Studies in Systems, Decision and Control

Volume 12

Series editor

Janusz Kacprzyk, Polish Academy of Sciences, Warsaw, Poland
e-mail: kacprzyk@ibspan.waw.pl

About this Series

The series "Studies in Systems, Decision and Control" (SSDC) covers both new developments and advances, as well as the state of the art, in the various areas of broadly perceived systems, decision making and control- quickly, up to date and with a high quality. The intent is to cover the theory, applications, and perspectives on the state of the art and future developments relevant to systems, decision making, control, complex processes and related areas, as embedded in the fields of engineering, computer science, physics, economics, social and life sciences, as well as the paradigms and methodologies behind them. The series contains monographs, textbooks, lecture notes and edited volumes in systems, decision making and control spanning the areas of Cyber-Physical Systems, Autonomous Systems, Sensor Networks, Control Systems, Energy Systems, Automotive Systems, Biological Systems, Vehicular Networking and Connected Vehicles, Aerospace Systems, Automation, Manufacturing, Smart Grids, Nonlinear Systems, Power Systems, Robotics, Social Systems, Economic Systems and other. Of particular value to both the contributors and the readership are the short publication timeframe and the world-wide distribution and exposure which enable both a wide and rapid dissemination of research output.

More information about this series at http://www.springer.com/series/13304

Ligang Wu · Xiaojie Su
Peng Shi

Fuzzy Control Systems with Time-Delay and Stochastic Perturbation

Analysis and Synthesis

Springer

Ligang Wu
School of Astronautics
Harbin Institute of Technology
Harbin
China

Xiaojie Su
College of Automation
Chongqing University
Chongqing
China

Peng Shi
School of Electrical and Electronic Engineering
The University of Adelaide
Adelaide, SA
Australia

and

School of Engineering and Science
Victoria University
Melbourne, VIC
Australia

ISSN 2198-4182 ISSN 2198-4190 (electronic)
ISBN 978-3-319-38492-4 ISBN 978-3-319-11316-6 (eBook)
DOI 10.1007/978-3-319-11316-6

Springer Cham Heidelberg New York Dordrecht London

Softcover reprint of the hardcover 1st edition 2015

Printed on acid-free paper

Springer is part of Springer Science+Business Media (www.springer.com)

To Jingyan and Zhixin

L.Wu

To My Family

X.Su

To My Family

P.Shi

Preface

Mathematical modeling of physical systems and processes can often lead to complex nonlinear systems, causing synthesis and analysis difficulties. Research of nonlinear systems is often problematic due to their complexities. One effective way of representing a complex nonlinear dynamic system is the so-called Takagi-Sugeno (T-S) fuzzy model, which is governed by a family of fuzzy IF-THEN rules that represent local linear input-output relations of the system. It incorporates a family of local linear models that smoothly blend together through fuzzy membership functions. This in essence, is a multi-model approach in which simple sub-models (typically linear models) are fuzzily combined to describe the global behavior of a nonlinear system. Within these fuzzy models, local dynamics in different state space regions are represented by linear models. An overall fuzzy model of the system is created by fuzzily 'blending' these linear models. Based on the fuzzy model, the control design is carried out by using the parallel distributed compensation scheme. The strategy is that a linear state feedback controller is designed for each local linear model. The obtained overall controller is nonlinear in general, and is again a fuzzy 'blending' of each individual linear controller.

Practical systems are commonly fraught with time-delays such as chemical processes and communication, generally lowering the system's performance and may lead to instability. The prevalent use of stochastic systems is largely contributed to the numerous applications stochastic modeling has in branches of science and engineering. Many important results have been reported for T-S fuzzy model, time-delay systems, and stochastic systems. When investigating T-S fuzzy systems that incorporate state-delay and stochastic perturbation terms, general control synthesis methodologies do not meet requirements. This monograph intends to present intends research developments and innovative methodologies on optimal synthesis of T-S fuzzy systems with time-delay and stochastic perturbation in a unified matrix inequality setting. Researchers exploring the areas of optimal synthesis of T-S fuzzy systems with time delay and stochastic perturbation will find valuable reference material within this text.

Stability analysis and stabilization, dynamic output feedback (DOF) control, full- and reduced-order filter design, fault detection and model reduction problems for a class of T-S fuzzy systems with time-delay and stochastic perturbation are all thoroughly investigated. Fresh novel techniques are applied to such systems which include the input-output method, the delay-partitioning method, the slack matrix method, and so on. This monograph is divided into three sections. First, we focus on optimal synthesis problems for discrete-time T-S fuzzy systems with time-varying delay. The main contents include 1) stability analysis and stabilization; 2) robust $\mathcal{H}_\infty$ DOF controller design; 3) full- and reduced- order filter design; and 4) reduced-order model, delay-free model and zero-order model design. Secondly, the theories and techniques developed in the previous part are extended to deal with T-S fuzzy stochastic systems with/without time-delay. Topics include 1) stability analysis and stabilization of discrete-time T-S fuzzy stochastic systems with time-delay; 2) $\mathcal{L}_2$-$\mathcal{L}_\infty$ DOF controller design for continuous-time T-S fuzzy stochastic systems with time-delay; 3) robust filter design for discrete-time T-S fuzzy stochastic systems with time-delay; 4) robust fault detection of continuous-time T-S fuzzy stochastic systems; and 5) model approximation for continuous-time T-S fuzzy stochastic systems. Finally, two real applications are presented to illustrate the feasibility and the effectiveness of the fuzzy control design schemes proposed in the previous parts. The first application is the fuzzy control of nonlinear electromagnetic suspension systems. A T-S fuzzy model for the considered nonlinear system is initially established, and used to design a fuzzy state feedback controller which ensures the closed-loop electromagnetic suspension system to be asymptotically stable with a mixed ℓ_2-ℓ_∞ performance. The second one is the robust $\mathcal{H}_\infty$ DOF control of longitudinal nonlinear model of flexible air-breathing hypersonic vehicles.

The main contents are suitable for a one-semester graduate course. This publication is a research reference whose intended audience include researchers, postgraduate and graduate students.

Harbin, China,
Chongqing, China,
Adelaide, Australia,

July 2014

Ligang Wu
Xiaojie Su
Peng Shi

Acknowledgements

There are numerous individuals without whose help this book will not have been completed. Special thanks go to Professor Yong-Duan Song from Chongqing University, Professor Sing Kiong Nguang from the University of Auckland, Professor James Lam from the University of Hong Kong, Professor Daniel W.C. Ho from the City University of Hong Kong, Professor Zidong Wang from the Brunel University, Professor Wei Xing Zheng from the University of Western Sydney, Dr. Hak-Keung Lam from King's College London, and Professor Huijun Gao from Harbin Institute of Technology, for their valuable suggestions, constructive comments and support. Our acknowledgments also go to our fellow colleagues who have offered invaluable support and encouragement throughout this research effort. In particular, we would like to acknowledge the contributions from Jianbin Qiu, Ming Liu, Rongni Yang, Guanghui Sun and Hongyi Li. Thanks go also to our students, Xiaoxiang Hu, Fanbiao Li, Xiaozhan Yang, Chunsong Han, Yongyang Xiong, Zhongrui Hu and Huiyan Zhang, for their commentary. The authors are especially grateful to their families for their encouragement and never-ending support when it was most required. Finally, we would like to thank the editors at Springer for their professional and efficient handling of this project.

The writing of this book was supported in part by the National Natural Science Foundation of China (61403048, 61174126, 61222301 and 61134001), the Fok Ying Tung Education Foundation (141059), the Heilongjiang Outstanding Youth Science Fund (JC201406), the Fundamental Research Funds for the Central Universities (HIT.BRETIV.201303), and the Australian Research Council (DP140102180).

Contents

List of Figures

List of Tables

Notations and Acronyms

$\blacksquare$	end of proof
$\blacklozenge$	end of remark
$\triangleq$	is defined as
$\in$	belongs to
$\forall$	for all
$\sum$	sum
$\mathbf{C}$	field of complex numbers
$\mathbf{R}$	field of real numbers
$\mathbf{R}^n$	space of n-dimensional real vectors
$\mathbf{R}^{n\times m}$	space of $n \times m$ real matrices
$\mathbf{Z}$	field of integral numbers
$\mathbf{Z}^+$	field of positive integral numbers
$\mathbf{E}\{\cdot\}$	mathematical expectation operator
lim	limit
max	maximum
min	minimum
sup	supremum
inf	infimum
rank$(\cdot)$	rank of a matrix
trace$(\cdot)$	trace of a matrix
$\lambda_{\min}(\cdot)$	minimum eigenvalue of a real square matrix
$\lambda_{\max}(\cdot)$	maximum eigenvalue of a real square matrix
$\sigma_{\min}(\cdot)$	minimum singular value of a real square matrix
$\sigma_{\max}(\cdot)$	maximum singular value of a real square matrix
I	identity matrix
I_n	$n \times n$ identity matrix
0	zero matrix
$0_{n\times m}$	zero matrix of dimension $n \times m$

X^T	transpose of matrix X
X^*	conjugate transpose of matrix X
X^{-1}	inverse of matrix X
$X^{\perp}$	full row rank matrix satisfying $X^{\perp}X = 0$ and $X^{\perp}X^{\perp T} > 0$
$X > (<)0$	X is real symmetric positive (negative) definite
$X \geq (\leq)0$	X is real symmetric positive (negative) semi-definite
$\mathcal{L}_2\{[0,\infty),[0,\infty)\}$	space of square summable sequences on $\{[0,\infty),[0,\infty)\}$ (continuous case)
$\ell_2\{[0,\infty),[0,\infty)\}$	space of square summable sequences on $\{[0,\infty),[0,\infty)\}$ (discrete case)
$\lvert\cdot\rvert$	Euclidean vector norm
$\lVert\cdot\rVert$	Euclidean matrix norm (spectral norm)
$\lVert\cdot\rVert_2$	$\mathcal{L}_2$-norm: $\sqrt{\int_0^\infty \lvert\cdot\rvert^2 dt}$ (continuous case) ℓ_2-norm: $\sqrt{\sum_0^\infty \lvert\cdot\rvert^2}$ (discrete case)
$\lVert\cdot\rVert_E$	$\mathbf{E}\{\lVert\cdot\rVert_2\}$
$\lVert\mathbf{T}\rVert_\infty$	$\mathcal{H}_\infty$ norm of transfer function $\mathbf{T}$: $\sup_{\omega\in[0,\infty)} \lVert\mathbf{T}(j\omega)\rVert$ (continuous case) $\sup_{\omega\in[0,2\pi)} \lVert\mathbf{T}(e^{j\omega})\rVert$ (discrete case)
diag	block diagonal matrix with blocks $\{X_1,\ldots,X_m\}$
$\text{diag}_n\{X\}$	block diagonal matrix diag $\{X,\ldots,X\}$ with n blocks
$\text{diag}_n^p\{X\}$	block diagonal matrix with n blocks, where the pth block is X and all others are zero matrices
$\text{col}_n\{X\}$	column matrix $\{X^T,\ldots,X^T\}^T$ with n blocks
$\star$	symmetric terms in a symmetric matrix
AHVs	air-breathing hypersonic vehicles
CCL	cone complementary linearization
CQLF	common quadratic Lyapunov function
DOF	dynamic output feedback
FLF	fuzzy Lyapunov function
PLF	piecewise Lyapunov function
LMI	linear matrix inequality
LKF	Lyapunov-Krasovskii functional
LTI	linear time-invariant
PDC	parallel distributed compensation
SOF	static output feedback
T-S	Takagi-Sugeno

Chapter 1
Introduction

Modeling practical physical systems frequently results in complex nonlinear systems, which poses great difficulties regarding system analysis and synthesis. Local linearization is a typical method used for the analysis and synthesis of nonlinear systems. However, it has been well recognized that the resulted local linearization model is valid only for a certain range of operating conditions, and can only guarantee the local stability of the original nonlinear system. Another approach, fuzzy control, emerged and developed following the first paper on fuzzy sets [243], has attracted great attention from both the academic and industrial communities. The reason lies much in its effectiveness in obtaining nonlinear control systems, especially when knowledge of the plant or even the precise control action of the situation is unknown. Thus, fuzzy control has even been found to have many applications in industrial systems and processes, see for example, [5, 7, 9, 12, 13, 14]. In fact, fuzzy control has proved to be a successful control approach for complex nonlinear systems. Fuzzy control has even been suggested as an alternative approach to conventional control techniques.

The basic structure of a fuzzy system consists of four conceptual components: knowledge base, fuzzification interface, inference engine, and defuzzification interface [50]. Fig. 1.1 shows the block diagram of a fuzzy system.

The past decades saw fuzzy rule-based modeling become an active research field due to its unique merits in solving complex nonlinear system identification and control problems. In an attempt to obtain more flexibility and more effective means of handling and processing uncertainties in complicated and ill-defined systems, Zadeh proposed a linguistic approach as the model of human thinking, introducing the fuzziness into systems theory [243]. Different from conventional modeling, fuzzy rule-based modeling is essentially a multi-model approach in which individual rules (where each rule acts like a 'local model') are combined to describe the global behavior of the system.

Among the array of model-based fuzzy systems, the Takagi-Sugeno (T-S) fuzzy system [192] is one of the most popular. T-S fuzzy systems effectively represent complex nonlinear systems in terms of fuzzy sets and fuzzy

L. Wu et al., *Fuzzy Control Systems with Time-Delay and Stochastic Perturbation*,
Studies in Systems, Decision and Control 12, DOI: 10.1007/978-3-319-11316-6_1

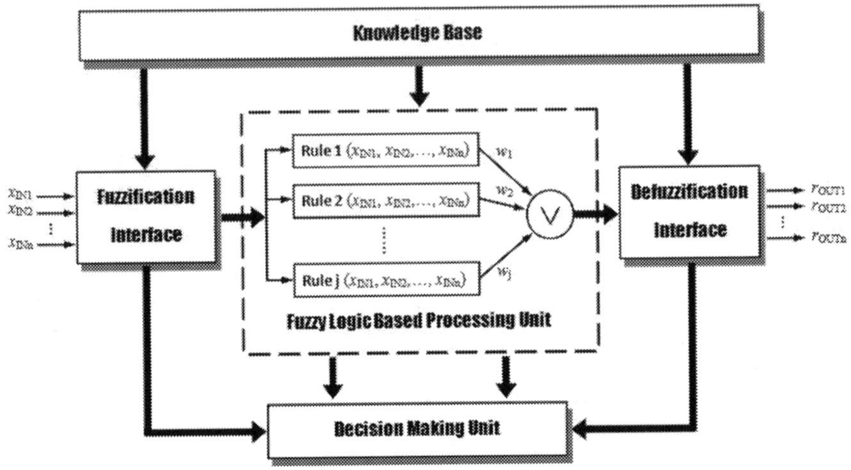

Fig. 1.1. Basic structure of fuzzy systems

reasoning applied to a set of linear input-output submodels. Using a T-S fuzzy plant model enables the description of a nonlinear system as a weighted sum of combined simple linear subsystems. This fuzzy model is made up of a family of fuzzy IF-THEN rules representing local linear input/output relations of the system. The overall fuzzy model of the system is achieved by smoothly blending these local linear models together through membership functions. Upon obtaining the fuzzy model, the control design is carried out via the parallel distributed compensation (PDC) approach [193, 206], which employs multiple linear controllers corresponding to the locally linear plant models with automatic scheduling performed via fuzzy rules. The T-S fuzzy model offers a fixed structure to some nonlinear systems and facilitates the related system analysis [50].

A continuous-time linear T-S fuzzy system can be modeled as

♦ **Plant Form:**

Rule i: IF $\theta_1(t)$ is $\mathcal{M}_{i1}$ and $\theta_2(t)$ is $\mathcal{M}_{i2}$ and $\cdots$ and $\theta_p(t)$ is $\mathcal{M}_{ip}$ THEN

$$\dot{x}(t) = A_i x(t) + B_i u(t), \quad i = 1, 2, \ldots, r,$$

where $x(t) \in \mathbf{R}^n$ is the state vector; $u(t) \in \mathbf{R}^m$ is the input vector. $\mathcal{M}_{ij}$ is the fuzzy set and r is the number of IF-THEN rules; $\theta(t) = \left[\theta_1(t)\ \theta_2(t) \cdots \theta_p(t)\right]^T$ is the premise variable vector. $A_i \in \mathbf{R}^{n\times n}$ and $B_i \in \mathbf{R}^{n\times m}$ are system parameter matrices.

It is assumed that the premise variables are not dependent on the input variables $u(t)$. This assumption is needed to avoid a complicated defuzzification process of fuzzy controllers [194]. Given a pair of $(x(t), u(t))$, the final output of the T-S fuzzy systems is inferred as follows:

$$\dot{x}(t) = \sum_{i=1}^{r} h_i(\theta(t)) \left[A_i x(t) + B_i u(t) \right], \tag{1.1}$$

where $h_i(\theta(t))$, sometimes denoted by $h_i(\theta)$ for simplicity, is the normalized membership function with

$$h_i(\theta(t)) = \frac{\nu_i(\theta(t))}{\sum_{i=1}^{r} \nu_i(\theta(t))}, \quad \nu_i(\theta(t)) = \prod_{j=1}^{p} \mathcal{M}_{ij}(\theta_j(t)),$$

where $\mathcal{M}_{ij}(\theta_j(t))$ is the grade of membership of $\theta_j(t)$ in $\mathcal{M}_{ij}$. It is assumed that

$$\nu_i(\theta(t)) \geq 0, \quad i = 1, 2, \ldots, r,$$
$$\sum_{i=1}^{r} \nu_i(\theta(t)) > 0, \quad \forall t \geq 0.$$

Therefore,

$$h_i(\theta(t)) \geq 0, \quad i = 1, 2, \ldots, r; \quad \sum_{i=1}^{r} h_i(\theta(t)) = 1.$$

Similarly, a discrete-time linear T-S fuzzy system can be described by

♦ **Plant Form:**

Rule i: IF $\theta_1(k)$ is $\mathcal{M}_{i1}$ and $\theta_2(k)$ is $\mathcal{M}_{i2}$ and $\cdots$ and $\theta_p(k)$ is $\mathcal{M}_{ip}$, THEN

$$x(k+1) = A_i x(k) + B_i u(k), \quad i = 1, 2, \ldots, r,$$

where $x(k) \in \mathbf{R}^n$ is the state vector; $u(k) \in \mathbf{R}^s$ is the input vector. $\mathcal{M}_{ij}$ is the fuzzy set and r is the number of IF-THEN rules; $\theta(k) = \left[\theta_1(k)\ \theta_2(k) \cdots \theta_p(k) \right]^T$ is the premise variables vector. A_i and B_i are known real constant matrices.

A more compact presentation of the discrete-time T-S fuzzy model can be given by

$$x(k+1) = \sum_{i=1}^{r} h_i(\theta(k)) \left[A_i x(k) + B_i u(k) \right]. \tag{1.2}$$

where $h_i(\theta(k))$, sometimes denoted by $h_i(\theta)$ for simplicity, is the normalized membership function with

$$h_i(\theta(k)) = \frac{\nu_i(\theta(k))}{\sum_{i=1}^{r} \nu_i(\theta(k))}, \quad \nu_i(\theta(k)) = \prod_{j=1}^{p} \mathcal{M}_{ij}(\theta_j(k)),$$

where $\mathcal{M}_{ij}(\theta_j(k))$ is the grade of membership of $\theta_j(k)$ in $\mathcal{M}_{ij}$. It is assumed that

$$\nu_i(\theta(k)) \geq 0, \quad i = 1, 2, \ldots, r,$$

$$\sum_{i=1}^{r} \nu_i(\theta(k)) > 0, \quad \forall k \geq 0.$$

Therefore,

$$h_i(\theta(k)) \geq 0, \quad i = 1, 2, \ldots, r; \quad \sum_{i=1}^{r} h_i(\theta(k)) = 1.$$

1.1 Stability and Synthesis of T-S Fuzzy Systems

1.1.1 Stability and Stabilization

One of the most important problems of fuzzy control system analysis is stability, as stability analysis results are vital to stabilization and synthesis problems. Over the past decade, the issues of stability and stabilization of T-S fuzzy systems have been considered extensively, see for example, with regards to stability analysis [16, 17, 27, 29, 47, 52, 100, 101, 102, 108, 130, 122, 123, 124, 125, 126, 139, 145, 155, 156, 157, 175, 176, 184, 190, 193, 209, 212, 217, 223, 252, 259], and for stabilization problems [31, 34, 35, 40, 42, 43, 45, 82, 109, 110, 116, 121, 127, 129, 149, 154, 161, 163, 195, 197, 206, 208, 228, 241, 258]. Stability analysis of T-S fuzzy systems was studied mainly based on Lyapunov stability theory, but with different kinds of Lyapunov functions. Examples include the common quadratic Lyapunov function (CQLF), the piecewise Lyapunov function (PLF), and the fuzzy Lyapunov function (FLF). Subsequently is a brief overview of existing stability analysis methods and a review of some stabilization results.

♣ Common Quadratic Lyapunov Function Approach

A well-known result on the stability analysis based on CQLF was presented in [193]. It was shown that if there exists a CQLF for all the subsystems, then the stability of the T-S fuzzy system can be guaranteed. Generally speaking, the existence of a CQLF is only sufficient for the asymptotic stability of the T-S fuzzy system, thus could be rather conservative.

For the continuous-time T-S fuzzy system in (1.1) with $u(t) = 0$ (i.e., open-loop system), by constructing a CQLF as $V(x) = x^T(t)Px(t)$, it is shown that the continuous-time T-S fuzzy system is asymptotically stable if there exists a common matrix $P > 0$ such that

$$PA_i + A_i^T P < 0, \quad i = 1, 2, \ldots, r.$$

Similarly, for discrete-time T-S fuzzy system in (1.2) with $u(k) = 0$, by constructing a CQLF as $V(x) = x^T(k)Px(k)$, it is shown that the discrete-time T-S fuzzy system is asymptotically stable if there exists a common matrix $P > 0$ such that

$$A_i^T P A_i - P < 0, \quad i = 1, 2, \ldots, r.$$

Now, consider the following continuous-time fuzzy controller:

♦ **Controller Form:**

Rule i: IF $\theta_1(t)$ is $\mathcal{M}_{i1}$ and $\theta_2(t)$ is $\mathcal{M}_{i2}$ and … and $\theta_p(t)$ is $\mathcal{M}_{ip}$, THEN

$$u(t) = K_i x(t), \quad i = 1, 2, \ldots, r,$$

where K_i is the gain matrix of the state feedback controller in each rule, and a compact form of the controller is given by

$$u(t) = \sum_{i=1}^{r} h_i(\theta) K_i x(t). \tag{1.3}$$

Under the above control, the closed-loop system of (1.1) can be described by

$$\dot{x}(t) = \sum_{i=1}^{r}\sum_{j=1}^{r} h_i(\theta) h_j(\theta) A_{ij} x(t),$$

where $A_{ij} \triangleq A_i + B_i K_j$.

Theorem 1.1. [194] *The continuous-time T-S fuzzy system in (1.1) is quadratically stabilizable via the fuzzy controller (1.3) if there exist matrices $X > 0$ and Y_i $(i = 1, 2, \ldots, r)$ such that*

$$\begin{aligned} -XA_i^T - A_i X - Y_i^T B_i^T - B_i Y_i &> 0, \\ -XA_i^T - A_i X - XA_j^T - A_j X - Y_j^T B_i^T - B_i Y_j - Y_i^T B_j^T - B_j Y_i &\geq 0, \\ i < j; \quad i, j = 1, 2, \ldots, r. \end{aligned}$$

Moreover, the matrix K_i in (1.3) can be computed by $K_i = Y_i X^{-1}$.

For discrete-time model (1.2), we design the following fuzzy controller:

♦ **Controller Form:**

Rule i: IF $\theta_1(k)$ is $\mathcal{M}_{i1}$ and $\theta_2(k)$ is $\mathcal{M}_{i2}$ and ... and $\theta_p(k)$ is $\mathcal{M}_{ip}$, THEN

$$u(k) = K_i x(k), \quad i = 1, 2, \ldots, r,$$

where K_i is the gain matrix of the state feedback controller in each rule, and a compact form of the controller is given by

$$u(k) = \sum_{i=1}^{r} h_i(\theta) K_i x(k). \tag{1.4}$$

Under the above control, the closed-loop system of (1.2) can be described by

$$x(k+1) = \sum_{i=1}^{r}\sum_{j=1}^{r} h_i(\theta) h_j(\theta) A_{ij} x(k).$$

Theorem 1.2. [194] *The discrete-time T-S fuzzy system in (1.2) is quadratically stabilizable via the fuzzy controller (1.4) if there exist matrices $X > 0$ and Y_i $(i = 1, 2, \ldots, r)$ such that*

$$\begin{bmatrix} X & XA_i^T + Y_i^T B_i^T \\ \star & X \end{bmatrix} > 0,$$

$$\begin{bmatrix} X & \frac{(A_i X + A_j X + B_i Y_j + B_j Y_i)^T}{2} \\ \star & X \end{bmatrix} \geq 0, \quad i < j;\ i, j = 1, 2, \ldots, r.$$

Moreover, the matrix K_i in (1.4) can be computed by $K_i = Y_i X^{-1}$.

♣ **Piecewise Lyapunov Function Approach.**

By the CQLF approach, it is required that a common positive definite matrix can be found to satisfy the Lyapunov equation or the LMI for all the local models. However, this is conservative since such a matrix might not exist in many cases, especially for highly nonlinear complex systems. Considering the conservativeness of CQLF approach, some attention has been paid to PLF approach, see for example, [16, 27, 45, 47, 52, 102, 197, 209] and the references therein. It is shown from the above-mentioned literature that the PLF is a much richer class of Lyapunov function candidates than the common Lyapunov function candidates and thus, it is able to deal with a larger class of fuzzy dynamic systems. In fact, the common Lyapunov function is a special case of the more general PLF [45].

Theorem 1.3. [45] *Consider the continuous-time T-S fuzzy system in (1.1) with $u(t) = 0$. If there exist a set of positive constants ε_i $(i = 1, 2, \ldots, r)$, and a symmetric matrix T such that with*

$$P_i = \left(F_i^T F_i\right)^{-1} F_i^T T F_i \left(F_i^T F_i\right)^{-1},$$

the following LMIs are satisfied:

$$P_i > 0,$$
$$\begin{bmatrix} P_i A_i^T + A_i P_i & P_i \\ \star & -\varepsilon_i I \end{bmatrix} < 0, \quad i = 1, 2, \ldots, r,$$

then the fuzzy system is globally exponentially stable, that is, $x(t)$ tends to the origin exponentially for every continuous piecewise trajectory in the state space.

Theorem 1.4. [47] *The discrete-time T-S fuzzy system in (1.2) with $u(k) = 0$ is globally exponentially stable if there exist a set of matrices $P_i > 0$ $(i = 1, 2, \ldots, r)$ such that*

$$\begin{bmatrix} A_i^T P_i A_i - P_i & A_i^T P_i \\ \star & -(I - P_i) \end{bmatrix} < 0,$$
$$\begin{bmatrix} A_i^T P_j A_i - P_i & A_i^T P_j \\ \star & -(I - P_j) \end{bmatrix} < 0, \quad i, j = 1, 2, \ldots, r.$$

Some results on the PLF approach to the stabilization problem for T-S fuzzy systems can be found in [27, 45, 197, 209].

♣ Fuzzy Lyapunov Function Approach.

Apart from the CQLF and PLF approaches to the stability analysis for T-S fuzzy systems, the FLF approach was also developed in [28, 153, 195, 196, 211, 218, 266]. For the continuous-time T-S fuzzy system in (1.1) with $u(t) = 0$, employ the following Lyapunov function:

$$V(x) = \sum_{i=1}^{r} h_i(\theta(t)) x^T(t) P_i x(t),$$

where $P_i > 0$ for $i = 1, 2, \ldots, r$, we have the following result.

Theorem 1.5. [195] *Assume that*

$$\left|\dot{h}_\rho(\theta(t))\right| \leq \phi_\rho, \quad \rho = 1, 2, \ldots, r,$$

where $\phi_\rho \geq 0$. The continuous-time T-S fuzzy system in (1.1) with $u(t) = 0$ is stable if there exist ϕ_ρ $(\rho = 1, 2, \ldots, r)$ such that

$$P_i > 0, \quad i = 1, 2, \ldots, r,$$

$$\sum_{\rho=1}^{r} \phi_\rho P_\rho + \frac{1}{2}\left(A_j^T P_i + P_i A_j + A_i^T P_j + P_j A_i\right) < 0, \quad i \le j;\ i, j = 1, 2, \ldots, r.$$

If the time derivatives of membership functions have the property of

$$\sum_{\rho=1}^{r} \dot{h}_\rho(\theta(t)) = 0, \quad \forall\, \theta(t),$$

that is,

$$\dot{h}_r(\theta(t)) = -\sum_{\rho=1}^{r-1} \dot{h}_\rho(\theta(t)),$$

the stability conditions given in Theorem 1.5 can be relaxed as follows.

Theorem 1.6. [195] *Assume that*

$$\left|\dot{h}_\rho(\theta(t))\right| \le \phi_\rho, \quad \rho = 1, 2, \ldots, r-1,$$

where $\phi_\rho \ge 0$. *The continuous-time T-S fuzzy system in (1.1) with* $u(t) = 0$ *is stable if there exist* ϕ_ρ $(\rho = 1, 2, \ldots, r-1)$ *such that*

$$P_i > 0, \quad i = 1, 2, \ldots, r,$$
$$P_\rho - P_r \ge 0, \quad \rho = 1, 2, \ldots, r-1,$$

$$\sum_{\rho=1}^{r-1} \phi_\rho (P_\rho - P_r) + \frac{1}{2}\left(A_j^T P_i + P_i A_j + A_i^T P_j + P_j A_i\right) < 0, \quad i \le j.$$

For discrete-time T-S fuzzy systems, we have the following results.

Theorem 1.7. [266] *The discrete-time T-S fuzzy system in (1.2) with* $u(k) = 0$ *is globally exponentially stable if there exist a set of matrices* $P_i > 0$ $(i = 1, 2, \ldots, r)$ *such that the following LMIs hold:*

$$A_i^T P_j A_i - P_i < 0, \quad i, j = 1, 2, \ldots, r.$$

Theorem 1.8. [194] *The discrete-time T-S fuzzy system in (1.2) is quadratically stabilizable via the fuzzy controller (1.4) if there exist matrices* $X_i > 0$, Y_i *and* Z_i $(i = 1, 2, \ldots, r)$ *such that*

$$\begin{bmatrix} X_i - Z_j^T - Z_j & Z_j^T A_i^T + Y_j^T B_i^T \\ \star & -X_l \end{bmatrix} < 0, \quad i, j, l = 1, 2, \ldots, r.$$

Moreover, the matrix K_i *in (1.4) can be computed by* $K_i = Y_i Z^{-1}$.

1.1.2 Optimal Synthesis Problems

Historically, considerable interest has been devoted to synthesis problems of T-S fuzzy systems, which include robust and optimal control, state estimation/filering, fault detection, and model approximation. In the following, we review some literature in this research field.

Extensive research into robust and optimal control problems for T-S fuzzy systems with/without time-delay over the last decade has yielded many important results [4, 19, 23, 24, 25, 26, 28, 36, 44, 46, 48, 67, 71, 88, 89, 94, 99, 111, 112, 113, 114, 115, 120, 131, 132, 133, 134, 140, 146, 167, 170, 171, 187, 199, 202, 207, 213, 220, 225, 227, 233, 237, 238, 246, 247, 249, 250, 261, 262, 266]. Formidable mentions include [44, 46, 48, 120, 171, 207, 247, 250] the piecewise Lyapunov function approach when applied to the $\mathcal{H}_\infty$ controller design; Choi and Park in [28] proposed a fuzzy weighting-dependent Lyapunov function approach to the $\mathcal{H}_\infty$ state-feedback controller design for discrete-time fuzzy systems; Kim and Park in [111, 112] presented a FLF approach to the $\mathcal{H}_\infty$ control design for fuzzy systems; and Zhou *et al.* in [266] established a basis-dependent Lyapunov function approach to the robust $\mathcal{H}_\infty$ control for discrete fuzzy systems. For uncertain fuzzy systems, Gassara *et al.* in [71] considered the observer-based robust $\mathcal{H}_\infty$ reliable controller design problem; Lee *et al.* in [131] studied the robust fuzzy control of nonlinear systems with parametric uncertainties; Li *et al.* in [134] addressed the robust $\mathcal{H}_\infty$ fuzzy control problem for a class of uncertain discrete fuzzy bilinear systems; Lo and Lin in [146] investigated the robust $\mathcal{H}_\infty$ control for Frobenius norm-bounded uncertain fuzzy systems; and Zhao *et al.* in [261] proposed a new approach to the guaranteed cost control of fuzzy systems with interval parameter uncertainties. Existing approaches to robust control of fuzzy time-delay systems were presented in [23, 115, 132, 140, 187, 246, 249, 262] for robust $\mathcal{H}_\infty$ control of fuzzy systems with time-varying delay; Also, the robust $\mathcal{H}_\infty$ control methods for fuzzy systems with interval delay were developed in [99, 113, 167], and fuzzy systems with infinite-distributed delays were presented in [213, 225]. In addition, considerable attention was paid to the study of the robust output feedback control problems, see for example, with regards to static output feedback (SOF) control problem [19, 25, 26, 89, 170], and for dynamic output feedback (DOF) control problems [4, 36, 132, 220, 233, 237].

Secondly, filtering is one of the fundamental problems in control systems and signal processing, which is the estimation of the state variables of a dynamical system through available noisy measurements. The celebrated Kalman filter [104] has been considered as the best possible (optimal) estimator for a large class of systems, which is an algorithm that uses a series of measurements observed over time, containing noise (random variations) and other inaccuracies, and produces estimates of unknown variables that tend to be more precise than those based on a single measurement alone. It is well known that the Kalman filtering is established on the assumptions that the system models are precisely known and the dynamic and measurement equations are additively affected by Gaussian noises [104]. These assumptions are often too strict for practical applications. The Kalman filtering scheme is no longer applicable when *a priori* information on the external noises is not precisely known. Therefore, the past two decades have witnessed significant progress on robust filtering involving various approaches [185, 188, 200, 201, 248, 251], and among them, the $\mathcal{H}_\infty$ filtering approach has drawn particular attention. One of its main advantages is the fact that it is insensitive to the exact knowledge of the statistics of the noise signals. To be specific, $\mathcal{H}_\infty$ filtering procedure ensures that the $\mathcal{L}_2$-induced gain from the noise input signals to the estimation error is less than a prescribed level, where the noise input is an arbitrary energy-bounded signal. The $\mathcal{H}_\infty$ filtering problem for T-S fuzzy systems has recently drawn a great deal of research interest, see for example, [1, 21, 49, 68, 90, 95, 138, 141, 169, 172, 236, 253, 254] and the references therein. For instance, Feng in [49] considered the robust $\mathcal{H}_\infty$ filtering problem for fuzzy dynamic systems, and three kinds of filtering design methods were proposed using quadratic stability theory and LMIs; An *et al.* in [1] proposed a delay-derivative-dependent fuzzy $\mathcal{H}_\infty$ filter design approach for T-S fuzzy time-delay systems; Chang in [21] investigated the robust nonfragile $\mathcal{H}_\infty$ filtering problem for fuzzy systems with linear fractional parametric uncertainties; Gao *et al.* in [68] addressed the $\mathcal{H}_\infty$ fuzzy filtering problem for nonlinear systems with intermittent measurements; Lin *et al.* in [141] considered $\mathcal{H}_\infty$ filter design for nonlinear systems with time-delay through T-S fuzzy model approach; Qiu *et al.* in [172] developed a nonsynchronized robust filter design scheme for continuous-time T-S fuzzy affine dynamic systems based on piecewise Lyapunov functions; and Zhang *et al.* in [251] studied the decentralized fuzzy $\mathcal{H}_\infty$ filtering problem for nonlinear interconnected systems with multiple time-delays.

Thirdly, the issues of fault detection and fault tolerant control are increasingly required in various kinds of practical complex systems for guaranteeing reliability and pursuing performance. The basic idea of fault detection is to construct a residual signal and, based on this, determine a residual evaluation function to compare with a predefined threshold. When the residual evaluation function has a value larger than the threshold, an alarm of faults is generated. Since accurate mathematical models are not always available, unavoidable modeling errors and external disturbances may seriously affect the performance of model-based fault detection systems. To overcome this, fault detection systems have to be robust to such modeling errors or disturbances. A system designed to provide both sensitivity to faults and robustness to modeling errors or disturbances is called a robust fault detection scheme. T-S fuzzy system fault detection has lately received much analysis. Exemplar studies include Gao *et al.* in [70] studied the sensor fault estimation problem for T-S fuzzy systems by fuzzy state/disturbance observer design; Jiang *et al.*

in [97] proposed an integrated fault estimation and accommodation design scheme for discrete-time T-S fuzzy systems with actuator faults; Wu and Ho in [219] considered the fault detection problem for Itô stochastic systems by using robust fuzzy filtering technique; Yang *et al.* in [239] presented a fault detection approach for T-S fuzzy discrete systems in finite-frequency domain; Zhang *et al.* in [256] developed a fault estimation approach for discrete-time T-S fuzzy systems based on piecewise Lyapunov functions; Zhao *et al.* in [260] investigated the fault detection problem for T-S fuzzy systems with intermittent measurements; and Zheng *et al.* in [263] addressed the T-S fuzzy-model-based fault detection problem for networked control systems with Markov delays. In addition, fault-tolerant control is a related issue that makes possible to develop a control feedback that allows keeping the required system performance in the case of faults. The fault-tolerant control problem for T-S fuzzy systems was also investigated, see for example, Jiang *et al.* in [96] proposed an adaptive fault-tolerant tracking control for near-space vehicles using T-S fuzzy model; Liu *et al.* in [143, 144] investigated the fuzzy-model-based fault detection and fault-tolerant control design for nonlinear stochastic systems; Shen *et al.* in [178, 179] considered the fault-tolerant control problem for T-S fuzzy systems with application to near-space hypersonic vehicles; and Zhang *et al.* in [255] studies the DOF fault tolerant control design for T-S fuzzy systems with actuator faults.

In addition, mathematical modeling of physical systems often results in complex high-order models, which bring serious difficulties to analysis and synthesis of the systems concerned. Therefore, in practical applications it is desirable to replace high-order models by reduced ones with respect to some given criterion, which is the model reduction problem. Model reduction has been a popular research area since it plays an important role in the process of control system design. Many important results on model reduction have been reported, which involve various efficient approaches such as the balanced truncation approach [83], Hankel-norm approach [73], Krylov projection approach [75], Padé reduction approach [6], and $\mathcal{H}_2$ approach [235], and $\mathcal{H}_\infty$ approach [221]. Please refer to [2] for a detailed survey of model reduction. Most model reduction techniques in these fields, however, aim at linear systems, and linear time-invariant systems in particular. The model reduction of nonlinear systems is still challenging. Considering that T-S fuzzy model is an effective way of representing a complex nonlinear dynamic system, the model reduction problem for nonlinear systems can be converted into a model reduction problem for T-S fuzzy systems. Wu *et al.* in [222] investigated the model reduction problem for discrete-time state-delay nonlinear systems in the T-S fuzzy framework; Su *et al.* in [186] studied the $\mathcal{H}_\infty$ model reduction problem for T-S fuzzy stochastic systems.

1.2 Stability Analysis of Time-Delay Systems: An LMI Approach

Time-delays are commonly recognized in various practical systems, such as communication, electronics, hydraulic, and chemical processes. Ignoring these time-delays may lead to degradation, instability and damage of systems [158, 173]. Therefore time-delay systems have been and continue to be widely investigated.

A continuous-time linear time-delay system can be modeled as

$$\dot{x}(t) = Ax(t) + A_d x(t-d), \tag{1.5a}$$

$$x(t) = \phi(t), \quad t \in [-d, 0]. \tag{1.5b}$$

Similarly, a discrete-time linear time-delay system can be described by

$$x(k+1) = Ax(k) + A_d x(k-d), \tag{1.6a}$$

$$x(k) = \phi(k), \quad k = -d, -d+1, \ldots, 0, \tag{1.6b}$$

where $x(\cdot) \in \mathbf{R}^n$ is the state vector; $\phi(\cdot) \in \mathbf{R}^n$ is the continuous initial condition; A and A_d are known real constant matrices; and d is a real constant representing time-delay.

If the time-delay considered is time-varying, the system is called linear time-varying delay system, and it can be formulated by the following equations for the continuous-time system:

$$\dot{x}(t) = Ax(t) + A_d x(t-d(t)), \tag{1.7a}$$

$$x(t) = \phi(t), \quad t \in [-d, 0]. \tag{1.7b}$$

and the following equations for the discrete-time system:

$$x(k+1) = Ax(k) + A_d x(k-d(k)), \tag{1.8a}$$

$$x(k) = \phi(k), \quad k = -d_2, -d_2+1, \ldots, 0, \tag{1.8b}$$

where the time-varying delay $d(t)$ in (1.7a) satisfies $0 \leq d(t) \leq d$ and $\dot{d}(t) \leq \tau$; and $d(k)$ in (1.8a) satisfies $1 \leq d_1 \leq d(k) \leq d_2$, and we define $\hat{d} = d_2 - d_1$.

Stability analysis is a fundamental and vital issue in studying time-delay systems, and the conservativeness of a stability condition is an important index to evaluate a stability result. Usually, stability conditions for time-delay systems can be classified into two types: delay-dependent and delay-independent stability conditions. The former include the information on the size of the delay, while the latter does not. Generally, delay-independent stability conditions are simpler to apply, while delay-dependent stability conditions are less conservative especially in the case when the time delay is small [234]. There have been a number of excellent survey papers on the stability analysis of time-delay systems, see for example, [79, 173, 234].

1.2.1 Delay-Independent Case

For continuous-time linear time-delay system (1.5), by constructing a Lyapunov-Krasovskii functional (LKF) as

$$V(t) = x^T(t)Px(t) + \int_{t-d}^{t} x^T(s)Qx(s)ds,$$

we have the following stability result:

Theorem 1.9. [234] *The continuous-time linear time-delay system in (1.5) is asymptotically stable if there matrices $P > 0$ and $Q > 0$ such that*

$$\begin{bmatrix} PA + A^T P + Q & PA_d \\ \star & -Q \end{bmatrix} < 0.$$

For discrete-time linear time-delay system (1.6), by constructing a LKF as

$$V(k) = x^T(k)Px(k) + \sum_{s=k-d}^{k-1} x^T(s)Qx(s),$$

we have the following stability result:

Theorem 1.10. *The discrete-time linear time-delay system in (1.6) is asymptotically stable if there matrices $P > 0$ and $Q > 0$ such that*

$$\begin{bmatrix} A^T PA - P + Q & A^T PA_d \\ \star & A_d^T PA_d - Q \end{bmatrix} < 0.$$

1.2.2 Delay-Dependent Case

Several methods have been proposed to develop delay-dependent stability conditions such as the model transformation approach (based on Newton-Leibniz formula) [135, 117], the descriptor system approach [57], the slack matrix approach [224, 232], the delay partitioning approach [74], and the input-output method (based on small gain theorem) [78]. In what follows, we summarize some recently developed delay-dependent approaches to the stability analysis for time-delay systems.

♣ Model Transformation Based on Newton-Leibniz Formula Approach

By using Newton-Leibniz formula and considering (1.5), we have

$$x(t-d) = x(t) - \int_{t-d}^{t} \dot{x}(s)ds$$

$$= x(t) - \int_{t-d}^{t} [Ax(t) + A_d x(t-d)]\, ds.$$

then, submitting the above equation to (1.5), it follows that

$$\dot{x}(t) = (A + A_d)\, x(t) + A_d \int_{t-d}^{t} [Ax(t) + A_d x(t-d)]\, ds, \tag{1.9a}$$

$$x(t) = \phi(t), \quad t \in [-d, 0]\,, \tag{1.9b}$$

It is well known that (1.5) is a special case of (1.9) and, thus, any solution of (1.5) is also a solution of (1.9). This means that the asymptotic stability of the time-delay system in (1.9) will ensure the asymptotic stability of the system in (1.5). For this reason, we now turn to study the stability of (1.9).

Theorem 1.11. [135] *The continuous-time linear time-delay system in (1.5) is asymptotically stable for any delay d satisfying $0 < d \le \bar{d}$ if there a matrix $P > 0$ and a scalar $\beta > 0$ such that*

$$\begin{bmatrix} P(A + A_d)^T + (A + A_d)P + \bar{d}A_d A_d^T & \bar{d}PA^T & \bar{d}PA_d^T \\ \star & -\bar{d}\beta I & 0 \\ \star & \star & -\bar{d}(1-\beta)I \end{bmatrix} < 0.$$

Constructing a LKF as

$$V(t) = x^T(t)P^{-1}x(t) + \int_{-d}^{0}\int_{t+\theta}^{t} x^T(s)A_d^T Q_1^{-1} A_d x(s)dsd\theta$$
$$+ \int_{-d}^{0}\int_{t-d+\theta}^{t} x^T(s)A_d^T Q_2^{-1} A_d x(s)dsd\theta,$$

where $P > 0$, $Q_1 > 0$ and $Q_2 > 0$.

The following stability result is based on the above LKF.

Theorem 1.12. [18] *The continuous-time linear time-delay system in (1.5) is asymptotically stable for any delay d satisfying $0 < d \le \bar{d}$ if there exist matrices $P > 0$, $Q_1 > 0$ and $Q_2 > 0$ such that*

$$\begin{bmatrix} P(A + A_d)^T + (A + A_d)P + A_d(Q_1 + Q_2)A_d^T & \bar{d}PA^T & \bar{d}PA_d^T \\ \star & -Q_1 & 0 \\ \star & \star & -Q_2 \end{bmatrix} < 0.$$

Remark 1.13. In Theorem 1.12, if we choose $Q_1 = \bar{d}\beta I$ and $Q_2 = \bar{d}(1-\beta)I$, the stability condition coincides with that in Theorem 1.11, which means that the stability condition in Theorem 1.12 is less conservative than that in Theorem 1.11. ♦

The results based on the model transformation approach can also be found in [33, 63, 117, 118, 136, 160, 159] and the references therein.

Remark 1.14. Note that by using the Newton-Leibniz formula-based model transformation approach, the new system in (1.9) is not equivalent to the original time-delay system in (1.5). As mentioned above that (1.5) is a special case of (1.9). Thus, there are some additional dynamics introduced through the model transformation, which will inevitably induce some conservativeness in the analysis of delay-dependent stability. ♦

♣ Bounding Techniques

In deriving the delay-dependent stability conditions in [18] and [135], the upper bound inequality for an inner product of two vectors is utilized, and it can be formulated as the following lemma.

Lemma 1.15. [210] *For any vectors* $a, b \in \mathbf{R}^n$ *and any positive definite matrix* $X \in \mathbf{R}^{n\times n}$*, it holds that*

$$-2a^T b \leq a^T X a + b^T X^{-1} b.$$

In [164], an improved upper bound for the inner product of two vectors was proposed.

Lemma 1.16. [164] (Park's Inequality) *Assume that* $a(\alpha) \in \mathbf{R}^{n_a}$ *and* $b(\alpha) \in \mathbf{R}^{n_b}$ *are given for* $\alpha \in \Omega$*. Then, for any positive definite matrices* $X \in \mathbf{R}^{n_a \times n_a}$ *and* $Y \in \mathbf{R}^{n_b \times n_b}$*, the following inequality holds:*

$$-2\int_{\Omega} b^T(\alpha)a(\alpha)d\alpha \leq \int_{\Omega} \begin{bmatrix} a(\alpha) \\ b(\alpha) \end{bmatrix}^T \times \begin{bmatrix} X & XY \\ \star & (Y^T X + I)\, X^{-1}\, (XY + I) \end{bmatrix} \begin{bmatrix} a(\alpha) \\ b(\alpha) \end{bmatrix} d\alpha.$$

Based on the new bounding technique, a new delay-dependent robust stability criterion was established, see the below theorem.

Theorem 1.17. [164] *The continuous-time linear time-delay system in (1.5) is asymptotically stable for any delay* d *satisfying* $0 < d \leq \bar{d}$ *if there exist matrices* $P > 0$*,* $Q > 0$*,* $V > 0$ *and* W *such that*

$$\begin{bmatrix} (1,1) & -W^T A_d & A^T A_d^T V & \bar{d}\left(W^T + P\right) \\ \star & -Q & A_d^T A_d^T V & 0 \\ \star & \star & -V & 0 \\ \star & \star & \star & -V \end{bmatrix} < 0,$$

where

$$(1,1) \triangleq (A + A_d)^T P + P\,(A + A_d) + W^T A_d + A_d^T W + Q.$$

In [151], Moon *et. al* further improved Park's Inequality, and presented a new bounding inequality, named Moon's Inequality.

Lemma 1.18. [151] (Moon's Inequality) *Assume that* $a(\alpha) \in \mathbf{R}^{n_a}$, $b(\alpha) \in \mathbf{R}^{n_b}$ *and* $\mathcal{N} \in \mathbf{R}^{n_a \times n_b}$ *are given for* $\alpha \in \Omega$. *Then, for any matrices* $X \in \mathbf{R}^{n_a \times n_a}$, $Y \in \mathbf{R}^{n_a \times n_b}$ *and* $Z \in \mathbf{R}^{n_b \times n_b}$, *the following inequality holds:*

$$-2\int_{\Omega} a^T(\alpha)\mathcal{N}b(\alpha)d\alpha \leq \int_{\Omega} \begin{bmatrix} a(\alpha) \\ b(\alpha) \end{bmatrix}^T \begin{bmatrix} X & Y-\mathcal{N} \\ \star & Z \end{bmatrix} \begin{bmatrix} a(\alpha) \\ b(\alpha) \end{bmatrix} d\alpha,$$

where

$$\begin{bmatrix} X & Y \\ \star & Z \end{bmatrix} \geq 0.$$

Based on the above bounding inequality, and choose the LKF as

$$V(t) = x^T(t)Px(t) + \int_{t-d}^{t} x^T(s)Qx(s)ds \\ + \int_{-d}^{0}\int_{t+\theta}^{t} \dot{x}^T(s)Z\dot{x}(s)dsd\theta,$$

where $P > 0$, $Q > 0$ and $Z > 0$, a new less conservative stability condition was presented in [151]. We re-state it as follows.

Theorem 1.19. [151] *The continuous-time linear time-delay system in (1.5) is asymptotically stable for any delay* d *satisfying* $0 < d \leq \bar{d}$ *if there exist matrices* $P > 0$, $Q > 0$, X, Y *and* Z *such that*

$$\begin{bmatrix} A^TP + PA + \bar{d}X + Y + Y^T + Q & -Y + PA_d & \bar{d}A^TZ \\ \star & -Q & \bar{d}A_{\underline{d}}^TZ \\ \star & \star & -\bar{d}Z \end{bmatrix} < 0,$$

$$\begin{bmatrix} X & Y \\ \star & Z \end{bmatrix} \geq 0.$$

Lemma 1.20. [77] (Jensen's Inequality) *For any positive definite matrix* $M \in \mathbf{R}^{n\times n}$, *scalars* a *and* b *satisfying* $a < b$, *and a vector function* $\omega : [a, b] \to \mathbf{R}^n$ *such that the integrations concerned are well defined, then the following inequality holds:*

$$\left(\int_a^b \omega(s)ds\right)^T M \left(\int_a^b \omega(s)ds\right) \leq (b-a)\int_a^b \omega^T(s)M\omega(s)ds.$$

Based on Jensen's Inequality, and choose the LKF as

$$V(t) = x^T(t)Px(t) + \int_{t-d}^{t} x^T(s)Qx(s)ds + d\int_{-d}^{0}\int_{t+\theta}^{t} \dot{x}^T(s)Z\dot{x}(s)dsd\theta,$$

where $P > 0$, $Q > 0$ and $Z > 0$. A new less conservative stability condition was presented in [74].

Theorem 1.21. [74] *The continuous-time linear time-delay system in (1.5) is asymptotically stable for any delay d satisfying $0 < d \leq \bar{d}$ if there exist matrices $P > 0$, $Q > 0$, X, Y and Z such that*

$$\begin{bmatrix} A^TP + PA + Q - Z & PA_d + Z & \bar{d}A^TZ \\ \star & -Q - Z & \bar{d}A_d^TZ \\ \star & \star & -Z \end{bmatrix} < 0.$$

Lemma 1.22. [98] *For any positive definite matrix $M \in \mathbf{R}^{n\times n}$, scalars a and b satisfying $a < b$, and a vector function w: $\{a, a+1, \ldots, b\} \to \mathbf{R}^n$ such that the integrations concerned are well defined, then the following inequality holds:*

$$-(b-a+1)\sum_{i=a}^{b} w^T(i)Mw(i) \leq -\left(\sum_{i=a}^{b} w^T(i)\right) M \left(\sum_{i=a}^{b} w(i)\right).$$

Theorem 1.23. *The discrete-time linear time-delay system in (1.6) is asymptotically stable for any delay d satisfying $0 < d \leq \bar{d}$ if there exist matrices $P > 0$, $Q > 0$ and $Z > 0$ such that*

$$\begin{bmatrix} A^TPA - P + Q - Z & A^TPA_d + Z & \bar{d}(A-I)^TZ \\ \star & A_d^TPA_d - Q - Z & \bar{d}A_d^TZ \\ \star & \star & -Z \end{bmatrix} < 0.$$

Proof. The result can be obtained by employing the discrete-time Jensen's Inequality, and choosing the following LKF:

$$V(k) = x^T(k)Px(k) + \sum_{s=k-d}^{k-1} x^T(s)Qx(s) + d\sum_{i=-d}^{-1}\sum_{s=k+i}^{k-1} \eta^T(s)Z\eta(s),$$

where $\eta(k) \triangleq x(k+1) - x(k)$, $P > 0$, $Q > 0$ and $Z > 0$. ∎

♣ Descriptor System Approach

Fridman and Shaked in [57, 58, 59, 60] proposed a new approach, namely descriptor system approach, to deal with time-delay systems. To use this approach, represent (1.5) in an equivalent descriptor form of

$$\begin{cases} \dot{x}(t) = y(t), \\ 0 = -y(t) + (A + A_d)x(t) - A_d \int_{t-d}^{t} y(s)ds. \end{cases}$$

Or equivalently,

$$E\dot{\xi}(t) = \bar{A}\xi(t) - \bar{A}_d \int_{t-d}^{t} y(s)ds, \tag{1.10}$$

where

$$\xi(t) \triangleq \begin{bmatrix} x(t) \\ y(t) \end{bmatrix}, \quad E \triangleq \begin{bmatrix} I & 0 \\ 0 & 0 \end{bmatrix}, \quad \bar{A} \triangleq \begin{bmatrix} 0 & I \\ A + A_d & -I \end{bmatrix}, \quad \bar{A}_d \triangleq \begin{bmatrix} 0 \\ A_d \end{bmatrix}.$$

The new model of (1.10) is equivalent to the original system in (1.5). Thus, the stability analysis for system (1.5) can be converted equivalently to that of (1.10). To this end, the following LKF is applied:

$$V(t) = \xi^T(t)EP\xi(t) + \int_{t-d}^{t} x^T(s)Qx(s)ds + \int_{-d}^{0}\int_{t+\theta}^{t} y^T(s)Ry(s)dsd\theta,$$

where $P \triangleq \begin{bmatrix} P_1 & 0 \\ P_2 & P_3 \end{bmatrix}$, $P_1 > 0$, $Q > 0$, $R > 0$, and $EP = P^T E \geq 0$.

Theorem 1.24. [59] *The continuous-time linear time-delay system in (1.5) is asymptotically stable for any delay d satisfying $0 < d \leq \bar{d}$ if there exist matrices $P \triangleq \begin{bmatrix} P_1 & 0 \\ P_2 & P_3 \end{bmatrix}$, $Z \triangleq \begin{bmatrix} Z_1 & Z_2 \\ \star & Z_3 \end{bmatrix}$, $Y \triangleq \begin{bmatrix} Y_1 & Y_2 \end{bmatrix}$, $P_1 > 0$, $Q > 0$, $R > 0$, such that*

$$\begin{bmatrix} \Omega + \bar{d}Z & P^T \begin{bmatrix} 0 \\ A_d \end{bmatrix} - Y^T \\ \star & -Q \end{bmatrix} < 0,$$

$$\begin{bmatrix} R & Y \\ \star & Z \end{bmatrix} \geq 0,$$

where

$$\Omega \triangleq P^T \begin{bmatrix} 0 & I \\ A & -I \end{bmatrix} + \begin{bmatrix} 0 & I \\ A & -I \end{bmatrix}^T P + \begin{bmatrix} Q & 0 \\ 0 & \bar{d}R \end{bmatrix} + \begin{bmatrix} Y \\ 0 \end{bmatrix} + \begin{bmatrix} Y \\ 0 \end{bmatrix}^T.$$

♣ Slack Matrix Approach

Considering the Newton-Leibniz formula:

$$x(t-d) = x(t) - \int_{t-d}^{t} \dot{x}(s)ds,$$

it follows that for any appropriately dimensioned matrices X and Y,

$$2\left[x^T(t)X - x^T(t-d)Y\right]\left[x(t) - x(t-d) - \int_{t-d}^{t} \dot{x}(s)ds\right] = 0,$$

where the slack matrices X and Y indicate the relationship between the terms in the Newton-Leibniz formula.

On the other hand, for any semi-positive definite matrix

$$W = \begin{bmatrix} W_{11} & W_{12} \\ \star & W_{22} \end{bmatrix} \geq 0,$$

the following holds:

$$d\xi^T(t)W\xi(t) - \int_{t-d}^{t} \xi^T W\xi(t)ds = 0, \tag{1.11}$$

where $\xi(t) \triangleq \begin{bmatrix} x(t) \\ x(t-d) \end{bmatrix}$.

By employing the following LKF:

$$V(x_t) = x^T(t)Px(t) + \int_{t-d}^{t} x^T(s)Qx(s)ds + \int_{-d}^{0}\int_{t+\theta}^{t} \dot{x}^T(s)Z\dot{x}(s)dsd\theta,$$

where $P > 0$, $Q > 0$ and $Z > 0$.

The following two theorems are based on the slack matrix approach and the above LKF.

Theorem 1.25. [224] *The continuous-time linear time-delay system in (1.5) is asymptotically stable for any delay d satisfying $0 < d \leq \bar{d}$ if there exist matrices $P > 0$, $Q > 0$, $Z > 0$, $W = \begin{bmatrix} W_{11} & W_{12} \\ \star & W_{22} \end{bmatrix} \geq 0$, X and Y such that*

$$\begin{bmatrix} PA + A^TP + X + X^T + Q + \bar{d}W_{11} & PA_d - X + Y^T + \bar{d}W_{12} & \bar{d}A^TZ \\ \star & -Q - Y - Y^T + \bar{d}W_{22} & \bar{d}A_d^TZ \\ \star & \star & -\bar{d}Z \end{bmatrix} < 0,$$

$$\begin{bmatrix} W_{11} & W_{12} & X \\ \star & W_{22} & Y \\ \star & \star & Z \end{bmatrix} \geq 0.$$

If don't consider (1.11) in the proof of Theorem 1.25, the stability result turn out to be the following.

Theorem 1.26. [232] *The continuous-time linear time-delay system in (1.5) is asymptotically stable for any delay d satisfying $0 < d \leq \bar{d}$ if there exist matrices $P > 0$, $Q > 0$, $Z > 0$, X and Y such that*

$$\begin{bmatrix} PA + A^T P + X + X^T + Q & PA_d - X + Y^T & -\bar{d}X & \bar{d}A^T Z \\ \star & -Q - Y - Y^T & -\bar{d}Y & \bar{d}A_d^T Z \\ \star & \star & -\bar{d}Z & 0 \\ \star & \star & \star & -\bar{d}Z \end{bmatrix} < 0.$$

♣ Delay Partitioning Approach

The delay partitioning technique was originally presented in [74]. This basic idea of this approach is to partition time-delay into several components evenly. By constructing a LKF when considering every delay component, it can be shown that the stability condition has been considerably improved. The results in [37, 148] were proved to be less and less conservative as the partitioning becomes increasingly thinner.

For the discrete-time time-varying delay system in (1.8), partition $d(k)$ into two parts: the constant part τm and the time-varying part $h(k)$, that is, $d(k) = \tau m + h(k)$, where $h(k)$ satisfies $0 \leq h(k) \leq d_2 - \tau m$. Define

$$\Upsilon(l) \triangleq \begin{bmatrix} x(l) \\ x(l-\tau) \\ \vdots \\ x(l - \tau m + \tau) \end{bmatrix}.$$

By applying the delay partitioning idea partially to the lower delay bound $d_1 = \tau m$ which gives m parts, it constructs the following LKF:

$$\begin{aligned} V(x_k) \triangleq\; & x^T(k)P(k)x(k) + \sum_{l=k-\tau}^{k-1} \Upsilon^T(l)Q_1\Upsilon(l) \\ & + \sum_{s=-d_2+1}^{-\tau m+1} \sum_{l=k-1+s}^{k-1} x^T(l)Rx(l) + \sum_{l=k-d_2}^{k-1} x^T(l)Q_2x(l) \\ & + \sum_{s=-\tau}^{-1} \sum_{l=k+s}^{k-1} \Big[x(l+1) - x(l)\Big]^T S_1(l) \Big[x(l+1) - x(l)\Big] \\ & + \sum_{s=-\bar{d}}^{-\tau m-1} \sum_{l=k+s}^{k-1} \Big[x(l+1) - x(l)\Big]^T S_2(l) \Big[x(l+1) - x(l)\Big], \end{aligned}$$

where $P > 0$, $Q_1 > 0$, $Q_2 > 0$, $R > 0$, $S_1 > 0$ and $S_2 > 0$.

Theorem 1.27. [148] *Given positive integers τ, m and d_2, the discrete-time linear time-varying delay system in (1.8) is asymptotically stable if there exist real matrices $P > 0$, $Q_1 > 0$, $Q_2 > 0$, $R > 0$, $S_1 > 0$, $S_2 > 0$, $M \geq 0$, $N \geq 0$, X, Y and Z satisfying*

$$\begin{bmatrix} \Phi + \Psi + \Psi^T & \Xi \\ \star & -\mathrm{diag}\{P, Q_2, R, S_1, S_2\} \end{bmatrix} < 0,$$
$$\begin{bmatrix} M & X \\ \star & S_1 \end{bmatrix} \geq 0,$$
$$\begin{bmatrix} N & Y \\ \star & S_2 \end{bmatrix} \geq 0,$$
$$\begin{bmatrix} N & Z \\ \star & S_2 \end{bmatrix} \geq 0,$$

where

$$\Phi \triangleq -\Xi_2^T P \Xi_2 + W_{Q_1}^T \bar{Q}_1 W_{Q_1} - W_R^T R W_R - W_{Q_2}^T Q_2 W_{Q_2} + \tau M + (d_2 - \tau m) N,$$
$$\Xi \triangleq \begin{bmatrix} \Xi_1^T P & \Xi_2^T P & \sqrt{d_2 - \tau m + 1}\Xi_2^T R & \sqrt{\tau}\Xi_3^T S_1 & \sqrt{d_2 - \tau m}\Xi_3^T S_2 \end{bmatrix},$$
$$\Psi \triangleq \begin{bmatrix} X & Y & Z \end{bmatrix} \left[\begin{array}{cccc} I_n & -I_n & 0_{n\times(m+1)n} & \\ \hline 0_{n\times mn} & I_n & -I_n & 0_n \\ \hline 0_{n\times(m+1)n} & I_n & -I_n & \end{array}\right], \quad \bar{Q}_1 \triangleq \begin{bmatrix} Q_1 & 0 \\ 0 & -Q_1 \end{bmatrix},$$
$$W_{Q_1} \triangleq \begin{bmatrix} I_{mn} & 0_{mn\times 3n} & \\ 0_{mn\times n} & I_{mn} & 0_{mn\times 2n} \end{bmatrix}, \quad W_{Q_2} \triangleq \begin{bmatrix} 0_{n\times(m+2)n} & I_n \end{bmatrix},$$
$$W_R \triangleq \begin{bmatrix} 0_{n\times(m+1)n} & I_n & 0_n \end{bmatrix}, \quad \Xi_3 \triangleq \Xi_1 - \Xi_2,$$
$$\Xi_1 \triangleq \begin{bmatrix} A & 0_{n\times mn} & A_d & 0_n \end{bmatrix}, \quad \Xi_2 \triangleq \begin{bmatrix} I_n & 0_{n\times(m+2)n} \end{bmatrix}.$$

Noted that with the delay partitioning approach, the delay partition $mn \times mn$ positive matrix is involved in Theorem 1.27 for asymptotically stability of the time-delay system. The proposed stability condition is much less conservative due to the introduced $mn \times mn$ positive matrix. However, the matrix often results in insolubility for analysis and synthesis of time-delay systems. Therefore, the $mn \times mn$ positive matrix is substituted for these $n \times n$ positive matrix by using the revised delay partitioning method.

Benefiting from this less conservative approach, many results for other complex systems are extended, such as time delay neural networks [152, 240], singular time-delay systems [51, 53], Markovian jump time delay systems [41, 69], and T-S fuzzy time-delay systems [187, 223].

♣ Input-Output Approach

The input-output technique, borrowed from the robust control theory [265], is one of the most effective ways to deal with time delay. The main procedures of this approach involves the model transformation of the concerned system into feedback interconnection formulation, which contains a constant time-delay forward subsystem and a delay "uncertainty" feedback subsystem. By applying the scaled small gain theorem when considering the feedback interconnection formulation, it can be shown that the new stability condition will reduce the conservativeness for the original system considerably, and the performance of the original system will be greatly improved.

In the past, many results [78, 137, 185, 258] on this approach prevailed in many research areas, such as linear time-invariant delay systems and time-varying delay systems, in which the method was employed to cope with the delay "uncertainty".

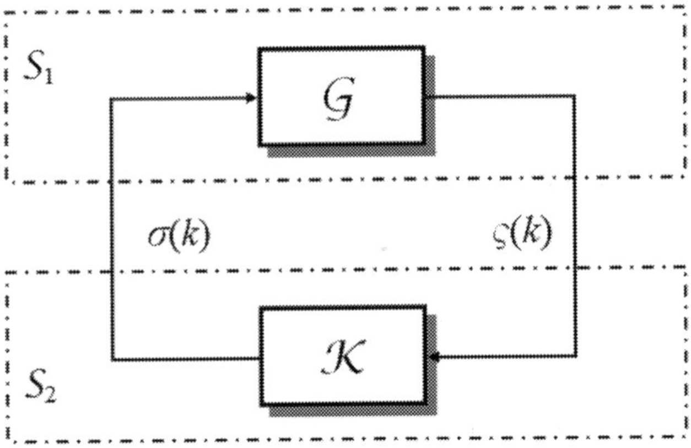

Fig. 1.2. Block diagram of closed-loop connected system

Consider an interconnection system, Fig. 1.2, consisting of two subsystems:

$$(\mathcal{S}_1): \quad \varsigma(k) = \mathcal{G}\sigma(k), \tag{1.12a}$$
$$(\mathcal{S}_2): \quad \sigma(k) = \mathcal{K}\varsigma(k), \tag{1.12b}$$

where the forward $(\mathcal{S}_1)$ is a known LTI system with operator $\mathcal{G}$ mapping $\sigma(k)$ to $\varsigma(k)$, the feedback $(\mathcal{S}_2)$ is an unknown linear time-varying one with operator $\mathcal{K} \in \mathbb{K}$ which has a block-diagonal structure, a mapping from $\varsigma(k)$ to $\sigma(k)$. The symbol $\mathbb{K}$ denotes a compact set of appropriately dimensioned time-varying matrices with a diagonal structure specified by

$$\mathbb{K} \triangleq \text{diag}\{\varepsilon_1(k)I_n, \varepsilon_2(k)I_n, \ldots, \varepsilon_s(k)I_n\}, \tag{1.13}$$

where $\varepsilon_i(k) \in \mathbf{R}$, $|\varepsilon_i(k)| \leq 1$, $i = 1, 2, \ldots, s$ (i is the position of repeated scalar). As a direct result of the small gain theorem [32, 265], a sufficient condition regarding robust asymptotic stability of the interconnection in (1.12) is given as follows.

Lemma 1.28. [137] *Assumed that* $(\mathcal{S}_1)$ *in (1.12) is internally stable, the closed-loop system of interconnection system described by (1.12) is robustly asymptotically stable for all* $\mathcal{K} \in \mathbb{K}$ *if there exist*

$$\Upsilon_0(\mathcal{G}) \times \Upsilon_0(\mathcal{K}) < 1, \tag{1.14}$$

where

$$\begin{aligned}\Upsilon_0(\mathcal{G}) &= \|T \circ \mathcal{G} \circ T^{-1}\|_\infty, \\ \Upsilon_0(\mathcal{K}) &= \|T^{-1} \circ \mathcal{K} \circ T\|_\infty, \\ T &= \text{diag}\{T_1, T_2, \ldots, T_s\} > 0.\end{aligned}$$

The objective of model transformation is to pull time-varying delay uncertainties out of the original time-varying delay system in (1.8) so that $(\mathcal{S}_1)$ is an LTI while $(\mathcal{S}_1)$ includes all the uncertainties.

To pull out the uncertainty in $d(k)$, estimate the time-varying $x(k - d(k))$ using its lower bound d_1 and upper bound d_2. The two-term approximation $\frac{x(k-d_1)+x(k-d_2)}{2}$ results in the estimation error:

$$\begin{aligned}\sigma(k) &= \frac{2}{d}\left\{x(k - d(k)) - \frac{1}{2}\left[x(k - d_1) + x(k - d_2)\right]\right\}, \\ &= \frac{1}{d}\left(\sum_{i=k-d_2}^{k-d_1-1} \beta(i)\varsigma(i)\right),\end{aligned} \tag{1.15}$$

where $d \triangleq d_2 - d_1$, $\varsigma(i) \triangleq x(i+1) - x(i)$ and

$$\beta(i) \triangleq \begin{cases} 1, & \text{when } i \leq k - d(k) - 1, \\ -1, & \text{when } i > k - d(k) - 1. \end{cases}$$

For brevity, operator $\sigma(k)$ denote

$$\mathcal{K}: \quad \varsigma(k) \to \sigma(k) = \frac{1}{d}\left(\sum_{i=k-d_2}^{k-d_1-1} \beta(i)\varsigma(i)\right), \tag{1.16}$$

to denote the relation $(\mathcal{S}_2)$ from $\varsigma(k)$ to $\sigma(k)$ in (1.12). The following result gives an upper bound of the ℓ_2 norm of $\mathcal{K}$.

Lemma 1.29. [137] *Operator* $\mathcal{K}$ *in (1.16) bears the property* $\|\mathcal{K}\|_\infty \leq 1$.

In view of Lemma 1.29, we can see that the ℓ_2 norm of $(\mathcal{S}_2)$ in (1.12) from input to output is bounded by one. Then based on Lemma 1.28, we focus on researching the scaled small gain of $(\mathcal{S}_1)$ for the interconnection frame (1.12).

Theorem 1.30. [137] *Given a scalar $\gamma > 0$, system $(\mathcal{S}_1)$ in (1.12) is asymptotically stable and satisfies $\|T \circ \mathcal{G} \circ T^{-1}\|_\infty < \gamma$ for some nonsingular $T \in \mathbf{R}^{n\times n}$ if either of the following two conditions holds:*

i) *There exist symmetric matrices $0 < P \in \mathbf{R}^{(d_2+1)n\times(d_2+1)n}$, $0 < S \in \mathbf{R}^{n\times n}$ such that*

$$\mathcal{G}^T \begin{bmatrix} P & 0 \\ 0 & S \end{bmatrix} \mathcal{G} - \begin{bmatrix} P & 0 \\ 0 & \gamma^2 S \end{bmatrix} < 0,$$

where

$$\mathcal{G} \triangleq \left[\begin{array}{ccccc|c} A & (\mathbf{0})_1 & \frac{1}{2}A_d & (\mathbf{0})_2 & \frac{1}{2}A_d & \frac{d}{2}A_d \\ & & I_{d_2 n} & & 0 & 0 \\ \hline A-I & (\mathbf{0})_1 & \frac{1}{2}A_d & (\mathbf{0})_2 & \frac{1}{2}A_d & \frac{d}{2}A_d \end{array}\right],$$

with $(\mathbf{0})_1 = 0_{n\times(d_1-1)n}$ and $(\mathbf{0})_2 = 0_{n\times(d-1)n}$.

ii) *There exist symmetric matrices $0 < P \in \mathbf{R}^{n\times n}$, $0 < Q_i \in \mathbf{R}^{n\times n}$, $0 < R_i \in \mathbf{R}^{n\times n}$, $i = 1, 2$, $0 < S \in \mathbf{R}^{n\times n}$ such that*

$$\begin{bmatrix} \Xi_1 & \begin{bmatrix} \Xi_2^T P & d_1\Xi_3^T R_1 & d_2\Xi_3^T R_2 & \Xi_3^T S \end{bmatrix} \\ \star & \mathrm{diag}\left\{-P, -R_1, -R_2, -S\right\} \end{bmatrix} < 0,$$

where

$$\Xi_1 \triangleq \begin{bmatrix} -P + Q_1 + Q_2 - R_1 - R_2 & \begin{bmatrix} R_1 & R_2 & 0 \end{bmatrix} \\ \star & -\mathrm{diag}\left\{(Q_1+R_1), (Q_2+R_2), \gamma^2 S\right\} \end{bmatrix},$$

$$\Xi_2 \triangleq \begin{bmatrix} A & \frac{1}{2}A_d & \frac{1}{2}A_d & \frac{d}{2}A_d \end{bmatrix}, \quad \Xi_3 \triangleq \begin{bmatrix} A-I & \frac{1}{2}A_d & \frac{1}{2}A_d & \frac{d}{2}A_d \end{bmatrix}.$$

Theorem 1.31. [137] *The discrete-time linear time-varying delay system in (1.8) is asymptotically stable for all $d_k \in \{d_1, d_1+1, \ldots, d_2\}$ if (i) or (ii) of Theorem 1.30 holds for $\gamma \leq 1$.*

With the idea of "pulling out uncertainties", it is easy to extend the proposed input-output results to multiple time-varying delay case, for which, the uncertainty of each delayed state will be pulled out and forms a feedback channel similar to the diagram in Fig. 1.2. Besides the uncertainties included in delay and system matrices, the input-output performance of the original system can also be considered in such a formulation, such as the standard ℓ_2-induced norm gain. This facilitates to extend these results to many other analysis and synthesis problems of time-delay systems.

♣ Reciprocally Convex Approach

The reciprocally convex approach [54, 142, 165, 189] suggests a lower bound lemma for such a linear combination of positive functions with inverses of convex parameters as the coefficients. Based on the lemma, we develop a stability criterion that directly handles the inversely weighted convex combination of quadratic terms of integral quantities, which achieves performance behavior identical to approaches based on the integral inequality lemma but with much less decision variables, comparable to those based on the Jensen inequality lemma.

It concerns a special type of function combinations, that is, a linear combination of positive functions with inverses of convex parameters as the coefficients, which is defined below.

Definition 1.32. [165] Let $\Psi_1, \Psi_2, \ldots, \Psi_N : \mathbf{R}^m \to \mathbf{R}^n$ be a given finite number of functions such that they have positive values in an open subset $\mathbf{D}$ of $\mathbf{R}^m$. Then, a reciprocally convex combination of these functions over $\mathbf{D}$ is a function of the form

$$\frac{1}{\vartheta_1}\Psi_1 + \frac{1}{\vartheta_2}\Psi_2 + \cdots + \frac{1}{\vartheta_N}\Psi_N : \quad \mathbf{D} \to \mathbf{R}^n, \tag{1.17}$$

where the real numbers ϑ_i satisfy $\vartheta_i > 0$ and $\sum_i \vartheta_i = 1$.

The following lemma suggests a lower bound for a reciprocally convex combination of scalar positive functions $\Psi_i = f_i$.

Lemma 1.33. [165] *Let $f_1, f_2, \ldots, f_N : \mathbf{R}^m \to \mathbf{R}$ have positive values in an open subset $\mathbf{D}$ of $\mathbf{R}^m$, then the reciprocally convex combination of f_i over $\mathbf{D}$ satisfies*

$$\min_{\{\vartheta_i | \vartheta_i > 0, \sum_i \vartheta_i = 1\}} \sum_i \frac{1}{\vartheta_i} f_i(\theta) = \sum_i f_i(\theta) + \max_{g_{i,j}(\theta)} \sum_{i \neq j} g_{i,j}(\theta),$$

subject to

$$\left\{ g_{i,j} : \quad \mathbf{R}^m \to \mathbf{R}, \quad g_{j,i}(\theta) = g_{i,j}(\theta), \quad \begin{bmatrix} f_i(\theta) & g_{i,j}(\theta) \\ g_{j,i}(\theta) & f_j(\theta) \end{bmatrix} \geq 0 \right\}.$$

The above lemma can be applied to handle the double integral terms of the following LKF for system (1.7):

$$\begin{aligned} V(x_t) &= x^T(t)Px(t) + \int_{t-d_1}^{t} x^T(s)Q_1x(s)ds + \int_{t-d_2}^{t} x^T(s)Q_2x(s)ds \\ &\quad + d_1 \int_{-d_1}^{0} \int_{t+\theta}^{t} \dot{x}^T(s)R_1\dot{x}(s)dsd\theta + d \int_{-d_2}^{-d_1} \int_{t+\theta}^{t} \dot{x}^T(s)R_2\dot{x}(s)dsd\theta, \end{aligned}$$

where $P > 0$, $Q_1 > 0$, $Q_2 > 0$, $R_1 > 0$ and $R_2 > 0$.

Theorem 1.34. [165] *The continuous-time linear time-varying delay system in (1.7) is asymptotically stable if there exist matrices $P > 0$, $Q_1 > 0$, $Q_2 > 0$, $R_1 > 0$, $R_2 > 0$ and S_{12} such that*

$$E_5 P E_1^T + E_1 P E_5^T + E_1 Q_2 E_1^T - E_3 Q_1 E_3^T - (E_1 - E_3) R_1 (E_1 - E_3)^T - \begin{bmatrix} E_3 - E_2 & E_2 - E_4 \end{bmatrix} \begin{bmatrix} R_2 & S_{12} \\ \star & R_2 \end{bmatrix} \begin{bmatrix} E_3^T - E_2^T \\ E_2^T - E_4^T \end{bmatrix} < 0,$$

$$\begin{bmatrix} R_2 & S_{12} \\ \star & R_2 \end{bmatrix} \geq 0,$$

where $E_5 \triangleq E_1 A^T + E_2 A_d^T$ and

$$E_1 \triangleq \begin{bmatrix} I \\ 0 \\ 0 \\ 0 \end{bmatrix}, \quad E_2 \triangleq \begin{bmatrix} 0 \\ I \\ 0 \\ 0 \end{bmatrix}, \quad E_3 \triangleq \begin{bmatrix} 0 \\ 0 \\ I \\ 0 \end{bmatrix}, \quad E_4 \triangleq \begin{bmatrix} 0 \\ 0 \\ 0 \\ I \end{bmatrix}.$$

1.3 Publication Contribution

This book represents the first of a few attempts to reflect the state-of-the-art of the research area for handling stability/performance analysis and optimal synthesis problems for T-S fuzzy systems with time-delay and stochastic perturbation. The content of this book can be divided into three parts. The first part will be focused on analysis and synthesis of T-S fuzzy time-delay systems. Some sufficient conditions are derived for the stability and some optimal performances by developing new techniques for the considered T-S fuzzy time-delay systems. The developed methodologies include the fuzzy LKF approach, the slack matrix approach, the delay-partitioning approach, the small gain theorem based input-output approach, and the reciprocally convex approach, etc. The main aim by using these advanced approaches is to effectively reduced the conservatism of the obtained results, thus facilitate the design subsequently. Then, some optimal synthesis problems, including the stabilization, the DOF controller design, the robust $\mathcal{H}_\infty$ filtering, and the model approximation, are investigated based on the analysis results. The second section focuses on the parallel theories and techniques developed in the previous part are extended to deal with T-S fuzzy stochastic systems (or say T-S fuzzy systems with stochastic perturbation). A unified framework under 'stochastic stability' is established for analyzing the considered T-S fuzzy stochastic systems. Specifically, in this part, the main focus is on stochastic stability analysis, stabilization, $\mathcal{L}_2$-$\mathcal{L}_\infty$ DOF control, $\mathcal{H}_\infty$ filtering, fault detection and model approximation problems for the considered T-S fuzzy stochastic systems. Sufficient conditions are established first for the stochastic stability and optimal performances (such as $\mathcal{H}_\infty$, $\mathcal{L}_2$-$\mathcal{L}_\infty$ and dissipativity) of the continuous- and discrete-time T-S fuzzy stochastic systems. Based on the obtained analysis results, the optimal synthesis issues are solved. In the third part, two fuzzy control applications are presented to illustrate the feasibility and the effectiveness of the fuzzy control design schemes proposed in the previous parts. The first one is the fuzzy control of nonlinear electromagnetic suspension systems. A T-S fuzzy model for the considered nonlinear system is firstly established, and then based on which a fuzzy state feedback controller is designed, which ensures the closed-loop electromagnetic suspension system to be asymptotically stable with a mixed ℓ_2-ℓ_∞ performance. The second one is the robust $\mathcal{H}_\infty$ DOF control of longitudinal nonlinear model of flexible air-breathing hypersonic vehicles (AHVs).

The features of this book can be highlighted as follows. 1) A unified framework is established for analysis and optimal synthesis of T-S fuzzy systems, where there are time-delay existing system states, and there are external stochastic perturbations. 2) A series of problems are solved with new approaches for analysis and synthesis of continuous- and discrete-time T-S fuzzy systems with time-delay and stochastic perturbation, including stability/performances analysis and stabilization, DOF control, robust filtering, fault detection, and model approximation. 3) Three advanced methods, namely the delay-partitioning approach, the small gain theorem based input-output approach, and the reciprocally convex approach, are developed to deal with T-S fuzzy time-delay systems. 4) A set of newly developed techniques (e.g., the fuzzy LKF method, the LMI technique, the cone complementary linearization (CCL) approach, the slack matrix approach, and the sums of squares technique) are exploited to handle the emerging mathematical/computational challenges.

This publication is a timely reflection of the developing area of system analysis and synthesis theories for T-S fuzzy systems with time-delay and stochastic perturbation. It is a collection of a series of latest research results and therefore serves as a useful textbook for senior and/or graduate students who are interested in knowing 1) the state-of-the-art of fuzzy systems area; 2) recent advances in time-delay systems; 3) recent advances in stochastic systems; and 4) recent advances in stability/performances analysis, stabilization, DOF control, robust filtering and model approximation problems. Readers will also benefit from some new concepts, new models and new methodologies with theoretical significance in system analysis and control synthesis. It can also be used as a practical research reference for engineers dealing with stabilization, optimal control and state estimation problems for T-S fuzzy systems, time-delay systems, stochastic perturbed systems, and nonlinear systems. The aim of this book is to close the gap in literature by providing a unified yet neat framework for stability/performances analysis and optimal synthesis of T-S fuzzy systems with time-delay and stochastic perturbation.

Generally, this is an advanced publication aimed at 3rd/4th-year undergraduates, postgraduates and academic researchers. Prerequisite knowledge includes linear algebra, matrix analysis, linear control system theory and stochastic systems.

Expected readers include 1) control engineers working on nonlinear control, fuzzy control and optimal control; 2) system engineers working on intelligent control and systems; 3) mathematicians and physician working on time-delay systems and stochastic systems; and 4) postgraduate students majoring on control engineering, system sciences and applied mathematics. This publication is also a useful reference for 1) mathematicians and physicians working on intelligent systems and nonlinear systems; 2) computer scientists working on algorithms and computational complexity; and 3) 3rd/4th-year students who are interested in advances in control theory and applications.

1.4 Publication Outline

The general layout of presentation of this monograph is divided into three parts. Part one focuses on the analysis and optimal synthesis for T-S fuzzy time-delay systems, whilst part two studies the analysis and optimal synthesis for T-S fuzzy stochastic systems. Lastly, part three presents some applications of fuzzy control methods. The organization structure of this monograph is shown in Fig. 1.3, and the main contents of this monograph are shown in Fig. 1.4.

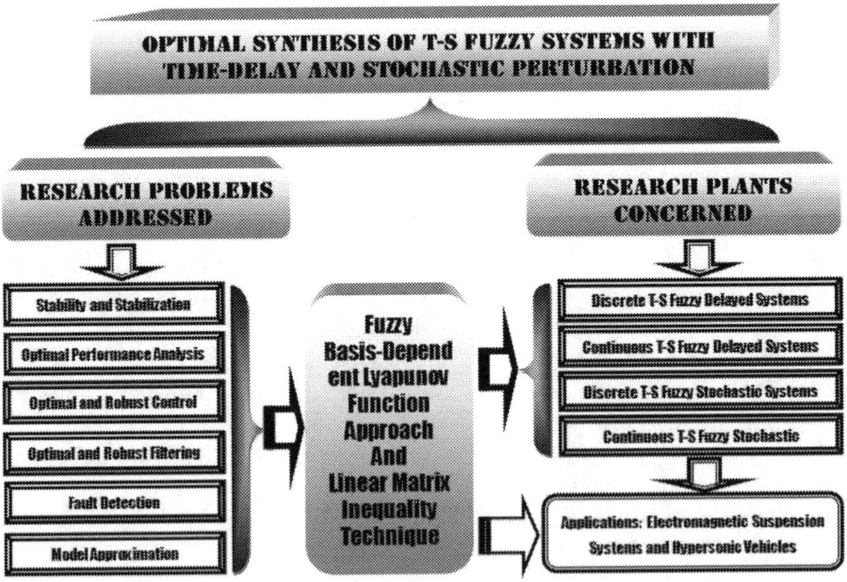

Fig. 1.3. Organizational structure of this publication

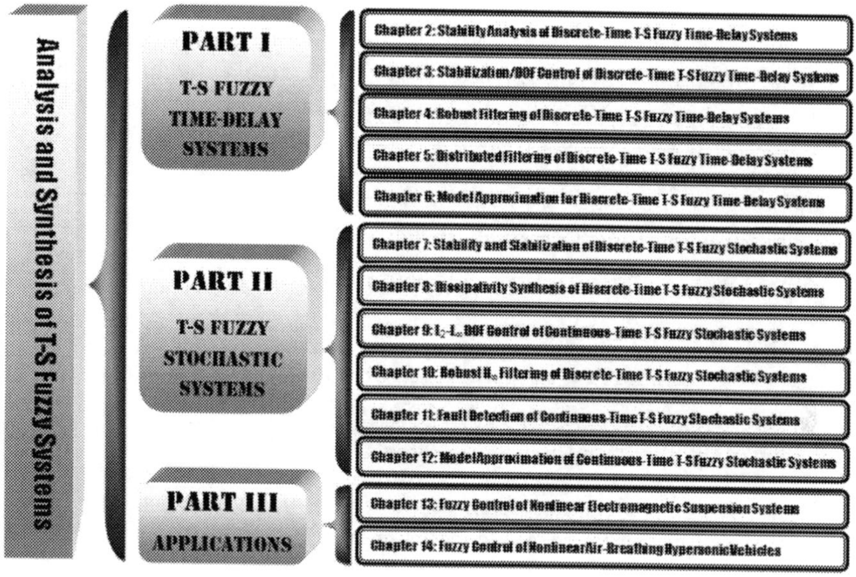

Fig. 1.4. Main contents of this publication

Chapter 1 presents the research background, motivations and research problems, which involve optimal analysis and synthesis of T-S fuzzy systems, time-delay systems, then the outline of the monograph is listed.

Part One focuses on the analysis and optimal synthesis for T-S fuzzy time-delay systems. Part One which begins with Chapter 2 consists of five chapters as follows.

Chapter 2 is concerned with some innovative methods combined with the construction of basis-dependent LKF, the delay partitioning method, the input-output method and the reciprocally convex method, to solve the stability analysis problem of discrete-time T-S fuzzy time-varying delay systems. A set of the parameter-dependent (delay-dependent and fuzzy-rule-dependent) conditions with less conservativeness are established in the form of LMIs.

Chapter 3 is focused on the problems of fuzzy state feedback control and $\mathcal{H}_\infty$ DOF control for discrete-time T-S fuzzy systems with time-varying delay. Based on the obtained less conservative stability results in Chapter 2, and combining with the delay partitioning and the input-output methods, the fuzzy state feedback control and the DOF control problems are solved by the construction of the basis-dependent LKF, which makes the corresponding closed-loop system stable with the specified performances.

Chapter 4 considers the system performance analysis and robust filter design for T-S fuzzy systems with time-varying delay. Sufficient conditions of the given performance are presented for the augmented filtering error system, based on which, the filter design problem is then solved. Furthermore, the obtained methods are extended to solve the problem of reliable filter design for T-S fuzzy systems with time-varying delay. All filter design conditions, which are in terms of strict LMIs, are obtained by employing the basis-dependent LKF method combined with the convex linearization technique.

Chapter 5 is devoted to studying the distributed fuzzy filtering problem for discrete-time T-S fuzzy systems with time-varying delay. Based on the proposed reliable filtering results in Chapter 4, the fuzzy distributed filtering problem is settled for sensor networks with the occurrence of incomplete information (including time-delay and sensor faults). The fuzzy distributed filter is designed by introducing the topological structure and based on the scale small gain theorem, which guarantees the corresponding filtering error system stable with a given distributed $\mathcal{H}_\infty$ performance. Moreover, the distributed filter with an average $\mathcal{H}_\infty$ performance is also designed in terms of the feasibility of a convex optimization problem.

Chapter 6 investigates the $\mathcal{H}_\infty$ model approximation problem for discrete-time T-S fuzzy time-delay systems. For a high-order T-S fuzzy system, our attention is focused on the construction of a reduced-order model which approximates the original system well in a specified $\mathcal{H}_\infty$ performance. By applying the delay partitioning approach, a delay-dependent sufficient condition is proposed for the asymptotic stability with an $\mathcal{H}_\infty$ performance for the approximation error system. Then, the $\mathcal{H}_\infty$ model approximation problem is solved by using the projection approach, which casts the model approximation into a sequential minimization problem subject to LMI constraints. Moreover, the model approximation results for some special structures, such as the delay-free model and the zero-order model, are also presented.

Part Two studies the analysis and optimal synthesis for T-S fuzzy stochastic systems. Part Two which begins with Chapter 7 consists of six chapters as follows.

Chapter 7 is concerned with the stability analysis and stabilization for discrete-time T-S fuzzy stochastic systems with time-varying delay. Our attention is focused on employing the novel idea of delay partitioning method combining with the basis-dependent LKF technique to obtain a less conservative sufficient stability condition, by which the stabilization can be solved via non-PDC scheme. The proposed design scheme is applied to stabilize a complex inverted pendulum system.

Chapter 8 studies the problems of dissipativity analysis and synthesis for discrete-time T-S fuzzy stochastic systems with time-varying delay. A new model transformation method is first introduced to pull time-delay uncertainty out of the original system. The uncertainty is confined to a subsystem and the approximated main system contains only constant delays. A sufficient condition of dissipativity is derived by LKF approach, and the dissipativity condition is finally converted into a set of LMIs. A fuzzy controller, which guarantees the closed-loop system to be dissipative, is then designed based on the obtained dissipativity condition.

Chapter 9 considers the $\mathcal{L}_2$-$\mathcal{L}_\infty$ DOF control for T-S fuzzy stochastic systems with time-varying delay. The slack matrix approach is used to derive a delay-dependent sufficient condition which guarantees the mean-square asymptotic stability with an $\mathcal{L}_2$-$\mathcal{L}_\infty$ performance for the closed-loop system. The corresponding solvability condition for a desired $\mathcal{L}_2$-$\mathcal{L}_\infty$ DOF controller is then established. These obtained conditions, which are not all expressed in terms of LMI, are cast into sequential minimization problems subject to LMI constraints by applying the CCL method.

Chapter 10 is devoted to studying the $\mathcal{H}_\infty$ filter design problem for discrete-time T-S fuzzy stochastic systems with time-varying delays. Firstly, a model transformation of the original system is studied by way of that of a comparison system consisting of two subsystems, which are a constant time-delay forward subsystem and a delay "uncertainty" feedback subsystem. The forward subsystem needs to be under consideration to ensure the stability of the original systems by applying the scaled small gain theorem. A sufficient condition for the mean-square asymptotically stability of the filter error system is obtained, while an $\mathcal{H}_\infty$ performance is guaranteed. The explicit expression of the desired filter parameters is also derived by applying a convex linearization approach, which casts the $\mathcal{H}_\infty$ desired filter design into a convex optimization problem.

Chapter 11 addresses the robust $\mathcal{H}_\infty$ fault detection problem for T-S fuzzy stochastic systems. By using a general observer-based fault detection filter as residual generator, the robust fault detection is formulated as a filtering problem. Attention is focused on the design of both the fuzzy-rule-independent and the fuzzy-rule-dependent fault detection filters, guaranteeing a prescribed noise attenuation level in an $\mathcal{H}_\infty$ sense. Sufficient conditions are proposed to guarantee the mean-square asymptotic stability with an $\mathcal{H}_\infty$ performance for the fault detection system. The corresponding solvability conditions for the desired fuzzy-rule-independent and fuzzy-rule-dependent fault detection filters are established.

Chapter 12 investigates the $\mathcal{H}_\infty$ model approximation problem for T-S fuzzy stochastic systems. At first, sufficient conditions are proposed in terms of LMIs for the existence of the admissible reduced-order models. Then, two different approaches are proposed to solve the considered $\mathcal{H}_\infty$

model approximation problem. One is the convex linearization approach, which casts the model reduction into a convex optimization problem, while the other is the projection approach, which casts the model reduction into a sequential minimization problem subject to LMI constraints by employing the CCL algorithm.

Part Three studies two applications of fuzzy control methods. Part three which begins with Chapter 13 consists of two chapters as follows.

Chapter 13 presents a T-S model-based fuzzy controller design approach for nonlinear electromagnetic suspension systems. The T-S fuzzy modeling approach is applied, and a new T-S fuzzy model is established to represent such nonlinear physical plants. Then, based on the obtained T-S fuzzy model, a fuzzy state feedback controller is designed, which ensures the closed-loop electromagnetic suspension system to be asymptotically stable with a mixed ℓ_2-ℓ_∞ performance. The controller is designed in a non-PDC scheme, and sufficient conditions for the existence of the desired controller are derived in terms of LMIs.

Chapter 14 studies T-S fuzzy robust $\mathcal{H}_\infty$ DOF control problem for longitudinal nonlinear model of flexible AHVs. The developed T-S fuzzy model of the flexible AHVs include uncertainties and external disturbances, and it is shown that the fuzzy model can approach the dynamics of flexible AHVs well. Based on the PDC scheme, a fuzzy DOF controller is designed to stabilize the closed-loop system considering that some components of the states of the flexible AHVs models are not available. Sufficient conditions for the existence of the desired $\mathcal{H}_\infty$ controllers are proposed in terms of LMIs, and the controller can be designed by solving a convex optimization problem.

Part I
Analysis and Synthesis of T-S Fuzzy Time-Delay Systems

Chapter 2
Stability Analysis of Discrete-Time T-S Fuzzy Time-Delay Systems

2.1 Introduction

New stability analysis methods are investigated for discrete-time T-S fuzzy time-delay systems, which include the basis-dependent LKF method, the delay partitioning method, the small gain theorem based input-output method, and the reciprocally convex method. Our main attention is focused on reducing the conservativeness of the stability conditions caused by the time-delays concerned. The stability analysis results developed in this chapter form an important theoretic foundation to optimal synthesis problems, such as robust/optimal control, robust filtering, fault detection, and model approximation, for T-S fuzzy time-delay systems in the subsequent chapters.

2.2 System Description and Preliminaries

Consider the following T-S fuzzy time-varying delay system:

♦ **Plant Form:**

Rule i: IF $\theta_1(k)$ is $\mathcal{M}_{i1}$ and $\theta_2(k)$ is $\mathcal{M}_{i2}$ and $\cdots$ and $\theta_p(k)$ is $\mathcal{M}_{ip}$, THEN

$$\begin{aligned} x(k+1) &= A_i x(k) + A_{di} x(k-d(k)), \\ x(k) &= \phi(k), \quad k = -d_2, d_2+1, \ldots, 0, \end{aligned}$$

where $i = 1, 2, \ldots, r$, and r is the number of IF-THEN rules; $\mathcal{M}_{ij}(i = 1, 2, \ldots, r; j = 1, 2, \ldots, p)$ are the fuzzy sets; $\theta = \begin{bmatrix} \theta_1(k) \ \theta_2(k) \ \cdots \ \theta_p(k) \end{bmatrix}^T$ is the premise variable vector; $x(k) \in \mathbf{R}^n$ is the state vector; $d(k)$ denotes the time-varying delay satisfying $1 \leqslant d_1 \leqslant d(k) \leqslant d_2$, where d_1 and d_2 are constant positive scalars representing its lower and upper bounds, respectively. A_i and A_{di} are real constant matrices, and $\phi(k)$ denotes the initial condition.

L. Wu et al., *Fuzzy Control Systems with Time-Delay and Stochastic Perturbation*, Studies in Systems, Decision and Control 12, DOI: 10.1007/978-3-319-11316-6_2

A more compact presentation of the T-S fuzzy delayed model is given by

$$x(k+1) = \bar{A}(k)x(k) + \bar{A}_d(k)x(k-d(k)), \tag{2.1}$$

where

$$\bar{A}(k) \triangleq \sum_{i=1}^{r} h_i(\theta)A_i, \quad \bar{A}_d(k) \triangleq \sum_{i=1}^{r} h_i(\theta)A_{di},$$

with $h_i(\theta)$, $i = 1, 2, \ldots, r$ are the normalized membership functions, which are defined as that of (1.2) in Chapter 1.

Consider the T-S fuzzy time-delay system in (2.1). To use the input-output method to analyze its stability, we estimate the time-delayed state vector $x(k-d(k))$ using the two-term approximation of $\frac{x(k-d_1)+x(k-d_2)}{2}$ at first, which results in the following estimation error:

$$\begin{aligned}\sigma(k) &= \frac{2}{d}\left\{x(k-d(k)) - \frac{1}{2}\left[x(k-d_1) + x(k-d_2)\right]\right\} \\ &= \frac{1}{d}\left[\sum_{i=k-d_2}^{k-d_1-1} \beta(i)\varsigma(i)\right],\end{aligned}$$

where $d \triangleq d_2 - d_1$, $\varsigma(i) \triangleq x(i+1) - x(i)$ and

$$\beta(i) \triangleq \begin{cases} 1, & \text{when } i \leq k-d(k)-1, \\ -1, & \text{when } i > k-d(k)-1. \end{cases}$$

The following auxiliary system is introduced to replace system (2.1):

$$x(k+1) = \bar{A}(k)x(k) + \frac{1}{2}\bar{A}_d(k)\left[x(k-d_1) + x(k-d_2)\right] + \frac{d}{2}\bar{A}_d(k)\sigma(k),$$

and it can be reformulated as

$$(\mathcal{S}_1): \begin{bmatrix} x(k+1) \\ \varsigma(k) \end{bmatrix} = \begin{bmatrix} \bar{A}(k) & \frac{1}{2}\bar{A}_d(k) & \frac{1}{2}\bar{A}_d(k) & \frac{d}{2}\bar{A}_d(k) \\ \bar{A}(k)-I & \frac{1}{2}\bar{A}_d(k) & \frac{1}{2}\bar{A}_d(k) & \frac{d}{2}\bar{A}_d(k) \end{bmatrix} \begin{bmatrix} x(k) \\ x(k-d_1) \\ x(k-d_2) \\ \sigma(k) \end{bmatrix}, \tag{2.2a}$$

$$(\mathcal{S}_2): \quad \sigma(k) = \mathcal{K}\varsigma(k), \tag{2.2b}$$

which forms the interconnection frame shown in Fig. 1.1. For brevity, we use the following operator:

$$\mathcal{K}: \quad \varsigma(k) \to \sigma(k) = \frac{1}{d}\left[\sum_{i=k-d_2}^{k-d_1-1} \beta(i)\varsigma(i)\right],$$

to denote the relation $(\mathcal{S}_2)$ from $\varsigma(k)$ to $\sigma(k)$ in Fig. 1.1.

Lemma 2.1. *Operator $\mathcal{K}$ in (2.2) bears the property that $\|\mathcal{K}\|_\infty \leq 1$.*

Proof. In view of the formulation $\sigma(k)$ in (2.2) and using Jensen inequality [77], we obtain the following inequality under the zero initial condition:

$$\begin{aligned}
\|\sigma(k)\|_2^2 = \sum_{i=0}^{\infty} \sigma^T(i)\sigma(i) &= \frac{1}{d^2}\sum_{i=0}^{\infty}\left[\sum_{i=k-d_2}^{k-d_1-1} \beta(i)\varsigma^T(i)\right]\left[\sum_{i=k-d_2}^{k-d_1-1} \beta(i)\varsigma(i)\right] \\
&\leq \frac{1}{d^2}\sum_{i=0}^{\infty}\left[(d_2-d_1)\sum_{i=k-d_2}^{k-d_1-1} \beta^2(i)\varsigma^T(i)\varsigma(i)\right] \\
&\leq \frac{1}{d}\sum_{j=-d_2}^{-d_1-1}\sum_{i=0}^{\infty} \varsigma^T(i)\varsigma(i) \\
&= \sum_{i=0}^{\infty} \varsigma^T(i)\varsigma(i) = \|\varsigma(k)\|_2^2,
\end{aligned}$$

which implies

$$\|\mathcal{K}\|_\infty = \sup_{\|\varsigma(k)\|_2 \neq 0} \frac{\|\sigma(k)\|_2}{\|\varsigma(k)\|_2} \leq 1.$$

This completes the proof. ∎

Notice from Lemma 2.1 that the norm of $(\mathcal{S}_2)$ in (2.2b) from input to output is bounded by one. In the follows, based on Lemma 1.28, we focus on the scaled small gain of $(\mathcal{S}_1)$ for the interconnection frame in (2.2a).

Lemma 2.2. *Assumed that the $(\mathcal{S}_1)$ is internally stable in (2.2a), the closed-loop system of the interconnection system described by (2.2) is asymptotically stable for $\mathcal{K}$ if there exist matrix $\bar{\mathcal{X}} > 0$ such that*

$$\left\|\bar{\mathcal{X}} \circ \mathcal{G} \circ \bar{\mathcal{X}}^{-1}\right\|_\infty < 1,$$

where

$$\mathcal{G} \triangleq \begin{bmatrix} \bar{A}(k) & \frac{1}{2}\bar{A}_d(k) & \frac{1}{2}\bar{A}_d(k) & \frac{d}{2}\bar{A}_d(k) \\ \bar{A}(k) - I & \frac{1}{2}\bar{A}_d(k) & \frac{1}{2}\bar{A}_d(k) & \frac{d}{2}\bar{A}_d(k) \end{bmatrix}.$$

Remark 2.3. Along the interconnection frame (2.2), the sufficient condition in Lemma 2.2 can be converted to another one, that is, assumed that $(\mathcal{S}_1)$ is internally stable in (2.2a), the closed-loop system of interconnection system described by (2.2) is asymptotically stable for $\mathcal{K}$ if there exist exists a matrix $\mathcal{X} \triangleq \bar{\mathcal{X}}^T\bar{\mathcal{X}}$ such that $\sum_{k=0}^{\infty}\left[\varsigma^T(k)\mathcal{X}\varsigma(k) - \sigma^T(k)\mathcal{X}\sigma(k)\right] < 0$. ♦

2.3 Main Results

In the following, by using some different approaches, we present some stability conditions for the T-S fuzzy time-delay system in (2.1).

2.3.1 Delay Partitioning Approach

To employ the delay partitioning approach, firstly, we partition $d(k)$ into two parts: constant part τm and time-varying part $h(k)$, that is, $d(k) = \tau m + h(k)$, where $h(k)$ satisfies $0 \leq h(k) \leq d_2 - \tau m \triangleq \tilde{d}$. Define

$$\begin{cases} P(k) \triangleq \sum_{i=1}^{r} h_i(\theta)P_i, & R(k) \triangleq \sum_{i=1}^{r} h_i(\theta)R_i, \\ R_1(k) \triangleq \sum_{i=1}^{r} h_i(\theta)R_{1i}, & R_2(k) \triangleq \sum_{i=1}^{r} h_i(\theta)R_{2i}, \\ S_1(k) \triangleq \sum_{i=1}^{r} h_i(\theta)S_{1i}, & S_2(k) \triangleq \sum_{i=1}^{r} h_i(\theta)S_{2i}, \\ Z_1(k) \triangleq \sum_{i=1}^{r} h_i(\theta)Z_{1i}, & Z_2(k) \triangleq \sum_{i=1}^{r} h_i(\theta)Z_{2i}, \\ F(k) \triangleq \sum_{i=1}^{r} h_i(\theta)F_i, & G(k) \triangleq \sum_{i=1}^{r} h_i(\theta)G_i, \\ X_v(k) \triangleq \sum_{i=1}^{r} h_i(\theta)X_{vi}, & Y_v(k) \triangleq \sum_{i=1}^{r} h_i(\theta)Y_{vi}, \\ B_v(k) \triangleq \sum_{i=1}^{r} h_i(\theta)B_{vi}, & D_v(k) \triangleq \sum_{i=1}^{r} h_i(\theta)D_{vi}, \\ E_v(k) \triangleq \sum_{i=1}^{r} h_i(\theta)E_{vi}, & Q_\alpha(k) \triangleq \sum_{i=1}^{r} h_i(\theta)Q_{Ni}, \end{cases} \tag{2.3}$$

where $P_i > 0$, $Z_{1i} > 0$, $Z_{2i} > 0$, $Q_{Ni} > 0$, $S_{1i} > 0$, $S_{2i} > 0$, $R_{1i} > 0$, $R_{2i} > 0$, $R_i > 0$, $i = 1, 2, \ldots, r$; $v = 0, 1, \ldots, m + 2$ and $\alpha = 0, 1, \ldots, m - 1$.

Construct the following LKF:

$$V(k) \triangleq \sum_{i=1}^{4} V_i(k), \tag{2.4}$$

with

$$\begin{cases} V_1(k) \triangleq x^T(k)\hat{P}(k)x(k) \\ V_2(k) \triangleq \sum_{s=-\tau}^{-1} \sum_{l=k+s}^{k-1} \eta^T(l)\hat{Z}_1(l)\eta(l) + \sum_{s=-d_2}^{-\tau m-1} \sum_{l=k+s}^{k-1} \eta^T(l)\hat{Z}_2(l)\eta(l) \\ V_3(k) \triangleq \sum_{\alpha=0}^{m-1} \sum_{l=k-\tau}^{k-1} x^T(l-\alpha\tau)\hat{Q}_\alpha(k,l)x(l-\alpha\tau) \\ \qquad + \sum_{l=k-d_2}^{k-1} x^T(l)\hat{S}_1(k,l)x(l) \\ V_4(k) \triangleq \sum_{s=-d_2+1}^{-\tau m+1} \sum_{l=k+s-1}^{k-1} x^T(l)\hat{S}_2(k,l)x(l) \\ \eta(l) \triangleq x(l+1) - x(l), \end{cases}$$

where $\hat{P}(k) \triangleq G^{-T}(k)P(k)G^{-1}(k)$, $\hat{Z}_1(k) \triangleq Z_1^{-1}(k)$, $\hat{Z}_2(k) \triangleq Z_2^{-1}(k)$, $\hat{S}_1(k,l) \triangleq F^{-T}(k)S_1(l)F^{-1}(k)$, $\hat{S}_2(k,l) \triangleq F^{-T}(k)S_2(l)F^{-1}(k)$ and $\hat{Q}_\alpha(k,l) \triangleq F^{-T}(k)Q_\alpha(l)F^{-1}(k)$, $\alpha = 0, 1, \ldots, m-1$.

Then, based on the fuzzy LKF in (2.4), the following result can be obtained.

Theorem 2.4. *Given positive integers τ, m and d_2, system (2.1) is asymptotically stable if there exist matrices $P(k) > 0$, $Z_1(k) > 0$, $Z_2(k) > 0$, $Q_\alpha(k) > 0$, $S_1(k) > 0$, $S_2(k) > 0$, $R_1(k) > 0$, $R_2(k) > 0$, $R(k) > 0$, $X_v(k)$, $Y_v(k)$, $B_v(k)$, $D_v(k)$, $E_v(k)$, ($v = 0, 1, \ldots, m+2$ and $\alpha = 0, 1, \ldots, m-1$), and nonsingular matrices $F(k)$ and $G(k)$, which are defined in (2.3), such that for any positive scalars k, s and ε, $F(k) = \varepsilon G(k)$ and the following inequalities hold:*

$$\begin{bmatrix} \Pi_{11}(k) & \Pi_{12}(k) & 0 & 0 & \cdots & 0 & 0 & \Pi_{17}(k) & 0 \\ \star & \Pi_{22}(k) & \Pi_{23}(k) & \Pi_{24}(k) & \cdots & \Pi_{25}(k) & \Pi_{26}(k) & \Pi_{27}(k) & \Pi_{28}(k) \\ \star & \star & \Pi_{33}(k) & \Pi_{34}(k) & \cdots & \Pi_{35}(k) & \Pi_{36}(k) & \Pi_{37}(k) & \Pi_{38}(k) \\ \star & \star & \star & \Pi_{44}(k) & \cdots & 0 & \Pi_{46}(k) & \Pi_{47}(k) & \Pi_{48}(k) \\ \star & \star & \star & \star & \ddots & \vdots & \vdots & \vdots & \vdots \\ \star & \star & \star & \star & \star & \Pi_{55}(k) & \Pi_{56}(k) & \Pi_{57}(k) & \Pi_{58}(k) \\ \star & \star & \star & \star & \star & \star & \Pi_{66}(k) & \Pi_{67}(k) & \Pi_{68}(k) \\ \star & \star & \star & \star & \star & \star & \star & \Pi_{77}(k) & \Pi_{78}(k) \\ \star & \star & \star & \star & \star & \star & \star & \star & \Pi_{88}(k) \end{bmatrix} < 0, \text{(2.5a)}$$

$$\begin{bmatrix} X_0(k) & 0 & 0 & 0 & Y_0(k) \\ \star & X_1(k) & 0 & 0 & Y_1(k) \\ \star & \star & \ddots & 0 & \vdots \\ \star & \star & \star & X_{m+2}(k) & Y_{m+2}(k) \\ \star & \star & \star & \star & \varepsilon G^T(k) + \varepsilon G(k) - R_1(k) \end{bmatrix} \geq 0, \text{(2.5b)}$$

$$\begin{bmatrix} E_0(k) & 0 & 0 & 0 & B_0(k) \\ \star & E_1(k) & 0 & 0 & B_1(k) \\ \star & \star & \ddots & 0 & \vdots \\ \star & \star & \star & E_{m+2}(k) & B_{m+2}(k) \\ \star & \star & \star & \star & \varepsilon G^T(k) + \varepsilon G(k) - R_2(k) \end{bmatrix} \geq 0, \quad (2.5c)$$

$$\begin{bmatrix} E_0(k) & 0 & 0 & 0 & D_0(k) \\ \star & E_1(k) & 0 & 0 & D_1(k) \\ \star & \star & \ddots & 0 & \vdots \\ \star & \star & \star & E_{m+2}(k) & D_{m+2}(k) \\ \star & \star & \star & \star & \varepsilon G^T(k) + \varepsilon G(k) - R_2(k) \end{bmatrix} \geq 0, \quad (2.5d)$$

$$Z_1(k) - R_1(s) < 0, \quad (2.5e)$$

$$Z_2(k) - R_2(s) < 0, \quad (2.5f)$$

$$R(k) - S_2(s) < 0, \quad (2.5g)$$

where

$$\Pi_{12}(k) \triangleq \begin{bmatrix} \left(\bar{A}(k) - I\right) G(k) \\ \left(\bar{A}(k) - I\right) G(k) \\ \bar{A}(k) G(k) \end{bmatrix}, \quad \Pi_{17}(k) \triangleq \begin{bmatrix} \varepsilon \bar{A}_d(k) G(k) \\ \varepsilon \bar{A}_d(k) G(k) \\ \varepsilon \bar{A}_d(k) G(k) \end{bmatrix},$$

$$\Pi_{11}(k) \triangleq \mathrm{diag}\left\{-\tau^{-1} Z_1(k), -\tilde{d}^{-1} Z_2(k), P(k+1) - G(k+1) - G^T(k+1)\right\},$$

$$\Pi_{22}(k) \triangleq \varepsilon^{-2}\left[Q_0(k) + S_1(k) + (\tilde{d}+1) S_2(k)\right] + \varepsilon^{-1}\left[Y_0(k) + Y_0^T(k)\right]$$
$$-P(k) + J_0(k), \quad J_\nu(k) \triangleq \tau X_\nu(k) + \tilde{d} E_\nu(k),$$

$$\Pi_{33}(k) \triangleq \varepsilon^2 J_1(k) - \varepsilon\left[Y_1(k) + Y_1^T(k)\right] + Q_1(k) - Q_0(k-\tau),$$

$$\Pi_{44}(k) \triangleq \varepsilon^2 J_2(k) + Q_2(k) - Q_1(k-\tau),$$

$$\Pi_{55}(k) \triangleq \varepsilon^2 J_{m-1}(k) + Q_{m-1}(k) - Q_{m-2}(k-\tau),$$

$$\Pi_{66}(k) \triangleq \varepsilon^2 J_m(k) + \varepsilon\left[B_m(k) + B_m^T(k)\right] - Q_{m-1}(k-\tau),$$

$$\Pi_{77}(k) \triangleq \varepsilon^2 J_{m+1}(k) - \varepsilon\left[B_{m+1}(k) + B_{m+1}^T(k)\right] + \varepsilon\left[D_{m+1}(k) + D_{m+1}^T(k)\right] - R(k),$$

$$\Pi_{88}(k) \triangleq \varepsilon^2 J_{m+2}(k) - \varepsilon\left[D_{m+2}(k) + D_{m+2}^T(k)\right] - S_1(k-d_2),$$

$$\Pi_{23}(k) \triangleq Y_1(k) - Y_0^T(k), \quad \Pi_{24}(k) \triangleq Y_2(k), \quad \Pi_{25}(k) \triangleq Y_{m-1}(k),$$

$$\Pi_{26}(k) \triangleq Y_m(k) + B_0^T(k), \quad \Pi_{28}(k) \triangleq Y_{m+2}(k) - D_0^T(k),$$

$$\Pi_{27}(k) \triangleq Y_{m+1}(k) - B_0^T(k) + D_0^T(k), \quad \Pi_{34}(k) \triangleq -\varepsilon Y_2(k),$$

$$\Pi_{35}(k) \triangleq -\varepsilon Y_{m-1}(k), \quad \Pi_{36}(k) \triangleq -\varepsilon Y_m(k) + \varepsilon B_1^T(k), \quad \Pi_{46}(k) \triangleq \varepsilon B_2^T(k),$$

$$\Pi_{37}(k) \triangleq -\varepsilon\left[Y_{m+1}(k) + B_1^T(k) - D_1^T(k)\right], \quad \Pi_{47}(k) \triangleq -\varepsilon B_2^T(k) + \varepsilon D_2^T(k),$$

$$\Pi_{38}(k) \triangleq -\varepsilon Y_{m+2}(k) - \varepsilon D_1^T(k), \ \Pi_{48}(k) \triangleq -\varepsilon D_2^T(k), \ \Pi_{58}(k) \triangleq -\varepsilon D_{m-1}^T(k),$$

$$\Pi_{56}(k) \triangleq \varepsilon B_{m-1}^T(k), \quad \Pi_{57}(k) \triangleq -\varepsilon\left[B_{m-1}(k) - D_{m-1}^T(k)\right],$$

$$\Pi_{67}(k) \triangleq \varepsilon B_{m+1}(k) - \varepsilon B_m^T(k) + D_m^T(k), \quad \Pi_{68}(k) \triangleq \varepsilon B_{m+2}(k) - \varepsilon D_m^T(k),$$

$$\Pi_{78}(k) \triangleq -\varepsilon\left[B_{m+2}(k) - D_{m+2}(k) + D_{m+1}^T(k)\right].$$

Proof. From the facts of

$$\begin{aligned}
[P(k+1)-G(k+1)]\,P^{-1}(k+1)\,[P(k+1)-G(k+1)]^T &\geq 0,\\
[R_1(k)-F(k)]^T\,\hat{R}_1(k)\,[R_1(k)-F(k)] &\geq 0,\\
[R_2(k)-F(k)]^T\,\hat{R}_2(k)\,[R_2(k)-F(k)] &\geq 0,
\end{aligned}$$

where $\hat{R}_1(k) \triangleq R_1^{-1}(k)$ and $\hat{R}_2(k) \triangleq R_2^{-1}(k)$, we have

$$\begin{aligned}
-\hat{P}^{-1}(k+1) &\leq -G(k+1)-G^T(k+1)+P(k+1),\\
F^T(k)\hat{R}_1(k)F(k) &\geq F(k)+F(k)-R_1(k),\\
F^T(k)\hat{R}_2(k)F(k) &\geq F(k)+F(k)-R_2(k).
\end{aligned}$$

Thus, it follows from (2.5a)–(2.5d) and $F(k)=\varepsilon G(k)$ that

$$\begin{bmatrix}
\hat{\Pi}_{11}(k) & \Pi_{12}(k) & 0 & 0 & \cdots & 0 & 0 & \Pi_{17}(k) & 0\\
\star & \Pi_{22}(k) & \Pi_{23}(k) & \Pi_{24}(k) & \cdots & \Pi_{25}(k) & \Pi_{26}(k) & \Pi_{27}(k) & \Pi_{28}(k)\\
\star & \star & \Pi_{33}(k) & \Pi_{34}(k) & \cdots & \Pi_{35}(k) & \Pi_{36}(k) & \Pi_{37}(k) & \Pi_{38}(k)\\
\star & \star & \star & \Pi_{44}(k) & \cdots & 0 & \Pi_{46}(k) & \Pi_{47}(k) & \Pi_{48}(k)\\
\star & \star & \star & \star & \ddots & \vdots & \vdots & \vdots & \vdots\\
\star & \star & \star & \star & \star & \Pi_{55}(k) & \Pi_{56}(k) & \Pi_{57}(k) & \Pi_{58}(k)\\
\star & \star & \star & \star & \star & \star & \Pi_{66}(k) & \Pi_{67}(k) & \Pi_{68}(k)\\
\star & \star & \star & \star & \star & \star & \star & \Pi_{77}(k) & \Pi_{78}(k)\\
\star & \star & \star & \star & \star & \star & \star & \star & \Pi_{88}(k)
\end{bmatrix} < 0, \quad (2.6)$$

$$\begin{bmatrix}
X_0(k) & 0 & 0 & 0 & Y_0(k)\\
\star & X_1(k) & 0 & 0 & Y_1(k)\\
\star & \star & \ddots & 0 & \vdots\\
\star & \star & \star & X_{m+2}(k) & Y_{m+2}(k)\\
\star & \star & \star & \star & F^T(k)\hat{R}_1(k)F(k)
\end{bmatrix} \geq 0, \quad (2.7)$$

$$\begin{bmatrix}
E_0(k) & 0 & 0 & 0 & B_0(k)\\
\star & E_1(k) & 0 & 0 & B_1(k)\\
\star & \star & \ddots & 0 & \vdots\\
\star & \star & \star & E_{m+2}(k) & B_{m+2}(k)\\
\star & \star & \star & \star & F^T(k)\hat{R}_2(k)F(k)
\end{bmatrix} \geq 0, \quad (2.8)$$

$$\begin{bmatrix}
E_0(k) & 0 & 0 & 0 & D_0(k)\\
\star & E_1(k) & 0 & 0 & D_1(k)\\
\star & \star & \ddots & 0 & \vdots\\
\star & \star & \star & E_{m+2}(k) & D_{m+2}(k)\\
\star & \star & \star & \star & F^T(k)\hat{R}_2(k)F(k)
\end{bmatrix} \geq 0, \quad (2.9)$$

where

$$\hat{\Pi}_{11}(k) \triangleq \text{diag}\left\{-\tau^{-1}Z_1(k), -\tilde{d}^{-1}Z_2(k), -\hat{P}^{-1}(k+1)\right\}.$$

Define the following matrices:

$$\begin{cases} \mathcal{T}_1(k) \triangleq \text{diag}\{I, I, I, G^{-1}(k), F^{-1}(k), \cdots, F^{-1}(k), F^{-1}(k)\}, \\ \mathcal{T}_2(k) \triangleq \text{diag}\{G^{-1}(k), G^{-1}(k), \cdots, G^{-1}(k), F^{-1}(k)\}, \\ \mathcal{T}_3(k) \triangleq \text{diag}\{G^{-1}(k), G^{-1}(k), \cdots, G^{-1}(k), F^{-1}(k)\}, \\ \mathcal{T}_4(k) \triangleq \text{diag}\{G^{-1}(k), G^{-1}(k), \cdots, G^{-1}(k), F^{-1}(k)\}. \end{cases}$$

Performing congruence transformations to (2.6)–(2.9) by matrices $\mathcal{T}_1(k)$, $\mathcal{T}_2(k)$, $\mathcal{T}_3(k)$ and $\mathcal{T}_4(k)$, respectively, and considering (2.3), we have

$$\begin{bmatrix} \hat{\Pi}_{11}(k) & \hat{\Pi}_{12}(k) & 0 & 0 & \cdots & 0 & 0 & \hat{\Pi}_{17}(k) & 0 \\ \star & \hat{\Pi}_{22}(k) & \hat{\Pi}_{23}(k) & \hat{\Pi}_{24}(k) & \cdots & \hat{\Pi}_{25}(k) & \hat{\Pi}_{26}(k) & \hat{\Pi}_{27}(k) & \hat{\Pi}_{28}(k) \\ \star & \star & \hat{\Pi}_{33}(k) & \hat{\Pi}_{34}(k) & \cdots & \hat{\Pi}_{35}(k) & \hat{\Pi}_{36}(k) & \hat{\Pi}_{37}(k) & \hat{\Pi}_{38}(k) \\ \star & \star & \star & \hat{\Pi}_{44}(k) & \cdots & 0 & \hat{\Pi}_{46}(k) & \hat{\Pi}_{47}(k) & \hat{\Pi}_{48}(k) \\ \star & \star & \star & \star & \ddots & \vdots & \vdots & \vdots & \vdots \\ \star & \star & \star & \star & \star & \hat{\Pi}_{55}(k) & \hat{\Pi}_{56}(k) & \hat{\Pi}_{57}(k) & \hat{\Pi}_{58}(k) \\ \star & \star & \star & \star & \star & \star & \hat{\Pi}_{66}(k) & \hat{\Pi}_{67}(k) & \hat{\Pi}_{68}(k) \\ \star & \star & \star & \star & \star & \star & \star & \hat{\Pi}_{77}(k) & \hat{\Pi}_{78}(k) \\ \star & \star & \star & \star & \star & \star & \star & \star & \hat{\Pi}_{88}(k) \end{bmatrix} < 0, \quad (2.10)$$

$$\Psi_1(k) \triangleq \begin{bmatrix} \hat{X}_0(k) & 0 & 0 & 0 & \hat{Y}_0(k) \\ \star & \hat{X}_1(k) & 0 & 0 & \hat{Y}_1(k) \\ \star & \star & \ddots & 0 & \vdots \\ \star & \star & \star & \hat{X}_{m+2}(k) & \hat{Y}_{m+2}(k) \\ \star & \star & \star & \star & \hat{R}_1(k) \end{bmatrix} \geq 0, \quad (2.11)$$

$$\Psi_2(k) \triangleq \begin{bmatrix} \hat{E}_0(k) & 0 & 0 & 0 & \hat{B}_0(k) \\ \star & \hat{E}_1(k) & 0 & 0 & \hat{B}_1(k) \\ \star & \star & \ddots & 0 & \vdots \\ \star & \star & \star & \hat{E}_{m+2}(k) & \hat{B}_{m+2}(k) \\ \star & \star & \star & \star & \hat{R}_2(k) \end{bmatrix} \geq 0, \quad (2.12)$$

$$\Psi_3(k) \triangleq \begin{bmatrix} \hat{E}_0(k) & 0 & 0 & 0 & \hat{D}_0(k) \\ \star & \hat{E}_1(k) & 0 & 0 & \hat{D}_1(k) \\ \star & \star & \ddots & 0 & \vdots \\ \star & \star & \star & \hat{E}_{m+2}(k) & \hat{D}_{m+2}(k) \\ \star & \star & \star & \star & \hat{R}_2(k) \end{bmatrix} \geq 0, \quad (2.13)$$

where $\hat{\Pi}_{11}(k)$ is defined in (2.6) and

$$\hat{\Pi}_{12}(k) \triangleq \begin{bmatrix} \bar{A}(k) - I \\ \bar{A}(k) - I \\ \bar{A}(k) \end{bmatrix}, \; \hat{\Pi}_{17}(k) \triangleq \begin{bmatrix} \bar{A}_d(k) \\ \bar{A}_d(k) \\ \bar{A}_d(k) \end{bmatrix},$$

$$\begin{aligned}
\hat{\Pi}_{22}(k) &\triangleq \hat{Y}_0^T(k) + \hat{Y}_0(k) + \hat{S}_1(k,k) + (\tilde{d}+1)\hat{S}_2(k,k) + \hat{Q}_0(k,k) - \hat{P}(k) + \hat{J}_0(k), \\
\hat{\Pi}_{33}(k) &\triangleq \hat{Q}_1(k,k) - \hat{Q}_0(k,k-\tau) + \hat{J}_1(k) - \hat{Y}_1^T(k) - \hat{Y}_1(k), \\
\hat{\Pi}_{44}(k) &\triangleq \hat{Q}_2(k,k) - \hat{Q}_1(k,k-\tau) + \hat{J}_2(k), \\
\hat{\Pi}_{55}(k) &\triangleq \hat{Q}_{m-1}(k,k) - \hat{Q}_{m-2}(k,k-\tau) + \hat{J}_{m-1}(k), \\
\hat{\Pi}_{66}(k) &\triangleq -\hat{Q}_{m-1}(k,k-\tau) + \hat{B}_m(k) + \hat{B}_m^T(k) + \hat{J}_m(k), \\
\hat{\Pi}_{77}(k) &\triangleq -\hat{B}_{m+1}(k) - \hat{B}_{m+1}^T(k) + \hat{D}_{m+1}(k) + \hat{D}_{m+1}^T(k) - \hat{R}(k,k) + \hat{J}_{m+1}(k), \\
\hat{\Pi}_{88}(k) &\triangleq -\hat{D}_{m+2}(k) - \hat{D}_{m+2}^T(k) - \hat{S}_1(k,k-d_2) + \hat{J}_{m+2}(k), \\
\hat{\Pi}_{23}(k) &\triangleq \hat{Y}_1(k) - \hat{Y}_0^T(k), \; \hat{J}_\nu(k) \triangleq \tau\hat{X}_\nu(k) + \tilde{d}\hat{E}_\nu(k), \\
\hat{\Pi}_{24}(k) &\triangleq \hat{Y}_2(k), \; \hat{\Pi}_{25}(k) \triangleq \hat{Y}_{m-1}(k), \; \hat{\Pi}_{26}(k) \triangleq \hat{Y}_m(k) + \hat{B}_0^T(k), \\
\hat{\Pi}_{27}(k) &\triangleq \hat{Y}_{m+1}(k) - \hat{B}_0^T(k) + \hat{D}_0^T(k), \; \hat{\Pi}_{28}(k) \triangleq \hat{Y}_{m+2}(k) - \hat{D}_0^T(k), \\
\hat{\Pi}_{34}(k) &\triangleq -\hat{Y}_2(k), \; \hat{\Pi}_{35}(k) \triangleq -\hat{Y}_{m-1}(k), \; \hat{\Pi}_{36}(k) \triangleq -\hat{Y}_m(k) + \hat{B}_1^T(k), \\
\hat{\Pi}_{37}(k) &\triangleq -\hat{Y}_{m+1}(k) - \hat{B}_1^T(k) + \hat{D}_1^T(k), \; \hat{\Pi}_{38}(k) \triangleq -\hat{Y}_{m+2}(k) - \hat{D}_1^T(k), \\
\hat{\Pi}_{46}(k) &\triangleq \hat{B}_2^T(k), \; \hat{\Pi}_{47}(k) \triangleq -\hat{B}_2^T(k) + \hat{D}_2^T(k), \; \hat{\Pi}_{48}(k) \triangleq -\hat{D}_2^T(k), \\
\hat{\Pi}_{56}(k) &\triangleq \hat{B}_{m-1}^T(k), \; \hat{\Pi}_{57}(k) \triangleq -\hat{B}_{m-1}^T(k) + \hat{D}_{m-1}^T(k), \; \hat{\Pi}_{58}(k) \triangleq -\hat{D}_{m-1}^T(k), \\
\hat{\Pi}_{67}(k) &\triangleq \hat{B}_{m+1}(k) - \hat{B}_m^T(k) + \hat{D}_m^T(k), \; \hat{\Pi}_{68}(k) \triangleq \hat{B}_{m+2}(k) - \hat{D}_m^T(k), \\
\hat{\Pi}_{78}(k) &\triangleq -\hat{B}_{m+2}(k) + \hat{D}_{m+2}(k) - \hat{D}_{m+1}^T(k).
\end{aligned}$$

with $\hat{R}(k,l) \triangleq F^{-T}(k)R(l)F^{-1}(k)$, $\hat{X}_\upsilon(k) \triangleq G^{-T}(k)X_\upsilon(k)G^{-1}(k)$, $\hat{Y}_\upsilon(k) \triangleq G^{-T}(k)\,Y_\upsilon(k)F^{-1}(k)$, $\hat{B}_\upsilon(k) \triangleq G^{-T}(k)B_\upsilon(k)F^{-1}(k)$, $\hat{D}_\upsilon(k) \triangleq G^{-T}(k)\,D_\upsilon(k)F^{-1}(k)$, and $\hat{E}_\upsilon(k) \triangleq G^{-T}(k)E_\upsilon(k)G^{-1}(k)$, $(\upsilon = 0, 1, \ldots, m+2)$.

By Schur complement, it follows that (2.10) is equivalent to

$$\hat{\Pi}(k) \triangleq \begin{bmatrix}
\tilde{\Pi}_{22}(k) & \hat{\Pi}_{23}(k) & \hat{\Pi}_{24}(k) & \cdots & \hat{\Pi}_{25}(k) & \hat{\Pi}_{26}(k) & \tilde{\Pi}_{27}(k) & \hat{\Pi}_{28}(k) \\
\star & \hat{\Pi}_{33}(k) & \hat{\Pi}_{34}(k) & \cdots & \hat{\Pi}_{35}(k) & \hat{\Pi}_{36}(k) & \hat{\Pi}_{37}(k) & \hat{\Pi}_{38}(k) \\
\star & \star & \hat{\Pi}_{44}(k) & \cdots & 0 & \hat{\Pi}_{46}(k) & \hat{\Pi}_{47}(k) & \hat{\Pi}_{48}(k) \\
\star & \star & \star & \ddots & \vdots & \vdots & \vdots & \vdots \\
\star & \star & \star & \star & \hat{\Pi}_{55}(k) & \hat{\Pi}_{56}(k) & \hat{\Pi}_{57}(k) & \hat{\Pi}_{58}(k) \\
\star & \star & \star & \star & \star & \hat{\Pi}_{66}(k) & \hat{\Pi}_{67}(k) & \hat{\Pi}_{68}(k) \\
\star & \star & \star & \star & \star & \star & \tilde{\Pi}_{77}(k) & \hat{\Pi}_{78}(k) \\
\star & \star & \star & \star & \star & \star & \star & \hat{\Pi}_{88}(k)
\end{bmatrix} < 0, \quad (2.14)$$

where

$$\begin{aligned}
\tilde{\Pi}_{22}(k) &\triangleq \hat{\Pi}_{33}(k) + \Sigma_1(k), \\
\tilde{\Pi}_{27}(k) &\triangleq \hat{\Pi}_{39}(k) + \Sigma_2(k), \\
\tilde{\Pi}_{77}(k) &\triangleq \hat{\Pi}_{99}(k) + \Sigma_3(k),
\end{aligned}$$

with

$$\Sigma_1(k) \triangleq \left[\bar{A}(k) - I\right]^T \left[\tau\hat{Z}_1(k) + \tilde{d}\hat{Z}_2(k)\right]\left[\bar{A}(k) - I\right] + \bar{A}^T(k)\hat{P}(k+1)\bar{A}(k),$$

$$\Sigma_2(k) \triangleq \left[\bar{A}(k) - I\right]^T \left[\tau\hat{Z}_1(k) + \tilde{d}\hat{Z}_2(k)\right]\bar{A}_d(k) + \bar{A}^T(k)\hat{P}(k+1)\bar{A}_d(k),$$

$$\Sigma_3(k) \triangleq \bar{A}_d^T(k)\left[\tau\hat{Z}_1(k) + \tilde{d}\hat{Z}_2(k)\right]\bar{A}_d(k) + \bar{A}_d^T(k)\hat{P}(k+1)\bar{A}_d(k).$$

Consider LKF (2.4), and along the trajectories of system (2.1), we have

$$\begin{aligned}
\Delta V_1(k) &= x^T(k+1)\hat{P}(k+1)x(k+1) - x^T(k)\hat{P}(k)x(k)\\
\Delta V_2(k) &\le \eta^T(k)\left[\tau\hat{Z}_1(k) + \tilde{d}\hat{Z}_2(k)\right]\eta(k) - \sum_{s=k-\tau}^{k-1}\eta^T(s)\hat{R}_1(k)\eta(s)\\
&\quad - \sum_{s=k-d(k)}^{k-\tau m-1}\eta^T(s)\hat{R}_2(k)\eta(s) - \sum_{s=k-d_2}^{k-d(k)-1}\eta^T(s)\hat{R}_2(k)\eta(s)\\
\Delta V_3(k) &= \sum_{N=0}^{m-1}\Big[x^T(k-N\tau)\hat{Q}_N(k,k)x(k-N\tau)\\
&\quad - x^T(k-(N+1)\tau)\hat{Q}_N(k,k-\tau)x(k-(N+1)\tau)\Big]\\
&\quad + x^T(k)\hat{S}_1(k,k)x(k) - x^T(k-d_2)\hat{S}_1(k,k-d_2)x(k-d_2)\\
\Delta V_4(k) &\le (\tilde{d}+1)x^T(k)\hat{S}_2(k,k)x(k) - x^T(k-d(k))\hat{R}(k,k)x(k-d(k)).
\end{aligned}$$

On the other hand, for any matrices

$$\mathbf{Y}(k) \triangleq \begin{bmatrix}\hat{Y}_0(k)\\ \hat{Y}_1(k)\\ \vdots\\ \hat{Y}_{m+2}(k)\end{bmatrix}, \quad \mathbf{B}(k) \triangleq \begin{bmatrix}\hat{B}_0(k)\\ \hat{B}_1(k)\\ \vdots\\ \hat{B}_{m+2}(k)\end{bmatrix}, \quad \mathbf{D}(k) \triangleq \begin{bmatrix}\hat{D}_0(k)\\ \hat{D}_1(k)\\ \vdots\\ \hat{D}_{m+2}(k)\end{bmatrix},$$

the following equations always hold:

$$2\hat{x}^T(k)\mathbf{Y}(k)\left[x(k) - x(k-\tau) - \sum_{s=k-\tau}^{k-1}\eta(s)\right] = 0,$$

$$2\hat{x}^T(k)\mathbf{B}(k)\left[x(k-\tau m) - x(k-d(k)) - \sum_{s=k-d(k)}^{k-\tau m-1}\eta(s)\right] = 0,$$

$$2\hat{x}^T(k)\mathbf{D}(k)\left[x(k-d(k)) - x(k-d_2) - \sum_{s=k-d_2}^{k-d(k)-1}\eta(s)\right] = 0,$$

where

$$\hat{x}(k) \triangleq \begin{bmatrix} x(k) \\ x(k-\tau) \\ \vdots \\ x(k-(m-1)\tau) \\ x(k-\tau m) \\ x(k-d(k)) \\ x(k-d_2) \end{bmatrix}.$$

Moreover, for any matrices $\mathbf{X}(k) \triangleq \mathrm{diag}\left\{\hat{X}_0(k), \hat{X}_1(k), \ldots, \hat{X}_{m+2}(k)\right\}$ and $\mathbf{E}(k) \triangleq \mathrm{diag}\left\{\hat{E}_0(k), \hat{E}_1(k), \ldots, \hat{E}_{m+2}(k)\right\}$, the following identities hold:

$$0 = \tilde{d}\hat{x}^T(k)\mathbf{E}(k)\hat{x}(k) - \sum_{s=k-d(k)}^{k-\tau m-1} \hat{x}^T(k)\mathbf{E}(k)\hat{x}(k) - \sum_{s=k-d_2}^{k-d(k)-1} \hat{x}^T(k)\mathbf{E}(k)\hat{x}(k),$$

$$0 = \tau\hat{x}^T(k)\mathbf{X}(k)\hat{x}(k) - \sum_{s=k-\tau}^{k-1} \hat{x}^T(k)\mathbf{X}(k)\hat{x}(k).$$

Then from the above analysis, we have

$$\begin{aligned} \Delta V(k) &= \Delta V_1(k) + \Delta V_2(k) + \Delta V_3(k) + \Delta V_4(k) \\ &\leq -\sum_{s=k-\tau}^{k-1} \begin{bmatrix} \hat{x}(k) \\ \eta(s) \end{bmatrix}^T \Psi_1(k) \begin{bmatrix} \hat{x}(k) \\ \eta(s) \end{bmatrix} - \sum_{s=k-d(k)}^{k-\tau m-1} \begin{bmatrix} \hat{x}(k) \\ \eta(s) \end{bmatrix}^T \Psi_2(k) \begin{bmatrix} \hat{x}(k) \\ \eta(s) \end{bmatrix} \\ &\quad - \sum_{s=k-d_2}^{k-d(k)-1} \begin{bmatrix} \hat{x}(k) \\ \eta(s) \end{bmatrix}^T \Psi_3(k) \begin{bmatrix} \hat{x}(k) \\ \eta(s) \end{bmatrix} + \hat{x}^T(k)\hat{\Pi}(k)\hat{x}(k), \end{aligned}$$

where $\Psi_i(k), i = 1, 2, 3$ and $\hat{\Pi}(k)$ are defined in (2.11)–(2.13) and (2.14), respectively. It is shown from inequalities (2.11)–(2.14) that $\Delta V(k) < 0$, which concludes from the Lyapunov stability theory that the T-S fuzzy time-delay system in (2.1) is asymptotically stable. ■

Note that Theorem 2.4, expressed in the form of time-dependent matrix inequalities, cannot be directly implemented for the stability analysis. Our next objective is to convert these inequalities into a set of LMIs, which thus can be readily solved using standard software. We have the following theorem.

Theorem 2.5. *Given positive integers τ, m and d_2, system (2.1) is asymptotically stable if there exist matrices $P_i > 0$, $Z_{1i} > 0$, $Z_{2i} > 0$, $Q_{\alpha i} > 0$, $S_{1i} > 0$, $S_{2i} > 0$, $R_{1i} > 0$, $R_{2i} > 0$, $R_i > 0$, X_{vi}, E_{vi}, Y_{vi}, B_{vi}, D_{vi}, and nonsingular matrices G_i and F_i ($i = 1, 2, \ldots, r$; $\alpha = 0, 1, \ldots, m-1$; $v = 0, 1, \ldots, m+2$) such that for any positive scalar ε, $F_i = \varepsilon G_i$ and*

$$\Phi_{ostii} < 0,$$
$$o, s, t, i = 1, 2, \ldots, r, \tag{2.15a}$$
$$\frac{1}{r-1}\Phi_{ostii} + \frac{1}{2}(\Phi_{ostij} + \Phi_{ostji}) < 0,$$
$$o, s, t, i, j = 1, 2, \ldots, r;\ 1 \le i \ne j \le r, \tag{2.15b}$$
$$\Psi_{1i} \triangleq \begin{bmatrix} X_{0i} & 0 & 0 & 0 & Y_{0i} \\ \star & X_{1i} & 0 & 0 & Y_{1i} \\ \star & \star & \ddots & 0 & \vdots \\ \star & \star & \star & X_{(m+2)i} & Y_{(m+2)i} \\ \star & \star & \star & \star & \varepsilon\left(G_i^T + G_i\right) - R_{1i} \end{bmatrix} \ge 0, \tag{2.15c}$$
$$\Psi_{2i} \triangleq \begin{bmatrix} E_{0i} & 0 & 0 & 0 & B_{0i} \\ \star & E_{1i} & 0 & 0 & B_{1i} \\ \star & \star & \ddots & 0 & \vdots \\ \star & \star & \star & B_{(m+2)i} & B_{(m+2)i} \\ \star & \star & \star & \star & \varepsilon\left(G_i^T + G_i\right) - R_{2i} \end{bmatrix} \ge 0, \tag{2.15d}$$
$$\Psi_{3i} \triangleq \begin{bmatrix} E_{0i} & 0 & 0 & 0 & D_{0i} \\ \star & E_{1i} & 0 & 0 & D_{1i} \\ \star & \star & \ddots & 0 & \vdots \\ \star & \star & \star & D_{(m+2)i} & D_{(m+2)i} \\ \star & \star & \star & \star & \varepsilon\left(G_i^T + G_i\right) - R_{2i} \end{bmatrix} \ge 0,$$
$$i = 1, 2, \ldots, r, \tag{2.15e}$$
$$Z_{1i} - R_{1j} < 0, \tag{2.15f}$$
$$Z_{2i} - R_{2j} < 0, \tag{2.15g}$$
$$R_i - S_{2j} < 0,$$
$$i, j = 1, 2, \ldots, r, \tag{2.15h}$$

where

$$\Phi_{ostij} \triangleq \begin{bmatrix} \sum_{11stij} & \cdots & \sum_{12ij} \\ \star & \ddots & \vdots \\ \star & \star & \sum_{22osi} \end{bmatrix},$$
$$\sum_{11stij} \triangleq \begin{bmatrix} -\tau^{-1}Z_{1i} & 0 & 0 & (A_i - I)G_j & 0 & 0 \\ \star & -\tilde{d}^{-1}Z_{2i} & 0 & (A_i - I)G_j & 0 & 0 \\ \star & \star & \Pi_{11t} & A_iG_j & 0 & 0 \\ \star & \star & \star & \Pi_{22i} & Y_{1i} - Y_{0i}^T & Y_{2i} \\ \star & \star & \star & \star & \Pi_{33si} & -\varepsilon Y_{2i} \\ \star & \star & \star & \star & \star & \Pi_{44si} \end{bmatrix},$$

$$\sum\nolimits_{12ij} \triangleq \begin{bmatrix} 0 & 0 & \varepsilon A_{di}G_j & 0 \\ 0 & 0 & \varepsilon A_{di}G_j & 0 \\ 0 & 0 & \varepsilon A_{di}G_j & 0 \\ Y_{(m-1)i} & \Pi_{26i} & \Pi_{27i} & \Pi_{28i} \\ -\varepsilon Y_{(m-1)i} & \Pi_{36i} & \Pi_{37i} & \Pi_{38i} \\ 0 & \varepsilon B_{2i}^T & \Pi_{47i} & -\varepsilon D_{2i}^T \end{bmatrix},$$

$$\sum\nolimits_{22osi} \triangleq \begin{bmatrix} \Pi_{55si} & \varepsilon B_{(m-1)i}^T & \Pi_{57i} & -\varepsilon D_{(m-1)i}^T \\ \star & \Pi_{66si} & \Pi_{67i} & \Pi_{68i} \\ \star & \star & \Pi_{77i} & \Pi_{78i} \\ \star & \star & \star & \Pi_{88oi} \end{bmatrix},$$

with

$$\begin{aligned}
\Pi_{11t} &\triangleq -G_t - G_t^T + P_t, \\
\Pi_{22i} &\triangleq \varepsilon^{-2}\left(Q_{0i} + S_{1i} + (\tilde{d}+1)S_{2i}\right) + \varepsilon^{-1}\left(Y_{0i} + Y_{0i}^T\right) - P_i + J_{0i}, \\
\Pi_{33si} &\triangleq \varepsilon^2 J_{1i} - \varepsilon\left(Y_{1i} + Y_{1i}^T\right) - Q_{0s} + Q_{1i}, \\
\Pi_{44si} &\triangleq \varepsilon^2 J_{2i} - Q_{1s} + Q_{2i}, \ J_{\nu i} \triangleq \tau X_{\nu i} + \tilde{d}E_{\nu i}, \\
\Pi_{55si} &\triangleq \varepsilon^2 J_{(m-1)i} + Q_{(m-1)i} - Q_{(m-2)s}, \\
\Pi_{66si} &\triangleq \varepsilon^2 J_{mi} - Q_{(m-1)s} + \varepsilon\left(B_{mi} + B_{mi}^T\right), \\
\Pi_{26i} &\triangleq Y_{mi} + B_{0i}^T, \\
\Pi_{27i} &\triangleq Y_{(m+1)i} - B_{0i}^T + D_{0i}^T, \\
\Pi_{28i} &\triangleq Y_{(m+2)i} - D_{0i}^T, \\
\Pi_{36i} &\triangleq -\varepsilon\left(Y_{mi} - B_{1i}^T\right), \\
\Pi_{37i} &\triangleq -\varepsilon\left(Y_{(m+1)i} + B_{1i}^T - D_{1i}^T\right), \\
\Pi_{38i} &\triangleq -\varepsilon\left(Y_{(m+2)i} + D_{1i}^T\right), \\
\Pi_{47i} &\triangleq -\varepsilon\left(B_{2i}^T - D_{2i}^T\right), \\
\Pi_{57i} &\triangleq -\varepsilon\left(B_{(m-1)i}^T - D_{(m-1)i}^T\right), \\
\Pi_{67i} &\triangleq \varepsilon\left(B_{(m+1)i} - B_{mi}^T + D_{mi}^T\right), \\
\Pi_{68i} &\triangleq \varepsilon\left(B_{(m+2)i} - D_{mi}^T\right), \\
\Pi_{78i} &\triangleq \varepsilon\left(D_{(m+2)i} - D_{(m+1)i}^T - B_{(m+2)i}\right), \\
\Pi_{77i} &\triangleq \varepsilon^2 J_{(m+1)i} - \varepsilon\left(B_{(m+1)i} + B_{(m+1)i}^T\right) + \varepsilon\left(D_{(m+1)i} + D_{(m+1)i}^T\right) - R_i, \\
\Pi_{88oi} &\triangleq \varepsilon^2 J_{(m+2)i} - \varepsilon\left(D_{(m+2)i} - D_{(m+2)i}^T\right) - S_{1o}.
\end{aligned}$$

Proof. With Φ_{ostij}, it is clear from (2.15a) that $\Pi_{22t} < 0$. Since $P_t > 0$, we have $G_t + G_t^T > 0$, which ensure that G_t^{-1} exists. Moreover, with $\Psi_{\iota i}$, it follows from (2.15c) that $\varepsilon(G_i^T + G_i) - R_{1i} \geq 0$, which imply $\varepsilon > 0$ since

$G_i > 0$ and $R_{1i} > 0$. The inequalities of (2.5a)–(2.5g) can be respectively written as

$$\sum_{o=1}^{r}\sum_{s=1}^{r}\sum_{t=1}^{r} h_o(\theta(k-d_2))h_s(\theta(k-\tau))h_t(\theta(k+1)) \times \sum_{i=1}^{r}\sum_{j=1}^{r} h_i(\theta)h_j(\theta)\Phi_{ostij} < 0, \tag{2.16}$$

$$\sum_{i=1}^{r} h_i(\theta)\Psi_{\iota i} \geq 0, \quad \iota = 1, 2, 3, \tag{2.17}$$

$$\sum_{i=1}^{r} h_i(\theta)Z_{1i} - \sum_{j=1}^{r} h_j(\theta(s))R_{1j} < 0, \tag{2.18}$$

$$\sum_{i=1}^{r} h_i(\theta)Z_{2i} - \sum_{j=1}^{r} h_j(\theta(s))R_{2j} < 0, \tag{2.19}$$

$$\sum_{i=1}^{r} h_i(\theta)R_i - \sum_{j=1}^{r} h_j(\theta(s))S_{2j} < 0. \tag{2.20}$$

According to [203], if conditions (2.15a)–(2.15b) hold, then (2.16) is fulfilled. Moreover, it is obvious that (2.17)–(2.20) are satisfied if the LMIs of (2.15c)–(2.15h) hold. Therefore, it follows from Theorem 2.4 that system (2.1) is asymptotically stable. Thus, the proof is completed. ■

Note that the delay partition positive matrix, $Q(k) \triangleq \text{diag}\,\{Q_0(k), Q_1(k), \ldots, Q_{m-1}(k)\} > 0$, is introduced in Theorem 2.4. The proposed stability condition is not much less conservative due to the special diagonal matrix $Q(k)$. In the following, this diagonal positive matrix $Q(k)$ is substituted for the general $mn \times mn$ positive matrix by using the delay partitioning method. Based on (2.3), we construct the following LKF:

$$\begin{aligned}
\bar{V}(k) &\triangleq x^T(k)\mathcal{P}(k)x(k) \\
&+ \sum_{l=k-\tau}^{k-1} \begin{bmatrix} x(l) \\ x(l-\tau) \\ \vdots \\ x(l-\tau m+\tau) \end{bmatrix}^T \mathcal{Q}_1(l) \begin{bmatrix} x(l) \\ x(l-\tau) \\ \vdots \\ x(l-\tau m+\tau) \end{bmatrix} \\
&+ \sum_{l=k-d_2}^{k-1} x^T(l)\mathcal{Q}_2(l)x(l) + \sum_{s=-d_2+1}^{-\tau m+1}\sum_{l=k-1+s}^{k-1} x^T(l)\mathcal{R}(l)x(l) \\
&+ \sum_{s=-\tau}^{-1}\sum_{l=k+s}^{k-1} \eta^T(l)\mathcal{X}_1(l)\eta(l) + \sum_{s=-d_2}^{-\tau m-1}\sum_{l=k+s}^{k-1} \eta^T(l)\mathcal{X}_2(l)\eta(l),
\end{aligned} \tag{2.21}$$

where

$$\mathcal{P}(k) \triangleq \sum_{i=1}^{r} h_i(\theta)\mathcal{P}_i, \; \mathcal{Q}_1(k) \triangleq \sum_{i=1}^{r} h_i(\theta)\mathcal{Q}_{1i}, \; \mathcal{Q}_2(k) \triangleq \sum_{i=1}^{r} h_i(\theta)\mathcal{Q}_{2i},$$
$$\mathcal{R}(k) \triangleq \sum_{i=1}^{r} h_i(\theta)\mathcal{R}_i, \; \mathcal{X}_1(k) \triangleq \sum_{i=1}^{r} h_i(\theta)\mathcal{X}_{1i}, \; \mathcal{X}_2(k) \triangleq \sum_{i=1}^{r} h_i(\theta)\mathcal{X}_{2i},$$

with $\mathcal{P}_i > 0$, $\mathcal{R}_i > 0$, $\mathcal{Q}_{1i} > 0$, $\mathcal{Q}_{2i} > 0$, $\mathcal{X}_{1i} > 0$, $\mathcal{X}_{2i} > 0$, $i = 1, 2, \ldots, r$.

Thus, based on the fuzzy LKF in (2.21), we have the following result.

Theorem 2.6. *Given positive integers τ, m and d_2, system (2.1) is asymptotically stable if there exist matrices $\mathcal{P}_i > 0$, $\mathcal{R}_i > 0$, $\mathcal{T}_i > 0$, $\mathcal{Q}_{1i} > 0$, $\mathcal{Q}_{2i} > 0$, $\mathcal{X}_{1i} > 0$, $\mathcal{X}_{2i} > 0$, $\mathcal{T}_{1i} > 0$, $\mathcal{T}_{2i} > 0$, $\mathcal{M}_i$, $\mathcal{N}_i$, $\mathcal{X}_i$, $\mathcal{Y}_i$ and $\mathcal{Z}_i$ $(i = 1, 2, \ldots, r)$ such that*

$$\begin{aligned}
\Gamma_{ostlii} &< 0, \quad o, s, t, l, i = 1, 2, \ldots, r, \\
\Gamma_{ostlij} + \Gamma_{ostlji} &< 0, \quad o, s, t, l, i, j = 1, 2, \ldots, r; \; 1 \le i \ne j \le r, \\
\begin{bmatrix} \mathcal{M}_i & \mathcal{X}_i \\ \star & \mathcal{T}_{1i} \end{bmatrix} &\ge 0, \\
\begin{bmatrix} \mathcal{N}_i & \mathcal{Y}_i \\ \star & \mathcal{T}_{2i} \end{bmatrix} &\ge 0, \\
\begin{bmatrix} \mathcal{N}_i & \mathcal{Z}_i \\ \star & \mathcal{T}_{2i} \end{bmatrix} &\ge 0, \quad i = 1, 2, \ldots, r, \\
\mathcal{T}_i - \mathcal{R}_j &< 0, \\
\mathcal{T}_{1i} - \mathcal{X}_{1j} &< 0, \\
\mathcal{T}_{2i} - \mathcal{X}_{2j} &< 0, \quad i, j = 1, 2, \ldots, r,
\end{aligned}$$

where

$$\begin{aligned}
\Gamma_{ostlij} \triangleq\; & \Xi_{1i}^T \mathcal{P}_t \Xi_{1i} + \Xi_2^T \left[\mathcal{Q}_{2j} + (\tilde{d}+1)\mathcal{R}_i\right] \Xi_2 + \Xi_{3i}^T \left[\tau \mathcal{X}_{1j} + \tilde{d}\mathcal{X}_{2j}\right] \Xi_{3i} \\
& - \Xi_2^T \mathcal{P}_i \Xi_2 - \mathscr{W}^T \mathcal{T}_s \mathscr{W} + \mathscr{W}_1^T \bar{\mathcal{Q}}_{1io} \mathscr{W}_1 - \mathscr{W}_2^T \mathcal{Q}_{2l} \mathscr{W}_2 \\
& + \Theta_i + \Theta_i^T + \tau \mathcal{M}_i + \tilde{d}\mathcal{N}_i,
\end{aligned}$$

with

$$\begin{aligned}
\Xi_{1i} &\triangleq \begin{bmatrix} A_i \; 0_{n\times mn} \; A_{di} \; 0_{n\times n} \end{bmatrix}, \; \Xi_2 \triangleq \begin{bmatrix} I_n \; 0_{n\times(m+2)n} \end{bmatrix}, \; \Xi_{3i} \triangleq \Xi_{1i} - \Xi_2, \\
\mathscr{W} &\triangleq \begin{bmatrix} 0_{n\times(m+1)n} & I_n & 0_{n\times n} \end{bmatrix}, \quad \mathscr{W}_2 \triangleq \begin{bmatrix} 0_{n\times(m+2)n} & I_n \end{bmatrix}, \\
\Theta_i &\triangleq \begin{bmatrix} \mathcal{X}_i & \mathcal{Y}_i & \mathcal{Z}_i \end{bmatrix} \begin{bmatrix} I_n & -I_n & 0_{n\times(m+1)n} & \\ 0_{n\times mn} & I_n & -I_n & 0_{n\times n} \\ 0_{n\times(m+1)n} & I_n & -I_n & \end{bmatrix}, \\
\bar{\mathcal{Q}}_{1io} &\triangleq \begin{bmatrix} \mathcal{Q}_{1i} & 0 \\ \star & -\mathcal{Q}_{1o} \end{bmatrix}, \quad \mathscr{W}_1 \triangleq \begin{bmatrix} I_{mn} & 0_{mn\times 3n} & \\ 0_{mn\times n} & I_{mn} & 0_{mn\times 2n} \end{bmatrix}.
\end{aligned}$$

2.3.2 Input-Output Approach

In this section, by using the input-output method and the delay partitioning technique, we present a stability condition for the T-S fuzzy time-delay system in (2.2). To this end, we construct the following LKF:

$$\begin{aligned}\mathscr{V}(k) &\triangleq x^T(k)\mathscr{P}(k)x(k) + \sum_{\kappa=0}^{m-1}\sum_{l=k-\tau}^{k-1} x^T(l-\kappa\tau)\mathscr{Q}_\kappa(l)x(l-\kappa\tau)\\ &\quad + \sum_{l=k-d_2}^{k-1} x^T(l)\mathscr{X}(l)x(l) + \sum_{s=-d_2+1}^{-\tau m+1}\sum_{l=k-1+s}^{k-1} x^T(l)\mathscr{R}(l)x(l)\\ &\quad + \sum_{s=-\tau}^{-1}\sum_{l=k+s}^{k-1} \varsigma^T(l)\mathscr{Z}_1(l)\varsigma(l) + \sum_{s=-d_2}^{-1}\sum_{l=k+s}^{k-1} \varsigma^T(l)\mathscr{Z}_2(l)\varsigma(l), \qquad (2.22)\end{aligned}$$

where for $i = 1, 2, \ldots, r$ and $\kappa = 0, 1, \ldots, m-1$,

$$\begin{aligned}\mathscr{P}(k) &\triangleq \sum_{i=1}^{r} h_i(\theta)\mathscr{P}_i, \ \mathscr{R}(k) \triangleq \sum_{i=1}^{r} h_i(\theta)\mathscr{R}_i, \ \mathscr{Q}_\kappa(k) \triangleq \sum_{i=1}^{r} h_i(\theta)\mathscr{Q}_{\kappa i},\\ \mathscr{X}(k) &\triangleq \sum_{i=1}^{r} h_i(\theta)\mathscr{X}_i, \ \mathscr{Z}_1(k) \triangleq \sum_{i=1}^{r} h_i(\theta)\mathscr{Z}_{1i}, \ \mathscr{Z}_2(k) \triangleq \sum_{i=1}^{r} h_i(\theta)\mathscr{Z}_{2i},\end{aligned}$$

with $\mathscr{P}_i > 0$, $\mathscr{R}_i > 0$, $\mathscr{X}_i > 0$, $\mathscr{Z}_{1i} > 0$, $\mathscr{Z}_{2i} > 0$, $\mathscr{Q}_{\kappa i} > 0$.

Thus, based on LKF in (2.22) we have the following result.

Theorem 2.7. *Given positive integers τ, m, $d \triangleq d_2 - d_1$ and $d_1 \triangleq \tau m$, system (2.2) is asymptotically stable if there exist matrices $\mathcal{X} > 0$, $\mathscr{P}_i > 0$, $\mathscr{Z}_{1i} > 0$, $\mathscr{Z}_{2i} > 0$, $\mathscr{X}_i > 0$, $\mathscr{R}_i > 0$, $\mathscr{T}_i > 0$, $\mathscr{T}_{1i} > 0$, $\mathscr{T}_{2i} > 0$, $\mathscr{Q}_{\kappa i} > 0$, $\mathscr{Y}_{\vartheta i} > 0$, $\mathscr{U}_{\vartheta i} > 0$, $\mathscr{M}_{\vartheta i}$ and $\mathscr{N}_{\vartheta i}$ such that for $s, t, l, i, j = 1, \ldots, r$; $\kappa = 0, \ldots, m-1$; $\vartheta = 0, \ldots, m+1$, the following inequalities hold:*

$$\begin{aligned}\Gamma_{ostlii} &< 0, \quad o, s, t, l, i = 1, 2, \ldots, r,\\ \frac{1}{r-1}\Gamma_{ostlii} + \frac{1}{2}\left(\Gamma_{ostlij} + \Gamma_{ostlji}\right) &< 0, \quad o, s, t, l, i, j = 1, 2, \ldots, r;\ 1 \le i \ne j \le r,\\ \begin{bmatrix} \Xi_{111i} & \Xi_{112i}\\ \star & \mathscr{T}_{1i}\end{bmatrix} &\ge 0,\\ \begin{bmatrix} \Xi_{211i} & \Xi_{212i}\\ \star & \mathscr{T}_{2i}\end{bmatrix} &\ge 0, \quad i = 1, 2, \ldots, r,\\ \mathscr{T}_i - \mathscr{R}_j &< 0,\\ \mathscr{T}_{1i} - \mathscr{Z}_{1j} &< 0,\\ \mathscr{T}_{2i} - \mathscr{Z}_{2j} &< 0, \quad i, j = 1, 2, \ldots, r,\end{aligned}$$

where

$$\Xi_{111i} \triangleq \mathrm{diag}\{\mathscr{Y}_{0i}, \mathscr{Y}_{1i}, \mathscr{Y}_{\hbar i}, \mathscr{Y}_{(m-1)i}, \mathscr{Y}_{mi}, \mathscr{Y}_{(m+1)i}\},$$

$$\Xi_{112i} \triangleq \begin{bmatrix} \mathscr{M}_{0i}^T & \mathscr{M}_{1i}^T & \mathscr{M}_{\hbar i}^T & \mathscr{M}_{(m-1)i}^T & \mathscr{M}_{mi}^T & \mathscr{M}_{(m+1)i}^T \end{bmatrix}^T,$$

$$\Xi_{211i} \triangleq \mathrm{diag}\{\mathscr{U}_{0i}, \mathscr{U}_{1i}, \mathscr{U}_{\hbar i}, \mathscr{U}_{(m-1)i}, \mathscr{U}_{mi}, \mathscr{U}_{(m+1)i}\},$$

$$\Xi_{212i} \triangleq \begin{bmatrix} \mathscr{N}_{0i}^T & \mathscr{N}_{1i}^T & \mathscr{N}_{\hbar i}^T & \mathscr{N}_{(m-1)i}^T & \mathscr{N}_{mi}^T & \mathscr{N}_{(m+1)i}^T \end{bmatrix}^T, \ \hbar = 2, \ldots, m-2,$$

$$\Gamma_{ostlij} \triangleq \begin{bmatrix} \Gamma_{11oij} & \Gamma_{12i} & \Gamma_{13i} & \Gamma_{14i} & \Gamma_{15oij} & \Gamma_{16oij} & d\Gamma_{17oij} \\ \star & \Gamma_{22is} & -\mathscr{M}_{\hbar i} & -\mathscr{M}_{(m-1)i} & -\mathscr{M}_{mi} & \Gamma_{26i} & 0 \\ \star & \star & \Gamma_{33is} & 0 & 0 & -\mathscr{N}_{\hbar i}^T & 0 \\ \star & \star & \star & \Gamma_{44is} & 0 & -\mathscr{N}_{(m-1)i}^T & 0 \\ \star & \star & \star & \star & \Gamma_{55ostij} & \Gamma_{56oij} & d\Gamma_{77oij} \\ \star & \star & \star & \star & \star & \Gamma_{66otij} & d\Gamma_{77oij} \\ \star & \star & \star & \star & \star & \star & d^2\Gamma_{77oij} - \mathcal{X} \end{bmatrix},$$

with

$$\begin{aligned}
\Gamma_{11oij} &\triangleq -\mathscr{P}_i + \mathscr{Q}_{0i} + \mathscr{X}_i + (d+1)\mathscr{R}_i + \mathscr{M}_{0i} + \mathscr{M}_{0i}^T + \mathscr{N}_{oi} + \mathscr{N}_{oi}^T + \tau\mathscr{Y}_{0i} \\
&\quad + d_2\mathscr{U}_{0i} + A_j^T\mathscr{P}_o A_j + (A_j - I)^T (\tau\mathscr{Z}_{1i} + d_2\mathscr{Z}_{2i} + \mathcal{X})(A_j - I), \\
\Gamma_{12i} &\triangleq -\mathscr{M}_{0i}^T + \mathscr{M}_{1i} + \mathscr{N}_{1i}, \quad \Gamma_{33is} \triangleq \mathscr{Q}_{\hbar i} - \mathscr{Q}_{(\hbar-1)s} + \tau\mathscr{Y}_{\hbar i} + d_2\mathscr{U}_{\hbar i}, \\
\Gamma_{13i} &\triangleq \mathscr{M}_{\hbar i} + \mathscr{N}_{\hbar i}, \quad \Gamma_{22is} \triangleq \mathscr{Q}_{1i} - \mathscr{Q}_{0s} - \mathscr{M}_{1i} - \mathscr{M}_{1i}^T + \tau\mathscr{Y}_{1i} + d_2\mathscr{U}_{1i}, \\
\Gamma_{44is} &\triangleq \mathscr{Q}_{(m-1)i} - \mathscr{Q}_{(m-2)s} + \tau\mathscr{Y}_{(m-1)i} + d_2\mathscr{U}_{(m-1)i}, \\
\Gamma_{14i} &\triangleq \mathscr{M}_{(m-1)i} + \mathscr{N}_{(m-1)i}, \quad \Gamma_{15oij} \triangleq \mathscr{M}_{mi} + \mathscr{N}_{mi} + \Gamma_{17oij}, \\
\Gamma_{26i} &\triangleq -\mathscr{M}_{(m+1)i} - \mathscr{N}_{1i}^T, \quad \Gamma_{56oij} \triangleq -\mathscr{N}_{mi}^T + \Gamma_{77oij}, \\
\Gamma_{16oij} &\triangleq \mathscr{M}_{(m+1)i} + \mathscr{N}_{(m+1)i} - \mathscr{N}_{0i}^T + \Gamma_{17oij}, \\
\Gamma_{17oij} &\triangleq \frac{1}{2}A_j^T\mathscr{P}_o A_{dj} + \frac{1}{2}(A_j - I)^T (\tau\mathscr{Z}_{1i} + d_2\mathscr{Z}_{2i} + \mathcal{X}) A_{dj}, \\
\Gamma_{55ostij} &\triangleq -\mathscr{Q}_{(m-1)s} + \tau\mathscr{Y}_{mi} + d_2\mathscr{U}_{mi} - \mathscr{T}_t + \Gamma_{77oij}, \\
\Gamma_{66olij} &\triangleq -\mathscr{X}_l - \mathscr{N}_{(m+1)i} - \mathscr{N}_{(m+1)i}^T + \tau\mathscr{Y}_{(m+1)i} + d_2\mathscr{U}_{(m+1)i} + \Gamma_{77oij}, \\
\Gamma_{77oij} &\triangleq \frac{1}{4}A_{dj}^T(\mathscr{P}_o + \tau\mathscr{Z}_{1i} + d_2\mathscr{Z}_{2i} + \mathcal{X}) A_{dj}.
\end{aligned}$$

With the previous results, the proof of Theorem 2.4 can be carried out in a straightforward way thus is omitted here.

Note that the special diagonal matrix $\mathscr{Q}_i \triangleq \mathrm{diag}\{\mathscr{Q}_{0i}, \mathscr{Q}_{1i}, \mathscr{Q}_{\hbar i}, \mathscr{Q}_{(m-1)i}\}$ is used to facilitate the proof of Theorem 2.7, which inevitably introduces some conservativeness into the proposed result. To overcome this, the block-diagonal matrix $\mathscr{Q}_i$ can be replaced by the matrix $\bar{\mathscr{Q}}_i \in R^{mn \times mn}$ with a general structure, and the corresponding result is proposed in Theorem 2.6. In the following, we further extend the obtained results in Theorem 2.6.

Construct the following LKF:

$$
\begin{aligned}
\bar{\mathscr{V}}(k) &\triangleq x^T(k)\mathscr{P}(k)x(k) \\
&+ \sum_{l=k-\tau}^{k-1} \begin{bmatrix} x(l) \\ x(l-\tau) \\ \vdots \\ x(l-\tau m+\tau) \end{bmatrix}^T \bar{\mathscr{Q}}(l) \begin{bmatrix} x(l) \\ x(l-\tau) \\ \vdots \\ x(l-\tau m+\tau) \end{bmatrix} \\
&+ \sum_{s=-d_2+1}^{-\tau m+1} \sum_{l=k-1+s}^{k-1} x^T(l)\mathscr{R}(l)x(l) + \sum_{l=k-d_2}^{k-1} x^T(l)\mathscr{T}(l)x(l) \\
&+ \sum_{s=-\tau}^{-1} \sum_{l=k+s}^{k-1} \eta^T(l)\mathscr{X}_1(l)\eta(l) + \sum_{s=-d_2}^{-\tau m-1} \sum_{l=k+s}^{k-1} \eta^T(l)\mathscr{X}_2(l)\eta(l), \quad (2.23)
\end{aligned}
$$

where

$$
\begin{aligned}
\mathscr{P}(k) &\triangleq \sum_{i=1}^{r} h_i(\theta)\mathscr{P}_i, \quad \mathscr{R}(k) \triangleq \sum_{i=1}^{r} h_i(\theta)\mathscr{R}_i, \quad \bar{\mathscr{Q}}(k) \triangleq \sum_{i=1}^{r} h_i(\theta)\bar{\mathscr{Q}}_i, \\
\mathscr{T}(k) &\triangleq \sum_{i=1}^{r} h_i(\theta)\mathscr{T}_i, \quad \mathscr{X}_1(k) \triangleq \sum_{i=1}^{r} h_i(\theta)\mathscr{X}_{1i}, \quad \mathscr{X}_2(k) \triangleq \sum_{i=1}^{r} h_i(\theta)\mathscr{X}_{2i},
\end{aligned}
$$

with $\mathscr{P}_i > 0$, $\mathscr{R}_i > 0$, $\mathscr{T}_i > 0$, $\mathscr{X}_{1i} > 0$, $\mathscr{X}_{2i} > 0$, $\bar{\mathscr{Q}}_i > 0$, $i = 1, 2, \ldots, r$.

Thus, based on fuzzy LKF (2.23), we have the following result.

Theorem 2.8. *Given positive integers τ, m, d_2 and $d_1 \triangleq \tau m$, the discrete-time T-S fuzzy time-delay system in (2.2) is asymptotically stable if there exist matrices $\mathscr{P}_i > 0$, $\mathscr{R}_i > 0$, $\bar{\mathscr{Q}}_i > 0$, $\mathscr{T}_i > 0$, $\mathscr{X}_{1i} > 0$, $\mathscr{X}_{2i} > 0$, $\mathscr{T}_{1i} > 0$, $\mathscr{T}_{2i} > 0$, $\mathcal{X} > 0$, $\mathscr{M}_i$, $\mathscr{N}_i$, $\mathscr{X}_i$ and $\mathscr{Y}_i$ $(i = 1, 2, \ldots, r)$ such that*

$$
\begin{aligned}
\Gamma_{ostlii} &< 0, \quad o, s, t, l, i = 1, 2, \ldots, r, \\
\Gamma_{ostlij} + \Gamma_{ostlji} &< 0, \quad o, s, t, l, i, j = 1, 2, \ldots, r; \ 1 \le i \ne j \le r, \\
&\begin{bmatrix} \mathscr{M}_i & \mathscr{X}_i \\ \star & \mathscr{T}_{1i} \end{bmatrix} \ge 0, \\
&\begin{bmatrix} \mathscr{N}_i & \mathscr{Y}_i \\ \star & \mathscr{T}_{2i} \end{bmatrix} \ge 0, \quad i = 1, 2, \ldots, r, \\
&\mathscr{T}_{1i} - \mathscr{X}_{1j} < 0, \\
&\mathscr{T}_{2i} - \mathscr{X}_{2j} < 0, \quad i, j = 1, 2, \ldots, r,
\end{aligned}
$$

where

$$
\begin{aligned}
\Gamma_{ostlij} &\triangleq \Xi_{1i}^T \mathscr{P}_t \Xi_{1i} + \Xi_2^T \left[\mathscr{T}_j + (\tilde{d}+1)\mathscr{R}_i\right] \Xi_2 + \Xi_{3i}^T \left[\tau \mathscr{X}_{1j} + \tilde{d}\mathscr{X}_{2j}\right] \Xi_{3i} \\
&-\Xi_4^T \mathcal{X} \Xi_4 + \Xi_{5i}^T \mathcal{X} \Xi_{5i} + \mathcal{I}_1^T(\Theta_i^T + \Theta_i)\mathcal{I}_1 + \mathcal{I}_2^T \hat{\mathscr{Q}}_{io}\mathcal{I}_2 - \Xi_2^T \mathscr{P}_i \Xi_2 \\
&-\mathcal{I}_3^T(\mathscr{T}_l + \mathscr{R}_l)\mathcal{I}_3 + \mathcal{I}_4^T \left[\tau \mathscr{M}_i + \tilde{d}\mathscr{N}_i\right] \mathcal{I}_4 - \mathcal{I}_5^T \mathscr{R}_s \mathcal{I}_5,
\end{aligned}
$$

with

$$\Theta_i \triangleq \begin{bmatrix} \mathscr{X}_i & \mathscr{Y}_i \end{bmatrix} \begin{bmatrix} I_n & -I_n & 0_{n\times mn} \\ 0_{n\times mn} & I_n & -I_n \end{bmatrix}, \quad \hat{\mathscr{Q}}_{io} \triangleq \begin{bmatrix} \bar{\mathscr{Q}}_i & 0 \\ \star & -\bar{\mathscr{Q}}_o \end{bmatrix},$$

$$\Xi_{1i} \triangleq \begin{bmatrix} A_i & 0_{n\times(m-1)n} & \frac{1}{2}A_{di} & \frac{1}{2}A_{di} & 0_{n\times n} \end{bmatrix}, \quad \Xi_2 \triangleq \begin{bmatrix} I_n & 0_{n\times(m+2)n} \end{bmatrix},$$

$$\Xi_4 \triangleq \begin{bmatrix} 0_{n\times(m+2)n} & I_n \end{bmatrix}, \quad \mathcal{I}_3 \triangleq \begin{bmatrix} 0_{n\times(m+1)n} & I_n & 0_{n\times n} \end{bmatrix},$$

$$\Xi_{5i} \triangleq \begin{bmatrix} A_i - I_n & 0_{n\times(m-1)n} & \frac{1}{2}A_{di} & \frac{1}{2}A_{di} & \frac{d}{2}A_{di} \end{bmatrix}, \quad \Xi_{3i} \triangleq \Xi_{1i} - \Xi_2,$$

$$\mathcal{I}_2 \triangleq \begin{bmatrix} I_{mn} & 0_{mn\times 3n} & \\ 0_{mn\times n} & I_{mn} & 0_{mn\times 2n} \end{bmatrix}, \quad \mathcal{I}_1 \triangleq \begin{bmatrix} I_{(m+2)n} & 0_{(m+2)n\times n} \end{bmatrix},$$

$$\mathcal{I}_5 \triangleq \begin{bmatrix} 0_{n\times mn} & I_n & 0_{n\times 2n} \end{bmatrix}, \quad \mathcal{I}_4 \triangleq \begin{bmatrix} I_{(m+2)n} & 0_{(m+2)n\times n} \end{bmatrix}.$$

2.3.3 Reciprocally Convex Approach

This section will explore the reciprocally convex approach to the stability analysis. Based on Lemma 1.33, we present the following by-product condition for the T-S fuzzy time-delay system in (2.1).

Theorem 2.9. *Given integers $1 \le d_1 \le d_2$, system (2.1) with time-varying delay $d(k)$ satisfying $d_1 \le d(k) \le d_2$ is asymptotically stable if there exist matrices $\bar{\mathcal{P}}_i > 0$, $\bar{\mathcal{X}}_{1i} > 0$, $\bar{\mathcal{X}}_{2i} > 0$, $\bar{\mathcal{Q}}_{1i} > 0$, $\bar{\mathcal{Q}}_{2i} > 0$, $\bar{\mathcal{Q}}_{3i} > 0$, $\bar{\mathcal{S}}_{1i} > 0$, $\bar{\mathcal{S}}_{2i} > 0$ and $\bar{\mathcal{M}}_i$ $(i = 1, 2, \ldots, r)$ such that the following LMIs hold:*

$$\begin{aligned}
\Phi_{ostlii} &< 0, \quad o, s, t, l, i = 1, 2, \ldots, r, \\
\Phi_{ostlij} + \Phi_{ostlji} &< 0, \quad 1 \le i \ne j \le r;\ o, s, t, l, i, j = 1, 2, \ldots, r, \\
\begin{bmatrix} \bar{\mathcal{S}}_{2i} & \bar{\mathcal{M}}_i^T \\ \star & \bar{\mathcal{S}}_{2i} \end{bmatrix} &\ge 0, \quad i = 1, 2, \ldots, r, \\
\bar{\mathcal{X}}_{1i} - \bar{\mathcal{S}}_{1j} &\ge 0, \\
\bar{\mathcal{X}}_{2i} - \bar{\mathcal{S}}_{2j} &\ge 0, \quad i, j = 1, 2, \ldots, r,
\end{aligned}$$

where

$$\Phi_{ostlji} \triangleq \begin{bmatrix} \Phi_{11tlji} & \bar{\mathcal{S}}_{1t} & \Phi_{13lji} & 0 \\ \star & -\left(\bar{\mathcal{Q}}_{1t} + \bar{\mathcal{S}}_{1t} + \bar{\mathcal{S}}_{2i}\right) & -\bar{\mathcal{M}}_i + \bar{\mathcal{S}}_{2i} & \bar{\mathcal{M}}_i \\ \star & \star & \Phi_{33slji} & -\bar{\mathcal{M}}_i + \bar{\mathcal{S}}_{2i} \\ \star & \star & \star & -\bar{\mathcal{Q}}_{2o} - \bar{\mathcal{S}}_{2i} \end{bmatrix},$$

$$\begin{aligned}
\Phi_{11tlji} &\triangleq \bar{\mathcal{Q}}_{1i} + \bar{\mathcal{Q}}_{2i} + (d+1)\bar{\mathcal{Q}}_{3i} - \bar{\mathcal{P}}_i - \bar{\mathcal{S}}_{1t} + A_j^T \bar{\mathcal{P}}_l A_j \\
&\quad + d_1^2 \left(A_j - I\right)^T \bar{\mathcal{X}}_{1i} \left(A_j - I\right) + d^2 \left(A_j - I\right)^T \bar{\mathcal{X}}_{2i} \left(A_j - I\right), \\
\Phi_{13lji} &\triangleq A_j^T \bar{\mathcal{P}}_l A_{dj} + d_1^2 \left(A_j - I\right)^T \bar{\mathcal{X}}_{1i} A_{dj} + d^2 \left(A_j - I\right)^T \bar{\mathcal{X}}_{2i} A_{dj}, \\
\Phi_{33slji} &\triangleq A_{dj}^T \bar{\mathcal{P}}_l A_{dj} + d_1^2 A_{dj}^T \bar{\mathcal{X}}_{1i} A_{dj} + d^2 A_{dj}^T \bar{\mathcal{X}}_{2i} A_{dj} - \bar{\mathcal{Q}}_{3s} - 2\bar{\mathcal{S}}_{2i} + \bar{\mathcal{M}}_i + \bar{\mathcal{M}}_i^T.
\end{aligned}$$

Proof. Choose the following fuzzy LKF:

$$\begin{aligned}\mathcal{V}(k) &\triangleq x^T(k)\bar{\mathcal{P}}(k)x(k) + \sum_{j=1}^{2}\sum_{i=k-d_j}^{k-1} x^T(i)\bar{\mathcal{Q}}_j(i)x(i)\\ &+ \sum_{i=k-d(k)}^{k-1} x^T(i)\bar{\mathcal{Q}}_3(i)x(i) + \sum_{j=-d_2+1}^{-d_1}\sum_{i=k+j}^{k-1} x^T(i)\bar{\mathcal{Q}}_3(i)x(i)\\ &+ \sum_{j=-d_1}^{-1}\sum_{i=k+j}^{k-1} d_1\eta^T(i)\bar{\mathcal{X}}_1(i)\eta(i) + \sum_{j=-d_2}^{-d_1-1}\sum_{i=k+j}^{k-1} d\eta^T(i)\bar{\mathcal{X}}_2(i)\eta(i),\end{aligned}$$

where

$$\begin{aligned}\bar{\mathcal{P}}(k) &\triangleq \sum_{i=1}^{r} h_i(\theta)\bar{\mathcal{P}}_i, \quad \bar{\mathcal{X}}_1(k) \triangleq \sum_{i=1}^{r} h_i(\theta)\bar{\mathcal{X}}_{1i}, \ \bar{\mathcal{X}}_2(k) \triangleq \sum_{i=1}^{r} h_i(\theta)\bar{\mathcal{X}}_{2i},\\ \bar{\mathcal{Q}}_1(k) &\triangleq \sum_{i=1}^{r} h_i(\theta)\bar{\mathcal{Q}}_{1i}, \ \bar{\mathcal{Q}}_2(k) \triangleq \sum_{i=1}^{r} h_i(\theta)\bar{\mathcal{Q}}_{2i}, \ \bar{\mathcal{Q}}_3(k) \triangleq \sum_{i=1}^{r} h_i(\theta)\bar{\mathcal{Q}}_{3i},\end{aligned}$$

with $\bar{\mathcal{P}}_i > 0$, $\bar{\mathcal{X}}_{1i} > 0$, $\bar{\mathcal{X}}_{2i} > 0$, $\bar{\mathcal{Q}}_{1i} > 0$, $\bar{\mathcal{Q}}_{2i} > 0$, $\bar{\mathcal{Q}}_{3i} > 0$, $i = 1, 2, \ldots, r$.

Then, the result can be easily derived by following the same lines in the proof of Theorem 2.4. ■

2.4 Illustrative Example

The effectiveness and superiority of the fore mentioned methods will now be demonstrated illustratively.

Example 2.10. (Conservativeness analysis): Consider the discrete-time T-S fuzzy time-delay system in (2.1) with the parameters given as follows:

$$\begin{aligned}A_1 &= \begin{bmatrix} -0.291 & 1 \\ 0 & 0.95 \end{bmatrix}, \quad A_{d1} = \begin{bmatrix} 0.012 & 0.014 \\ 0 & 0.015 \end{bmatrix},\\ A_2 &= \begin{bmatrix} -0.1 & 0 \\ 1 & -0.2 \end{bmatrix}, \quad A_{d2} = \begin{bmatrix} 0.01 & 0 \\ 0.01 & 0.015 \end{bmatrix},\end{aligned}$$

which has been considered in [66]. In this example, $d(k)$ presents a time-varying state delay. We obtain the upper delay bound by using the method proposed in Theorems 2.5, 2.6, 2.7, 2.8, 2.9, respectively. A detailed comparison is given in Table 2.1, where the achieved upper bounds of time-delay in the above system are listed for their respective lower bounds. It can be seen that the methods proposed in this chapter are better than the recently published results in [66].

Notice from Table 2.1 that by introducing the novel model transformation technique and the input-output method in Theorem 2.8, the allowable upper

Table 2.1. Allowable upper bound of d_2 for different values of d_1

For different d_1	$d_1 = 3$	$d_1 = 5$	$d_1 = 10$
Theorem 2.5	$d_2 = 6$ $(m = 3)$	$d_2 = 7$ $(m = 5)$	$d_2 = 11$ $(m = 5)$
Theorem 2.7	$d_2 = 13$ $(m = 3)$	$d_2 = 14$ $(m = 5)$	$d_2 = 19$ $(m = 5)$
Theorem 1 of [66]	$d_2 = 14$	$d_2 = 16$	$d_2 = 20$
Theorem 2.6	$d_2 = 23$ $(m = 3)$	$d_2 = 25$ $(m = 5)$	$d_2 = 29$ $(m = 5)$
Theorem 2.8	$d_2 = 100$ $(m = 3)$	$d_2 = 102$ $(m = 5)$	$d_2 = 107$ $(m = 5)$
Theorem 2.9	$d_2 = 26$	$d_2 = 28$	$d_2 = 33$

bound d_2 of the time-varying delay $d(k)$ is the largest compared to the results with the other methods, which means that the conservativeness of the result in Theorem 2.8 is smallest. However, the computational load for the conditions in Theorem 2.8 is more than the other ones.

2.5 Conclusion

New methods were proposed in combination with the construction of basis-dependent LKF, the delay partitioning method, the input-output method and the reciprocally convex method, to solve the stability analysis problem of discrete-time T-S fuzzy time-varying delay systems, and obtains the parameter-dependent (delay-dependent and fuzzy-rule-dependent) conditions with further less conservativeness. These obtained stability analysis results with less conservativeness are the foundation for the following system analysis and synthesis in following chapters of Part II.

Chapter 3
Stabilization and DOF Control of Discrete-Time T-S Fuzzy Time-Delay Systems

3.1 Introduction

In this chapter, the stabilization and DOF control problems are investigated for discrete-time T-S fuzzy time-delay systems. By utilizing a novel idea of delay partitioning technique, a delay-dependent stability condition with less conservativeness is proposed at first. Then, based on the stability result, the stabilization problem via the non-PDC scheme is addressed with the gain matrix of the stabilization state feedback controller, which can be obtained by solving a set of LMIs. Furthermore, we consider the DOF control problem in case some state components are not available. By applying the scaled small gain theorem and the delay partitioning method, a sufficient condition is proposed, which guarantees that the closed-loop system is asymptotically stable and has an induced ℓ_2 performance. Then, the desired DOF controller can be designed by the convex linearization approach, and a solvability condition for the DOF controller is also established in terms of LMIs.

3.2 System Description and Preliminaries

Consider the following T-S fuzzy time-varying delay system:

♦ **Plant Form:**

Rule i: IF $\theta_1(k)$ is $\mathcal{M}_{i1}$ and $\theta_2(k)$ is $\mathcal{M}_{i2}$ and … and $\theta_p(k)$ is $\mathcal{M}_{ip}$, THEN

$$x(k+1) = A_i x(k) + A_{di} x(k - d(k)) + B_i u(k) \tag{3.1a}$$

$$x(k) = \phi(k), \quad k = -d_2, -d_2+1, \ldots, 0, \tag{3.1b}$$

where $i = 1, 2, \ldots, r$, and r is the number of IF-THEN rules; $\mathcal{M}_{ij}(i = 1, 2, \ldots, r; j = 1, 2, \ldots, p)$ are the fuzzy sets; $\theta = \left[\theta_1(k)\ \theta_2(k) \cdots \theta_p(k)\right]^T$ is the premise variable vector. $x(k) \in \mathbf{R}^n$ is the system state vector; $u(k) \in \mathbf{R}^p$ is the control input and $d(k)$ is the time-varying delay satisfying $1 \leqslant d_1 \leqslant d(k) \leqslant d_2$, where d_1 and d_2 are positive constants representing

L. Wu et al., *Fuzzy Control Systems with Time-Delay and Stochastic Perturbation*, Studies in Systems, Decision and Control 12, DOI: 10.1007/978-3-319-11316-6_3

the lower and upper bounds, respectively. The lower bound d_1 can be described by $d_1 = \tau m$, where τ and m are two integers. A_i, A_{di} and B_i are known real constant matrices with appropriate dimensions; $\phi(k)$ denotes the initial condition.

It is assumed that the premise variables do not depend on the input variable $u(k)$ explicitly. Given a pair of $(x(k), u(k))$, a more compact presentation of the discrete T-S fuzzy time-varying delay model can be given by

$$x(k+1) = \sum_{i=1}^{r} h_i(\theta)\Big[A_i x(k) + A_{di} x(k-d(k)) + B_i u(k)\Big], \tag{3.2}$$

where $h_i(\theta)$, $i = 1, 2, \ldots, r$ are the normalized membership functions, which are defined as that of (1.2) in Chapter 1.

A more compact presentation of the T-S fuzzy model is given by

$$x(k+1) = \bar{A}(k)x(k) + \bar{A}_d(k)x(k-d(k)) + \bar{B}(k)u(k), \tag{3.3}$$

where

$$\bar{A}(k) \triangleq \sum_{i=1}^{r} h_i(\theta)A_i, \quad \bar{A}_d(k) \triangleq \sum_{i=1}^{r} h_i(\theta)A_{di}, \quad \bar{B}(k) \triangleq \sum_{i=1}^{r} h_i(\theta)B_i.$$

Assume that the premise variable of the fuzzy model θ is available for feedback which implies that $h_i(\theta)$ is available for feedback. Suppose that the controller's premise variable is the same as the plant's premise variable. The parallel distributed compensation strategy is utilized and the fuzzy state feedback controller obeys the following rules:

♦ **Controller Form:**

Rule i: IF $\theta_1(k)$ is $\mathcal{M}_{i1}$ and $\theta_2(k)$ is $\mathcal{M}_{i2}$ and $\cdots$ and $\theta_p(k)$ is $\mathcal{M}_{ip}$, THEN

$$u(k) = K_i x(k), \tag{3.4}$$

for $i = 1, 2, \ldots, r$, where K_i is the gain matrix of the state-feedback controller. Thus, the controller in (3.4) can also be represented by the following form:

$$u(k) = \bar{K}(k)x(k),$$

where

$$\bar{K}(k) = \sum_{i=1}^{r} h_i(\theta)K_i. \tag{3.5}$$

Therefore, the closed-loop system can be obtained as

$$x(k+1) = \sum_{i=1}^{r}\sum_{j=1}^{r} h_i(\theta)h_j(\theta)\Big[(A_i + B_iK_j)\,x(k) + A_{di}x(k-d(k))\Big],$$

and its compact form is given by

$$x(k+1) = \Big[\bar{A}(k) + \bar{B}(k)K(k)\Big]x(k) + \bar{A}_d(k)x(k-d(k)). \tag{3.6}$$

In practice, some system state components can not be accessible, thus the state feedback control can not be implemented. In this case, we consider the DOF control problem. To this end, we introduce the following system:

♦ **Plant Form:**

Rule i: IF $\theta_1(k)$ is $\mathcal{M}_{i1}$ and $\theta_2(k)$ is $\mathcal{M}_{i2}$ and $\cdots$ and $\theta_p(k)$ is $\mathcal{M}_{ip}$, THEN

$$x(k+1) = A_ix(k) + A_{di}x(k-d(k)) + B_iu(k) + D_i\omega(k), \tag{3.7a}$$
$$y(k) = C_ix(k) + C_{di}x(k-d(k)) + F_i\omega(k), \tag{3.7b}$$
$$z(k) = E_ix(k) + E_{di}x(k-d(k)) + G_iu(k) + H_i\omega(k), \tag{3.7c}$$

where $y(k) \in \mathbf{R}^p$ is the measured output; $\omega(k) \in \mathbf{R}^l$ denotes the disturbance input belonging to $\ell_2[0,\infty)$; $z(k) \in \mathbf{R}^q$ is the controlled output. D_i, C_i, C_{di}, F_i, E_i, E_{di}, G_i and H_i are known real constant matrices.

A more compact presentation of (3.7) can be given by

$$x(k+1) = \sum_{i=1}^{r} h_i(\theta)\left[A_ix(k) + A_{di}x(k-d(k)) + B_iu(k) + D_i\omega(k)\right], \tag{3.8a}$$

$$y(k) = \sum_{i=1}^{r} h_i(\theta)\left[C_ix(k) + C_{di}x(k-d(k)) + F_i\omega(k)\right], \tag{3.8b}$$

$$z(k) = \sum_{i=1}^{r} h_i(\theta)\left[E_ix(k) + E_{di}x(k-d(k)) + G_iu(k) + H_i\omega(k)\right]. \tag{3.8c}$$

Assume that the premise variable of the fuzzy model θ is available, thus $h_i(\theta)$ is also available for feedback. In addition, suppose the controller's premise variable be the same as that of the plant. Therefore, based on the PDC technique, we consider the following DOF controllers for system in (3.8):

♦**DOF Controller Form:**

Rule i: IF $\theta_1(k)$ is $\mathcal{M}_{i1}$ and $\theta_2(k)$ is $\mathcal{M}_{i2}$ and $\cdots$ and $\theta_p(k)$ is $\mathcal{M}_{ip}$, THEN

$$\hat{x}(k+1) = A_{Ki}\hat{x}(k) + B_{Ki}y(k),$$
$$u(k) = C_{Ki}\hat{x}(k),$$

where $i = 1, 2, \ldots, r$, and r is the number of IF-THEN rules; $\hat{x}(k) \in \mathbf{R}^k$ is the controller state, A_{Ki}, B_{Ki} and C_{Ki} are matrices to be determined. A compact form of the DOF controller is given as

$$\hat{x}(k+1) = \sum_{i=1}^{r} h_i(\theta) \left[A_{Ki}\hat{x}(k) + B_{Ki}y(k) \right], \tag{3.9a}$$

$$u(k) = \sum_{i=1}^{r} h_i(\theta) C_{Ki}\hat{x}(k). \tag{3.9b}$$

Considering (3.8) and (3.9), the closed-loop system can be described by

$$\xi(k+1) = \bar{A}_c(k)\xi(k) + \bar{A}_{cd}(k)\xi(k-d(k)) + \bar{D}_c(k)\omega(k), \tag{3.10a}$$
$$z(k) = \bar{E}_c(k)\xi(k) + \bar{E}_{cd}(k)\xi(k-d(k)) + \bar{H}_c(k)\omega(k), \tag{3.10b}$$

where $\xi(k) \triangleq \begin{bmatrix} x(k) \\ \hat{x}(k) \end{bmatrix}$ and

$$\bar{A}_c(k) \triangleq \sum_{i=1}^{r}\sum_{j=1}^{r} h_i(\theta)h_j(\theta)\bar{A}_{cij}, \quad \bar{A}_{cd}(k) \triangleq \sum_{i=1}^{r}\sum_{j=1}^{r} h_i(\theta)h_j(\theta)\bar{A}_{cdij},$$
$$\bar{D}_c(k) \triangleq \sum_{i=1}^{r}\sum_{j=1}^{r} h_i(\theta)h_j(\theta)\bar{D}_{cij}, \quad \bar{E}_{cd}(k) \triangleq \sum_{i=1}^{r} h_i(\theta)\bar{E}_{cdi},$$
$$\bar{E}_c(k) \triangleq \sum_{i=1}^{r}\sum_{j=1}^{r} h_i(\theta)h_j(\theta)\bar{E}_{cij}, \quad \bar{H}_c(k) \triangleq \sum_{i=1}^{r} h_i(\theta)\bar{H}_{ci}.$$

with

$$\bar{A}_{cij} \triangleq \begin{bmatrix} A_i & B_iC_{Kj} \\ B_{Kj}C_i & A_{Kj} \end{bmatrix}, \quad \bar{A}_{cdij} \triangleq \begin{bmatrix} A_{di} & 0 \\ B_{Kj}C_{di} & 0 \end{bmatrix}, \quad \bar{D}_{cij} \triangleq \begin{bmatrix} D_i \\ B_{Kj}F_i \end{bmatrix},$$
$$\bar{E}_{cij} \triangleq \begin{bmatrix} E_i & G_iC_{Kj} \end{bmatrix}, \quad \bar{E}_{cdi} \triangleq \begin{bmatrix} E_{di} & 0 \end{bmatrix}, \quad \bar{H}_{ci} \triangleq H_i.$$

Definition 3.1. The closed-loop system in (3.10) with $\omega(k) = 0$ is said to be asymptotically stable if

$$\lim_{k\to\infty} \|\xi(k)\| = 0.$$

Definition 3.2. Given a scalar $\gamma > 0$, the closed-loop system in (3.10) is said to be asymptotically stable with an induced ℓ_2 performance γ if it is asymptotically stable and satisfies

$$\|z(k)\|_2 < \gamma \|\omega(k)\|_2, \quad \forall 0 \neq \omega(k) \in \ell_2[0, \infty), \tag{3.11}$$

where

$$\|z(k)\|_2 \triangleq \sqrt{\sum_{k=0}^{\infty} z^T(k)z(k)}.$$

3.3 Stabilization

In this section, we consider the stabilization problem based on the proposed stability analysis in Theorem 2.5 of Chapter 2. Assume that all of the states are available for feedback control design. As in [82], the following non-PDC controller is considered for system (3.3):

$$u(k) = \bar{K}(k)\bar{G}^{-1}(k)x(k), \tag{3.12}$$

where

$$\bar{G}(k) = \sum_{i=1}^{r} h_i(\theta)G_i,$$

and $\bar{K}(k)$ is defined in (3.5). Obviously, if we take $G_i = G$ then (3.12) becomes a PDC controller. Substituting (3.12) into (3.3), the closed-loop system can be described by

$$x(k+1) = \bar{A}_c(k)x(k) + \bar{A}_d(k)x(k - d(k)), \tag{3.13}$$

where $\bar{A}_c(k) = \bar{A}(k) + \bar{B}(k)\bar{K}(k)\bar{G}^{-1}(k)$. In addition, when $d_1 = d_2$, we can obtain the T-S fuzzy closed-loop system with constant delay as follows:

$$x(k+1) = \hat{A}_c(k)x(k) + \bar{A}_d(k)x(k - d),$$

where $\hat{A}_c(k) = \bar{A}(k) + \bar{B}(k)\bar{K}(k)\bar{G}^{-1}(k)$.

Theorem 3.3. *For given positive integers τ, m and d_2, the closed-loop system in (3.13) is asymptotically stable if there exist matrices $P_i > 0$, $Z_{1i} > 0$, $Z_{2i} > 0$, $Q_{Ni} > 0$, $S_{1i} > 0$, $S_{2i} > 0$, $R_{1i} > 0$, $R_{2i} > 0$, $R_i > 0$, X_{Mi}, E_{Mi}, Y_{Mi}, B_{Mi}, D_{Mi}, G_i and F_i, $i = 1, 2, \ldots, r$; $N = 0, 1, \ldots, m-1$; $M = 0, 1, \ldots, m+2$, which ensure that G_i^{-1} and F_i^{-1} exist and $F_i = \varepsilon G_i$, such that for any positive scalar ε, LMIs (2.15a)–(2.15h) (shown in Theorem 2.5 of Chapter 2) hold, in this case, Φ_{ostij} is given by*

$$\Phi_{ostij} \triangleq \begin{bmatrix} \hat{\sum}_{11ostij} & \cdots & \sum_{12ostij} \\ \star & \ddots & \vdots \\ \star & \star & \sum_{22ostij} \end{bmatrix},$$

where

$$\sum\nolimits_{11ostij} \triangleq \begin{bmatrix} -\tau^{-1}Z_{1i} & 0 & 0 & (A_i - I)G_j + B_iK_j & 0 & 0 \\ \star & -\tilde{d}^{-1}Z_{2i} & 0 & (A_i - I)G_j + B_iK_j & 0 & 0 \\ \star & \star & -G_t - G_t^T + P_t & A_iG_j + B_iK_j & 0 & 0 \\ \star & \star & \star & \Pi_{22i} & Y_{1i} - Y_{0i}^T & Y_{2i} \\ \star & \star & \star & \star & \Pi_{33si} & -\varepsilon Y_{2i} \\ \star & \star & \star & \star & \star & \Pi_{44si} \end{bmatrix},$$

$\sum_{12ostij}$, $\sum_{22ostij}$, Π_{22i}, Π_{33i} and Π_{44i} are defined in Theorem 2.5 of Chapter 2. Moreover, there exists a fuzzy controller of the form (3.12).

With the previous results, the proof of Theorem 3.3 can be carried out in a straightforward way, thus it is omitted here.

3.4 Dynamic Output Feedback Control

In this section, a model transformation approach is introduced, and dynamic output control is designed based on the input-output method. Firstly, consider the following auxiliary system for the closed-loop system in (3.10):

$$\xi(k+1) = \bar{A}_c(k)\xi(k) + \bar{A}_{cd}(k)\xi(k - d(k)) + \gamma^{-1}\bar{D}_c(k)\hat{\omega}(k), \quad (3.14a)$$
$$z(k) = \bar{E}_c(k)\xi(k) + \bar{E}_{cd}(k)\xi(k - d(k)) + \gamma^{-1}\bar{H}_c(k)\hat{\omega}(k). \quad (3.14b)$$

It is clear that the induced ℓ_2 performance in (3.11) is equivalent to

$$\|z(k)\|_2 < \|\hat{\omega}(k)\|_2, \quad \forall\, 0 \neq \hat{\omega}(k) \in \ell_2[0, \infty). \quad (3.15)$$

The auxiliary system (3.14) is introduced to obtain the following model transformation and apply the input-output technique. It is not difficult to see that the induced ℓ_2 performance γ of the auxiliary system (3.14) is transferred to systems parameter, and is equivalent to the induced ℓ_2 performance in (3.11) for the original closed-loop system (3.10). It is the premise condition that the induced ℓ_2 performance γ of the auxiliary system (3.14) embodies in system parameters. Thus, based on the input-output approach, the scale small gain theorem can be used directly in the following model transformation.

In the following, considering the closed-loop system in (3.14), we now estimate the time-varying $\xi(k - d(k))$ using its lower bound d_1 and upper bound d_2. The two-term approximation $\frac{\xi(k-d_1)+\xi(k-d_2)}{2}$ results in the estimation error:

$$\begin{aligned} \sigma(k) &= \frac{2}{d}\left\{\xi(k - d(k)) - \frac{1}{2}\left[\xi(k - d_1) + \xi(k - d_2)\right]\right\} \\ &= \frac{1}{d}\left[\sum_{i=k-d_2}^{k-d_1-1} \beta(i)\varsigma(i)\right], \end{aligned} \quad (3.16)$$

where $d \triangleq d_2 - d_1$, $\varsigma(i) \triangleq \xi(i+1) - \xi(i)$ and

$$\beta(i) \triangleq \begin{cases} 1, & \text{when } i \leq k - d(k) - 1, \\ -1, & \text{when } i > k - d(k) - 1. \end{cases}$$

To employ the input-output approach, the following auxiliary system is introduced to replace system (3.14):

$$\begin{aligned} \xi(k+1) &= \bar{A}_c(k)\xi(k) + \frac{1}{2}\bar{A}_{cd}(k)\left[\xi(k-d_1) + \xi(k-d_2)\right] \\ &\quad + \frac{d}{2}\bar{A}_{cd}(k)\sigma(k) + \gamma^{-1}\bar{D}_c(k)\hat{\omega}(k), \end{aligned} \tag{3.17a}$$

$$\begin{aligned} z(k) &= \bar{E}_c(k)\xi(k) + \frac{1}{2}\bar{E}_{cd}(k)\left[\xi(k-d_1) + \xi(k-d_2)\right] \\ &\quad + \frac{d}{2}\bar{E}_{cd}(k)\sigma(k) + \gamma^{-1}\bar{H}_c(k)\hat{\omega}(k). \end{aligned} \tag{3.17b}$$

Now, based on (3.17), the following model reformulate system (3.17) into the interconnection frame in Fig. 1.1:

$$(\mathcal{S}_1): \begin{bmatrix} \xi(k+1) \\ \varsigma(k) \\ z(k) \end{bmatrix} = \begin{bmatrix} \Sigma_1(k) & \frac{d}{2}\bar{A}_{cd}(k) & \gamma^{-1}\bar{D}_c(k) \\ \Sigma_2(k) & \frac{d}{2}\bar{A}_{cd}(k) & \gamma^{-1}\bar{D}_c(k) \\ \Sigma_3(k) & \frac{d}{2}\bar{E}_{cd}(k) & \gamma^{-1}\bar{H}_c(k) \end{bmatrix} \begin{bmatrix} \xi(k) \\ \xi(k-d_1) \\ \xi(k-d_2) \\ \sigma(k) \\ \hat{\omega}(k) \end{bmatrix}, \tag{3.18a}$$

$$(\mathcal{S}_2): \quad \sigma(k) = \mathcal{K}\varsigma(k), \tag{3.18b}$$

where

$$\begin{aligned} \Sigma_1(k) &\triangleq \begin{bmatrix} \bar{A}_c(k) & \frac{1}{2}\bar{A}_{cd}(k) & \frac{1}{2}\bar{A}_{cd}(k) \end{bmatrix}, \\ \Sigma_2(k) &\triangleq \begin{bmatrix} \bar{A}_c(k) - I & \frac{1}{2}\bar{A}_{cd}(k) & \frac{1}{2}\bar{A}_{cd}(k) \end{bmatrix}, \\ \Sigma_3(k) &\triangleq \begin{bmatrix} \bar{E}_c(k) & \frac{1}{2}\bar{E}_{cd}(k) & \frac{1}{2}\bar{E}_{cd}(k) \end{bmatrix}. \end{aligned}$$

For brevity, let us use the following operator:

$$\mathcal{K}: \quad \varsigma(k) \to \sigma(k) = \frac{1}{d}\left[\sum_{i=k-d_2}^{k-d_1-1} \beta(i)\varsigma(i)\right], \tag{3.19}$$

to denote the relation $(\mathcal{S}_2)$ from $\varsigma(k)$ to $\sigma(k)$ in Fig. 1.1. The following result gives an upper bound of the ℓ_2 norm of $\mathcal{K}$.

Lemma 3.4. *Operator $\mathcal{K}$ in (3.19) bears the property $\|\mathcal{K}\|_\infty \leq 1$.*

By Lemma 3.4, we can see that the ℓ_2 norm of $(\mathcal{S}_2)$ in (3.18b) from input to output is bounded by one. Then, based on Lemma 3.4, we focus on researching the scaled small gain of $(\mathcal{S}_1)$ for the interconnection frame (3.18a).

Lemma 3.5. *Assumed that the $(\mathcal{S}_1)$ is internally stable in (3.18a), the closed-loop system of the interconnection system described by (3.18) is asymptotically stable and has an induced ℓ_2 performance level γ for $\mathcal{K}$ if there exist matrix $\hat{X} = \text{diag}\{\bar{X}, I\} > 0$ such that*

$$\|\hat{X} \circ \mathcal{G} \circ \hat{X}^{-1}\|_\infty < 1, \tag{3.20}$$

where

$$\mathcal{G} \triangleq \begin{bmatrix} \Sigma_1(k) & \frac{d}{2}\bar{A}_{cd}(k) & \gamma^{-1}\bar{D}_c(k) \\ \Sigma_2(k) & \frac{d}{2}\bar{A}_{cd}(k) & \gamma^{-1}\bar{D}_c(k) \\ \Sigma_3(k) & \frac{d}{2}\bar{E}_{cd}(k) & \gamma^{-1}\bar{H}_c(k) \end{bmatrix}.$$

Proof. From (3.18) and (3.20), we have

$$\|\varsigma(k)\|_2^2 + \|z(k)\|_2^2 < \|\sigma(k)\|_2^2 + \|\hat{\omega}(k)\|_2^2.$$

This together with Lemma 3.5 yields (3.11) and (3.15). Moreover, based on Lemmas 3.4 and 3.5, we can see that the result can be resulted. ■

Remark 3.6. By Lemma 3.5 and supposed that $(\mathcal{S}_1)$ is internally stable in (3.18), the closed-loop system of interconnection system in (3.18) is asymptotically stable and has an induced ℓ_2 performance level γ for $\mathcal{K}$ if there exist exists a matrix $X \triangleq \bar{X}^T\bar{X}$ such that

$\mathcal{J} \triangleq \sum_{k=0}^{\infty} \left[\varsigma^T(k)X\varsigma(k) - \sigma^T(k)X\sigma(k) + z^T(k)z(k) - \hat{\omega}^T(k)\hat{\omega}(k)\right] < 0.$ ♦

By applying the new model transformation and the instrumental idea of delay partitioning, we derive a LMI formulation of induced ℓ_2 bound for closed-loop system (3.18). Let

$$Q_\vartheta(k) \triangleq \sum_{i=1}^{r} h_i(\theta)Q_{\vartheta i}, \ R(k) \triangleq \sum_{i=1}^{r} h_i(\theta)R_i, \ S(k) \triangleq \sum_{i=1}^{r} h_i(\theta)S_i,$$

where $S_i > 0$, $R_i > 0$, $Q_{\vartheta i} > 0$, $i = 1, \ldots, r$, $\vartheta = 0, \ldots, m-1$, are $(n+k) \times (n+k)$ matrices.

Thus, we construct the following fuzzy LKF:

$$V(k) \triangleq \sum_{i=1}^{4} V_i(k), \tag{3.21}$$

with

$$V_1(k) \triangleq \xi^T(k)P\xi(k),$$
$$V_2(k) \triangleq \sum_{\vartheta=0}^{m-1}\sum_{l=k-\tau}^{k-1} \xi^T(l-\vartheta\tau)Q_\vartheta(l)\xi(l-\vartheta\tau),$$
$$V_3(k) \triangleq \sum_{l=k-d_2}^{k-1} \xi^T(l)S(l)\xi(l) + \sum_{s=-d_2+1}^{-\tau m+1}\sum_{l=k-1+s}^{k-1} \xi^T(l)R(l)\xi(l),$$
$$V_4(k) \triangleq \sum_{i=-\tau}^{-1}\sum_{j=k+i}^{k-1} \varsigma^T(j)Z_1\varsigma(j) + \sum_{i=-d_2}^{-1}\sum_{j=k+i}^{k-1} \varsigma^T(j)Z_2\varsigma(j).$$

Based on the fuzzy LKF in (3.21), we have the following result.

Theorem 3.7. *Given positive integers τ, m, $d = d_2 - d_1$ and $d_1 = \tau m$, the discrete-time T-S fuzzy time-varying delay system in (3.18) is asymptotically stable and has an induced ℓ_2 performance, if there exist matrices $P > 0$, $X > 0$, $Z_1 > 0$, $Z_2 > 0$, $S_i > 0$, $R_i > 0$, $T_i > 0$, $Q_{\vartheta i} > 0$, $Y_{\kappa i} > 0$, $U_{\kappa i} > 0$, $M_{\kappa i}$ and $N_{\kappa i}$, such that for $s,t,l,i,j = 1,\ldots,r$, $\vartheta = 0,\ldots,m-1$, $\kappa = 0,\ldots,m+1$, $\iota = 2,3,\ldots,m-2$,*

$$\left.\begin{aligned} \frac{1}{r-1}\Pi_{stlii} + \frac{1}{2}\left(\Pi_{stlij} + \Pi_{stlji}\right) < 0, \quad 1 \le i \ne j \le r, \\ \Pi_{stlii} < 0, \end{aligned}\right\} \tag{3.22a}$$

$$\begin{bmatrix} \Psi_{111i} & \Psi_{112i} \\ \star & Z_1 \end{bmatrix} \ge 0, \tag{3.22b}$$

$$\begin{bmatrix} \Psi_{211i} & \Psi_{212i} \\ \star & Z_2 \end{bmatrix} \ge 0, \tag{3.22c}$$

$$T_i - R_j < 0, \tag{3.22d}$$

where

$$\Pi_{stlij} \triangleq \begin{bmatrix} \Pi_{11} & \Pi_{12ij} & 0 & \frac{1}{2}\Pi_{14ij} & \frac{1}{2}\Pi_{15ij} \\ \star & \Pi_{22i} & \Pi_{23i} & \Pi_{24i} & 0 \\ \star & \star & \Pi_{33is} & \Pi_{34i} & 0 \\ \star & \star & \star & \Pi_{44istl} & 0 \\ \star & \star & \star & \star & \Pi_{55} \end{bmatrix},$$

$$\Psi_{111i} \triangleq \mathrm{diag}\left\{Y_{0i}, Y_{1i}, Y_{\iota i}, Y_{(m-1)i}, Y_{mi}, Y_{(m+1)i}\right\},$$
$$\Psi_{211i} \triangleq \mathrm{diag}\left\{U_{0i}, U_{1i}, U_{\iota i}, U_{(m-1)i}, U_{mi}, U_{(m+1)i}\right\},$$
$$\Psi_{112i} \triangleq \begin{bmatrix} M_{0i}^T & M_{1i}^T & M_{\iota i}^T & M_{(m-1)i}^T & M_{mi}^T & M_{(m+1)i}^T \end{bmatrix}^T,$$
$$\Psi_{212i} \triangleq \begin{bmatrix} N_{0i}^T & N_{1i}^T & N_{\iota i}^T & N_{(m-1)i}^T & N_{mi}^T & N_{(m+1)i}^T \end{bmatrix}^T,$$

with

$$\Pi_{11} \triangleq \operatorname{diag}\left\{-P^{-1}, -\tau^{-1}Z_1^{-1}, -d_2^{-1}Z_2^{-1}, -X^{-1}, -I\right\},$$
$$\Pi_{22i} \triangleq S_i - P + Q_{0i} + (d+1)R_i + M_{0i} + M_{0i}^T + N_{0i} + N_{0i}^T + J_{0i},$$
$$\Pi_{33is} \triangleq \begin{bmatrix} G_{1is} - M_{1i} - M_{1i}^T + J_{1i} & -M_{\iota i} & -M_{(m-1)i} \\ \star & G_{\iota is} + J_{\iota i} & 0 \\ \star & \star & G_{(m-1)is} + J_{(m-1)i} \end{bmatrix},$$
$$\Pi_{44istl} \triangleq \begin{bmatrix} -Q_{(m-1)s} - T_t + J_{mi} & -N_{mi}^T \\ \star & -S_l - N_{(m+1)i} - N_{(m+1)i}^T + J_{(m+1)i} \end{bmatrix},$$
$$\Pi_{55} \triangleq \begin{bmatrix} -X & 0 \\ \star & -\gamma^2 I \end{bmatrix}, \quad J_{\nu i} \triangleq \tau Y_{\nu i} + d_2 U_{\nu i}, \quad G_{\nu is} \triangleq Q_{\nu i} - Q_{(\nu-1)s},$$
$$\Pi_{12ij} \triangleq \begin{bmatrix} \bar{A}_{cij} \\ \bar{A}_{cij} - I \\ \bar{A}_{cij} - I \\ \bar{A}_{cij} - I \\ \bar{E}_{cij} \end{bmatrix}, \quad \Pi_{34i} \triangleq \begin{bmatrix} -M_{mi} & -M_{(m+1)i} - N_{1i}^T \\ 0 & -N_{\iota i}^T \\ 0 & -N_{(m-1)i}^T \end{bmatrix},$$
$$\Pi_{14ij} \triangleq \begin{bmatrix} \bar{A}_{cdij} & \bar{A}_{cdij} \\ \bar{A}_{cdij} & \bar{A}_{cdij} \\ \bar{A}_{cdij} & \bar{A}_{cdij} \\ \bar{A}_{cdij} & \bar{A}_{cdij} \\ \bar{E}_{cdi} & \bar{E}_{cdi} \end{bmatrix}, \quad \Pi_{15ij} \triangleq \begin{bmatrix} d\bar{A}_{cdij} & 2\bar{D}_{cij} \\ d\bar{A}_{cdij} & 2\bar{D}_{cij} \\ d\bar{A}_{cdij} & 2\bar{D}_{cij} \\ d\bar{A}_{cdij} & 2\bar{D}_{cij} \\ d\bar{E}_{cdi} & 2\bar{H}_{ci} \end{bmatrix},$$
$$\Pi_{23i} \triangleq \begin{bmatrix} -M_{0i}^T + M_{1i} + N_{1i} & M_{\iota i} + N_{\iota i} & M_{(m-1)i} + N_{(m-1)i} \end{bmatrix},$$
$$\Pi_{24i} \triangleq \begin{bmatrix} M_{mi} + N_{mi} & M_{(m+1)i} + N_{(m+1)i} - N_{0i}^T \end{bmatrix}.$$

With the previous results, the proof of Theorem 3.7 can be carried out in a straightforward way thus is omitted here.

In the following, we are in a position to present a solution to the induced ℓ_2 DOF control problem based on Theorem 3.7, and give the following result.

Theorem 3.8. *Given positive integers τ, m, $d = d_2 - d_1$ and $d_1 \triangleq \tau m$, the discrete-time T-S fuzzy time-varying delay system in (3.18) is asymptotically stable and has an induced ℓ_2 performance, if there exist matrices $\mathscr{P} > 0$, $\mathscr{W} > 0$, $\bar{X} \triangleq \begin{bmatrix} X_1 & X_2 \\ \star & X_3 \end{bmatrix} > 0$, $\bar{Z}_1 \triangleq \begin{bmatrix} Z_{11} & Z_{12} \\ \star & Z_{13} \end{bmatrix} > 0$, $\bar{Z}_2 \triangleq \begin{bmatrix} Z_{21} & Z_{22} \\ \star & Z_{23} \end{bmatrix} > 0$, $\bar{S}_i \triangleq \begin{bmatrix} S_{i1} & S_{i2} \\ \star & S_{i3} \end{bmatrix} > 0$, $\bar{R}_i \triangleq \begin{bmatrix} R_{i1} & R_{i2} \\ \star & R_{i3} \end{bmatrix} > 0$, $\bar{T}_i \triangleq \begin{bmatrix} T_{i1} & T_{i2} \\ \star & T_{i3} \end{bmatrix} > 0$, $\bar{Q}_{\vartheta i} \triangleq \begin{bmatrix} Q_{\vartheta i1} & Q_{\vartheta i2} \\ \star & Q_{\vartheta i3} \end{bmatrix} > 0$, $\bar{Y}_{\kappa i} \triangleq \begin{bmatrix} Y_{\kappa i1} & Y_{\kappa i2} \\ \star & Y_{\kappa i3} \end{bmatrix} > 0$, $\bar{U}_{\kappa i} \triangleq \begin{bmatrix} U_{\kappa i1} & U_{\kappa i2} \\ \star & U_{\kappa i3} \end{bmatrix} > 0$, $\bar{M}_{\kappa i} \triangleq \begin{bmatrix} M_{\kappa i1} & M_{\kappa i2} \\ M_{\kappa i3} & M_{\kappa i4} \end{bmatrix}$, $\bar{N}_{\kappa i} \triangleq \begin{bmatrix} N_{\kappa i1} & N_{\kappa i2} \\ N_{\kappa i3} & N_{\kappa i4} \end{bmatrix}$, $\mathscr{A}_{ij}$, $\mathscr{A}_{dij}$, $\mathscr{B}_i$ and $\mathscr{C}_i$ such that for $s, t, l, i, j = 1, \ldots, r$, $\vartheta = 0, \ldots, m-1$, $\kappa = 0, \ldots, m+1$, $\iota = 2, 3, \ldots, m-2$,*

$$\left.\begin{aligned} \frac{1}{r-1}\Gamma_{stlii} + \frac{1}{2}\left(\Gamma_{stlij} + \Gamma_{stlji}\right) < 0, \quad 1 \le i \ne j \le r, \\ \Gamma_{stlii} < 0, \end{aligned}\right\} \tag{3.23a}$$

$$\begin{bmatrix} \Phi_{111i} & 0 & 0 & \Phi_{114i} \\ \star & \Phi_{122i} & 0 & \Phi_{124i} \\ \star & \star & \Phi_{133i} & \Phi_{134i} \\ \star & \star & \star & \bar{Z}_1 \end{bmatrix} \geq 0, \qquad (3.23b)$$

$$\begin{bmatrix} \Phi_{211i} & 0 & 0 & \Phi_{214i} \\ \star & \Phi_{222i} & 0 & \Phi_{224i} \\ \star & \star & \Phi_{233i} & \Phi_{234i} \\ \star & \star & \star & \bar{Z}_2 \end{bmatrix} \geq 0, \qquad (3.23c)$$

$$\begin{bmatrix} T_{1i} - R_{1j} & T_{2i} - R_{2j} \\ \star & T_{3i} - R_{3j} \end{bmatrix} < 0, \qquad (3.23d)$$

where

$$\Phi_{111i} \triangleq \begin{bmatrix} Y_{0i1} & Y_{0i2} & 0 & 0 \\ \star & Y_{0i3} & 0 & 0 \\ \star & \star & Y_{1i1} & Y_{1i2} \\ \star & \star & \star & Y_{1i3} \end{bmatrix}, \ \Phi_{122i} \triangleq \begin{bmatrix} Y_{\iota i1} & Y_{\iota i2} & 0 & 0 \\ \star & Y_{\iota i3} & 0 & 0 \\ \star & \star & Y_{(m-1)i1} & Y_{(m-1)i2} \\ \star & \star & \star & Y_{(m-1)i3} \end{bmatrix},$$

$$\Phi_{133i} \triangleq \begin{bmatrix} Y_{mi1} & Y_{mi2} & 0 & 0 \\ \star & Y_{mi3} & 0 & 0 \\ \star & \star & Y_{(m+1)i1} & Y_{(m+1)i2} \\ \star & \star & \star & Y_{(m+1)i3} \end{bmatrix}, \ \Phi_{114i} \triangleq \begin{bmatrix} M_{0i1} & M_{0i2} \\ M_{0i3} & M_{0i4} \\ M_{1i1} & M_{1i2} \\ M_{1i3} & M_{1i4} \end{bmatrix},$$

$$\Phi_{124i} \triangleq \begin{bmatrix} M_{\iota i1} & M_{\iota i2} \\ M_{\iota i3} & M_{\iota i4} \\ M_{(m-1)i1} & M_{(m-1)i2} \\ M_{(m-1)i3} & M_{(m-1)i4} \end{bmatrix}, \ \Phi_{134i} \triangleq \begin{bmatrix} M_{mi1} & M_{mi2} \\ M_{mi3} & M_{mi4} \\ M_{(m+1)i1} & M_{(m+1)i2} \\ M_{(m+1)i3} & M_{(m+1)i4} \end{bmatrix},$$

$$\Phi_{211i} \triangleq \begin{bmatrix} U_{0i1} & U_{0i2} & 0 & 0 \\ \star & U_{0i3} & 0 & 0 \\ \star & \star & U_{1i1} & U_{1i2} \\ \star & \star & \star & U_{1i3} \end{bmatrix}, \ \Phi_{222i} \triangleq \begin{bmatrix} U_{\iota i1} & U_{\iota i2} & 0 & 0 \\ \star & U_{\iota i3} & 0 & 0 \\ \star & \star & U_{(m-1)i1} & U_{(m-1)i2} \\ \star & \star & \star & U_{(m-1)i3} \end{bmatrix},$$

$$\Phi_{233i} \triangleq \begin{bmatrix} U_{mi1} & U_{mi2} & 0 & 0 \\ \star & U_{mi3} & 0 & 0 \\ \star & \star & U_{(m+1)i1} & U_{(m+1)i2} \\ \star & \star & \star & U_{(m+1)i3} \end{bmatrix}, \ \Phi_{214i} \triangleq \begin{bmatrix} N_{0i1} & N_{0i2} \\ N_{0i3} & N_{0i4} \\ N_{1i1} & N_{1i2} \\ N_{1i3} & N_{1i4} \end{bmatrix},$$

$$\Phi_{224i} \triangleq \begin{bmatrix} N_{\iota i1} & N_{\iota i2} \\ N_{\iota i3} & N_{\iota i4} \\ N_{(m-1)i1} & N_{(m-1)i2} \\ N_{(m-1)i3} & N_{(m-1)i4} \end{bmatrix}, \ \Phi_{234i} \triangleq \begin{bmatrix} N_{mi1} & N_{mi2} \\ N_{mi3} & N_{mi4} \\ N_{(m+1)i1} & N_{(m+1)i2} \\ N_{(m+1)i3} & N_{(m+1)i4} \end{bmatrix},$$

$$\Gamma_{stlij} \triangleq \begin{bmatrix} \Gamma_1 & \Gamma_{2ij} & 0 & \frac{1}{2}\Gamma_{3ij} & \frac{1}{2}\Gamma_{4ij} \\ \star & \Gamma_{5i} & \Gamma_{6i} & \Gamma_{7i} & 0 \\ \star & \star & \Gamma_{8is} & \Gamma_{9i} & 0 \\ \star & \star & \star & \Gamma_{0istl} & 0 \\ \star & \star & \star & \star & \Gamma_{11} \end{bmatrix},$$

with

$$\varGamma_1 \triangleq \mathrm{diag}\{-\varGamma_{111}, \tau^{-1}\varGamma_{122}, d_2^{-1}\varGamma_{133}, \varGamma_{144}, -I\},$$

$$\varGamma_{2ij} \triangleq \begin{bmatrix} \varGamma_{\mathrm{Ac}ij} \\ \varGamma_{\mathrm{Ac}ij} \\ \varGamma_{\mathrm{Ac}ij} \\ \varGamma_{\mathrm{Ac}ij} \\ \varGamma_{\mathrm{Ec}ij} \end{bmatrix}, \quad \varGamma_{3ij} \triangleq \begin{bmatrix} \varGamma_{\mathrm{Acd}ij} & \varGamma_{\mathrm{Acd}ij} \\ \varGamma_{\mathrm{Acd}ij} & \varGamma_{\mathrm{Acd}ij} \\ \varGamma_{\mathrm{Acd}ij} & \varGamma_{\mathrm{Acd}ij} \\ \varGamma_{\mathrm{Acd}ij} & \varGamma_{\mathrm{Acd}ij} \\ \varGamma_{\mathrm{Ecd}i} & \varGamma_{\mathrm{Ecd}i} \end{bmatrix}, \quad \varGamma_{4ij} \triangleq \begin{bmatrix} d\varGamma_{\mathrm{Acd}ij} & 2\varGamma_{\mathrm{Dc}ij} \\ d\varGamma_{\mathrm{Acd}ij} & 2\varGamma_{\mathrm{Dc}ij} \\ d\varGamma_{\mathrm{Acd}ij} & 2\varGamma_{\mathrm{Dc}ij} \\ d\varGamma_{\mathrm{Acd}ij} & 2\varGamma_{\mathrm{Dc}ij} \\ d\varGamma_{\mathrm{Ecd}i} & 2H_i \end{bmatrix},$$

$$\varGamma_{5i} \triangleq \begin{bmatrix} \varGamma_{511i} & \varGamma_{512i} \\ \star & \varGamma_{513i} \end{bmatrix}, \quad \varGamma_{6i} \triangleq \begin{bmatrix} \varGamma_{611i} & \varGamma_{612i} & \varGamma_{613i} \\ \varGamma_{621i} & \varGamma_{622i} & \varGamma_{623i} \end{bmatrix}, \quad \varGamma_{7i} \triangleq \begin{bmatrix} \varGamma_{711i} & \varGamma_{712i} \\ \varGamma_{721i} & \varGamma_{722i} \end{bmatrix},$$

$$\varGamma_{8is} \triangleq \begin{bmatrix} \varGamma_{811is} & \varGamma_{812is} & -M_{\iota i1} & -M_{\iota i2} & -M_{(m-1)i1} & -M_{(m-1)i2} \\ \star & \varGamma_{822is} & -M_{\iota i3} & -M_{\iota i4} & -M_{(m-1)i3} & -M_{(m-1)i4} \\ \star & \star & \varGamma_{833is} & \varGamma_{834is} & 0 & 0 \\ \star & \star & \star & \varGamma_{844is} & 0 & 0 \\ \star & \star & \star & \star & \varGamma_{855is} & \varGamma_{856is} \\ \star & \star & \star & \star & \star & \varGamma_{866is} \end{bmatrix},$$

$$\varGamma_{9i} \triangleq \begin{bmatrix} -M_{mi1} & -M_{mi2} & -M_{(m+1)i1} - N_{1i1}^T & -M_{(m+1)i2} - N_{1i3}^T \\ -M_{mi3} & -M_{mi4} & -M_{(m+1)i3} - N_{1i2}^T & -M_{(m+1)i4} - N_{1i4}^T \\ 0 & 0 & -N_{\iota i1}^T & -N_{\iota i3}^T \\ 0 & 0 & -N_{\iota i2}^T & -N_{\iota i4}^T \\ 0 & 0 & -N_{(m-1)i1}^T & -N_{(m-1)i3}^T \\ 0 & 0 & -N_{(m-1)i2}^T & -N_{(m-1)i4}^T \end{bmatrix},$$

$$\varGamma_{0istl} \triangleq \begin{bmatrix} \varGamma_{011ist} & \varGamma_{012ist} & -N_{mi1}^T & -N_{mi3}^T \\ \star & \varGamma_{022ist} & -N_{mi2}^T & -N_{mi4}^T \\ \star & \star & \varGamma_{033il} & \varGamma_{034il} \\ \star & \star & \star & \varGamma_{044il} \end{bmatrix}, \quad \varGamma_{11} \triangleq \begin{bmatrix} -X_1 & -X_2 & 0 \\ \star & -X_3 & 0 \\ \star & \star & -\gamma^2 I \end{bmatrix},$$

$$\varGamma_{122} \triangleq \begin{bmatrix} Z_{11} - \mathscr{P} - \mathscr{P}^T & Z_{12} - 2I \\ \star & Z_{13} - \mathscr{W} - \mathscr{W}^T \end{bmatrix}, \quad \varGamma_{111} \triangleq \begin{bmatrix} \mathscr{P} & I \\ \star & \mathscr{W} \end{bmatrix},$$

$$\varGamma_{133} \triangleq \begin{bmatrix} Z_{21} - \mathscr{P} - \mathscr{P}^T & Z_{22} - 2I \\ \star & Z_{23} - \mathscr{W} - \mathscr{W}^T \end{bmatrix}, \quad \varGamma_{\mathrm{Dc}ij} \triangleq \begin{bmatrix} \mathscr{P} D_i + \mathscr{B}_j F_i \\ D_i \end{bmatrix},$$

$$\varGamma_{144} \triangleq \begin{bmatrix} X_1 - \mathscr{P} - \mathscr{P}^T & X_2 - 2I \\ \star & X_3 - \mathscr{W} - \mathscr{W}^T \end{bmatrix}, \quad \begin{array}{l} \varGamma_{\mathrm{Ec}ij} \triangleq \begin{bmatrix} E_i & E_i \mathscr{W} + G_i \mathscr{C}_j \end{bmatrix}, \\ \varGamma_{\mathrm{Ecd}i} \triangleq \begin{bmatrix} E_{di} & E_{di} \mathscr{W} \end{bmatrix}, \end{array}$$

$$\varGamma_{\mathrm{Ac}ij} \triangleq \begin{bmatrix} \mathscr{P} A_i + \mathscr{B}_j C_i & \mathscr{A}_{ij} \\ A_i & A_i \mathscr{W} + B_i \mathscr{C}_j \end{bmatrix}, \quad \varGamma_{\mathrm{Acd}ij} \triangleq \begin{bmatrix} \mathscr{P} A_{di} + \mathscr{B}_j C_{di} & \mathscr{A}_{dij} \\ A_{\mathrm{d}i} & A_{\mathrm{d}i} \mathscr{W} \end{bmatrix},$$

$$\varGamma_{511i} \triangleq -\mathscr{P} + Q_{0i1} + S_{i1} + (d+1) R_{i1} + M_{0i1} + M_{0i1}^T + N_{0i1} + N_{0i1}^T + J_{0i1},$$

$$\varGamma_{512i} \triangleq -I + Q_{0i2} + S_{i2} + (d+1) R_{i2} + M_{0i2} + M_{0i3}^T + N_{0i2} + N_{0i3}^T + J_{0i2},$$

$$\varGamma_{513i} \triangleq -\mathscr{W} + Q_{0i3} + S_{i3} + (d+1) R_{i3} + M_{0i4} + M_{0i4}^T + N_{0i4} + N_{0i4}^T + J_{0i3},$$

$$\varGamma_{611i} \triangleq \begin{bmatrix} -M_{0i1}^T + M_{1i1} + N_{1i1} & -M_{0i3}^T + M_{1i2} + N_{1i2} \end{bmatrix},$$

$$\varGamma_{621i} \triangleq \begin{bmatrix} -M_{0i2}^T + M_{1i3} + N_{1i3} & -M_{0i4}^T + M_{1i4} + N_{1i4} \end{bmatrix},$$

$$\varGamma_{613i} \triangleq \begin{bmatrix} M_{(m-1)i1} + N_{(m-1)i1} & M_{(m-1)i2} + N_{(m-1)i2} \end{bmatrix},$$

$$\begin{aligned}
\Gamma_{623i} &\triangleq \left[M_{(m-1)i3} + N_{(m-1)i3} \;\; M_{(m-1)i4} + N_{(m-1)i4} \right],\\
\Gamma_{612i} &\triangleq \left[M_{\iota i1}+N_{\iota i1} \;\; M_{\iota i2}+N_{\iota i2} \right], \; \Gamma_{622i} \triangleq \left[M_{\iota i3} + N_{\iota i3} \;\; M_{\iota i4} + N_{\iota i4} \right],\\
\Gamma_{711i} &\triangleq \left[M_{mi1} + N_{mi1} \;\; M_{mi2} + N_{mi2} \right],\\
\Gamma_{721i} &\triangleq \left[M_{mi3} + N_{mi3} \;\; M_{mi4} + N_{mi4} \right],\\
\Gamma_{712i} &\triangleq \left[M_{(m+1)i1} + N_{(m+1)i1} - N_{0i1}^T \;\; M_{(m+1)i2} + N_{(m+1)i2} - N_{0i3}^T \right],\\
\Gamma_{722i} &\triangleq \left[M_{(m+1)i3} + N_{(m+1)i3} - N_{0i2}^T \;\; M_{(m+1)i4} + N_{(m+1)i4} - N_{0i4}^T \right],\\
\Gamma_{811is} &\triangleq G_{1is1} + J_{1i1} - M_{1i1} - M_{1i1}^T, \quad \Gamma_{833is} \triangleq G_{\iota is1} + J_{\iota i1},\\
\Gamma_{812is} &\triangleq G_{1is2} + J_{1i2} - M_{1i2} - M_{1i3}^T, \quad \Gamma_{834is} \triangleq G_{\iota is2} + J_{\iota i2},\\
\Gamma_{822is} &\triangleq G_{1is3} + J_{1i3} - M_{1i4} - M_{1i4}^T, \quad \Gamma_{844is} \triangleq G_{\iota is3} + J_{\iota i3},\\
\Gamma_{855is} &\triangleq G_{(m-1)is1} + J_{(m-1)i1}, \quad \Gamma_{856is} \triangleq G_{(m-1)is2} + J_{(m-1)i2},\\
\Gamma_{866is} &\triangleq G_{(m-1)is3} + J_{(m-1)i3}, \quad \Gamma_{011ist} \triangleq -Q_{(m-1)s1} - T_{t1} + J_{mi1},\\
\Gamma_{012ist} &\triangleq -Q_{(m-1)s2} - T_{t2} + J_{mi2}, \quad \Gamma_{022ist} \triangleq -Q_{(m-1)s3} - T_{t3} + J_{mi3},\\
\Gamma_{033il} &\triangleq -S_{l1} + J_{(m+1)i1} - N_{(m+1)i1} - N_{(m+1)i1}^T,\\
\Gamma_{034il} &\triangleq -S_{l2} + J_{(m+1)i2} - N_{(m+1)i2} - N_{(m+1)i3}^T,\\
\Gamma_{044il} &\triangleq -S_{l3} + J_{(m+1)i3} - N_{(m+1)i4} - N_{(m+1)i4}^T,\\
J_{\nu i1} &\triangleq \tau Y_{\nu i1} + d_2 U_{\nu i1}, \; G_{\nu is1} \triangleq Q_{\nu i1} - Q_{(\nu-1)s1}, \; J_{\nu i3} \triangleq \tau Y_{\nu i3} + d_2 U_{\nu i3},\\
J_{\nu i2} &\triangleq \tau Y_{\nu i2} + d_2 U_{\nu i2}, \; G_{\nu is2} \triangleq Q_{\nu i2} - Q_{(\nu-1)s2}, \; G_{\nu is3} \triangleq Q_{\nu i3} - Q_{(\nu-1)s3}.
\end{aligned}$$

Moreover, if the above conditions have feasible solutions then the matrices for the desired DOF controller in the form of (3.9) are given by

$$\begin{cases} A_{Kj} \triangleq \mathscr{U}^{-1} \left(\mathscr{A}_{ij} - \mathscr{P} A_i \mathscr{W} - \mathscr{B}_j C_i \mathscr{W} - \mathscr{P} B_i \mathscr{C}_j \right) \mathscr{V}^{-T}, \\ B_{Kj} \triangleq \mathscr{U}^{-1} \mathscr{B}_j, \\ C_{Kj} \triangleq \mathscr{C}_j \mathscr{V}^{-T}, \end{cases} \tag{3.24}$$

where $\mathscr{U}$ and $\mathscr{V}$ are any nonsingular matrices satisfying

$$\mathscr{U} \mathscr{V}^T = I - \mathscr{P} \mathscr{W}.$$

Proof. According to Theorem 3.7, it is easy to prove that the closed-loop system (3.10) is asymptotically stable and has an induced ℓ_2 performance if there exist matrices $P > 0$, $X > 0$, $Z_1 > 0$, $Z_2 > 0$, $S_i > 0$, $R_i > 0$, $T_i > 0$, $Q_{\vartheta i} > 0$, $Y_{\kappa i} > 0$, $U_{\kappa i} > 0$, $M_{\kappa i}$ and $N_{\kappa i}$, ($i = 1, 2, \ldots, r$; $\vartheta = 0, \ldots, m-1$; $\kappa = 0, \ldots, m+1$) satisfying (3.22b)–(3.22d) and the following inequalities:

$$\left.\begin{aligned} \frac{1}{r-1} \bar{\Pi}_{stlii} + \frac{1}{2} \left(\bar{\Pi}_{stlij} + \bar{\Pi}_{stlji} \right) < 0, \quad 1 \leq i \neq j \leq r, \\ \bar{\Pi}_{stlii} < 0, \end{aligned}\right\} \tag{3.25}$$

where

$$\bar{\Pi}_{stlij} \triangleq \begin{bmatrix} \bar{\Pi}_1 & \bar{\Pi}_{2ij} & 0 & \frac{1}{2}\bar{\Pi}_{3ij} & \frac{1}{2}\bar{\Pi}_{4ij} \\ \star & \Pi_{5i} & \Pi_{6i} & \Pi_{7i} & 0 \\ \star & \star & \Pi_{8is} & \Pi_{9i} & 0 \\ \star & \star & \star & \Pi_{0istl} & 0 \\ \star & \star & \star & \star & \Pi_{11} \end{bmatrix}, \quad \bar{\Pi}_{2ij} \triangleq \begin{bmatrix} P\bar{A}_{cij} \\ P(\bar{A}_{cij} - I) \\ P(\bar{A}_{cij} - I) \\ P(\bar{A}_{cij} - I) \\ \bar{E}_{cij} \end{bmatrix},$$

$$\bar{\Pi}_{3ij} \triangleq \begin{bmatrix} P\bar{A}_{cdij} & P\bar{A}_{cdij} \\ P\bar{A}_{cdij} & P\bar{A}_{cdij} \\ P\bar{A}_{cdij} & P\bar{A}_{cdij} \\ P\bar{A}_{cdij} & P\bar{A}_{cdij} \\ \bar{E}_{cdij} & \bar{E}_{cdij} \end{bmatrix}, \quad \bar{\Pi}_{4ij} \triangleq \begin{bmatrix} dP\bar{A}_{cdij} & 2P\bar{D}_{cij} \\ dP\bar{A}_{cdij} & 2P\bar{D}_{cij} \\ dP\bar{A}_{cdij} & 2P\bar{D}_{cij} \\ dP\bar{A}_{cdij} & 2P\bar{D}_{cij} \\ d\bar{E}_{cdij} & 2\bar{H}_{cij} \end{bmatrix},$$

$$\bar{\Pi}_1 \triangleq \mathrm{diag}\left\{-P, \tau^{-1}(Z_1 - 2P), d_2^{-1}(Z_2 - 2P), (X - 2P), -I\right\}.$$

Let the matrix P and $W = P^{-1}$ be partitioned respectively as

$$P \triangleq \begin{bmatrix} P_1 & P_2 \\ \star & P_3 \end{bmatrix}, \quad W \triangleq \begin{bmatrix} W_1 & W_2 \\ \star & W_3 \end{bmatrix}. \tag{3.26}$$

Without loss of generality, we assume P_2 and W_2 are nonsingular (if not, P_2 and W_2 may be perturbed by ΔP_2 and ΔW_2 with sufficiently small norm such that $P_2+\Delta P_2$ and $W_2+\Delta W_2$ are nonsingular and satisfy (3.25), respectively). Define the following nonsingular matrices:

$$\mathcal{J}_P \triangleq \begin{bmatrix} P_1 & I \\ P_2^T & 0 \end{bmatrix}, \quad \mathcal{J}_W \triangleq \begin{bmatrix} I & W_1 \\ 0 & W_2^T \end{bmatrix}. \tag{3.27}$$

Notice that $P\mathcal{J}_W = \mathcal{J}_P$, $W\mathcal{J}_P = \mathcal{J}_W$ and $P_1W_1 + P_2W_2^T = I$.

Define the following matrices which are also nonsingular:

$$\begin{cases} \mathscr{P} \triangleq P_1, \quad \mathscr{W} \triangleq W_1, \quad \mathscr{U} \triangleq P_2, \quad \mathscr{V} \triangleq W_2, \\ Z_1 \triangleq \mathcal{J}_W^{-T}\bar{Z}_1\mathcal{J}_W^{-1}, \quad Z_2 \triangleq \mathcal{J}_W^{-T}\bar{Z}_2\mathcal{J}_W^{-1}, \\ M_{\kappa i} \triangleq \mathcal{J}_W^{-T}\bar{M}_{\kappa i}\mathcal{J}_W^{-1}, \quad N_{\kappa i} \triangleq \mathcal{J}_W^{-T}\bar{N}_{\kappa i}\mathcal{J}_W^{-1}, \\ Y_{\kappa i} \triangleq \mathcal{J}_W^{-T}\bar{Y}_{\kappa i}\mathcal{J}_W^{-1}, \quad U_{\kappa i} \triangleq \mathcal{J}_W^{-T}\bar{U}_{\kappa i}\mathcal{J}_W^{-1}, \\ Q_{\vartheta i} \triangleq \mathcal{J}_W^{-T}\bar{Q}_{\vartheta i}\mathcal{J}_W^{-1}, \quad R_i \triangleq \mathcal{J}_W^{-T}\bar{R}_i\mathcal{J}_W^{-1}, \\ S_i \triangleq \mathcal{J}_W^{-T}\bar{S}_i\mathcal{J}_W^{-1}, \quad T_i \triangleq \mathcal{J}_W^{-T}\bar{T}_i\mathcal{J}_W^{-1}, \\ X \triangleq \mathcal{J}_W^{-T}\bar{X}\mathcal{J}_W^{-1}, \end{cases}$$

and

$$\begin{cases} \mathscr{A}_{ij} \triangleq P_1 A_i W_1 + P_2 B_{Kj} C_i W_1 + P_1 B_i C_{Kj} W_2^T + P_2 A_{Kj} W_2^T, \\ \mathscr{A}_{\mathrm{d}ij} \triangleq P_1 A_{di} W_1 + P_2 B_{Kj} C_{di} W_1, \\ \mathscr{B}_j \triangleq \mathscr{U} B_{\mathrm{K}j}, \\ \mathscr{C}_j \triangleq C_{\mathrm{K}j} \mathscr{V}^T. \end{cases} \tag{3.28}$$

Performing congruence transformations to (3.22b)–(3.22d) and (3.25) by $\mathrm{diag}\{\mathcal{J}_W, \mathcal{J}_W, \ldots, \mathcal{J}_W, \mathcal{J}_W, \mathcal{J}_W, \mathcal{J}_W\}$, $\mathrm{diag}\{\mathcal{J}_W, \mathcal{J}_W, \ldots, \mathcal{J}_W, \mathcal{J}_W, \mathcal{J}_W, \mathcal{J}_W\}$, $\mathcal{J}_W$ and $\mathrm{diag}\{\mathcal{J}_W, \mathcal{J}_W, \mathcal{J}_W, \mathcal{J}_W, I, \mathcal{J}_W, \mathcal{J}_W, \ldots, \mathcal{J}_W, \mathcal{J}_W, \mathcal{J}_W, \mathcal{J}_W, I\}$, respectively, and considering (3.26)–(3.28), we can obtain (3.23a)–(3.23d). The conditions in (3.28) yields (3.24). This completes the proof. ■

Remark 3.9. Note that Theorem 3.8 provides a sufficient condition for the solvability of the induced ℓ_2 DOF controller design for T-S fuzzy time-varying delay system (3.8). Since the obtained conditions are all in terms of strict LMIs, a desired DOF controller can be determined by solving the following convex optimization problem:

$$\min \delta \quad \text{subject to (3.23a)–(3.23d) with } \delta = \gamma^2.$$ ♦

3.5 Illustrative Example

Example 3.10. Consider the following Henon system:

$$\begin{aligned} x_1(k+1) &= -\left[\mathcal{C}x_1(k) + (1-\mathcal{C})x_1(k-d(k))\right]^2 + 0.3x_2(k) + 1.4 + u(k), \\ x_2(k+1) &= \mathcal{C}x_1(k) + (1-\mathcal{C})x_1(k-d(k)), \end{aligned}$$

where the constant $c \in [0, 1]$ is the retarded coefficient.

Let $\theta = cx_1(k) + (1-c)x_1(k-d(k))$. Assume that $\theta \in [-\mathcal{M}, \mathcal{M}], \mathcal{M} > 0$. By using the same procedure as in [194], the nonlinear term θ^2 can be exactly represented as

$$\theta^2(k) = h_1(\theta)(-m)\theta + h_2(\theta)m\theta,$$

where $h_1(\theta), h_2(\theta) \in [0, 1]$, and $h_1(\theta) + h_2(\theta) = 1$. By solving the equations, the membership functions $h_1(\theta)$ and $h_2(\theta)$ are obtained as

$$h_1(\theta) = \frac{1}{2}\left(1 - \frac{\theta}{\mathcal{M}}\right), \quad h_2(\theta) = \frac{1}{2}\left(1 + \frac{\theta}{\mathcal{M}}\right).$$

It can be seen from the aforementioned expressions that $h_1(\theta) = 1$ and $h_2(\theta) = 0$ when θ is $-\mathcal{M}$ and that $h_1(\theta) = 0$ and $h_2(\theta) = 1$ when θ is $\mathcal{M}$. Then, the above nonlinear system can be approximately represented by the following T-S fuzzy model:

♦ **Plant Form:**

Rule 1: IF θ is $-m$, THEN

$$x(k+1) = A_1 x(k) + A_{d1} x(k - d(k)) + B_1 u^\star(k),$$

Rule 2: IF θ is m, THEN

$$x(k+1) = A_2 x(k) + A_{d2} x(k - d(k)) + B_2 u^\star(k),$$

where $u^\star(k) = 1.4 + u(k)$ and

$$A1 = \begin{bmatrix} \mathcal{C}\mathcal{M} & 0.3 \\ \mathcal{C} & 0 \end{bmatrix}, \quad A_{d1} = \begin{bmatrix} (1-\mathcal{C})\mathcal{M} & 0 \\ 1-\mathcal{C} & 0 \end{bmatrix}, \quad B_1 = \begin{bmatrix} 1 \\ 0 \end{bmatrix},$$

$$A2 = \begin{bmatrix} -\mathcal{C}\mathcal{M} & 0.3 \\ c & 0 \end{bmatrix}, \quad A_{d2} = \begin{bmatrix} -(1-\mathcal{C})\mathcal{M} & 0 \\ 1-\mathcal{C} & 0 \end{bmatrix}, \quad B_2 = \begin{bmatrix} 1 \\ 0 \end{bmatrix}.$$

◊ **Fuzzy State Feedback Controller:**

In the example, $x(k) = \begin{bmatrix} x_1(k) \\ x_2(k) \end{bmatrix}$, $\mathcal{C} = 0.8$, $\mathcal{M} = 2$ and $d(k)$ represents a time-varying state delay. For simulation purpose, the initial condition is assumed to be $\phi(k) = \begin{bmatrix} e^{k/d_2} & 0 \end{bmatrix}^T$ for all $k = -d_2, -d_2+1, \ldots, 0$. Here, our purpose is to design a state feedback controller in the form of (3.12) such that the resulting closed-loop system is asymptotically stable.

With the choice of $\varepsilon = 10$, it is shown that the above system is asymptotically stable for all $d_2 \leq 8$. Let $d_2 = 6$. By using Theorem 3.7, we obtain the fuzzy controller gains as follows:

$$G_1 = 10^{-3} \times \begin{bmatrix} 0.0070 & -0.0001 \\ 0.0016 & 0.1383 \end{bmatrix}, \quad K_1 = 10^{-4} \times \begin{bmatrix} -0.1047 & -0.3921 \end{bmatrix},$$

$$G_2 = 10^{-3} \times \begin{bmatrix} 0.0069 & 0.0000 \\ 0.0016 & 0.1397 \end{bmatrix}, \quad K_2 = 10^{-4} \times \begin{bmatrix} 0.1189 & -0.4218 \end{bmatrix}.$$

In the simulation, let the delay $d(k)$ change randomly between $d_1 = 3$ and $d_2 = 6$ (see Fig. 3.1). Fig. 3.2 depicts the states of the open-loop system, and the states of the closed-loop system are shown in Fig. 3.3.

◊ **Fuzzy Dynamic Output Feedback Controller:**

For simulation, we add some disturbance terms and a regulated output. The above nonlinear system becomes

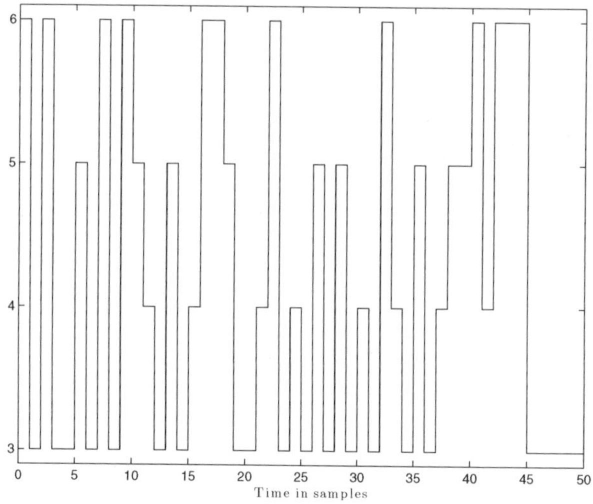

Fig. 3.1. Time-varying delays $d(k)$

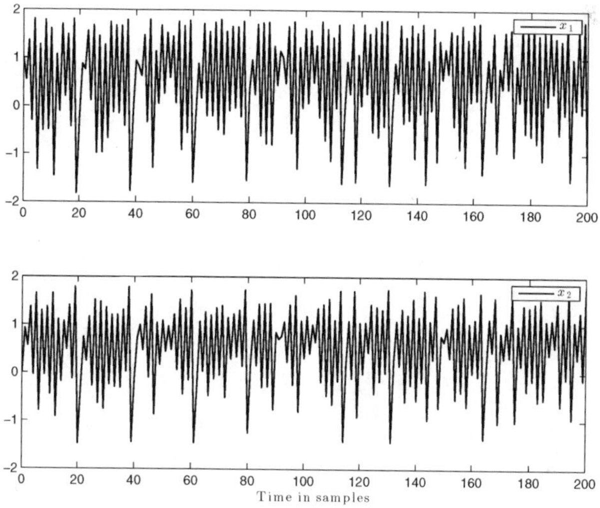

Fig. 3.2. States of the open-loop system

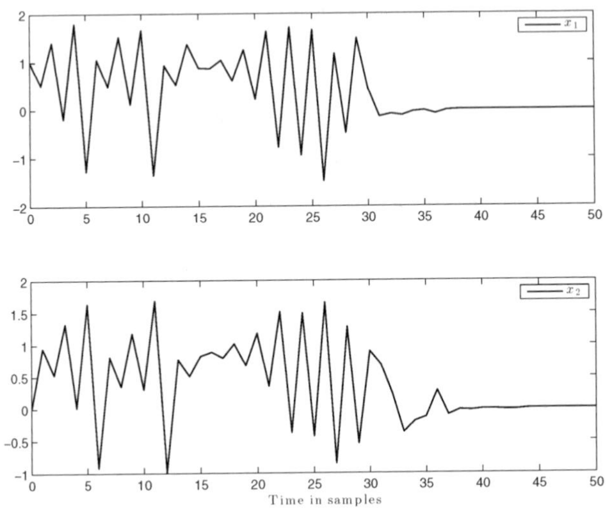

Fig. 3.3. States of the closed-loop system by state feedback control

♦ **Plant Form:**

Rule 1: IF θ is $-\mathcal{M}$, THEN

$$\begin{cases} x(k+1) = A_1x(k) + A_{d1}x(k-d(k)) + B_1u^\star(k) + D_1\omega(k), \\ y(k) = C_1x(k) + C_{d1}x(k-d(k)) + F_1\omega(k), \\ z(k) = E_1x(k), \end{cases}$$

Rule 2: IF θ is $\mathcal{M}$, THEN

$$\begin{cases} x(k+1) = A_2x(k) + A_{d2}x(k-d(k)) + B_2u^\star(k) + D_2\omega(k), \\ y(k) = C_2x(k) + C_{d2}x(k-d(k)) + F_2\omega(k), \\ z(k) = E_2x(k), \end{cases}$$

where

$$C_1 = C_2 = \begin{bmatrix} \mathcal{C} & 0 \end{bmatrix}, \quad C_{d1} = C_{d2} = \begin{bmatrix} 1-\mathcal{C} & 0 \end{bmatrix}, \quad F_1 = 1, \quad F_2 = 0.5,$$
$$D_1 = D_2 = \begin{bmatrix} 1 & 0 \end{bmatrix}^T, \quad E_1 = E_2 = \begin{bmatrix} 1 & 0 \end{bmatrix}.$$

In the example, $x(k) = \begin{bmatrix} x_1(k) \\ x_2(k) \end{bmatrix}$, $\mathcal{C} = 0.8$, $\mathcal{M} = 0.9$ and $d(k)$ represents a time-varying state delay. For simulation purpose, the initial condition is assumed to be $\phi(k) = \begin{bmatrix} e^{k/d_2} & 0 \end{bmatrix}^T$ for all $k = -d_2, -d_2+1, \ldots, 0$. Here, our purpose is to design DOF controllers in the form of (3.9) such that the resulting closed-loop system is asymptotically stable and has an induced ℓ_2 performance. By solve the LMI conditions in Theorem 3.8, we have that the achieved performance level is $\gamma = 19.8033$ and the corresponding desired DOF controller parameters are as follows:

$$\begin{aligned} A_{K1} &= \begin{bmatrix} -0.5054 & -0.1262 \\ 0.1452 & 0.0363 \end{bmatrix}, \quad B_{K1} = \begin{bmatrix} -5.9473 \\ 14.0525 \end{bmatrix}, \\ A_{K2} &= \begin{bmatrix} 0.2305 & 0.0576 \\ -0.0844 & -0.0211 \end{bmatrix}, \quad B_{K2} = \begin{bmatrix} -0.3637 \\ 11.9290 \end{bmatrix}, \\ C_{K1} &= \begin{bmatrix} 0.1671 & 0.0189 \end{bmatrix}, \quad C_{K2} = \begin{bmatrix} -0.0833 & -0.0436 \end{bmatrix}. \end{aligned}$$

In the following, we will present the simulation results to illustrate the DOF induced ℓ_2 controller. Suppose the disturbance input $\omega(k)$ be

$$\omega(k) = \frac{3\sin(0.85k)}{(0.55k)^2 + 1}.$$

The simulation results are shown in Figs. 3.4–3.7. Among them, Fig. 3.4 shows the time-varying delay $d(k)$ which changes randomly between $d_1 = 3$ and $d_2 = 6$; Fig. 3.5 plots the states of the open-loop system; Fig. 3.6 shows the states $x_1(k)$ (solid line) and $x_2(k)$ (dash-dot line) of the closed-loop system; and Fig. 3.7 shows the control input $u(k)$.

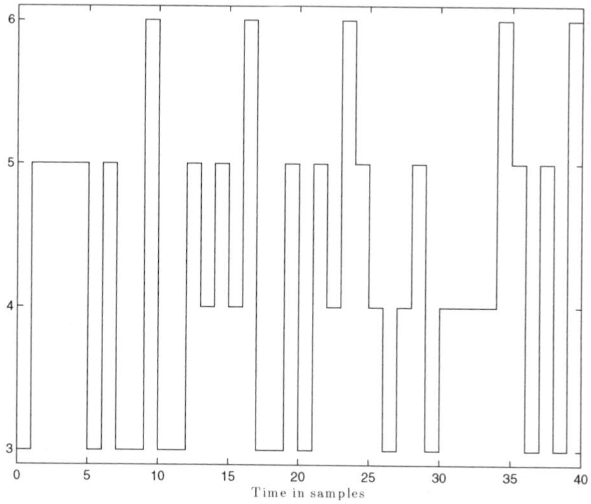

Fig. 3.4. Time-varying delays $d(k)$

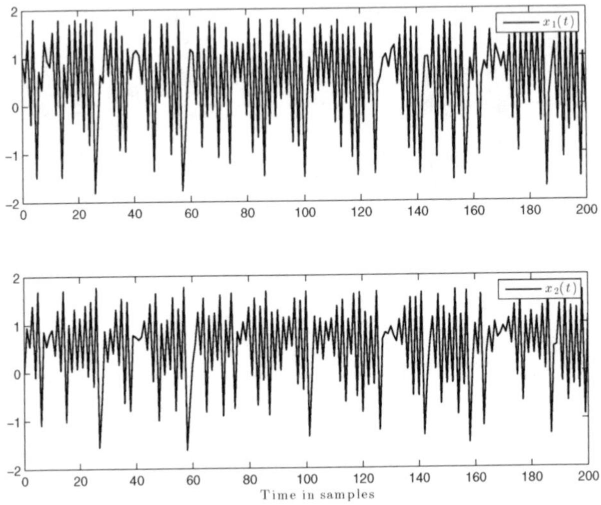

Fig. 3.5. States of the open-loop system

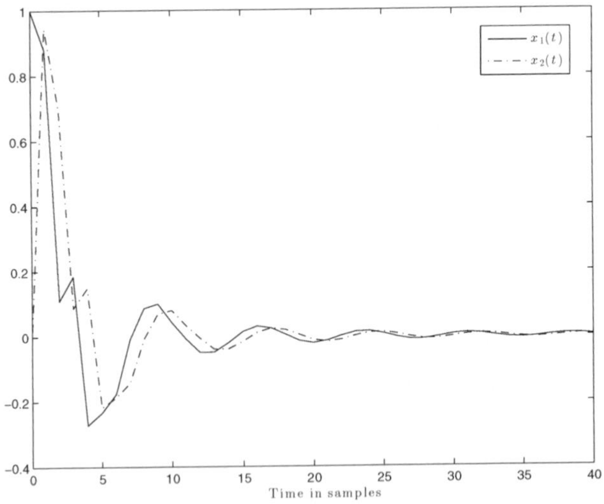

Fig. 3.6. States of the closed-loop system by DOF control

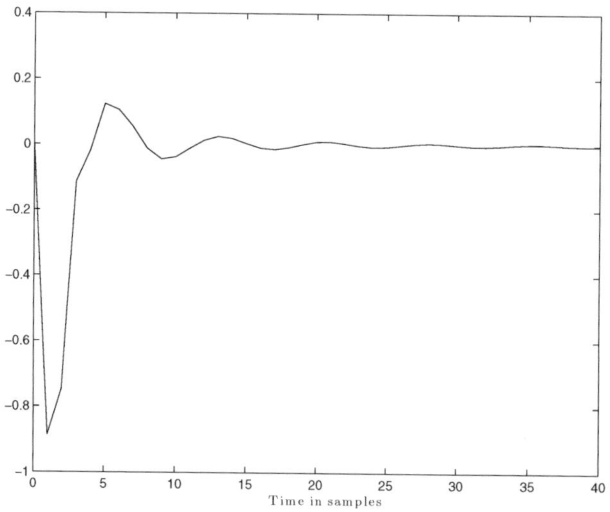

Fig. 3.7. DOF control input $u(k)$

3.6 Conclusion

In this chapter, the stabilization and DOF control problems have been studied for discrete T-S fuzzy systems with time-varying delay. Firstly, the non-PDC control law is proposed to stabilize the resulting closed-loop fuzzy system with time-varying state by delay partitioning method. Moreover, by introducing scaled small gain theorem and model transformation method, the desired DOF controller can be designed by optimization techniques. Finally, an illustrative example is given to show the effectiveness of the design schemes.

Chapter 4
Robust Filtering of Discrete-Time T-S Fuzzy Time-Delay Systems

4.1 Introduction

This chapter is concerned with system performance analysis and filter design for T-S fuzzy systems with time-varying delay. Sufficient conditions of stability analysis satisfying the given performance level are presented for the augmented filtering error system by the input-output approach. Based on these conditions, the filtering problem for the concerned T-S fuzzy systems can be solved efficiently. Furthermore, the reliable filtering problem is studied for T-S fuzzy time-delay systems with incomplete sensor information. Based on the extension of reciprocally convex idea to the construction of basis-dependent LKF, the desired reliable filter is also obtained, which makes the corresponding filtering error system stable with strict dissipativity.

4.2 $\mathcal{H}_\infty$ Filter Design

4.2.1 System Description and Preliminaries

Consider the following discrete-time T-S fuzzy time-delay system:

♦**Plant Form:**

Rule i: IF $\theta_1(k)$ is $\mathcal{M}_{i1}$ and $\theta_2(k)$ is $\mathcal{M}_{i2}$ and $\cdots$ and $\theta_p(k)$ is $\mathcal{M}_{ip}$, THEN

$$\begin{aligned} x(k+1) &= A_i x(k) + A_{di} x(k-d(k)) + B_i \omega(k), \\ y(k) &= C_i x(k) + C_{di} x(k-d(k)) + D_i \omega(k), \\ z(k) &= L_i x(k) + L_{di} x(k-d(k)) + F_i \omega(k), \\ x(k) &= \phi(k), \quad k = -d_2, -d_2+1, \ldots, 0, \end{aligned}$$

where $i = 1, 2, \ldots, r$, and r is the number of IF-THEN rules; $\mathcal{M}_{ij}(i = 1, 2, \ldots, r; j = 1, 2, \ldots, p)$ are the fuzzy sets; $\theta = \left[\theta_1(k)\ \theta_2(k)\ \cdots\ \theta_p(k)\right]^T$ is

L. Wu et al., *Fuzzy Control Systems with Time-Delay and Stochastic Perturbation*, Studies in Systems, Decision and Control 12, DOI: 10.1007/978-3-319-11316-6_4

the premise variable vector. $x(k) \in \mathbf{R}^n$ is the system state vector; $y(k) \in \mathbf{R}^p$ is the measured output; $\omega(k) \in \mathbf{R}^l$ is the disturbance input that belongs to $\ell_2[0, \infty)$; $z(k) \in \mathbf{R}^q$ is the signal to be estimated, and $d(k)$ is the time-varying delay satisfying $1 \leqslant d_1 \leqslant d(k) \leqslant d_2$, where d_1 and d_2 are positive constants representing the lower and upper bounds, respectively. A_i, A_{di}, B_i, C_i, C_{di}, D_i, L_i, L_{di} and F_i are known real constant matrices; $\phi(k)$ denotes the initial condition.

A more compact form of the above system can be described by

$$x(k+1) = \sum_{i=1}^{r} h_i(\theta) \left[A_i x(k) + A_{di} x(k - d(k)) + B_i \omega(k)\right], \tag{4.1a}$$

$$y(k) = \sum_{i=1}^{r} h_i(\theta) \left[C_i x(k) + C_{di} x(k - d(k)) + D_i \omega(k)\right], \tag{4.1b}$$

$$z(k) = \sum_{i=1}^{r} h_i(\theta) \left[L_i x(k) + L_{di} x(k - d(k)) + F_i \omega(k)\right], \tag{4.1c}$$

where $h_i(\theta)$, $i = 1, 2, \ldots, r$ are the normalized membership functions, which are defined as that of (1.2) in Chapter 1.

Here, we shall design a full-order and a reduced-order filters of general structure described by:

$$\hat{x}(k+1) = A_f \hat{x}(k) + B_f y(k), \tag{4.2a}$$

$$\hat{z}(k) = L_f \hat{x}(k) + D_f y(k), \tag{4.2b}$$

where $\hat{x}(k) \in \mathbf{R}^k$ is the state vector of the filter with $k \leq n$; $\hat{z}(k) \in \mathbf{R}^q$ is an estimation of $z(k)$; and A_f, B_f, L_f and D_f are filter parameter matrices to be determined.

Remark 4.1. In fact, there exist two kinds of filters, that is, fuzzy-rule-independent filter and fuzzy-rule-dependent one. In the fuzzy-rule-independent case, the premise variable of the original fuzzy model θ is usually supposed to be unavailable in filter implementation, thus the filter structure will have to be independent of the fuzzy rules. In other words, a fixed filter is to be designed. While, if it is assumed that the premise variable of the fuzzy model θ is available for feedback, that is, $h_i(\theta)$ is available for feedback. Also suppose that the filter and the plant have the same premise variable. In this case, the fuzzy-rule-dependent filter can be designed. Generally speaking, a fuzzy-rule-dependent filter, due to the face that it takes the fuzzy rule into account, has less conservativeness than a fuzzy-rule-independent one. In this chapter, we only considered the design of a fuzzy-rule-independent filter, see (4.2), however, the obtained results can be easily extended to the design of a fuzzy-rule-dependent filter. ♦

Then, augmenting the model of (4.1) to include the filter model of (4.2), we obtain the filtering error system as

$$\xi(k+1) = \sum_{i=1}^{r} h_i(\theta)\left[\bar{A}_i\xi(k) + \bar{A}_{di}\xi(k-d(k)) + \bar{B}_i\omega(k)\right], \quad (4.3a)$$

$$e(k) = \sum_{i=1}^{r} h_i(\theta)\left[\bar{L}_i\xi(k) + \bar{L}_{di}\xi(k-d(k)) + \bar{F}_i\omega(k)\right], \quad (4.3b)$$

where $\xi(k) \triangleq \begin{bmatrix} x(k) \\ \hat{x}(k) \end{bmatrix}$, $e(k) \triangleq z(k) - \hat{z}(k)$ and

$$\bar{A}_i \triangleq \begin{bmatrix} A_i & 0 \\ B_fC_i & A_f \end{bmatrix}, \quad \bar{A}_{di} \triangleq \begin{bmatrix} A_{di} & 0 \\ B_fC_{di} & 0 \end{bmatrix}, \quad \bar{B}_i \triangleq \begin{bmatrix} B_i \\ B_fD_i \end{bmatrix},$$
$$\bar{F}_i \triangleq F_i - D_fD_i, \quad \bar{L}_i \triangleq \begin{bmatrix} L_i - D_fC_i & -L_f \end{bmatrix}, \quad \bar{L}_{di} \triangleq \begin{bmatrix} L_{di} - D_fC_{di} & 0 \end{bmatrix}.$$

Moreover, we define

$$\bar{A}(k) \triangleq \sum_{i=1}^{r} h_i(\theta)\bar{A}_i, \quad \bar{A}_d(k) \triangleq \sum_{i=1}^{r} h_i(\theta)\bar{A}_{di}, \quad \bar{B}(k) \triangleq \sum_{i=1}^{r} h_i(\theta)\bar{B}_i,$$
$$\bar{F}(k) \triangleq \sum_{i=1}^{r} h_i(\theta)\bar{F}_i, \quad \bar{L}(k) \triangleq \sum_{i=1}^{r} h_i(\theta)\bar{L}_i, \quad \bar{L}_d(k) \triangleq \sum_{i=1}^{r} h_i(\theta)\bar{L}_{di}.$$

Definition 4.2. The filtering error system in (4.3) with $\omega(k) = 0$ is said to be asymptotically stable if

$$\lim_{k\to\infty} \|\xi(k)\| = 0.$$

Definition 4.3. Given a scalar $\gamma > 0$, the filtering error system in (4.3) is said to be asymptotically stable with an $\mathcal{H}_\infty$ performance level γ if it is asymptotically stable under $\omega = 0$, and satisfies

$$\|e(k)\|_2 < \gamma\|\omega(k)\|_2, \quad \forall\, 0 \neq \omega(k) \in \ell_2[0,\infty), \quad (4.4)$$

where $\|e(k)\|_2 \triangleq \sqrt{\sum_{k=0}^{\infty} e^T(k)e(k)}$.

Define $\hat{\omega}(k) \triangleq \gamma\omega(k)$, and consider an auxiliary system for the filtering error system (4.3):

$$\xi(k+1) = \sum_{i=1}^{r} h_i(\theta)\left[\bar{A}_i\xi(k) + \bar{A}_{di}\xi(k-d(k)) + \gamma^{-1}\bar{B}_i\hat{\omega}(k)\right], \quad (4.5a)$$

$$e(k) = \sum_{i=1}^{r} h_i(\theta)\left[\bar{L}_i\xi(k) + \bar{L}_{di}\xi(k-d(k)) + \gamma^{-1}\bar{F}_i\hat{\omega}(k)\right], \quad (4.5b)$$

It is clear that $\mathcal{H}_\infty$ performance (4.4) is equivalent to

$$\|e(k)\|_2 < \|\hat{\omega}(k)\|_2, \quad \forall\, 0 \neq \hat{\omega}(k) \in \ell_2[0,\infty).$$

Therefore, our objective in this chapter is to determine the filter matrices (A_f, B_f, L_f, D_f) in (4.2) such that the filtering error system in (4.3) is asymptotically stable and has a guaranteed $\mathcal{H}_\infty$ performance level γ.

To employ the input-output approach in Chapter 1, the following auxiliary system is introduced to replace system (4.5) based on the two-term approximation in (3.16):

$$\begin{aligned} \xi(k+1) = \sum_{i=1}^{r} h_i(\theta) \Big\{ &\frac{1}{2}\bar{A}_{di}\left[\xi(k-d_1) + \xi(k-d_2)\right] + \bar{A}_i\xi(k) \\ &+ \frac{d}{2}\bar{A}_{di}\sigma(k) + \gamma^{-1}\bar{B}_i\hat{\omega}(k) \Big\}, \end{aligned} \tag{4.6a}$$

$$\begin{aligned} e(k) = \sum_{i=1}^{r} h_i(\theta) \Big\{ &\frac{1}{2}\bar{L}_{di}\left[\xi(k-d_1) + \xi(k-d_2)\right] + \bar{L}_i\xi(k) \\ &+ \frac{d}{2}\bar{L}_{di}\sigma(k) + \gamma^{-1}\bar{F}_i\hat{\omega}(k) \Big\}. \end{aligned} \tag{4.6b}$$

The following model can formulate system (4.6) in the interconnection frame in Fig. 1.1:

$$(\mathcal{S}_1): \quad \begin{bmatrix} \xi(k+1) \\ \varsigma(k) \\ e(k) \end{bmatrix} = \begin{bmatrix} \Sigma_1(k) & \frac{d}{2}\bar{A}_d(k) & \gamma^{-1}\bar{B}(k) \\ \Sigma_2(k) & \frac{d}{2}\bar{A}_d(k) & \gamma^{-1}\bar{B}(k) \\ \Sigma_3(k) & \frac{d}{2}\bar{L}_d(k) & \gamma^{-1}\bar{F}(k) \end{bmatrix} \begin{bmatrix} \bar{\xi}(k) \\ \sigma(k) \\ \hat{\omega}(k) \end{bmatrix}, \tag{4.7a}$$

$$(\mathcal{S}_2): \quad \sigma(k) = \mathcal{K}\varsigma(k), \tag{4.7b}$$

where $\bar{\xi}(k) \triangleq \begin{bmatrix} \xi(k) \\ \xi(k-d_1) \\ \xi(k-d_2) \end{bmatrix}$ and

$$\begin{aligned} \Sigma_1(k) &\triangleq \begin{bmatrix} \bar{A}(k) & \frac{1}{2}\bar{A}_d(k) & \frac{1}{2}\bar{A}_d(k) \end{bmatrix}, \\ \Sigma_2(k) &\triangleq \begin{bmatrix} (\bar{A}(k) - I) & \frac{1}{2}\bar{A}_d(k) & \frac{1}{2}\bar{A}_d(k) \end{bmatrix}, \\ \Sigma_3(k) &\triangleq \begin{bmatrix} \bar{L}(k) & \frac{1}{2}\bar{L}_d(k) & \frac{1}{2}\bar{L}_d(k) \end{bmatrix}. \end{aligned}$$

For brevity, let us use the following operator:

$$(\mathcal{K}): \quad \varsigma(k) \to \sigma(k) = \frac{1}{d}\left[\sum_{i=k-d_2}^{k-d_1-1} \beta(i)\varsigma(i)\right],$$

to denote the relation $(\mathcal{S}_2)$ from $\varsigma(k)$ to $\sigma(k)$ as the same in Fig. 1.1. By Lemma 3.4, we can obtain $\|\mathcal{K}\|_\infty \leq 1$.

Thus, we can see that the ℓ_2 norm of $(\mathcal{S}_2)$ in (4.7) from input to output is bounded by one. Then, based on Lemma 3.4, we focus on researching the scaled small gain of $(\mathcal{S}_1)$ for the interconnection frame (4.7).

Lemma 4.4. *Assume $(\mathcal{S}_1)$ is internally stable in (4.7), the filtering error system of interconnection system described by (4.7) is asymptotically stable and has a guaranteed $\mathcal{H}_\infty$ performance level γ for $(\mathcal{K})$ if there exists a matrix* $\hat{X} = \text{diag}\{\bar{X}, I\} > 0$ *such that*

$$\|\hat{X} \circ \mathcal{G} \circ \hat{X}^{-1}\|_\infty < 1,$$

where

$$\mathcal{G} \triangleq \begin{bmatrix} \Sigma_1(k) & \frac{d}{2}\bar{A}_d(k) & \gamma^{-1}\bar{B}(k) \\ \Sigma_2(k) & \frac{d}{2}\bar{A}_d(k) & \gamma^{-1}\bar{B}(k) \\ \Sigma_3(k) & \frac{d}{2}\bar{L}_d(k) & \gamma^{-1}\bar{F}(k) \end{bmatrix}.$$

Remark 4.5. Along the interconnection frame of (4.7), the sufficient condition in Lemma 4.4 is equivalent to the following condition: Assume the $\mathcal{S}_1$ is internally stable in (4.7), the closed loop system of interconnection system described by (4.7) is asymptotically stable and has a guaranteed $\mathcal{H}_\infty$ performance level γ for $\mathcal{K}$ if there exist exists a matrix $X \triangleq \bar{X}^T\bar{X}$ such that

$$\mathcal{J} \triangleq \sum_{k=0}^{\infty}\left[\varsigma^T(k)X\varsigma(k) - \sigma^T(k)X\sigma(k) + e^T(k)e(k) - \hat{\omega}^T(k)\hat{\omega}(k)\right] < 0. \quad \blacklozenge$$

4.2.2 Main Results

In the following, by applying the input-output approach, we will derive a LMI formulation of $\mathcal{H}_\infty$ bound for system (4.7). Firstly, we let $d = d_2 - d_1$ and

$$\bar{Q}_1(k) \triangleq \sum_{i=1}^{r} h_i(\theta)Q_{1i}, \quad \bar{Q}_2(k) \triangleq \sum_{i=1}^{r} h_i(\theta)Q_{2i},$$

where $Q_{1i} > 0$, $Q_{2i} > 0$, $i = 1, 2, \ldots, r$, are all $(n+k) \times (n+k)$ matrices. Thus, we construct the following fuzzy LKF:

$$V(k) \triangleq \sum_{i=1}^{3} V_i(k), \tag{4.8}$$

with

$$
\begin{aligned}
V_1(k) &\triangleq \xi^T(k)P\xi(k), \\
V_2(k) &\triangleq \sum_{i=k-d_1}^{k-1} \xi^T(i)\bar{Q}_1(i)\xi(i) + \sum_{i=k-d_2}^{k-1} \xi^T(i)\bar{Q}_2(i)\xi(i), \\
V_3(k) &\triangleq \sum_{i=-d_1}^{-1} \sum_{j=k+i}^{k-1} \varsigma^T(j)Z_1\varsigma(j) + \sum_{i=-d_2}^{-1} \sum_{j=k+i}^{k-1} \varsigma^T(j)Z_2\varsigma(j).
\end{aligned}
$$

Then, based on the above LKF, we can obtain the following result.

Theorem 4.6. *The filtering error system in (4.7) is asymptotically stable and has a guaranteed $\mathcal{H}_\infty$ performance level γ if there exist matrices $P > 0$, $X > 0$, $Z_1 > 0$, $Z_2 > 0$, $Q_{1i} > 0$, $Q_{2i} > 0$, $S_{1i} > 0$, $S_{2i} > 0$, $S_{3i} > 0$, $T_{1i} > 0$, $T_{2i} > 0$, $T_{3i} > 0$, M_{1i}, M_{2i}, M_{3i}, N_{1i}, N_{2i}, and N_{3i}, such that for $i, j, s = 1, 2, \ldots, r$,*

$$
\begin{bmatrix}
\Pi_5 & \Pi_{6i} & \Pi_{7i} & \Pi_{7i} & d\Pi_{7i} & \Pi_{10i} \\
\star & \Pi_{66i} & \Pi_{67i} & \Pi_{68i} & 0 & 0 \\
\star & \star & \Pi_{77ij} & \Pi_{78i} & 0 & 0 \\
\star & \star & \star & \Pi_{88is} & 0 & 0 \\
\star & \star & \star & \star & -X & 0 \\
\star & \star & \star & \star & \star & -\gamma^2 I
\end{bmatrix} < 0, \tag{4.9a}
$$

$$
\begin{bmatrix}
S_{1i} & 0 & 0 & M_{1i} \\
\star & S_{2i} & 0 & M_{2i} \\
\star & \star & S_{3i} & M_{3i} \\
\star & \star & \star & Z_1
\end{bmatrix} \geq 0, \tag{4.9b}
$$

$$
\begin{bmatrix}
T_{1i} & 0 & 0 & N_{1i} \\
\star & T_{2i} & 0 & N_{2i} \\
\star & \star & T_{3i} & N_{3i} \\
\star & \star & \star & Z_2
\end{bmatrix} \geq 0, \tag{4.9c}
$$

where

$$
\begin{aligned}
\Pi_5 &\triangleq \mathrm{diag}\left\{-P^{-1},\ -d_1^{-1}Z_1^{-1},\ -d_2^{-1}Z_2^{-1},\ -X^{-1},\ -I\right\}, \\
\Pi_{6i} &\triangleq \begin{bmatrix} \bar{A}_i \\ \bar{A}_i - I \\ \bar{A}_i - I \\ \bar{A}_i - I \\ \bar{L}_i \end{bmatrix}, \quad
\Pi_{7i} \triangleq \frac{1}{2} \begin{bmatrix} \bar{A}_{di} \\ \bar{A}_{di} \\ \bar{A}_{di} \\ \bar{A}_{di} \\ \bar{L}_{di} \end{bmatrix}, \quad
\Pi_{10i} \triangleq \begin{bmatrix} \bar{B}_i \\ \bar{B}_i \\ \bar{B}_i \\ \bar{B}_i \\ \bar{F}_i \end{bmatrix}, \\
\Pi_{66i} &\triangleq -P + Q_{1i} + Q_{2i} + M_{1i} + M_{1i}^T + N_{1i} + N_{1i}^T + d_1 S_{1i} + d_2 T_{1i}, \\
\Pi_{67i} &\triangleq -M_{1i}^T + M_{2i} + N_{2i}, \\
\Pi_{68i} &\triangleq M_{3i} + N_{3i} - N_{1i}^T,
\end{aligned}
$$

$$\Pi_{78i} \triangleq -M_{3i} - N_{2i}^T,$$
$$\Pi_{77ij} \triangleq -Q_{1j} - M_{2i} - M_{2i}^T + d_1 S_{2i} + d_2 T_{2i},$$
$$\Pi_{88is} \triangleq -Q_{2s} - N_{3i} - N_{3i}^T + d_1 S_{3i} + d_2 T_{3i}.$$

With the previous results, the proof of Theorem 4.6 can be carried out in a straight forward way thus is omitted here.

Remark 4.7. In Theorem 4.6, we applied the input-output method combining with a novel LKF in (4.8) to analyze the stability and $\mathcal{H}_\infty$ performance of the filtering error system. The main attention was focused on reduction of the conservativeness, such that the filter synthesis problems in the full- and reduced-order cases have feasible solutions. To use the small scale gain theorem, first we transformed the time-varying delay in the original system into the uncertainties, and then the original system was transformed into a comparison system consisting of two subsystems (that is, a constant time-delay forward subsystem and a delayed "uncertainty" feedback one). Then, we can apply the Lyapunov functional in (4.8) to establish the $\mathcal{H}_\infty$ performance criterion. Finally, we can obtain the desired filters for the original system based on Theorem 4.6. Here, it should be pointed out that to get better results, we can introduce some recently developed techniques such as the delay partitioning method. ♦

In Theorem 4.6, we presented the $\mathcal{H}_\infty$ performance analysis result for the filtering error system in (4.7). In the following part, based on Theorem 4.6, we will investigate the full- and reduced-filter designs.

Firstly, we are in a position to present a solution to the full-order $\mathcal{H}_\infty$ filtering problem for the T-S fuzzy time-varying delay system in (4.1).

Theorem 4.8. *The filtering error system (4.3) is asymptotically stable and has a guaranteed $\mathcal{H}_\infty$ performance level γ if there exist matrices $\mathscr{P} > 0$, $\mathscr{X} > 0$, $\bar{X} \triangleq \begin{bmatrix} X_1 & X_2 \\ \star & X_3 \end{bmatrix} > 0$, $\bar{Z}_1 \triangleq \begin{bmatrix} Z_{11} & Z_{12} \\ \star & Z_{13} \end{bmatrix} > 0$, $\bar{Z}_2 \triangleq \begin{bmatrix} Z_{21} & Z_{22} \\ \star & Z_{23} \end{bmatrix} > 0$, $\bar{Q}_{1i} \triangleq \begin{bmatrix} Q_{1i1} & Q_{1i2} \\ \star & Q_{1i3} \end{bmatrix} > 0$, $\bar{Q}_{2i} \triangleq \begin{bmatrix} Q_{2i1} & Q_{2i2} \\ \star & Q_{2i3} \end{bmatrix} > 0$, $\bar{S}_{mi} \triangleq \begin{bmatrix} S_{mi1} & S_{mi2} \\ \star & S_{mi3} \end{bmatrix} > 0$, $\bar{T}_{mi} \triangleq \begin{bmatrix} T_{mi1} & T_{mi2} \\ \star & T_{mi3} \end{bmatrix} > 0$, $\bar{M}_{mi} \triangleq \begin{bmatrix} M_{mi1} & M_{mi2} \\ M_{mi3} & M_{mi4} \end{bmatrix}$, $\bar{N}_{mi} \triangleq \begin{bmatrix} N_{mi1} & N_{mi2} \\ N_{mi3} & N_{mi4} \end{bmatrix}$, $\mathscr{A}$, $\mathscr{B}$, $\mathscr{L}$, $\mathscr{D}$, W_1 and W_2 such that the following inequalities hold for $i, j, s = 1, 2, \ldots, r$, $m = 1, 2, 3$,*

$$\begin{bmatrix} \Gamma_5 & \Gamma_{6i} & \frac{1}{2}\Gamma_{7i} & \frac{1}{2}\Gamma_{7i} & \frac{d}{2}\Gamma_{7i} & \Gamma_{10i} \\ \star & \Gamma_{66i} & \Gamma_{67i} & \Gamma_{68i} & 0 & 0 \\ \star & \star & \Gamma_{77ij} & \Gamma_{78i} & 0 & 0 \\ \star & \star & \star & \Gamma_{88is} & 0 & 0 \\ \star & \star & \star & \star & -\bar{X} & 0 \\ \star & \star & \star & \star & \star & -\gamma^2 I \end{bmatrix} < 0, \qquad (4.10a)$$

$$\begin{bmatrix} S_{1i1} & S_{1i2} & 0 & 0 & 0 & 0 & M_{1i1} & M_{1i2} \\ \star & S_{1i3} & 0 & 0 & 0 & 0 & M_{1i3} & M_{1i4} \\ \star & \star & S_{2i1} & S_{2i2} & 0 & 0 & M_{2i1} & M_{2i2} \\ \star & \star & \star & S_{2i3} & 0 & 0 & M_{2i3} & M_{2i4} \\ \star & \star & \star & \star & S_{3i1} & S_{3i2} & M_{3i1} & M_{3i2} \\ \star & \star & \star & \star & \star & S_{3i3} & M_{3i3} & M_{3i4} \\ \star & \star & \star & \star & \star & \star & Z_{11} & Z_{12} \\ \star & \star & \star & \star & \star & \star & \star & Z_{13} \end{bmatrix} \geq 0, \tag{4.10b}$$

$$\begin{bmatrix} T_{1i1} & T_{1i2} & 0 & 0 & 0 & 0 & N_{1i1} & N_{1i2} \\ \star & T_{1i3} & 0 & 0 & 0 & 0 & N_{1i3} & N_{1i4} \\ \star & \star & T_{2i1} & T_{2i2} & 0 & 0 & N_{2i1} & N_{2i2} \\ \star & \star & \star & T_{2i3} & 0 & 0 & N_{2i3} & N_{2i4} \\ \star & \star & \star & \star & T_{3i1} & T_{3i2} & N_{3i1} & N_{3i2} \\ \star & \star & \star & \star & \star & T_{3i3} & N_{3i3} & N_{3i4} \\ \star & \star & \star & \star & \star & \star & Z_{21} & Z_{22} \\ \star & \star & \star & \star & \star & \star & \star & Z_{23} \end{bmatrix} \geq 0, \tag{4.10c}$$

where

$$\Gamma_5 \triangleq \mathrm{diag}\{-\Gamma_{11}, d_1^{-1}\Gamma_{22}, d_2^{-1}\Gamma_{33}, \Gamma_{44}, -I\},$$

$$\Gamma_{22} \triangleq \begin{bmatrix} Z_{11} - W_1 - W_1^T & Z_{12} - W_2 - \mathscr{L} \\ \star & Z_{13} - \mathscr{L} - \mathscr{L}^T \end{bmatrix},$$

$$\Gamma_{33} \triangleq \begin{bmatrix} Z_{21} - W_1 - W_1^T & Z_{22} - W_2 - \mathscr{L} \\ \star & Z_{23} - \mathscr{L} - \mathscr{L}^T \end{bmatrix},$$

$$\Gamma_{44} \triangleq \begin{bmatrix} X_1 - W_1 - W_1^T & X_2 - W_2 - \mathscr{L} \\ \star & X_3 - \mathscr{L} - \mathscr{L}^T \end{bmatrix},$$

$$\Gamma_{6i} \triangleq \begin{bmatrix} \mathscr{P}A_i + \mathscr{B}C_i & \mathscr{A} \\ \mathscr{L}^T A_i + \mathscr{B}C_i & \mathscr{A} \\ W_1^T A_i + \mathscr{B}C_i - W_1^T & \mathscr{A} - \mathscr{L} \\ W_2^T A_i + \mathscr{B}C_i - W_2^T & \mathscr{A} - \mathscr{L} \\ W_1^T A_i + \mathscr{B}C_i - W_1^T & \mathscr{A} - \mathscr{L} \\ W_2^T A_i + \mathscr{B}C_i - W_2^T & \mathscr{A} - \mathscr{L} \\ W_1^T A_i + \mathscr{B}C_i - W_1^T & \mathscr{A} - \mathscr{L} \\ W_2^T A_i + \mathscr{B}C_i - W_2^T & \mathscr{A} - \mathscr{L} \\ L_i - \mathscr{D}C_i & -\mathscr{L} \end{bmatrix}, \quad \Gamma_{7i} \triangleq \begin{bmatrix} \mathscr{P}A_{di} + \mathscr{B}C_{di} & 0 \\ \mathscr{L}^T A_{di} + \mathscr{B}C_{di} & 0 \\ W_1^T A_{di} + \mathscr{B}C_{di} & 0 \\ W_2^T A_{di} + \mathscr{B}C_{di} & 0 \\ W_1^T A_{di} + \mathscr{B}C_{di} & 0 \\ W_2^T A_{di} + \mathscr{B}C_{di} & 0 \\ W_1^T A_{di} + \mathscr{B}C_{di} & 0 \\ W_2^T A_{di} + \mathscr{B}C_{di} & 0 \\ L_{di} - \mathscr{D}C_{di} & 0 \end{bmatrix},$$

$$\Gamma_{10i} \triangleq \begin{bmatrix} \mathscr{P}B_i + \mathscr{B}D_i \\ \mathscr{L}^T B_i + \mathscr{B}D_i \\ W_1^T B_i + \mathscr{B}D_i \\ W_2^T B_i + \mathscr{B}D_i \\ W_1^T B_i + \mathscr{B}D_i \\ W_2^T B_i + \mathscr{B}D_i \\ W_1^T B_i + \mathscr{B}D_i \\ W_2^T B_i + \mathscr{B}D_i \\ F_i - \mathscr{D}D_i \end{bmatrix}, \quad \begin{aligned} \Gamma_{11} &\triangleq \begin{bmatrix} \mathscr{P} & \mathscr{L} \\ \star & \mathscr{L}^T \end{bmatrix}, \\ \Gamma_{66i} &\triangleq \begin{bmatrix} \Gamma_{66i1} & \Gamma_{66i2} \\ \star & \Gamma_{66i3} \end{bmatrix}, \\ \Gamma_{77ij} &\triangleq \begin{bmatrix} \Gamma_{77ij1} & \Gamma_{77ij2} \\ \star & \Gamma_{77ij3} \end{bmatrix}, \\ \Gamma_{88is} &\triangleq \begin{bmatrix} \Gamma_{88is1} & \Gamma_{88is2} \\ \star & \Gamma_{88is3} \end{bmatrix}, \end{aligned}$$

$$\Gamma_{67i} \triangleq \begin{bmatrix} -M_{1i1}^T + M_{2i1} + N_{2i1} & -M_{1i3}^T + M_{2i2} + N_{2i2} \\ -M_{1i2}^T + M_{2i3} + N_{2i3} & -M_{1i4}^T + M_{2i4} + N_{2i4} \end{bmatrix},$$

$$\Gamma_{68i} \triangleq \begin{bmatrix} M_{3i1} + N_{3i1} - N_{1i1}^T & M_{3i2} + N_{3i2} - N_{1i3}^T \\ M_{3i3} + N_{3i3} - N_{1i2}^T & M_{3i4} + N_{3i4} - N_{1i4}^T \end{bmatrix},$$

$$\Gamma_{78i} \triangleq \begin{bmatrix} -M_{3i1} - N_{2i1}^T & -M_{3i2} - N_{2i3}^T \\ -M_{3i3} - N_{2i2}^T & -M_{3i4} - N_{2i4}^T \end{bmatrix},$$

$$\Gamma_{66i1} \triangleq -\mathscr{P} + Q_{1i1} + Q_{2i1} + M_{1i1} + M_{1i1}^T + N_{1i1} + N_{1i1}^T + d_1 S_{1i1} + d_2 T_{1i1},$$

$$\Gamma_{66i2} \triangleq -\mathscr{X} + Q_{1i2} + Q_{2i2} + M_{1i2} + M_{1i3}^T + N_{1i2} + N_{1i3}^T + d_1 S_{1i2} + d_2 T_{1i2},$$

$$\Gamma_{66i3} \triangleq -\mathscr{X}^T + Q_{1i3} + Q_{2i3} + M_{1i4} + M_{1i4}^T + N_{1i4} + N_{1i4}^T + d_1 S_{1i3} + d_2 T_{1i3},$$

$$\Gamma_{77ij1} \triangleq -Q_{1j1} - M_{2i1} - M_{2i1}^T + d_1 S_{2i1} + d_2 T_{2i1},$$

$$\Gamma_{77ij2} \triangleq -Q_{1j2} - M_{2i2} - M_{2i3}^T + d_1 S_{2i2} + d_2 T_{2i2},$$

$$\Gamma_{77ij3} \triangleq -Q_{1j3} - M_{2i4} - M_{2i4}^T + d_1 S_{2i3} + d_2 T_{2i3},$$

$$\Gamma_{88is1} \triangleq -Q_{2s1} - N_{3i1} - N_{3i1}^T + d_1 S_{3i1} + d_2 T_{3i1},$$

$$\Gamma_{88is2} \triangleq -Q_{2s2} - N_{3i2} - N_{3i3}^T + d_1 S_{3i2} + d_2 T_{3i2},$$

$$\Gamma_{88is3} \triangleq -Q_{2s3} - N_{3i4} - N_{3i4}^T + d_1 S_{3i3} + d_2 T_{3i3}.$$

Moreover, if the above conditions have feasible solutions then the matrices for an admissible full-order $\mathcal{H}_\infty$ filter in the form of (4.2) are given by

$$\begin{bmatrix} A_f & B_f \\ L_f & D_f \end{bmatrix} = \begin{bmatrix} \mathscr{X}^{-1} & 0 \\ 0 & I \end{bmatrix} \begin{bmatrix} \mathscr{A} & \mathscr{B} \\ \mathscr{L} & \mathscr{D} \end{bmatrix}. \tag{4.11}$$

Proof. According to Theorem 4.6, it is easy to prove that the filtering error system (4.3) is asymptotically stable and has a guaranteed $\mathcal{H}_\infty$ performance γ if there exist matrices $P > 0$, $X > 0$, $Z_1 > 0$, $Z_2 > 0$, $Q_{1i} > 0$, $Q_{2i} > 0$, $S_{1i} > 0$, $S_{2i} > 0$, $S_{3i} > 0$, $T_{1i} > 0$, $T_{2i} > 0$, $T_{3i} > 0$, M_{1i}, M_{2i}, M_{3i}, N_{1i}, N_{2i}, and N_{3i}, $(i = 1, 2, \ldots, r)$, and W satisfying (4.9b)–(4.9c) and the following inequality

$$\begin{bmatrix} \bar{\Pi}_5 & \bar{\Pi}_{6i} & \frac{1}{2}\bar{\Pi}_{7i} & \frac{1}{2}\bar{\Pi}_{7i} & \frac{d}{2}\bar{\Pi}_{7i} & \bar{\Pi}_{10i} \\ \star & \Pi_{66i} & \Pi_{67i} & \Pi_{68i} & 0 & 0 \\ \star & \star & \Pi_{77ij} & \Pi_{78i} & 0 & 0 \\ \star & \star & \star & \Pi_{88is} & 0 & 0 \\ \star & \star & \star & \star & -X & 0 \\ \star & \star & \star & \star & \star & -\gamma^2 I \end{bmatrix} < 0, \tag{4.12}$$

where

$$\bar{\Pi}_{6i} \triangleq \begin{bmatrix} P\bar{A}_i \\ W^T(\bar{A}_i - I) \\ W^T(\bar{A}_i - I) \\ W^T(\bar{A}_i - I) \\ \bar{L}_i \end{bmatrix}, \quad \bar{\Pi}_{7i} \triangleq \begin{bmatrix} P\bar{A}_{di} \\ W^T\bar{A}_{di} \\ W^T\bar{A}_{di} \\ W^T\bar{A}_{di} \\ \bar{L}_{di} \end{bmatrix}, \quad \bar{\Pi}_{10i} \triangleq \begin{bmatrix} P\bar{B}_i \\ W^T\bar{B}_i \\ W^T\bar{B}_i \\ W^T\bar{B}_i \\ \bar{F}_i \end{bmatrix},$$

$\bar{\Pi}_5 \triangleq \text{diag}\left\{-P, d_1^{-1}\left(Z_1 - W - W^T\right), d_2^{-1}\left(Z_2 - W - W^T\right), \left(X - W - W^T\right), -I\right\}.$

Let the matrix P be partitioned as

$$P \triangleq \begin{bmatrix} P_1 & P_2 \\ \star & P_3 \end{bmatrix} > 0,$$

where $P_1 \in \mathbf{R}^{n\times n}$ and $P_3 \in \mathbf{R}^{n\times n}$ are symmetric positive definite matrices, and $P_2 \in \mathbf{R}^{n\times n}$. Without loss of generality, we assume P_2 is nonsingular (if not, P_2 may be perturbed by ΔP_2 with sufficiently small norm such that $P_2 + \Delta P_2$ is nonsingular and satisfies (4.12)).

Define the following matrices which are also nonsingular:

$$\begin{cases} \mathscr{F} \triangleq \begin{bmatrix} I & 0 \\ 0 & P_3^{-1}P_2^T \end{bmatrix}, & W \triangleq \begin{bmatrix} W_1 & W_2 P_2^{-T} P_3 \\ P_2^T & P_3 \end{bmatrix}, \\ \mathscr{Z} \triangleq P_2 P_3^{-1} P_2^T, & X \triangleq \mathscr{F}^{-T} \bar{X} \mathscr{F}^{-1}, \\ Z_1 \triangleq \mathscr{F}^{-T} \bar{Z}_1 \mathscr{F}^{-1}, & M_{mi} \triangleq \mathscr{F}^{-T} \bar{M}_{mi} \mathscr{F}^{-1}, \\ Z_2 \triangleq \mathscr{F}^{-T} \bar{Z}_2 \mathscr{F}^{-1}, & N_{mi} \triangleq \mathscr{F}^{-T} \bar{N}_{mi} \mathscr{F}^{-1}, \\ Q_{1i} \triangleq \mathscr{F}^{-T} \bar{Q}_{1i} \mathscr{F}^{-1}, & S_{mi} \triangleq \mathscr{F}^{-T} \bar{S}_{mi} \mathscr{F}^{-1}, \\ Q_{2i} \triangleq \mathscr{F}^{-T} \bar{Q}_{2i} \mathscr{F}^{-1}, & T_{mi} \triangleq \mathscr{F}^{-T} \bar{T}_{mi} \mathscr{F}^{-1}, \end{cases} \tag{4.13}$$

and $\mathscr{P} \triangleq P_1$,

$$\begin{bmatrix} \mathscr{A} & \mathscr{B} \\ \mathscr{L} & \mathscr{D} \end{bmatrix} \triangleq \begin{bmatrix} P_2 & 0 \\ 0 & I \end{bmatrix} \begin{bmatrix} A_f & B_f \\ L_f & D_f \end{bmatrix} \begin{bmatrix} P_3^{-1} P_2^T & 0 \\ 0 & I \end{bmatrix}. \tag{4.14}$$

Performing congruence transformations to (4.9b)–(4.9c) and (4.12) by matrices $\text{diag}\{\mathscr{F}, \mathscr{F}, \mathscr{F}, \mathscr{F}\}$, $\text{diag}\{\mathscr{F}, \mathscr{F}, \mathscr{F}, \mathscr{F}\}$ and $\text{diag}\{\mathscr{F}, \mathscr{F}, \mathscr{F}, \mathscr{F}, I, \mathscr{F}, \mathscr{F}, \mathscr{F}, \mathscr{F}, I\}$, respectively, and considering (4.13)–(4.14), we can obtain (4.10a)–(4.10c). Moreover, note that (4.14) is equivalent to

$$\begin{aligned} \begin{bmatrix} A_f & B_f \\ L_f & D_f \end{bmatrix} &= \begin{bmatrix} P_2^{-1} & 0 \\ 0 & I \end{bmatrix} \begin{bmatrix} \mathscr{A} & \mathscr{B} \\ \mathscr{L} & \mathscr{D} \end{bmatrix} \begin{bmatrix} P_2^{-T} P_3 & 0 \\ 0 & I \end{bmatrix} \\ &= \begin{bmatrix} \left(P_2^{-T} P_3\right)^{-1} \mathscr{Z}^{-1} & 0 \\ 0 & I \end{bmatrix} \begin{bmatrix} \mathscr{A} & \mathscr{B} \\ \mathscr{L} & \mathscr{D} \end{bmatrix} \begin{bmatrix} P_2^{-T} P_3 & 0 \\ 0 & I \end{bmatrix}. \end{aligned} \tag{4.15}$$

Notice also that the matrices A_f, B_f, L_f and D_f in (4.2) can be written as (4.15), which implies that $P_2^{-T}P_3$ can be viewed as a similarity transformation on the state-space realization of the filter and, as such, has no effect on the filter mapping from y to $\hat{z}$. Without loss of generality, we may set $P_2^{-T}P_3 = I$, thus obtain (4.11). Therefore, the full-order filter mode in (4.2) can be constructed by (4.11). This completes the proof. ∎

Remark 4.9. Theorem 4.8 provides a sufficient condition for the solvability of $\mathcal{H}_\infty$ filter, by which the desired full-order filter can be determined by solving the following convex optimization problem:

min δ subject to (4.10a)–(4.10c) with $\delta = \gamma^2$. ♦

Now, we consider the reduced-order filter design problem. Based upon Theorems 4.6 and 4.8, a solution to the $\mathcal{H}_\infty$ reduced-order filtering problem for system (4.1) is presented as follows:

Theorem 4.10. *The filtering error system in (4.3) is asymptotically stable and has a guaranteed $\mathcal{H}_\infty$ performance γ if there exist matrices $\mathscr{P} > 0$, $\mathscr{X} > 0$, $\bar{X} \triangleq \begin{bmatrix} X_1 & X_2 \\ \star & X_3 \end{bmatrix} > 0$, $\bar{Z}_1 \triangleq \begin{bmatrix} Z_{11} & Z_{12} \\ \star & Z_{13} \end{bmatrix} > 0$, $\bar{Z}_2 \triangleq \begin{bmatrix} Z_{21} & Z_{22} \\ \star & Z_{23} \end{bmatrix} > 0$, $\bar{Q}_{1i} \triangleq \begin{bmatrix} Q_{1i1} & Q_{1i2} \\ \star & Q_{1i3} \end{bmatrix} > 0$, $\bar{Q}_{2i} \triangleq \begin{bmatrix} Q_{2i1} & Q_{2i2} \\ \star & Q_{2i3} \end{bmatrix} > 0$, $\bar{S}_{mi} \triangleq \begin{bmatrix} S_{mi1} & S_{mi2} \\ \star & S_{mi3} \end{bmatrix} > 0$, $\bar{T}_{mi} \triangleq \begin{bmatrix} T_{mi1} & T_{mi2} \\ \star & T_{mi3} \end{bmatrix} > 0$, $\bar{M}_{mi} \triangleq \begin{bmatrix} M_{mi1} & M_{mi2} \\ M_{mi3} & M_{mi4} \end{bmatrix}$, $\bar{N}_{mi} \triangleq \begin{bmatrix} N_{mi1} & N_{mi2} \\ N_{mi3} & N_{mi4} \end{bmatrix}$, $\mathscr{A}$, $\mathscr{B}$, $\mathscr{L}$, $\mathscr{D}$, W_1 and W_2 such that (4.10b)–(4.10c) and the following inequalities hold for $i, j, s = 1, 2, \ldots, r$, $m = 1, 2, 3$,*

$$\begin{bmatrix} \hat{\Gamma}_5 & \hat{\Gamma}_{6i} & \frac{1}{2}\hat{\Gamma}_{7i} & \frac{1}{2}\hat{\Gamma}_{7i} & \frac{d}{2}\hat{\Gamma}_{7i} & \hat{\Gamma}_{10i} \\ \star & \hat{\Gamma}_{66i} & \Gamma_{67i} & \Gamma_{68i} & 0 & 0 \\ \star & \star & \Gamma_{77ij} & \Gamma_{78i} & 0 & 0 \\ \star & \star & \star & \Gamma_{88is} & 0 & 0 \\ \star & \star & \star & \star & -\bar{X} & 0 \\ \star & \star & \star & \star & \star & -\gamma^2 I \end{bmatrix} < 0, \tag{4.16}$$

where

$$\begin{aligned} \hat{\Gamma}_5 &\triangleq \text{diag}\{-\hat{\Gamma}_{11}, d_1^{-1}\hat{\Gamma}_{22}, d_2^{-1}\hat{\Gamma}_{33}, \hat{\Gamma}_{44}, -I\}, \\ \hat{\Gamma}_{22} &\triangleq \begin{bmatrix} Z_{11} - W_1 - W_1^T & Z_{12} - W_2 - \mathscr{H}\mathscr{X} \\ \star & Z_{13} - \mathscr{X} - \mathscr{X}^T \end{bmatrix}, \quad \hat{\Gamma}_{11} \triangleq \begin{bmatrix} \mathscr{P} & \mathscr{H}\mathscr{X} \\ \star & \mathscr{X}^T \end{bmatrix}, \\ \hat{\Gamma}_{33} &\triangleq \begin{bmatrix} Z_{21} - W_1 - W_1^T & Z_{22} - W_2 - \mathscr{H}\mathscr{X} \\ \star & Z_{23} - \mathscr{X} - \mathscr{X}^T \end{bmatrix}, \quad \mathscr{H} \triangleq \begin{bmatrix} I_{k\times k} \\ 0_{(n-k)\times k} \end{bmatrix}, \\ \hat{\Gamma}_{44} &\triangleq \begin{bmatrix} X_1 - W_1 - W_1^T & X_2 - W_2 - \mathscr{H}\mathscr{X} \\ \star & X_3 - \mathscr{X} - \mathscr{X}^T \end{bmatrix}, \quad \hat{\Gamma}_{66i} \triangleq \begin{bmatrix} \Gamma_{66i1} & \widetilde{\Gamma}_{66i2} \\ \star & \Gamma_{66i3} \end{bmatrix}, \end{aligned}$$

$$\hat{\Gamma}_{6i} \triangleq \begin{bmatrix} \mathscr{P}A_i + \mathscr{H}\mathscr{B}C_i & \mathscr{H}\mathscr{A} \\ \mathscr{Z}^T\mathscr{H}^T A_i + \mathscr{B}C_i & \mathscr{A} \\ W_1^T A_i + \mathscr{H}\mathscr{B}C_i - W_1^T & \mathscr{H}(\mathscr{A} - \mathscr{Z}) \\ W_2^T A_i + \mathscr{B}C_i - W_2^T & \mathscr{A} - \mathscr{Z} \\ W_1^T A_i + \mathscr{H}\mathscr{B}C_i - W_1^T & \mathscr{H}(\mathscr{A} - \mathscr{Z}) \\ W_2^T A_i + \mathscr{B}C_i - W_2^T & \mathscr{A} - \mathscr{Z} \\ W_1^T A_i + \mathscr{H}\mathscr{B}C_i - W_1^T & \mathscr{H}(\mathscr{A} - \mathscr{Z}) \\ W_2^T A_i + \mathscr{B}C_i - W_2^T & \mathscr{A} - \mathscr{Z} \\ L_i - \mathscr{D}C_i & -\mathscr{L} \end{bmatrix},$$

$$\hat{\Gamma}_{7i} \triangleq \begin{bmatrix} \mathscr{P}A_{di} + \mathscr{H}\mathscr{B}C_{di} & 0 \\ \mathscr{Z}^T\mathscr{H}^T A_{di} + \mathscr{B}C_{di} & 0 \\ W_1^T A_{di} + \mathscr{H}\mathscr{B}C_{di} & 0 \\ W_2^T A_{di} + \mathscr{B}C_{di} & 0 \\ W_1^T A_{di} + \mathscr{H}\mathscr{B}C_{di} & 0 \\ W_2^T A_{di} + \mathscr{B}C_{di} & 0 \\ W_1^T A_{di} + \mathscr{H}\mathscr{B}C_{di} & 0 \\ W_2^T A_{di} + \mathscr{B}C_{di} & 0 \\ L_{di} - \mathscr{D}C_{di} & 0 \end{bmatrix}, \quad \hat{\Gamma}_{10i} \triangleq \begin{bmatrix} \mathscr{P}B_i + \mathscr{H}\mathscr{B}D_i \\ \mathscr{Z}^T\mathscr{H}^T B_i + \mathscr{B}D_i \\ W_1^T B_i + \mathscr{H}\mathscr{B}D_i \\ W_2^T B_i + \mathscr{B}D_i \\ W_1^T B_i + \mathscr{H}\mathscr{B}D_i \\ W_2^T B_i + \mathscr{B}D_i \\ W_1^T B_i + \mathscr{H}\mathscr{B}D_i \\ W_2^T B_i + \mathscr{B}D_i \\ F_i - \mathscr{D}D_i \end{bmatrix},$$

$$\widetilde{\Gamma}_{66i2} \triangleq -\mathscr{H}\mathscr{Z} + Q_{1i2} + Q_{2i2} + M_{1i2} + M_{1i3}^T + N_{1i2} + N_{1i3}^T + d_1 S_{1i2} + d_2 T_{1i2},$$

and Γ_{66i1}, Γ_{66i3}, Γ_{67i}, Γ_{68i}, Γ_{77ij}, Γ_{78i} and Γ_{88is} are defined in Theorem 4.8. Moreover, if the above conditions have feasible solutions then the matrices for an admissible $\mathcal{H}_\infty$ reduced-order filter in the form of (4.2) are given by (4.11).

Proof. According to the proof of Theorem 4.8, partition P is given

$$P \triangleq \begin{bmatrix} P_1 & P_2 \\ \star & P_3 \end{bmatrix}, \quad P_2 \triangleq \begin{bmatrix} P_4 \\ 0_{(n-k)\times k} \end{bmatrix},$$

where $0 < P_1 \in \mathbf{R}^{n\times n}$, $0 < P_3 \in \mathbf{R}^{k\times k}$ and $P_4 \in \mathbf{R}^{k\times k}$. Without loss of generality, we assume P_4 is nonsingular. To see this, let the matrix $\mathscr{M} \triangleq P + \alpha\mathscr{N}$, where α is a positive scalar and

$$\mathscr{N} \triangleq \begin{bmatrix} 0_{n\times n} & \mathscr{H} \\ \star & 0_{n\times n} \end{bmatrix}, \quad \mathscr{M} \triangleq \begin{bmatrix} \mathscr{M}_1 & \mathscr{M}_2 \\ \star & \mathscr{M}_3 \end{bmatrix}, \quad \mathscr{M}_2 \triangleq \begin{bmatrix} \mathscr{M}_4 \\ 0_{(n-k)\times k} \end{bmatrix}.$$

Observe that since $P > 0$, we have that $\mathscr{M} > 0$ for $\alpha > 0$ in the neighborhood of the origin. Thus, it can be easily verified that there exists an arbitrarily small $\alpha > 0$ such that $\mathscr{M}_4$ is nonsingular and (4.12) is feasible with P replaced by $\mathscr{M}$. Since $\mathscr{M}_4$ is nonsingular, we thus conclude that there is no loss of generality to assume the matrix P_4 to be nonsingular.

Define the following matrices which are also nonsingular:

$$\mathscr{F} \triangleq \begin{bmatrix} I & 0 \\ 0 & P_3^{-1}P_4^T \end{bmatrix}, \quad W \triangleq \begin{bmatrix} W_1 & W_2 P_4^{-T} P_3 \\ (\mathscr{H}P_4)^T & P_3 \end{bmatrix},$$

$$\begin{bmatrix} \mathscr{A} & \mathscr{B} \\ \mathscr{L} & \mathscr{D} \end{bmatrix} \triangleq \begin{bmatrix} P_4 & 0 \\ 0 & I \end{bmatrix} \begin{bmatrix} A_f & B_f \\ L_f & D_f \end{bmatrix} \begin{bmatrix} P_3^{-1}P_4^T & 0 \\ 0 & I \end{bmatrix}, \quad \mathscr{X} \triangleq P_4P_3^{-1}P_4^T,$$

and X, Z_1, Z_2, Q_{1i}, Q_{2i}, S_{mi}, T_{mi}, M_{mi}, N_{mi} and $\mathscr{P}$ are defined in (4.13).

The reminder of the proof can also follow the same lines as in the proof of Theorem 4.8. This completes the proof. ■

Remark 4.11. Theorem 4.10 presents a sufficient solvability condition for the $\mathcal{H}_\infty$ reduced-order filtering problem, thus, a desired reduced-order filter can be determined by solving the following convex optimization problem:

min δ subject to (4.10b)–(4.10c) and (4.16) with $\delta = \gamma^2$. ♦

4.2.3 Illustrative Example

Example 4.12. Consider the following Henon mapping system with time-varying delay:

$$\begin{aligned} x_1(k+1) &= -\left[\mathcal{C}x_1(k) + (1-\mathcal{C})x_1(k-d(k))\right]^2 + 0.3x_2(k) + \omega(k), \\ x_2(k+1) &= \mathcal{C}x_1(k) + (1-\mathcal{C})x_1(k-d(k)), \\ y(k) &= \mathcal{C}x_1(k) + (1-\mathcal{C})x_1(k-d(k)) + \omega(k), \\ z(k) &= x_1(k), \end{aligned}$$

where $\omega(k)$ is the disturbance input. The constant $\mathcal{C} \in [0,1]$ is the retarded coefficient. Let $\theta = \mathcal{C}x_1(k) + (1-\mathcal{C})x_1(k-d)$. Assume that $\theta \in [-\mathcal{M}, \mathcal{M}]$, $\mathcal{M} > 0$. By using the same procedure as in [203], the nonlinear term $\theta^2(k)$ can be exactly represented by $\theta^2(k) = h_1(\theta)(-\mathcal{M})\theta + h_2(\theta)\mathcal{M}\theta$, where $h_1(\theta), h_2(\theta) \in [0,1]$, and $h_1(\theta) + h_2(\theta) = 1$. By this, the membership functions $h_1(\theta)$ and $h_2(\theta)$ can be chosen as

$$h_1(\theta) = \frac{1}{2}\left(1 - \frac{\theta}{\mathcal{M}}\right), \quad h_2(\theta) = \frac{1}{2}\left(1 + \frac{\theta}{\mathcal{M}}\right).$$

It can be seen from the aforementioned expressions that $h_1(\theta) = 1$ and $h_2(\theta) = 0$ when θ is $-\mathcal{M}$ and that $h_1(\theta) = 0$ and $h_2(\theta) = 1$ when θ is $\mathcal{M}$. Then, the above nonlinear system can be approximately represented by the following T-S fuzzy model:

♦ **Plant Form:**

Rule 1: IF θ is $-\mathcal{M}$, THEN

$$\begin{aligned} x(k+1) &= A_1x(k) + A_{d1}x(k-d(k)) + B_1\omega(k), \\ y(k) &= C_1x(k) + C_{d1}x(k-d(k)) + D_1\omega(k), \\ z(k) &= L_1x(k), \end{aligned}$$

Rule 2: IF θ is $\mathcal{M}$, THEN

$$\begin{aligned} x(k+1) &= A_2 x(k) + A_{d2} x(k-d(k)) + B_2 \omega(k), \\ y(k) &= C_2 x(k) + C_{d2} x(k-d(k)) + D_2 \omega(k), \\ z(k) &= L_2 x(k), \end{aligned}$$

where

$$A_1 = \begin{bmatrix} \mathcal{CM} & 0.3 \\ \mathcal{C} & 0 \end{bmatrix}, \quad A_{d1} = \begin{bmatrix} (1-\mathcal{C})\mathcal{M} & 0 \\ 1-\mathcal{C} & 0 \end{bmatrix}, \quad B_1 = B_2 = \begin{bmatrix} 1 \\ 0 \end{bmatrix},$$
$$A_2 = \begin{bmatrix} -\mathcal{CM} & 0.3 \\ \mathcal{C} & 0 \end{bmatrix}, \quad A_{d2} = \begin{bmatrix} -(1-\mathcal{C})\mathcal{M} & 0 \\ 1-\mathcal{C} & 0 \end{bmatrix}, \quad D_1 = 1, \quad D_2 = 0.5,$$
$$C_1 = C_2 = \begin{bmatrix} \mathcal{C} & 0 \end{bmatrix}, \quad C_{d1} = C_{d2} = \begin{bmatrix} 1-\mathcal{C} & 0 \end{bmatrix}, \quad L_1 = L_2 = \begin{bmatrix} 1 & 0 \end{bmatrix}.$$

In the example, $x(k) = \begin{bmatrix} x_1(k) \\ x_2(k) \end{bmatrix}$, $\mathcal{C} = 0.8$, $\mathcal{M} = 0.2$ and $1 \le d(k) \le 3$ represents time-varying delay. Then, by solving the conditions in Theorems 4.8 and 4.10, the obtained results for the full- and reduced-order filtering cases are as follows:

Case 1: with $k = 2$, in this case we obtain $\gamma_{\min} = 2.0403$:

$$\begin{aligned} A_f &= \begin{bmatrix} 0.7376 & 0.1511 \\ -0.0584 & 0.4706 \end{bmatrix}, \quad B_f = \begin{bmatrix} 0.2556 \\ -0.1479 \end{bmatrix}, \\ L_f &= \begin{bmatrix} -0.0020 & -0.0010 \end{bmatrix}, \quad D_f = 0.5816. \end{aligned} \tag{4.18}$$

Case 2: with $k = 1$, in this case we obtain $\gamma_{\min} = 2.0411$:

$$A_f = 0.7683, \quad B_f = 0.4373, \quad L_f = -0.0010, \quad D_f = 0.5817. \tag{4.19}$$

In the following, we will present the simulation results to illustrate the effectiveness of the designed full- and reduced-order $\mathcal{H}_\infty$ filters. Let the initial condition be zero, that is, $x(0) = 0$ and $\hat{x}(0) = 0$. Suppose the disturbance input $\omega(k) = \frac{3\sin(0.85k)}{(0.55k)^2+1}$. The simulation results are shown in Figs. 4.1–4.3. Among them, Fig. 4.1 shows the time-varying delay $d(k)$ which changes randomly between $d_1 = 1$ and $d_2 = 3$. Fig. 4.2 plots the signal $z(k)$ (solid line), and its estimations $\hat{z}(k)$ with the full-order filter of (4.18) (dash-dot line) and with the reduced-order filter of (4.19) (dotted line). The corresponding estimation errors $e(k)$ are shown in Fig. 4.3.

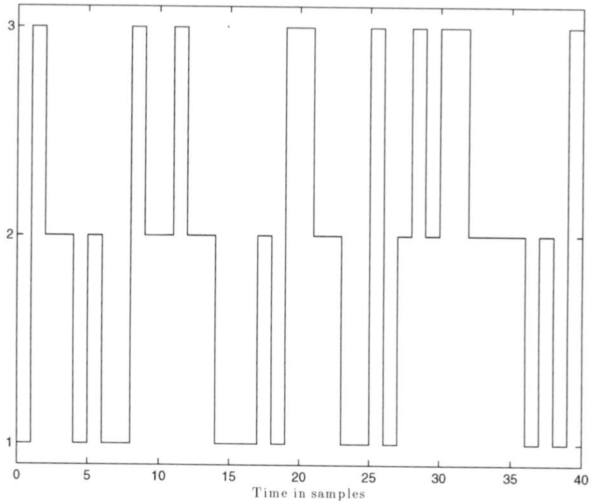

Fig. 4.1. Time-varying delays $d(k)$

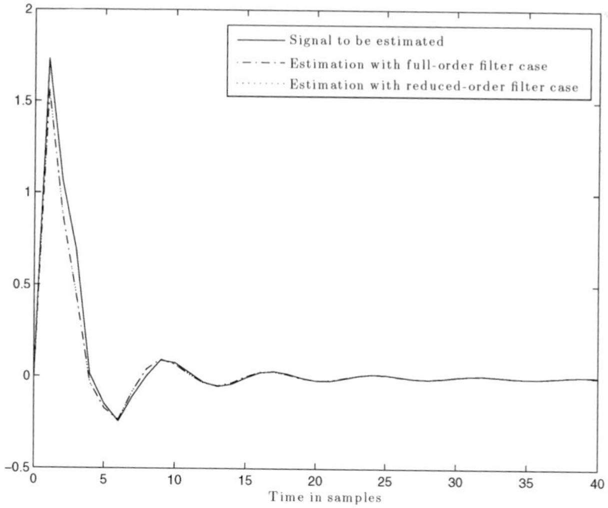

Fig. 4.2. Signal $z(k)$ and its estimations $\hat{z}(k)$ of the full- and reduced-order filters

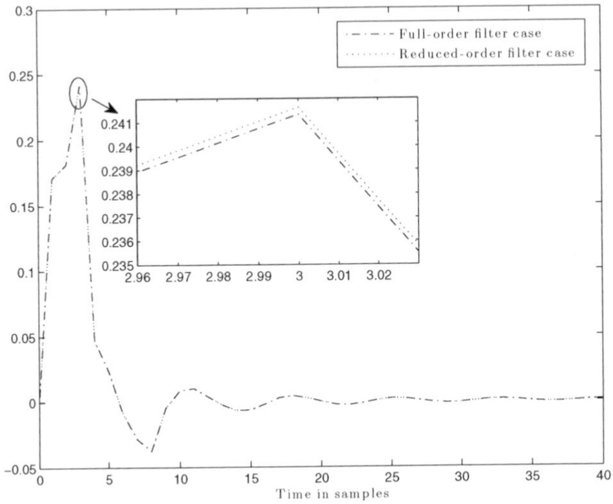

Fig. 4.3. Estimation error $e(k)$

4.3 Reliable Filter Design

4.3.1 System Description and Preliminaries

Consider the following discrete-time T-S fuzzy time-varying delay system:

♦ **Plant Form:**

Rule i: IF $\theta_1(k)$ is $\mathcal{M}_{i1}$ and $\theta_2(k)$ is $\mathcal{M}_{i2}$ and ... and $\theta_p(k)$ is $\mathcal{M}_{ip}$, THEN

$$\begin{aligned} x(k+1) &= A_i x(k) + A_{di} x(k-d(k)) + B_i \omega(k), \\ y(k) &= C_i x(k) + C_{di} x(k-d(k)) + D_i \omega(k), \\ z(k) &= L_i x(k) + L_{di} x(k-d(k)) + F_i \omega(k), \\ x(k) &= \phi(k), \quad k = -d_2, -d_2+1, \ldots, 0, \end{aligned}$$

where $i = 1, 2, \ldots, r$, and r is the number of IF-THEN rules; $\mathcal{M}_{ij}(i = 1, 2, \ldots, r; j = 1, 2, \ldots, p)$ are the fuzzy sets; $\theta = \left[\theta_1(k)\ \theta_2(k)\ \cdots\ \theta_p(k)\right]^T$ is the premise variable vector. $x(k) \in \mathbf{R}^n$ is the system state vector; $y(k) \in \mathbf{R}^m$ is the measured output; $\omega(k) \in \mathbf{R}^l$ is the disturbance input that belongs to $\ell_2[0, \infty)$; $z(k) \in \mathbf{R}^q$ is the signal to be estimated, and $d(k)$ is the time-varying delay satisfying $1 \leqslant d_1 \leqslant d(k) \leqslant d_2$, where d_1 and d_2 are positive constants representing the lower and upper bounds, respectively. A_i, A_{di}, B_i, C_i, C_{di}, D_i, L_i, L_{di} and F_i are known real constant matrices; and $\phi(k)$ denotes the initial condition.

A more compact presentation of the discrete-time T-S fuzzy time-varying delay model can be described by

$$x(k+1) = \sum_{i=1}^{r} h_i(\theta)\left[A_i x(k) + A_{di} x(k-d(k)) + B_i \omega(k)\right], \tag{4.20a}$$

$$y(k) = \sum_{i=1}^{r} h_i(\theta)\left[C_i x(k) + C_{di} x(k-d(k)) + D_i \omega(k)\right], \tag{4.20b}$$

$$z(k) = \sum_{i=1}^{r} h_i(\theta)\left[L_i x(k) + L_{di} x(k-d(k)) + F_i \omega(k)\right], \tag{4.20c}$$

where $h_i(\theta)$, $i = 1, 2, \ldots, r$ are the normalized membership functions, which are defined as that of (1.2) in Chapter 1.

Here, we design a desired filter in the following form by taking sensor failures into account:

$$\hat{x}(k+1) = A_f \hat{x}(k) + B_f \hat{y}(k), \tag{4.21a}$$

$$\hat{z}(k) = C_f \hat{x}(k) + D_f \hat{y}(k), \tag{4.21b}$$

where $\hat{x}(k) \in \mathbf{R}^k$ is the state vector of the filter system (4.21) with $k \leq n$; $\hat{z}(k) \in \mathbf{R}^q$ is an estimation of $z(k)$; A_f, B_f, C_f and D_f are parameter matrices to be determined, and $\hat{y}(k)$ denotes the signal from the sensor that may be faulty.

The following failure model from [54] is adopted here

$$\hat{y}_j(k) = \beta_{\varepsilon j} y_j(k), \quad j = 1, 2, \ldots, m,$$

where

$$0 \leq \underline{\beta}_{\varepsilon j} \leq \beta_{\varepsilon j} \leq \bar{\beta}_{\varepsilon j}, \quad j = 1, 2, \ldots, m,$$

with $0 \leq \beta_{\varepsilon j} \leq 1$ in which the variables $\beta_{\varepsilon j}$ quantify the failures of the sensors. Then, we have

$$\hat{y}(k) = B_\varepsilon y(k), \quad B_\varepsilon = \mathrm{diag}\{\beta_{\varepsilon 1}, \beta_{\varepsilon 2}, \ldots, \beta_{\varepsilon m}\}.$$

Remark 4.13. In the model mentioned above, when $\underline{\beta}_{\varepsilon j} = \bar{\beta}_{\varepsilon j}$, it is the normal fully operating case, $y_i^F(k) = y_i(k)$; when $\underline{\beta}_{\varepsilon j} = 0$, then it contains the outage case in [205]; when $\underline{\beta}_{\varepsilon j} \neq 0$ and $\bar{\beta}_{\varepsilon j} \neq 1$, then it corresponds to the case where the intensity of the feedback signal from actuator may variate. ♦

Define

$$\bar{B}_\varepsilon \triangleq \mathrm{diag}\{\bar{\beta}_{\varepsilon 1}, \bar{\beta}_{\varepsilon 2}, \ldots, \bar{\beta}_{\varepsilon j}, \ldots, \bar{\beta}_{\varepsilon m}\},$$

$$\underline{B}_\varepsilon \triangleq \mathrm{diag}\{\underline{\beta}_{\varepsilon 1}, \underline{\beta}_{\varepsilon 2}, \ldots, \underline{\beta}_{\varepsilon j}, \ldots, \underline{\beta}_{\varepsilon m}\},$$

$$
\begin{aligned}
B_{\varepsilon 0} &\triangleq \mathrm{diag}\{\beta_{\varepsilon 01}, \beta_{\varepsilon 02}, \ldots, \beta_{\varepsilon 0j}, \ldots, \beta_{\varepsilon 0m}\},\\
\Lambda &\triangleq \mathrm{diag}\{\alpha_1, \alpha_2, \ldots, \alpha_j, \ldots, \alpha_m\},\\
E_\varepsilon &\triangleq \mathrm{diag}\{\epsilon_{\varepsilon 1}, \epsilon_{\varepsilon 2}, \ldots, \epsilon_{\varepsilon j}, \ldots, \epsilon_{\varepsilon m}\},
\end{aligned}
$$

where $\beta_{\varepsilon 0j} \triangleq \frac{\bar{\beta}_{\varepsilon j} + \underline{\beta}_{\varepsilon j}}{2}$ and $\alpha_j \triangleq \frac{\bar{\beta}_{\varepsilon j} - \underline{\beta}_{\varepsilon j}}{2}$. Then, we have

$$
B_\varepsilon = B_{\varepsilon 0} + E_\varepsilon, \quad |\epsilon_{\varepsilon j}| \leq \alpha_j. \tag{4.22}
$$

Augmenting the model in (4.20) to include the filter system in (4.21), we obtain the filtering error system as:

$$
\xi(k+1) = \sum_{i=1}^{r} h_i(\theta)\left[\bar{A}_i \xi(k) + \bar{A}_{di}\xi(k - d(k)) + \bar{B}_i \omega(k)\right], \tag{4.23a}
$$

$$
e(k) = \sum_{i=1}^{r} h_i(\theta)\left[\bar{L}_i \xi(k) + \bar{L}_{di}\xi(k - d(k)) + \bar{F}_i \omega(k)\right], \tag{4.23b}
$$

where $\xi(k) \triangleq \begin{bmatrix} x(k) \\ \hat{x}(k) \end{bmatrix}$, $e(k) \triangleq z(k) - \hat{z}(k)$ and

$$
\begin{aligned}
\bar{A}_i &\triangleq \begin{bmatrix} A_i & 0 \\ B_f B_\varepsilon C_i & A_f \end{bmatrix}, \quad \bar{A}_{di} \triangleq \begin{bmatrix} A_{di} & 0 \\ B_f B_\varepsilon C_{di} & 0 \end{bmatrix},\\
\bar{B}_i &\triangleq \begin{bmatrix} B_i \\ B_f B_\varepsilon D_i \end{bmatrix}, \quad \bar{L}_{di} \triangleq \begin{bmatrix} L_{di} - D_f B_\varepsilon C_{di} & 0 \end{bmatrix},\\
\bar{L}_i &\triangleq \begin{bmatrix} L_i - D_f B_\varepsilon C_i & -C_f \end{bmatrix}, \quad \bar{F}_i \triangleq F_i - D_f B_\varepsilon D_i.
\end{aligned}
$$

Moreover, we define

$$
\begin{aligned}
\bar{A}(k) &\triangleq \sum_{i=1}^{r} h_i(\theta)\bar{A}_i, \quad \bar{A}_d(k) \triangleq \sum_{i=1}^{r} h_i(\theta)\bar{A}_{di},\\
\bar{B}(k) &\triangleq \sum_{i=1}^{r} h_i(\theta)\bar{B}_i, \quad \bar{F}(k) \triangleq \sum_{i=1}^{r} h_i(\theta)\bar{F}_i,\\
\bar{L}(k) &\triangleq \sum_{i=1}^{r} h_i(\theta)\bar{L}_i, \quad \bar{L}_d(k) \triangleq \sum_{i=1}^{r} h_i(\theta)\bar{L}_{di}.
\end{aligned}
$$

Definition 4.14. The filtering error system in (4.23) with $\omega(k) = 0$ is said to be asymptotically stable if

$$
\lim_{k\to\infty} \|\xi(k)\| = 0.
$$

In this following, discussions on dissipative systems are introduced. Dissipative systems can be regarded as a generalization of passive systems with more general internal and supplied energies [257]. A system is called "dissipative" if "power dissipation" exists in the system. Dissipative systems are those that cannot store more energy than that supplied by the environment and/or by other systems connected to them, i.e., dissipative systems can only dissipate but not generate energy [174]. Based on [87], associated with the discrete-time T-S fuzzy time-varying delay system in (4.23) is a real valued function $\mathcal{G}(\omega(k), e(k))$ called the *supply rate* which is formally defined as follows.

Definition 4.15. (Supply Rate) The supply rate is a real valued function, $\mathcal{G}(\omega(k), e(k)) : \Omega \times Z \to \mathbf{R}$, which is assumed to be locally Lebesgue integrable independently of the input and the initial conditions, i.e., for any $\omega(k) \in \Omega$, $e(k) \in Z$ and $T^\star \geq 0$, it holds that

$$\sum_{k=0}^{T^\star} \mathcal{G}(\omega(k), z(k)) < +\infty.$$

The classical form of dissipativity in [87] is obviously applicable to the discrete-time T-S fuzzy time-varying delay system in (4.23) in the following.

Definition 4.16. (Dissipative System) The discrete-time T-S fuzzy time-varying delay system (4.23) with supply rate $\mathcal{G}(\omega(k), e(k))$ is said to be dissipative if there exists a nonnegative function $V(x(k)) : X \to \mathbf{R}$, called the storage function, such that the following dissipation inequality holds:

$$V(x(T)) - V(x(0)) \leq \sum_{k=0}^{T^\star} \mathcal{G}(\omega(k), e(k)), \tag{4.24}$$

for all initial condition $\psi(k) \in X$, input $\omega(k) \in \Omega$ and $T^\star \geq 0$ (or said differently: for all admissible inputs $\omega(k)$ that drive the state from $x(0)$ to $x(T^\star)$ on the interval $[0, T^\star]$, where $x(T^\star)$ is the state variable at time $t = T^\star$).

Remark 4.17. Note that inequality (4.24) is known as the dissipation inequality and it possesses the property that the increase in stored energy is always less than the amount of energy supplied by the environment. Passive systems are a special class of dissipative systems that have a bilinear supply rate, i.e. $\mathcal{G}(\omega, e) = e^T\omega$. If a system with a constant *positive* feed forward of $\mathcal{X}$ is passive, then the process is dissipative with respect to the supply rate $\mathcal{G}(\omega, e) = e^T\omega + \omega^T\mathcal{X}\omega$, where $\mathcal{X} = \mathcal{X}^T \in \mathbf{R}^{p\times p}$. Similarly, if a system with a constant *negative* feedback of $\mathcal{Z}$ is passive, then the process is dissipative with respect to the supply rate $\mathcal{G}(\omega, e) = e^T\mathcal{Z}e + e^T\omega$, where $\mathcal{Z} = \mathcal{Z}^T \in \mathbf{R}^{p\times p}$. ♦

Based on the above analysis, a more general supply rate is presented in the following definition.

Definition 4.18. Given matrices $\mathcal{Z} \in \mathbf{R}^{q\times q}$, $\mathcal{X} \in \mathbf{R}^{p\times p}$, $\mathcal{Y} \in \mathbf{R}^{q\times p}$ with $\mathcal{Z}$ and $\mathcal{X}$ being symmetric, the discrete-time T-S fuzzy time-varying delay system in (4.23) is said to be dissipative if for some real function $\varrho(\cdot)$ with $\varrho(0) = 0$,

$$\sum_{k=0}^{T^*} \begin{bmatrix} e(k) \\ \omega(k) \end{bmatrix}^T \begin{bmatrix} \mathcal{Z} & \mathcal{Y} \\ \star & \mathcal{X} \end{bmatrix} \begin{bmatrix} e(k) \\ \omega(k) \end{bmatrix} + \varrho(\psi(0)) \geq 0, \quad \forall T^* \geq 0.$$

Furthermore, if for some scalar $\delta > 0$,

$$\sum_{k=0}^{T^*} \begin{bmatrix} e(k) \\ \omega(k) \end{bmatrix}^T \begin{bmatrix} \mathcal{Z} & \mathcal{Y} \\ \star & \mathcal{X} \end{bmatrix} \begin{bmatrix} e(k) \\ \omega(k) \end{bmatrix} + \varrho(\psi(0)) \geq \delta \sum_{k=0}^{T^*} \omega^T(k)\omega(k), \quad \forall T^* \geq 0,$$

then the system in (4.23) is said to be strictly dissipative.

We assume that $\mathcal{Z} \leq 0$. Then we can get $-\mathcal{Z} = (\mathcal{Z}_{-}^{\frac{1}{2}})^2$, for some $\mathcal{Z}_{-}^{\frac{1}{2}} \geq 0$.

Remark 4.19. The theory of dissipative systems generalizes the system theory, including the bounded real (small gain) theorem, passivity theorem, circle criterion, and sector bounded nonlinearity. To show this, a few special cases can be immediately got by setting the $(\mathcal{Z}, \mathcal{Y}, \mathcal{X})$ parameters. Some special cases are stated as follows:

- If $\mathcal{Z} = -I$, $\mathcal{Y} = 0$ and $\mathcal{X} = \gamma^2 I (\gamma > 0)$, strictly dissipative reduces to the $\mathcal{H}_\infty$ performance constraint.
- If $\mathcal{Z} = 0$, $\mathcal{Y} = I$ and $\mathcal{X} = 0$, strictly dissipative reduces to the positive real performance.
- If $\mathcal{Z} = -\theta I$, $\mathcal{Y} = 1 - \theta$ and $\mathcal{X} = \theta\gamma^2 I$ $(\gamma > 0, \theta \in [0, 1])$, strictly dissipative reduces to the mixed performance.
- If $\mathcal{Z} = -I$, $\mathcal{Y} = \frac{1}{2}(\mathcal{K}_1 + \mathcal{K}_2)^T$ and $\mathcal{X} = -\frac{1}{2}(\mathcal{K}_1^T \mathcal{K}_2 + \mathcal{K}_2^T \mathcal{K}_1)(\gamma > 0$, for some constant matrices $\mathcal{K}_1, \mathcal{K}_2)$, strictly dissipative reduces to the sector bounded performance. ♦

4.3.2 Main Results

Now, we will apply the reciprocally convex approach combining with the LKF technique to investigate the strict dissipativity and the stability for the filter error augmented system in (4.23). Firstly, let $d = d_2 - d_1$ and

$$Q_1(k) \triangleq \sum_{i=1}^{r} h_i(\theta) Q_{1i}, \quad Q_2(k) \triangleq \sum_{i=1}^{r} h_i(\theta) Q_{2i}, \quad Q_3(k) \triangleq \sum_{i=1}^{r} h_i(\theta) Q_{3i},$$

where $Q_{1i} > 0$, $Q_{2i} > 0$, $Q_{3i} > 0$, $i = 1, 2, \ldots, r$, are all $(n + k) \times (n + k)$ matrices.

Theorem 4.20. *Given matrices $0 \geq \mathcal{Z} \in \mathbf{R}^{q \times q}$, $\mathcal{X} \in \mathbf{R}^{p \times p}$, $\mathcal{Y} \in \mathbf{R}^{q \times p}$ with $\mathcal{Z}$ and $\mathcal{X}$ being symmetric, and scalar $\delta > 0$, suppose that there exist matrices*

$0 < P \in \mathbf{R}^{(n+k)\times(n+k)}$, $0 < Q_{1i} \in \mathbf{R}^{(n+k)\times(n+k)}$, $0 < Q_{2i} \in \mathbf{R}^{(n+k)\times(n+k)}$, $0 < Q_{3i} \in \mathbf{R}^{(n+k)\times(n+k)}$, $0 < S_1 \in \mathbf{R}^{(n+k)\times(n+k)}$, $0 < S_2 \in \mathbf{R}^{(n+k)\times(n+k)}$ *and* $M \in \mathbf{R}^{(n+k)\times(n+k)}$ *such that for* $i, j, s, t = 1, \ldots, r$,

$$\Xi_{ijst} \triangleq \begin{bmatrix} \Xi_{11i} & S_1 & 0 & 0 & -\bar{L}_i^T\mathcal{Y} & \Xi_{16i} \\ \star & \Xi_{22j} & -M+S_2 & M & 0 & 0 \\ \star & \star & \Xi_{33s} & -M+S_2 & -\bar{L}_{di}^T\mathcal{Y} & \Xi_{36i} \\ \star & \star & \star & -Q_{2t}-S_2 & 0 & 0 \\ \star & \star & \star & \star & \Xi_{55i} & \Xi_{56i} \\ \star & \star & \star & \star & \star & \Xi_{66} \end{bmatrix} < 0, \quad (4.25a)$$

$$\begin{bmatrix} S_2 & M^T \\ \star & S_2 \end{bmatrix} \geq 0, \quad (4.25b)$$

where

$$\begin{aligned} \Xi_{11i} &\triangleq -P + Q_{1i} + Q_{2i} + (d+1)Q_{3i} - S_1, \\ \Xi_{22j} &\triangleq -Q_{1j} - S_1 - S_2, \\ \Xi_{33s} &\triangleq -Q_{3s} - 2S_2 + M + M^T, \\ \Xi_{55i} &\triangleq -\bar{F}_i^T\mathcal{Y} - \mathcal{Y}^T\bar{F}_i - \mathcal{X} + \delta I, \\ \Xi_{16i} &\triangleq \begin{bmatrix} \bar{A}_i^T & d_1(\bar{A}_i^T - I) & d(\bar{A}_i^T - I) & \bar{L}_i^T\mathcal{Z}_-^{\frac{1}{2}} \end{bmatrix}, \\ \Xi_{36i} &\triangleq \begin{bmatrix} \bar{A}_{di}^T & d_1\bar{A}_{di}^T & d\bar{A}_{di}^T & \bar{L}_{di}^T\mathcal{Z}_-^{\frac{1}{2}} \end{bmatrix}, \\ \Xi_{56i} &\triangleq \begin{bmatrix} \bar{B}_i^T & d_1\bar{B}_i^T & d\bar{B}_i^T & \bar{F}_i^T\mathcal{Z}_-^{\frac{1}{2}} \end{bmatrix}, \\ \Xi_{66} &\triangleq \mathrm{diag}\left\{-P^{-1}, -S_1^{-1}, -S_2^{-1}, -I\right\}, \end{aligned}$$

then the filter error system in (4.23) with sensor failure is asymptotically stable and strictly dissipative in the sense of Definition 4.18.

Proof. Based on the fuzzy basis functions, from (4.25a) we obtain

$$\sum_{i=1}^{r}\sum_{j=1}^{r}\sum_{s=1}^{r}\sum_{t=1}^{r} h_i(\theta)h_j(\theta(k-d_1))h_s(\theta(k-d(k)))h_t(\theta(k-d_2))\Xi_{ijst} < 0.$$

A more compact presentation of the above equalities is given by

$$\begin{bmatrix} \Xi_{11}(k) & S_1 & 0 & 0 & -\bar{L}^T(k)\mathcal{Y} & \Xi_{16}(k) \\ \star & \Xi_{22}(k) & -M+S_2 & M & 0 & 0 \\ \star & \star & \Xi_{33}(k) & -M+S_2 & -\bar{L}_d^T(k)\mathcal{Y} & \Xi_{36}(k) \\ \star & \star & \star & \Xi_{44}(k) & 0 & 0 \\ \star & \star & \star & \star & \Xi_{55}(k) & \Xi_{56}(k) \\ \star & \star & \star & \star & \star & \Xi_{66} \end{bmatrix} < 0, (4.26)$$

where

$$\begin{aligned}
\Xi_{11}(k) &\triangleq -P + Q_1(k) + Q_2(k) + (d+1)Q_3(k) - S_1, \\
\Xi_{22}(k) &\triangleq -Q_1(k-d_1) - S_1 - S_2, \\
\Xi_{33}(k) &\triangleq -Q_3(k-d(k)) - 2S_2 + M + M^T, \\
\Xi_{44}(k) &\triangleq -Q_2(k-d_2) - S_2, \\
\Xi_{55}(k) &\triangleq -\bar{F}^T(k)\mathcal{Y} - \mathcal{Y}^T\bar{F}(k) - \mathcal{X} + \delta I, \\
\Xi_{16}(k) &\triangleq \left[\bar{A}^T(k) \;\; d_1\left(\bar{A}^T(k) - I\right) \;\; d\left(\bar{A}^T(k) - I\right) \;\; \bar{L}^T(k)\mathcal{Z}_-^{\frac{1}{2}} \right], \\
\Xi_{36}(k) &\triangleq \left[\bar{A}_d^T(k) \;\; d_1\bar{A}_d^T(k) \;\; d\bar{A}_d^T(k) \;\; \bar{L}_d^T(k)\mathcal{Z}_-^{\frac{1}{2}} \right], \\
\Xi_{56}(k) &\triangleq \left[\bar{B}^T(k) \;\; d_1\bar{B}^T(k) \;\; d\bar{B}^T(k) \;\; \bar{F}^T(k)\mathcal{Z}_-^{\frac{1}{2}} \right],
\end{aligned}$$

By Schur complement, inequality (4.26) implies

$$\bar{\Xi}(k) \triangleq \begin{bmatrix} \bar{\Xi}_{11}(k) & S_1 & \bar{\Xi}_{13}(k) & 0 & \bar{\Xi}_{15}(k) \\ \star & \Xi_{22}(k) & -M + S_2 & M & 0 \\ \star & \star & \bar{\Xi}_{33}(k) & -M + S_2 & \bar{\Xi}_{35}(k) \\ \star & \star & \star & \Xi_{44}(k) & 0 \\ \star & \star & \star & \star & \bar{\Xi}_{55}(k) \end{bmatrix} < 0, \quad (4.27)$$

where

$$\begin{aligned}
\bar{\Xi}_{11}(k) &\triangleq \bar{A}^T(k)P\bar{A}(k) - P + Q_1(k) + Q_2(k) + (d+1)Q_3(k) - \bar{L}^T(k)\mathcal{Z}\bar{L}(k) \\
&\quad - S_1 + d_1^2\left[\bar{A}(k) - I\right]^T S_1\left[\bar{A}(k) - I\right] + d^2\left[\bar{A}(k) - I\right]^T S_2\left[\bar{A}(k) - I\right], \\
\bar{\Xi}_{13}(k) &\triangleq \bar{A}^T(k)P\bar{A}_d(k) - \bar{L}^T(k)\mathcal{Z}\bar{L}_d(k) + d_1^2\left[\bar{A}(k) - I\right]^T S_1\bar{A}_d(k) \\
&\quad + d^2\left[\bar{A}(k) - I\right]^T S_2\bar{A}_d(k), \\
\bar{\Xi}_{15}(k) &\triangleq \bar{A}^T(k)P\bar{B}(k) - \bar{L}^T(k)\mathcal{Z}\bar{F}(k) - \bar{L}^T(k)\mathcal{Y} + d_1^2\left[\bar{A}(k) - I\right]^T S_1\bar{B}(k) \\
&\quad + d^2\left[\bar{A}(k) - I\right]^T S_2 B(k), \\
\bar{\Xi}_{33}(k) &\triangleq \bar{A}_d^T(k)P\bar{A}_d(k) - Q_3(k-d(k)) - 2S_2 + M + M^T - \bar{L}_d^T(k)\mathcal{Z}\bar{L}_d(k) \\
&\quad + d_1^2\bar{A}_d^T(k)S_1\bar{A}_d(k) + d^2\bar{A}_d^T(k)S_2\bar{A}_d(k), \\
\bar{\Xi}_{35}(k) &\triangleq \bar{A}_d^T(k)P\bar{B}(k) - \bar{L}_d^T(k)\mathcal{Y} - \bar{L}_d^T(k)\mathcal{Z}\bar{F}(k) + d_1^2\bar{A}_d^T(k)S_1\bar{B}(k) \\
&\quad + d^2\bar{A}_d^T(k)S_2 B(k), \\
\bar{\Xi}_{55}(k) &\triangleq \bar{B}^T(k)P\bar{B}(k) - \bar{F}^T(k)\mathcal{Z}\bar{F}(k) - \bar{F}^T(k)\mathcal{Y} - \mathcal{Y}^T\bar{F}(k) - \mathcal{X} + \delta I \\
&\quad + d_1^2\bar{B}^T(k)S_1\bar{B}(k) + d^2\bar{B}^T(k)S_2 B(k).
\end{aligned}$$

In the following, we construct the following fuzzy LKF:

$$V(k) \triangleq \sum_{i=1}^{5} V_i(k), \quad (4.28)$$

where

$$
\begin{aligned}
V_1(k) &\triangleq \xi^T(k)P\xi(k),\\
V_2(k) &\triangleq \sum_{j=1}^{2}\sum_{i=k-d_j}^{k-1} \xi^T(i)Q_j(i)\xi(i),\\
V_3(k) &\triangleq \sum_{i=k-d(k)}^{k-1} \xi^T(i)Q_3(i)\xi(i) + \sum_{j=-d_2+1}^{-d_1}\sum_{i=k+j}^{k-1} \xi^T(i)Q_3(i)\xi(i),\\
V_4(k) &\triangleq \sum_{j=-d_1}^{-1}\sum_{i=k+j}^{k-1} d_1\zeta^T(i)S_1\zeta(i),\\
V_5(k) &\triangleq \sum_{j=-d_2}^{-d_1-1}\sum_{i=k+j}^{k-1} d\zeta^T(i)S_2\zeta(i),\\
\zeta(k) &\triangleq \xi(k+1) - \xi(k).
\end{aligned}
$$

Along the trajectories of the filter error system in (4.23), and considering the difference of the fuzzy LKF in (4.28), we have

$$
\Delta V(k) \triangleq V(k+1) - V(k) = \sum_{i=1}^{5} \Delta V_i(k),
$$

where

$$
\begin{aligned}
\Delta V_1(k) &= \xi^T(k+1)P\xi(k+1) - \xi^T(k)P\xi(k),\\
\Delta V_2(k) &= \xi^T(k)Q_1(k)\xi(k) - \xi^T(k-d_1)Q_1(k-d_1)\xi(k-d_1)\\
&\quad +\xi^T(k)Q_2(k)\xi(k) - \xi^T(k-d_2)Q_2(k-d_2)\xi(k-d_2),\\
\Delta V_3(k) &= (d+1)\xi^T(k)Q_3(k)\xi(k) - \xi^T(k-d(k))Q_3(k-d(k))\xi(k-d(k))\\
&\quad + \sum_{i=k-d(k+1)+1}^{k-1} \xi^T(i)Q_3(i)\xi(i) - \sum_{i=k-d(k)+1}^{k-1} \xi^T(i)Q_3(i)\xi(i)\\
&\quad - \sum_{i=k-d_2+1}^{k-d_1} \xi^T(i)Q_3(i)\xi(i)\\
&\le (d+1)\xi^T(k)Q_3(k)\xi(k) - \xi^T(k-d(k))Q_3(k-d(k))\xi(k-d(k)),\\
\Delta V_4(k) &= d_1^2\zeta^T(k)S_1\zeta(k) - d_1\sum_{i=k-d_1}^{k-1} \zeta^T(i)S_1\zeta(i)\\
&\le -\left(\sum_{i=k-d_1}^{k-1}\zeta(i)\right)^T S_1\left(\sum_{i=k-d_1}^{k-1}\zeta^T(i)\right) + d_1^2\zeta^T(k)S_1\zeta(k)\\
&= -\left[\xi(k) - \xi(k-d_1)\right] S_1 \left[\xi(k) - \xi(k-d_1)\right] + d_1^2\zeta^T(k)S_1\zeta(k).
\end{aligned}
$$

Since $\begin{bmatrix} S_2 & M^T \\ \star & S_2 \end{bmatrix} \geq 0$, the following inequality holds:

$$\begin{bmatrix} \sqrt{\frac{\vartheta_1}{\vartheta_2}}\zeta_1(k) \\ -\sqrt{\frac{\vartheta_2}{\vartheta_1}}\zeta_2(k) \end{bmatrix}^T \begin{bmatrix} S_2 & M^T \\ \star & S_2 \end{bmatrix} \begin{bmatrix} \sqrt{\frac{\vartheta_1}{\vartheta_2}}\zeta_1(k) \\ -\sqrt{\frac{\vartheta_2}{\vartheta_1}}\zeta_2(k) \end{bmatrix} \geq 0,$$

where

$$\vartheta_1 \triangleq \frac{d_2 - d(k)}{d}, \quad \zeta_1(k) \triangleq \xi(k - d(k)) - \xi(k - d_2),$$
$$\vartheta_2 \triangleq \frac{d(k) - d_1}{d}, \quad \zeta_2(k) \triangleq \xi(k - d_1) - \xi(k - d(k)).$$

Then by employing Lemma 1.33, for $d_1 \leq d(k) \leq d_2$ we have

$$\begin{aligned} \Delta V_5(k) &= d^2\zeta^T(k)S_2\zeta(k) - d\sum_{i=k-d_2}^{k-d(k)-1}\zeta^T(i)S_2\eta(i) - d\sum_{i=k-d(k)}^{k-d_1-1}\zeta^T(i)S_2\zeta(i) \\ &\leq -\frac{d}{d_2 - d(k)}\left[\sum_{i=k-d_2}^{k-d(k)-1}\zeta(i)\right]^T S_2 \left[\sum_{i=k-d_2}^{k-d(k)-1}\zeta(i)\right] \\ &\quad -\frac{d}{d(k) - d_1}\left[\sum_{i=k-d(k)}^{k-d_1-1}\zeta(i)\right]^T S_2 \left[\sum_{i=k-d(k)}^{k-d_1-1}\zeta(i)\right] + d^2\zeta^T(k)S_2\zeta(k) \\ &\leq -\begin{bmatrix} \zeta_1(k) \\ \zeta_2(k) \end{bmatrix}^T \begin{bmatrix} S_2 & M^T \\ \star & S_2 \end{bmatrix} \begin{bmatrix} \zeta_1(k) \\ \zeta_2(k) \end{bmatrix} + d^2\zeta^T(k)S_2\zeta(k) \\ &= -\begin{bmatrix} \xi(k-d_1) \\ \xi(k-d(k)) \end{bmatrix}^T \begin{bmatrix} S_2 & -S_2 \\ \star & S_2 \end{bmatrix} \begin{bmatrix} \xi(k-d_1) \\ \xi(k-d(k)) \end{bmatrix} \\ &\quad -\begin{bmatrix} \xi(k-d(k)) \\ \xi(k-d_2) \end{bmatrix}^T \begin{bmatrix} S_2 & -S_2 \\ \star & S_2 \end{bmatrix} \begin{bmatrix} \xi(k-d(k)) \\ \xi(k-d_2) \end{bmatrix} \\ &\quad -\begin{bmatrix} \xi(k-d_1) \\ \xi(k-d(k)) \\ \xi(k-d_2) \end{bmatrix}^T \begin{bmatrix} 0 & M & -M \\ \star & -M - M^T & M \\ \star & \star & 0 \end{bmatrix} \begin{bmatrix} \xi(k-d_1) \\ \xi(k-d(k)) \\ \xi(k-d_2) \end{bmatrix} \\ &\quad +d^2\zeta^T(k)S_2\zeta(k). \end{aligned}$$

Note that when $d(k) = d_1$ or $d(k) = d_2$, it yields $\zeta_1(k) = 0$ or $\zeta_2(k) = 0$. Hence, the inequality in $\Delta V_5(k)$ still holds. Thus, the following conditions can be obtained:

$$\Delta V(k) = \bar{\zeta}^T(k)\hat{\Xi}(k)\bar{\zeta}(k), \tag{4.29}$$

where

$$\hat{\Xi}(k) \triangleq \begin{bmatrix} \hat{\Xi}_{11}(k) & S_1 & \hat{\Xi}_{13} & 0 & \hat{\Xi}_{15}(k) \\ \star & \Xi_{22}(k) & -M+S_2 & M & 0 \\ \star & \star & \hat{\Xi}_{33}(k) & -M+S_2 & \hat{\Xi}_{35}(k) \\ \star & \star & \star & \Xi_{44}(k) & 0 \\ \star & \star & \star & \star & \hat{\Xi}_{55}(k) \end{bmatrix}, \ \zeta(k) \triangleq \begin{bmatrix} \xi(k) \\ \xi(k-d_1) \\ \xi(k-d(k)) \\ \xi(k-d_2) \\ \omega(k) \end{bmatrix},$$

$$\begin{aligned}
\hat{\Xi}_{11}(k) &\triangleq \bar{A}^T(k)P\bar{A}(k) - P + Q_1(k) + Q_2(k) + (d+1)Q_3(k) - S_1 \\
&\quad + d_1^2 \left[\bar{A}(k) - I\right]^T S_1 \left[\bar{A}(k) - I\right] + d^2 \left[\bar{A}(k) - I\right]^T S_2 \left[\bar{A}(k) - I\right], \\
\hat{\Xi}_{13}(k) &\triangleq \bar{A}^T(k)P\bar{A}_d(k) + d_1^2 \left[\bar{A}(k)-I\right]^T S_1\bar{A}_d(k) + d^2 \left[\bar{A}(k)-I\right]^T S_2\bar{A}_d(k), \\
\hat{\Xi}_{15}(k) &\triangleq \bar{A}^T(k)P\bar{B}(k) + d_1^2 \left[\bar{A}(k) - I\right]^T S_1\bar{B}(k) + d^2 \left[\bar{A}(k) - I\right]^T S_2B(k), \\
\hat{\Xi}_{33}(k) &\triangleq \bar{A}_d^T(k)P\bar{A}_d(k) - Q_3(k-d(k)) - 2S_2 + M + M^T \\
&\quad + d_1^2\bar{A}_d^T(k)S_1\bar{A}_d(k) + d^2\bar{A}_d^T(k)S_2\bar{A}_d(k), \\
\hat{\Xi}_{35}(k) &\triangleq \bar{A}_d^T(k)P\bar{B}(k) + d_1^2\bar{A}_d^T(k)S_1\bar{B}(k) + d^2\bar{A}_d^T(k)S_2B(k), \\
\hat{\Xi}_{55}(k) &\triangleq \bar{B}^T(k)P\bar{B}(k) + d_1^2\bar{B}^T(k)S_1\bar{B}(k) + d^2\bar{B}^T(k)S_2B(k).
\end{aligned}$$

By considering (4.25b), (4.27), (4.29) and the zero inputs $\omega(k) = 0$, it follows that $\Delta V(k) < 0$, thus the filter error system (4.23) is asymptotically stable.

Next, we show the strict dissipativity of system (4.23). To this end, we define

$$\mathcal{J}(T^\star) \triangleq \sum_{k=0}^{T^\star} \begin{bmatrix} e(k) \\ \omega(k) \end{bmatrix}^T \begin{bmatrix} \mathcal{Z} & \mathcal{Y} \\ \star & \mathcal{X} \end{bmatrix} \begin{bmatrix} e(k) \\ \omega(k) \end{bmatrix} - \delta \sum_{k=0}^{T^\star} \omega^T(k)\omega(k), \quad \forall T^* \geq 0.$$

Then under the zero initial condition, that is, $\xi(k) = 0$ for $k = -d_2, -d_2 + 1, \ldots, 0$, it can be shown that for any non-zero $\omega(k) \in \ell_2[0, \infty)$,

$$\begin{aligned}
V(T^\star+1)-V(0)-\mathcal{J}(T^\star) &= \sum_{k=0}^{T^\star} \left\{ \Delta V(k) - \begin{bmatrix} e(k) \\ \omega(k) \end{bmatrix}^T \begin{bmatrix} \mathcal{Z} & \mathcal{Y} \\ \star & \mathcal{X} - \delta I \end{bmatrix} \begin{bmatrix} e(k) \\ \omega(k) \end{bmatrix} \right\} \\
&= \bar{\zeta}^T(k)\bar{\Xi}(k)\bar{\zeta}(k) < 0.
\end{aligned}$$

Based on the above inequality and $V(T^\star + 1) > 0$, we have $\mathcal{J}(T^\star) > 0$. By Definition 4.18, we can conclude that the filter error system in (4.23) with sensor failure is strictly dissipative. This completes the proof. ■

Now, our attention will be devoted to design a filter in (4.21) such that the filtering error system in (4.23) subject to possible actuator failures is strictly dissipative. Based on the result of Theorem 4.20, the reliable filter design with strict dissipativity method for T-S fuzzy time-varying delay system in (4.20) is given in the following theorems. We first consider the case that the filtering error system in (4.23) has known sensor failure parameters.

Theorem 4.21. *Given matrices $0 \geq \mathcal{Z} \in \mathbf{R}^{q\times q}$, $\mathcal{X} \in \mathbf{R}^{p\times p}$, $\mathcal{Y} \in \mathbf{R}^{q\times p}$ with $\mathcal{Z}$ and $\mathcal{X}$ being symmetric, and a scalar $\delta > 0$, if there exist matrices*

$\bar{Q}_{1i} \triangleq \begin{bmatrix} Q_{1i1} & Q_{2i2} \\ \star & Q_{1i4} \end{bmatrix} > 0,\ \bar{Q}_{2i} \triangleq \begin{bmatrix} Q_{2i1} & Q_{2i2} \\ \star & Q_{2i4} \end{bmatrix} > 0,\ \bar{Q}_{3i} \triangleq \begin{bmatrix} Q_{3i1} & Q_{3i2} \\ \star & Q_{3i4} \end{bmatrix} > 0,$
$\bar{S}_1 \triangleq \begin{bmatrix} S_{11} & S_{12} \\ \star & S_{14} \end{bmatrix} > 0,\ \bar{S}_2 \triangleq \begin{bmatrix} S_{21} & S_{22} \\ \star & S_{24} \end{bmatrix} > 0,\ \bar{M} \triangleq \begin{bmatrix} M_1 & M_2 \\ M_3 & M_4 \end{bmatrix},\ \mathcal{O} > 0,\ \mathcal{L} > 0,$
$\mathcal{W}_1$, $\mathcal{W}_2$, $\mathcal{A}_f$, $\mathcal{B}_f$, $\mathcal{C}_f$ *and* $\mathcal{D}_f$ *such that for* $i, j, s, t = 1, \ldots, r$,

$$\begin{bmatrix} \Upsilon_{11i} & \bar{S}_1 & 0 & 0 & \Upsilon_{15i} & \Upsilon_{16i} \\ \star & \Upsilon_{22j} & \Upsilon_{23} & \bar{M} & 0 & 0 \\ \star & \star & \Upsilon_{33s} & \Upsilon_{34} & \Upsilon_{35i} & \Upsilon_{36i} \\ \star & \star & \star & \Upsilon_{44t} & 0 & 0 \\ \star & \star & \star & \star & \Upsilon_{55i} & \Upsilon_{56i} \\ \star & \star & \star & \star & \star & \Upsilon_{66} \end{bmatrix} < 0, \tag{4.30a}$$

$$\begin{bmatrix} S_{21} & S_{22} & M_1^T & M_3^T \\ \star & S_{24} & M_2^T & M_4^T \\ \star & \star & S_{21} & S_{22} \\ \star & \star & \star & S_{24} \end{bmatrix} \geq 0, \tag{4.30b}$$

where

$$\Upsilon_{22j} \triangleq \begin{bmatrix} -Q_{1j1} - S_{11} - S_{21} & -Q_{1j2} - S_{12} - S_{22} \\ \star & -Q_{1j4} - S_{14} - S_{24} \end{bmatrix}, \quad \mathcal{I} \triangleq \begin{bmatrix} I_{k\times k} \\ 0_{(n-k)\times k} \end{bmatrix},$$
$$\Upsilon_{11i} \triangleq \begin{bmatrix} \Upsilon_{11i1} & \Upsilon_{11i2} \\ \star & \Upsilon_{11i4} \end{bmatrix}, \quad \Upsilon_{33s} \triangleq \begin{bmatrix} \Upsilon_{33s1} & \Upsilon_{33s2} \\ \star & \Upsilon_{33s4} \end{bmatrix},$$
$$\Upsilon_{44t} \triangleq \begin{bmatrix} -Q_{2t1} - S_{21} & -Q_{2t2} - S_{22} \\ \star & -Q_{2t4} - S_{24} \end{bmatrix},$$
$$\Upsilon_{23} \triangleq \begin{bmatrix} -M_1 + S_{21} & -M_2 + S_{22} \\ -M_3 + S_{22}^T & -M_4 + S_{24} \end{bmatrix}, \quad \Upsilon_{15i} \triangleq \begin{bmatrix} -L_i^T\mathcal{Y} + C_i^T B_\varepsilon^T \mathcal{D}_f^T \mathcal{Y} \\ \mathcal{C}_f^T \mathcal{Y} \end{bmatrix},$$
$$\Upsilon_{34} \triangleq \begin{bmatrix} -M_1 + S_{21} & -M_2 + S_{22} \\ -M_3 + S_{22}^T & -M_4 + S_{24} \end{bmatrix}, \quad \Upsilon_{35i} \triangleq \begin{bmatrix} -L_{di}^T\mathcal{Y} + C_{di}^T B_\varepsilon^T \mathcal{D}_f^T \mathcal{Y} \\ 0 \end{bmatrix},$$
$$\Upsilon_{55i} \triangleq -(F_i^T\mathcal{Y} - D_i^T B_\varepsilon^T \mathcal{D}_f^T \mathcal{Y}) - (F_i^T\mathcal{Y} - D_i^T B_\varepsilon^T \mathcal{D}_f^T \mathcal{Y})^T - \mathcal{X} + \delta I,$$
$$\Upsilon_{11i1} \triangleq Q_{1i1} + Q_{2i1} + (d+1)Q_{3i1} - S_{11} - \mathcal{O},$$
$$\Upsilon_{11i2} \triangleq Q_{1i2} + Q_{2i2} + (d+1)Q_{3i2} - S_{12} - \mathcal{I}\mathcal{L},$$
$$\Upsilon_{11i4} \triangleq Q_{1i4} + Q_{2i4} + (d+1)Q_{3i4} - S_{14} - \mathcal{L}^T,$$
$$\Upsilon_{33s1} \triangleq -Q_{3s1} - S_{21} - S_{21}^T + M_1 + M_1^T,$$
$$\Upsilon_{33s2} \triangleq -Q_{3s2} - S_{22} - S_{22}^T + M_2 + M_3^T,$$
$$\Upsilon_{33s4} \triangleq -Q_{3s4} - S_{24} - S_{24}^T + M_4 + M_4^T,$$
$$\Upsilon_{16i} \triangleq \begin{bmatrix} \Upsilon_{16i1} & d_1\Upsilon_{16i2} & d\Upsilon_{16i3} & \Upsilon_{16i4} \end{bmatrix},$$
$$\Upsilon_{36i} \triangleq \begin{bmatrix} \Upsilon_{36i1} & d_1\Upsilon_{36i2} & d\Upsilon_{36i3} & \Upsilon_{36i4} \end{bmatrix},$$
$$\Upsilon_{56i} \triangleq \begin{bmatrix} \Upsilon_{56i1} & d_1\Upsilon_{56i2} & d\Upsilon_{56i3} & \Upsilon_{56i4} \end{bmatrix},$$
$$\Upsilon_{66i} \triangleq \mathrm{diag}\{\Upsilon_{661}, \Upsilon_{662}, \Upsilon_{663}, -I\},$$

$$\Upsilon_{16i1} \triangleq \begin{bmatrix} A_i^T \mathcal{O} + C_i^T B_\varepsilon^T \mathcal{B}_f^T \mathcal{I}^T & A_i^T \mathcal{I}\mathcal{L} + C_i^T B_\varepsilon^T \mathcal{B}_f^T \\ \mathcal{A}_f^T \mathcal{I}^T & \mathcal{A}_f^T \end{bmatrix},$$

$$\Upsilon_{16i2} \triangleq \begin{bmatrix} A_i^T \mathcal{W}_1 + C_i^T B_\varepsilon^T \mathcal{B}_f^T \mathcal{I}^T - \mathcal{W}_1 & A_i^T \mathcal{W}_2 + C_i^T B_\varepsilon^T \mathcal{B}_f^T - \mathcal{W}_2 \\ \mathcal{A}_f^T \mathcal{I}^T - \mathcal{L}^T \mathcal{I}^T & \mathcal{A}_f^T - \mathcal{L}^T \end{bmatrix},$$

$$\Upsilon_{16i3} \triangleq \begin{bmatrix} A_i^T \mathcal{W}_1 + C_i^T B_\varepsilon^T \mathcal{B}_f^T \mathcal{I}^T - \mathcal{W}_1 & A_i^T \mathcal{W}_2 + C_i^T B_\varepsilon^T \mathcal{B}_f^T - \mathcal{W}_2 \\ \mathcal{A}_f^T \mathcal{I}^T - \mathcal{L}^T \mathcal{I}^T & \mathcal{A}_f^T - \mathcal{L}^T \end{bmatrix},$$

$$\Upsilon_{16i4} \triangleq \begin{bmatrix} L_i^T \mathcal{Z}_-^{\frac{1}{2}} - C_i^T B_\varepsilon^T \mathcal{D}_f^T \mathcal{Z}_-^{\frac{1}{2}} \\ -\mathcal{C}_f^T \mathcal{Z}_-^{\frac{1}{2}} \end{bmatrix},$$

$$\Upsilon_{36i1} \triangleq \begin{bmatrix} A_{di}^T \mathcal{O} + C_{di}^T B_\varepsilon^T \mathcal{B}_f^T \mathcal{I}^T & A_{di}^T \mathcal{I}\mathcal{L} + C_{di}^T B_\varepsilon^T \mathcal{B}_f^T \\ 0 & 0 \end{bmatrix},$$

$$\Upsilon_{36i2} \triangleq \begin{bmatrix} A_{di}^T \mathcal{W}_1 + C_{di}^T B_\varepsilon^T \mathcal{B}_f^T \mathcal{I}^T & A_{di}^T \mathcal{W}_2 + C_{di}^T B_\varepsilon^T \mathcal{B}_f^T \\ 0 & 0 \end{bmatrix},$$

$$\Upsilon_{36i3} \triangleq \begin{bmatrix} A_{di}^T \mathcal{W}_1 + C_{di}^T B_\varepsilon^T \mathcal{B}_f^T \mathcal{I}^T & A_{di}^T \mathcal{W}_2 + C_{di}^T B_\varepsilon^T \mathcal{B}_f^T \\ 0 & 0 \end{bmatrix},$$

$$\Upsilon_{36i4} \triangleq \begin{bmatrix} L_{di}^T \mathcal{Z}_-^{\frac{1}{2}} - C_{di}^T B_\varepsilon^T \mathcal{D}_f^T \mathcal{Z}_-^{\frac{1}{2}} \\ 0 \end{bmatrix},$$

$$\Upsilon_{56i1} \triangleq \begin{bmatrix} B_i^T \mathcal{O} + D_i^T B_\varepsilon^T \mathcal{B}_f^T \mathcal{I}^T & B_i^T \mathcal{I}\mathcal{L} + D_i^T B_\varepsilon^T \mathcal{B}_f^T \end{bmatrix},$$

$$\Upsilon_{56i2} \triangleq \begin{bmatrix} B_i^T \mathcal{W}_1 + D_i^T B_\varepsilon^T \mathcal{B}_f^T \mathcal{I}^T & B_i^T \mathcal{W}_2 + D_i^T B_\varepsilon^T \mathcal{B}_f^T \end{bmatrix},$$

$$\Upsilon_{56i3} \triangleq \begin{bmatrix} B_i^T \mathcal{W}_1 + D_i^T B_\varepsilon^T \mathcal{B}_f^T \mathcal{I}^T & B_i^T \mathcal{W}_2 + D_i^T B_\varepsilon^T \mathcal{B}_f^T \end{bmatrix},$$

$$\Upsilon_{56i4} \triangleq F_i^T \mathcal{Z}_-^{\frac{1}{2}} - D_i^T B_\varepsilon^T \mathcal{D}_f^T \mathcal{Z}_-^{\frac{1}{2}},$$

$$\Upsilon_{661} \triangleq \begin{bmatrix} -\mathcal{O} & -\mathcal{I}\mathcal{L} \\ \star & -\mathcal{L}^T \end{bmatrix},$$

$$\Upsilon_{662} \triangleq \begin{bmatrix} S_{11} - \mathcal{W}_1 - \mathcal{W}_1^T & S_{12} - \mathcal{W}_2 - (\mathcal{L}^T \mathcal{I}^T)^T \\ \star & S_{14} - \mathcal{L}^T - \mathcal{L} \end{bmatrix},$$

$$\Upsilon_{663} \triangleq \begin{bmatrix} S_{21} - \mathcal{W}_1 - \mathcal{W}_1^T & S_{22} - \mathcal{W}_2 - (\mathcal{L}^T \mathcal{I}^T)^T \\ \star & S_{24} - \mathcal{L}^T - \mathcal{L} \end{bmatrix},$$

then the filter error system in (4.23) with sensor failure is asymptotically stable and strictly dissipative in the sense of Definition 4.18. Moreover, if the above conditions have feasible solutions then the parameter matrices for the desired filter in the form of (4.21) are given by

$$A_f = \mathcal{L}^{-1} \mathcal{A}_f, \quad B_f = \mathcal{L}^{-1} \mathcal{B}_f, \quad C_f = \mathcal{C}_f, \quad D_f = \mathcal{D}_f. \tag{4.31}$$

Proof. According to Theorem 4.20, it is easy to show that the filtering error system in (4.23) is asymptotically stable and strictly dissipative if there exist matrices $0 < P \in \mathbf{R}^{(n+k)\times(n+k)}$, $0 < Q_{1i} \in \mathbf{R}^{(n+k)\times(n+k)}$, $0 < Q_{2i} \in \mathbf{R}^{(n+k)\times(n+k)}$, $0 < Q_{3i} \in \mathbf{R}^{(n+k)\times(n+k)}$, $(i = 1, 2, \ldots, r)$,

$0 < S_1 \in \mathbf{R}^{(n+k)\times(n+k)}$, $0 < S_2 \in \mathbf{R}^{(n+k)\times(n+k)}$, $M \in \mathbf{R}^{(n+k)\times(n+k)}$ and $\mathcal{W} \in \mathbf{R}^{(n+k)\times(n+k)}$ satisfying (4.25b) and the following inequality

$$\begin{bmatrix} \Xi_{11i} & S_1 & 0 & 0 & \Xi_{15i} & \bar{\Xi}_{16i} \\ \star & \Xi_{22j} & -M+S_2 & M & 0 & 0 \\ \star & \star & \Xi_{33s} & -M+S_2 & \Xi_{35i} & \bar{\Xi}_{36i} \\ \star & \star & \star & \Xi_{44t} & 0 & 0 \\ \star & \star & \star & \star & \Xi_{55i} & \bar{\Xi}_{56i} \\ \star & \star & \star & \star & \star & \bar{\Xi}_{66} \end{bmatrix} < 0, \tag{4.32}$$

where

$$\begin{aligned} \bar{\Xi}_{16i} &\triangleq \left[\bar{A}_i^T P \;\; d_1(\bar{A}_i^T - I)\mathcal{W} \;\; d(\bar{A}_i^T - I)\mathcal{W} \;\; \bar{L}_i^T \mathcal{Z}_-^{\frac{1}{2}} \right], \\ \bar{\Xi}_{36i} &\triangleq \left[\bar{A}_{di}^T P \;\; d_1 \bar{A}_{di}^T \mathcal{W} \;\; d\bar{A}_{di}^T \mathcal{W} \;\; \bar{L}_{di}^T \mathcal{Z}_-^{\frac{1}{2}} \right], \\ \bar{\Xi}_{56i} &\triangleq \left[\bar{B}_i^T P \;\; d_1 \bar{B}_i^T \mathcal{W} \;\; d\bar{B}_i^T \mathcal{W} \;\; \bar{F}_i^T \mathcal{Z}_-^{\frac{1}{2}} \right], \\ \bar{\Xi}_{66} &\triangleq \mathrm{diag}\left\{ -P, S_1 - \mathcal{W} - \mathcal{W}^T, S_2 - \mathcal{W} - \mathcal{W}^T, -I \right\}. \end{aligned}$$

Let the matrix P be partitioned as

$$P \triangleq \begin{bmatrix} P_1 & P_2 \\ \star & P_3 \end{bmatrix} > 0, \quad P_2 \triangleq \begin{bmatrix} P_4 \\ 0_{(n-k)\times k} \end{bmatrix},$$

where $0 < P_1 \in \mathbf{R}^{n\times n}$, $0 < P_3 \in \mathbf{R}^{k\times k}$ and $P_4 \in \mathbf{R}^{k\times k}$.

Define the following matrices which are also nonsingular:

$$\begin{cases} \mathcal{F} \triangleq \begin{bmatrix} I & 0 \\ 0 & P_3^{-1}P_4^T \end{bmatrix}, & \mathcal{W} \triangleq \begin{bmatrix} \mathcal{W}_1 & \mathcal{W}_2 P_4^{-T} P_3 \\ (\mathcal{I}P_4)^T & P_3 \end{bmatrix}, \\ \mathcal{L} \triangleq P_4 P_3^{-1} P_4^T, & M \triangleq \mathcal{F}^{-T} \bar{M} \mathcal{F}^{-1}, \\ S_1 \triangleq \mathcal{F}^{-T} \bar{S}_1 \mathcal{F}^{-1}, & S_2 \triangleq \mathcal{F}^{-T} \bar{S}_2 \mathcal{F}^{-1}, \\ Q_{1i} \triangleq \mathcal{F}^{-T} \bar{Q}_{1i} \mathcal{F}^{-1}, & Q_{3i} \triangleq \mathcal{F}^{-T} \bar{Q}_{3i} \mathcal{F}^{-1}, \\ \mathcal{O} \triangleq P_1, & Q_{2i} \triangleq \mathcal{F}^{-T} \bar{Q}_{2i} \mathcal{F}^{-1}, \end{cases} \tag{4.33}$$

and

$$\begin{cases} \mathcal{A}_f \triangleq P_4 A_f P_3^{-1} P_4^T, \\ \mathcal{B}_f \triangleq P_4 B_f, \\ \mathcal{C}_f \triangleq C_f P_3^{-1} P_4^T, \\ \mathcal{D}_f \triangleq D_f. \end{cases} \tag{4.34}$$

Then, performing congruence transformations to (4.25b) and (4.32) by matrices $\mathrm{diag}\{\mathcal{F}, \mathcal{F}\}$ and $\mathrm{diag}\{\mathcal{F}, \mathcal{F}, \mathcal{F}, \mathcal{F}, I, \mathcal{F}, \mathcal{F}, \mathcal{F}, I\}$, respectively, and considering (4.33)–(4.34), we can obtain inequalities (4.30a)–(4.30b). Furthermore, notice that (4.31) is equivalent to

$$\begin{cases} A_f \triangleq P_4^{-1}\mathcal{A}_f P_4^{-T} P_3 = \left(P_4^{-T}P_3\right)^{-1}\mathcal{L}^{-1}\mathcal{A}_f P_4^{-T}P_3, \\ B_f \triangleq P_4^{-1}\mathcal{B}_f = (P_4^{-T}P_3)^{-1}\mathcal{L}^{-1}\mathcal{B}_f, \\ C_f \triangleq \mathcal{C}_f P_4^{-T}P_3, \\ D_f \triangleq \mathcal{D}_f. \end{cases}$$

Notice also that the matrices A_f, B_f, C_f and D_f in (4.21) can be written as the above equations, which implies that $P_4^{-T}P_3$ can be viewed as a similarity transformation on the state-space realization of the filter and, as such, has no effect on the filter mapping from y to $\hat{e}$. Without loss of generality, we may set $P_4^{-T}P_3 = I$, thus obtain (4.31). Therefore, the filter in the form of (4.21) can be constructed by (4.31). This completes the proof. ■

Remark 4.22. Note that the conditions of strictly dissipative filter design have been presented when $B_\varepsilon = 1$ in Theorem 4.21. If $\mathcal{Z} = -I$, $\mathcal{Y} = 0$, $\mathcal{X} = \gamma^2 I$, it is easy to obtain the conditions of the special case for dissipativity, that is, $\mathcal{H}_\infty$ performance. In Examples, we give the simulation results for Henon mapping model to illustrate the effectiveness of the proposed $\mathcal{H}_\infty$ filter design method. ♦

With known sensor failure parameter, Theorem 4.21 provides a delay-dependent sufficient condition for the existence of strictly dissipative filter for discrete-time T-S fuzzy systems with time-varying delay. In the following, based on Theorem 4.21, an approach for designing the reliable filter with strict dissipativity will be shown in the case that the sensor failure parameter matrix is unknown but satisfies the constraint in (4.22).

Theorem 4.23. *Given matrices $0 \geq \mathcal{Z} \in \mathbf{R}^{q\times q}$, $\mathcal{X} \in \mathbf{R}^{p\times p}$, $\mathcal{Y} \in \mathbf{R}^{q\times p}$ with $\mathcal{Z}$ and $\mathcal{X}$ being symmetric, and scalars $\delta > 0$, suppose that there exist matrices $0 < \mathcal{O} \in \mathbf{R}^{n\times n}$, $0 < \mathcal{L} \in \mathbf{R}^{k\times k}$, $0 < \bar{Q}_{1i} \in \mathbf{R}^{(n+k)\times(n+k)}$, $0 < \bar{Q}_{2i} \in \mathbf{R}^{(n+k)\times(n+k)}$, $0 < \bar{Q}_{3i} \in \mathbf{R}^{(n+k)\times(n+k)}$, $0 < \bar{S}_1 \in \mathbf{R}^{(n+k)\times(n+k)}$, $0 < \bar{S}_2 \in \mathbf{R}^{(n+k)\times(n+k)}$, $\bar{M} \in \mathbf{R}^{(n+k)\times(n+k)}$, $\mathcal{W}_1 \in \mathbf{R}^{n\times n}$, $\mathcal{W}_2 \in \mathbf{R}^{n\times k}$, $\mathcal{A}_f \in \mathbf{R}^{k\times k}$, $\mathcal{B}_f \in \mathbf{R}^{k\times p}$, $\mathcal{C}_f \in \mathbf{R}^{q\times k}$, $\mathcal{D}_f \in \mathbf{R}^{q\times p}$, and $\pi > 0$ such that (4.30b) and the following inequality hold for $i, j, s, t = 1, 2, \ldots, r$,*

$$\left[\begin{array}{cccccc|c|c} \Upsilon_{11i} & \Upsilon_{12} & 0 & 0 & \hat{\Upsilon}_{15i} & \hat{\Upsilon}_{16i} & \begin{bmatrix} C_i^T \\ 0 \end{bmatrix} & 0 \\ \hline \star & \Upsilon_{22j} & \Upsilon_{23} & \Upsilon_{24} & \Upsilon_{25} & \Upsilon_{26} & 0 & 0 \\ \hline \star & \star & \Upsilon_{33s} & \Upsilon_{34} & \hat{\Upsilon}_{35i} & \hat{\Upsilon}_{36i} & \begin{bmatrix} C_i^T \\ 0 \end{bmatrix} & 0 \\ \hline \star & \star & \star & \Upsilon_{44t} & \Upsilon_{45} & \Upsilon_{46} & 0 & 0 \\ \star & \star & \star & \star & \hat{\Upsilon}_{55i} & \hat{\Upsilon}_{56i} & D_i^T & \pi(D_f^T\mathcal{Y})^T \\ \star & \star & \star & \star & \star & \Upsilon_{66} & 0 & \pi\hat{\Upsilon}_3 \\ \star & \star & \star & \star & \star & \star & -\pi\Lambda^{-2} & 0 \\ \star & \star & \star & \star & \star & \star & \star & -\pi I \end{array}\right] < 0,$$

where

$$\hat{\Upsilon}_{15i} \triangleq \begin{bmatrix} -L_i^T\mathcal{Y} + C_i^T B_{\varepsilon 0}^T \mathcal{D}_f^T \mathcal{Y} \\ \mathcal{C}_f^T \mathcal{Y} \end{bmatrix}, \quad \hat{\Upsilon}_{35i} \triangleq \begin{bmatrix} -L_{di}^T\mathcal{Y} + C_{di}^T B_{\varepsilon 0}^T \mathcal{D}_f^T \mathcal{Y} \\ 0 \end{bmatrix},$$

$$\hat{\Upsilon}_{55i} \triangleq -\left(F_i^T\mathcal{Y} - D_i^T B_{\varepsilon 0}^T \mathcal{D}_f^T \mathcal{Y}\right) - \left(F_i^T\mathcal{Y} - D_i^T B_{\varepsilon 0}^T \mathcal{D}_f^T \mathcal{Y}\right)^T - \mathcal{X} + \delta I,$$

$$\hat{\Upsilon}_3 \triangleq \begin{bmatrix} \left[\mathcal{B}_f^T \mathcal{I}^T \;\; \mathcal{B}_f^T\right] \\ \left[\mathcal{B}_f^T \mathcal{I}^T \;\; \mathcal{B}_f^T\right] \\ \left[\mathcal{B}_f^T \mathcal{I}^T \;\; \mathcal{B}_f^T\right] \\ -\mathcal{D}_f^T \mathcal{Z}_-^{\frac{1}{2}} \end{bmatrix}, \quad \begin{aligned} \hat{\Upsilon}_{16i} &\triangleq \left[\hat{\Upsilon}_{16i1} \;\; d_1\hat{\Upsilon}_{16i2} \;\; d\hat{\Upsilon}_{16i3} \;\; \hat{\Upsilon}_{16i4}\right], \\ \hat{\Upsilon}_{36i} &\triangleq \left[\hat{\Upsilon}_{36i1} \;\; d_1\hat{\Upsilon}_{36i2} \;\; d\hat{\Upsilon}_{36i3} \;\; \hat{\Upsilon}_{36i4}\right], \\ \hat{\Upsilon}_{56i} &\triangleq \left[\hat{\Upsilon}_{56i1} \;\; d_1\hat{\Upsilon}_{56i2} \;\; d\hat{\Upsilon}_{56i3} \;\; \hat{\Upsilon}_{56i4}\right], \end{aligned}$$

$$\hat{\Upsilon}_{16i1} \triangleq \begin{bmatrix} A_i^T\mathcal{O} + C_i^T B_{\varepsilon 0}^T \mathcal{B}_f^T \mathcal{I}^T & A_i^T \mathcal{I}\mathcal{L} + C_i^T B_{\varepsilon 0}^T \mathcal{B}_f^T \\ \mathcal{A}_f^T \mathcal{I}^T & \mathcal{A}_f^T \end{bmatrix},$$

$$\hat{\Upsilon}_{16i2} \triangleq \begin{bmatrix} A_i^T\mathcal{W}_1 + C_i^T B_{\varepsilon 0}^T \mathcal{B}_f^T \mathcal{I}^T - \mathcal{W}_1 & A_i^T \mathcal{W}_2 + C_i^T B_{\varepsilon 0}^T \mathcal{B}_f^T - \mathcal{W}_2 \\ \mathcal{A}_f^T \mathcal{I}^T - \mathcal{L}^T \mathcal{I}^T & \mathcal{A}_f^T - \mathcal{L}^T \end{bmatrix},$$

$$\hat{\Upsilon}_{16i3} \triangleq \begin{bmatrix} A_i^T\mathcal{W}_1 + C_i^T B_{\varepsilon 0}^T \mathcal{B}_f^T \mathcal{I}^T - \mathcal{W}_1 & A_i^T \mathcal{W}_2 + C_i^T B_{\varepsilon 0}^T \mathcal{B}_f^T - \mathcal{W}_2 \\ \mathcal{A}_f^T - \mathcal{L}^T & \mathcal{A}_f^T - \mathcal{L}^T \end{bmatrix},$$

$$\hat{\Upsilon}_{16i4} \triangleq \begin{bmatrix} L_i^T \mathcal{Z}_-^{\frac{1}{2}} - C_i^T B_{\varepsilon 0}^T \mathcal{D}_f^T \mathcal{Z}_-^{\frac{1}{2}} \\ -\mathcal{C}_f^T \mathcal{Z}_-^{\frac{1}{2}} \end{bmatrix},$$

$$\hat{\Upsilon}_{36i1} \triangleq \begin{bmatrix} A_{di}^T\mathcal{O} + C_{di}^T B_{\varepsilon 0}^T \mathcal{B}_f^T \mathcal{I}^T & A_{di}^T \mathcal{I}\mathcal{L} + C_{di}^T B_{\varepsilon 0}^T \mathcal{B}_f^T \\ 0 & 0 \end{bmatrix},$$

$$\hat{\Upsilon}_{36i2} \triangleq \begin{bmatrix} A_{di}^T\mathcal{W}_1 + C_{di}^T B_{\varepsilon 0}^T \mathcal{B}_f^T \mathcal{I}^T & A_{di}^T \mathcal{W}_2 + C_{di}^T B_{\varepsilon 0}^T \mathcal{B}_f^T \\ 0 & 0 \end{bmatrix},$$

$$\hat{\Upsilon}_{36i3} \triangleq \begin{bmatrix} A_{di}^T\mathcal{W}_1 + C_{di}^T B_{\varepsilon 0}^T \mathcal{B}_f^T \mathcal{I}^T & A_{di}^T \mathcal{W}_2 + C_{di}^T B_{\varepsilon 0}^T \mathcal{B}_f^T \\ 0 & 0 \end{bmatrix},$$

$$\hat{\Upsilon}_{36i4} \triangleq \begin{bmatrix} L_{di}^T \mathcal{Z}_-^{\frac{1}{2}} - C_{di}^T B_{\varepsilon 0}^T \mathcal{D}_f^T \mathcal{Z}_-^{\frac{1}{2}} \\ 0 \end{bmatrix},$$

$$\hat{\Upsilon}_{56i1} \triangleq \left[B_i^T\mathcal{O} + D_i^T B_{\varepsilon 0}^T \mathcal{B}_f^T \mathcal{I}^T \;\; B_i^T \mathcal{I}\mathcal{L} + D_i^T B_{\varepsilon 0}^T \mathcal{B}_f^T\right],$$

$$\hat{\Upsilon}_{56i2} \triangleq \left[B_i^T\mathcal{W}_1 + D_i^T B_{\varepsilon 0}^T \mathcal{B}_f^T \mathcal{I}^T \;\; B_i^T \mathcal{W}_2 + D_i^T B_{\varepsilon 0}^T \mathcal{B}_f^T\right],$$

$$\hat{\Upsilon}_{56i3} \triangleq \left[B_i^T\mathcal{W}_1 + D_i^T B_{\varepsilon 0}^T \mathcal{B}_f^T \mathcal{I}^T \;\; B_i^T \mathcal{W}_2 + D_i^T B_{\varepsilon 0}^T \mathcal{B}_f^T\right],$$

$$\hat{\Upsilon}_{56i4} \triangleq F_i^T \mathcal{Z}_-^{\frac{1}{2}} - D_i^T B_{\varepsilon 0}^T \mathcal{D}_f^T \mathcal{Z}_-^{\frac{1}{2}},$$

and Υ_{11i}, Υ_{12}, Υ_{22j}, Υ_{23}, Υ_{24}, Υ_{33s}, Υ_{34}, Υ_{44t}, *and* Υ_{66} *are defined in Theorem 4.21. Then the filter error system in (4.23) with sensor failure is asymptotically stable and strictly dissipative in the sense of Definition 4.18. Moreover, the matrices for an admissible filter in the form of (4.21) can be obtained by (4.31).*

Remark 4.24. Note that the conditions in Theorem 4.23 are in terms of strict LMIs. Therefore, the reliable filter design with strict dissipativity problem can be solved by using convex optimization algorithms. The parameter matrices of the desired filter can be easily obtained by using the standard software like Matlab. ♦

4.3.3 Illustrative Example

Example 4.25. Consider the Henon mapping system with time-varying delay, which can be described by the following equations:

$$\begin{aligned} x_1(k+1) &= -\left[\mu x_1(k) + (1-\mu)x_1(k-d(k))\right]^2 + 0.3x_2(k) + \omega(k), \\ x_2(k+1) &= \mu x_1(k) + (1-\mu)x_1(k-d(k)), \\ y(k) &= \mu x_1(k) + (1-\mu)x_1(k-d(k)) + \omega(k), \\ z(k) &= x_1(k), \end{aligned}$$

where $\omega(k)$ is the disturbance input. The constant $\mu \in [0,1]$ is the retarded coefficient.

Let $\theta = \mu x_1(k) + (1-\mu)x_1(k-d)$. Assume that $\theta \in [-\nu, \nu]$, $\nu > 0$. By using the same procedure as in [194], the nonlinear term $\theta^2(k)$ can be exactly represented as

$$\theta^2(k) = h_1(\theta)(-\nu)\theta + h_2(\theta)\nu\theta,$$

where $h_1(\theta), h_2(\theta) \in [0,1]$, and $h_1(\theta) + h_2(\theta) = 1$. By solving the equations, the membership functions $h_1(\theta)$ and $h_2(\theta)$ are obtained as

$$h_1(\theta) = \frac{1}{2}\left(1 - \frac{\theta}{\nu}\right), \quad h_2(\theta) = \frac{1}{2}\left(1 + \frac{\theta}{\nu}\right).$$

It can be seen from the aforementioned expressions that $h_1(\theta) = 1$ and $h_2(\theta) = 0$ when θ is $-\nu$ and that $h_1(\theta) = 0$ and $h_2(\theta) = 1$ when θ is ν. Then, the above nonlinear system can be approximately represented by the following T-S fuzzy model:

♦ **Plant Form:**

Rule 1: IF θ is $-\nu$, THEN

$$\begin{cases} x(k+1) = A_1 x(k) + A_{d1} x(k-d(k)) + B_1 \omega(k), \\ \quad y(k) = C_1 x(k) + C_{d1} x(k-d(k)) + D_1 \omega(k), \\ \quad z(k) = L_1 x(k), \end{cases}$$

Rule 2: IF θ is ν, THEN

$$\begin{cases} x(k+1) = A_2 x(k) + A_{d2} x(k - d(k)) + B_2 \omega(k), \\ y(k) = C_2 x(k) + C_{d2} x(k - d(k)) + D_2 \omega(k), \\ z(k) = L_2 x(k), \end{cases}$$

where

$$A_1 = \begin{bmatrix} \mu\nu & 0.3 \\ \mu & 0 \end{bmatrix}, \quad A_{d1} = \begin{bmatrix} (1-\mu)\nu & 0 \\ 1-\mu & 0 \end{bmatrix}, \quad B_1 = B_2 = \begin{bmatrix} 1 \\ 0 \end{bmatrix},$$
$$A_2 = \begin{bmatrix} -\mu\nu & 0.3 \\ c & 0 \end{bmatrix}, \quad A_{d2} = \begin{bmatrix} -(1-\mu)\nu & 0 \\ 1-\mu & 0 \end{bmatrix}, \quad L_1 = L_2 = \begin{bmatrix} 1 & 0 \end{bmatrix},$$
$$C_1 = C_2 = \begin{bmatrix} \mu & 0 \end{bmatrix}, \quad C_{d1} = C_{d2} = \begin{bmatrix} 1-\mu & 0 \end{bmatrix}, \quad D_1 = 1, \quad D_2 = 0.5.$$

In the example, $x(k) \triangleq \begin{bmatrix} x_1(k) \\ x_2(k) \end{bmatrix}$, $\mu = 0.8$, $\nu = 0.2$ and $1 \leq d(k) \leq 3$ represents time-varying state delay. Then by solving the conditions in Theorems 4.21 and 4.23, the obtained results for the desired filtering cases are as follows:

- $\mathcal{H}_\infty$ performance case: $\mathcal{Z} = -I$, $\mathcal{Y} = 0$, $\mathcal{X} = \gamma^2 I$, and $B_\varepsilon = 1$. By solving the conditions in Theorem 4.21, we obtain $\gamma_{\min} = 1.3774$, and the corresponding desired filter matrices are as follows:

$$\begin{cases} A_f = \begin{bmatrix} 1.1457 & 0.3504 \\ -0.3086 & 0.4303 \end{bmatrix}, \quad B_f = \begin{bmatrix} 0.7705 \\ -1.0511 \end{bmatrix}, \\ C_f = \begin{bmatrix} -0.7191 & -0.2939 \end{bmatrix}, \quad D_f = 0.5592. \end{cases} \tag{4.36}$$

- Strictly dissipative case: $\mathcal{Z} = -0.25$, $\mathcal{Y} = -0.2$, $\mathcal{X} = 1$, $\underline{B}_\varepsilon = 0.8$ and $\bar{B}_\varepsilon = 0.9$. By solving the LMI conditions in Theorem 4.23, the corresponding reliable filter matrices are as follows:

$$\begin{cases} A_f = \begin{bmatrix} 0.8147 & 0.0378 \\ -0.0847 & 0.6813 \end{bmatrix}, \quad B_f = \begin{bmatrix} 0.6635 \\ -0.9081 \end{bmatrix}, \\ C_f = \begin{bmatrix} -0.0258 & -0.0031 \end{bmatrix}, \quad D_f = 0.7645. \end{cases} \tag{4.37}$$

In the following, we will present the simulation results to illustrate the effectiveness of the designed $\mathcal{H}_\infty$ filter and the reliable one with strict dissipativity. Let the initial conditions be zero, that is, $x(0) = 0$ and $\hat{x}(0) = 0$, and suppose the disturbance input be $\omega(k) = \frac{3\sin(0.9k)}{(0.75k)^2+3.5}$. The simulation results for the designed $\mathcal{H}_\infty$ filter and the reliable one with strict dissipativity are shown in Figs. 4.5–4.6 and Figs. 4.7–4.8, respectively. Fig. 4.4 shows the time-varying delay $d(k)$ which changes randomly between $d_1 = 1$ and $d_2 = 3$. Figs. 4.5 and 4.7 plot, separately, the signal $z(k)$ (solid line), and its estimations $\hat{z}(k)$ with the each designed filter (dash-dot line). The corresponding estimation errors $e(k)$ are shown in Figs. 4.6 and 4.8. From them,

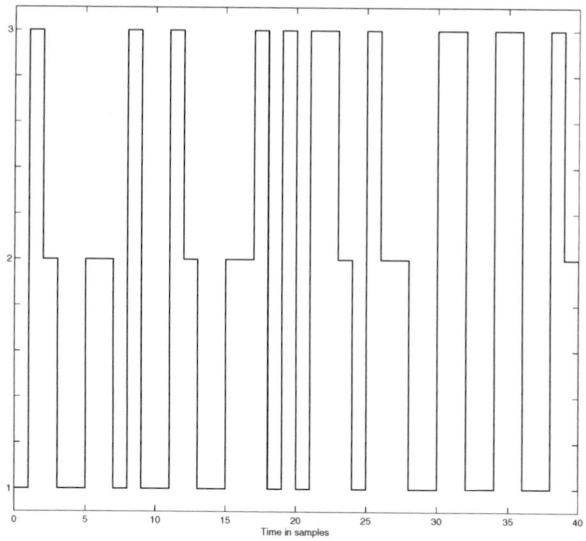

Fig. 4.4. Time-varying delays $d(k)$

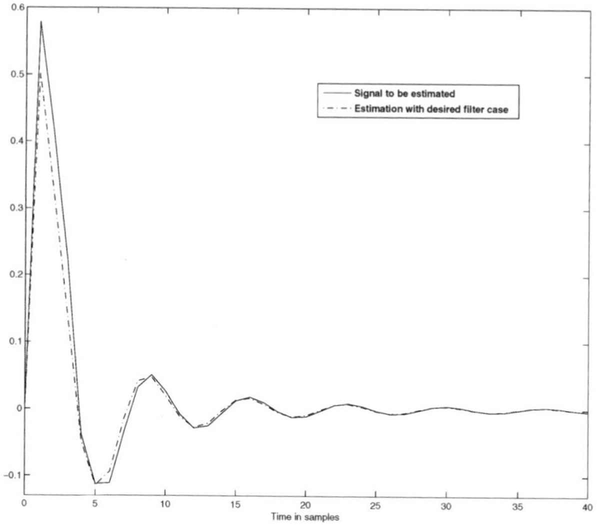

Fig. 4.5. Signal $z(k)$ and its estimation $\hat{z}(k)$ of the $\mathcal{H}_\infty$ filter

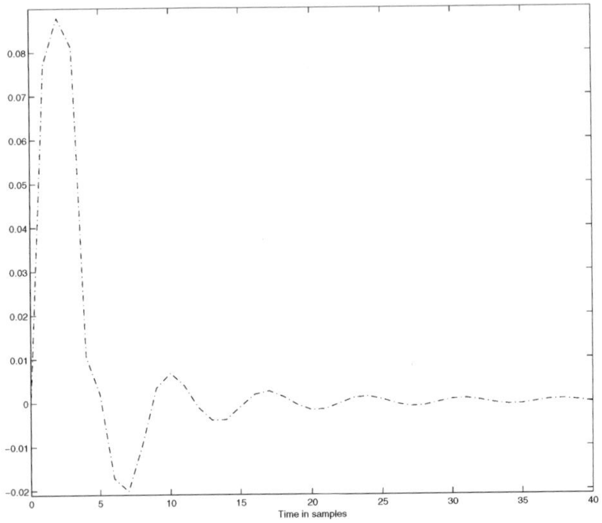

Fig. 4.6. Estimation error $e(k)$ for the $\mathcal{H}_\infty$ performance case

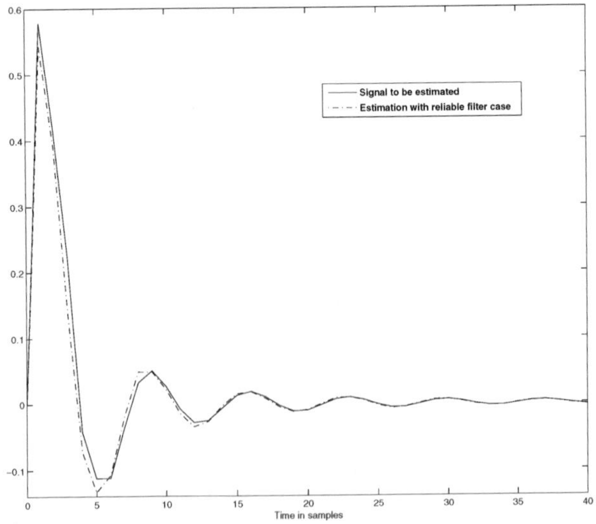

Fig. 4.7. Signal $z(k)$ and its estimation $\hat{z}(k)$ of the dissipative reliable filter

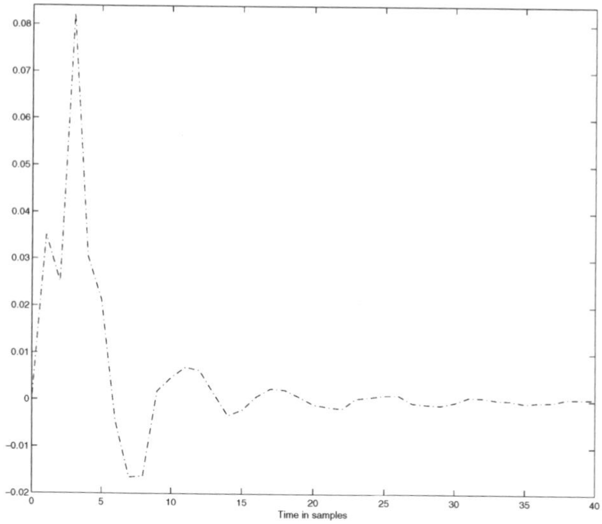

Fig. 4.8. Estimation error $e(k)$ for the dissipative case

it is obvious that the estimation error of the reliable filter with strict dissipativity with (4.37) is smaller than the obtained $\mathcal{H}_\infty$ filter with (4.36). This is consistent with common sense, since the $\mathcal{H}_\infty$ performance is a special case of strict dissipativity and has much more restricted conditions.

4.4 Conclusion

The focus of this chapter has been system performance analysis and filter design for T-S fuzzy systems with time-varying delay. Firstly, sufficient conditions of stability analysis satisfying the given performances have been presented for the augmented error system by the delay partitioning approach in combination with the input-output approach. Based on these conditions, the filtering problem for the concerned systems can be solved efficiently. Then, the obtained conditions have been extended to solve the problem of reliable filter design. Since all the filter design conditions are in terms of a set of strict LMIs, the desired filters can be obtained by solving optimization problems.

Chapter 5
Distributed Filtering of Discrete-Time T-S Fuzzy Time-Delay Systems

5.1 Introduction

This chapter is concerned with the distributed fuzzy filter design problem for a class of sensor networks described by discrete-time T-S fuzzy systems with time-varying delays and multiple probabilistic packet losses. In the sensor networks, each individual sensor can receive not only the data packets from its own measurement but also from its neighboring sensors' measurements according to the sensor networked topology. Our attention is focused on the design of distributed fuzzy filters to guarantee the filtering error dynamical system to be mean-square asymptotically stable with an average $\mathcal{H}_\infty$ performance. Sufficient conditions for the obtained filtering error dynamics are proposed by applying a comparison model and the scaled small gain theorem. Based on the measurements and estimates of the system states and its neighbors for each sensor, the solution of the parameters of the distributed fuzzy filters is characterized in terms of the feasibility of a convex optimization problem. An illustrative example will be provided to illustrate the effectiveness of the proposed approaches.

5.2 System Description and Preliminaries

The distributed fuzzy filtering problem in sensor networks is shown in Fig. 5.1. In this figure, each sensor can transfer the information from both the plant and its neighbouring sensors according to the sensor networked topology. The information received by the pth sensor node from the plant is transmitted via communication cables that are of limited capacity, and therefore may encounter the phenomena of random link failures and data losses.

This chapter assumes that the sensor network has n sensor nodes, which are distributed in the space according to a specific interconnection topology characterized by a directed graph $\mathscr{G} = (\mathcal{V}, \mathcal{E}, \mathcal{L})$, where $\mathcal{V} = \{1, 2, \ldots, n\}$ denotes the set of sensor nodes, $\mathcal{E} \subseteq \mathcal{V} \times \mathcal{V}$ is the set of edges, and $\mathcal{L} =$

L. Wu et al., *Fuzzy Control Systems with Time-Delay and Stochastic Perturbation*, Studies in Systems, Decision and Control 12, DOI: 10.1007/978-3-319-11316-6_5

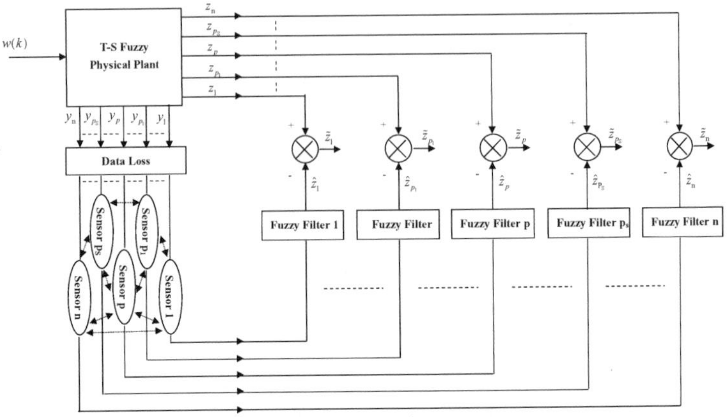

Fig. 5.1. Block diagram of the distributed fuzzy filtering in a sensor network

$(l_{pq})_{n\times n}$ is the nonnegative adjacency matrix associated with the edges of the graph, i.e., $l_{pq} > 0 \Leftrightarrow \text{edge}(p,q) \in \mathcal{E}$, which means that there is information transmission from the qth sensor node to the pth sensor node. Moreover, it is also assumed that $l_{pp} = 1$ for all $p \in \mathcal{V}$, *i.e.*, the sensors are self-connected. If $(p,q) \in \mathcal{E}$, then node q is called one of the neighbors of node p. For all $q \in \mathcal{V}$, denote $\mathcal{N}_q \triangleq \{q \in \mathcal{V} | (p,q) \in \mathcal{E}\}$, which means that in the sensor network, the pth sensor node can receive the information from its neighboring nodes $q \in \mathcal{N}_q$ according to the given network topology.

The plant is described by the nonlinear time-delay setting, which can be expressed as a set of linear systems in local operating regions and represented by the following T-S fuzzy time-varying delay model:

♦ **Plant Form:**

Rule i: IF $\theta_1(k)$ is $\mathcal{M}_{i1}$ and $\theta_2(k)$ is $\mathcal{M}_{i2}$ and … and $\theta_p(k)$ is $\mathcal{M}_{ip}$, THEN

$$\begin{aligned} x(k+1) &= A_i x(k) + A_{di} x(k-d(k)) + B_i \omega(k), \\ z(k) &= L_i x(k) + L_{\omega i}\omega(k), \\ x(k) &= \phi(k), \quad k = -d_2, -d_2+1, \ldots, 0, \end{aligned}$$

where $i = 1, 2, \ldots, r$, $x(k) \in \mathbf{R}^{n_x}$ represents the state vector; $\omega(k) \in \mathbf{R}^{n_\omega}$ is the disturbance input belonging to $\ell_2[0,\infty)$; $z(k) \in \mathbf{R}^{n_z}$ is the signal to be estimated, and $d(k)$ is the time-varying delay which satisfies $1 \leqslant d_1 \leqslant d(k) \leqslant d_2$, where d_1 and d_2 are two constant positive scalars representing its lower and upper bounds, respectively. r is the number of IF-THEN rules; $\mathcal{M}_{ij} (i = 1, 2, \ldots, r; j = 1, 2, \ldots, p)$ are the fuzzy sets; $\theta = \begin{bmatrix} \theta_1(k)\ \theta_2(k) \cdots \theta_p(k) \end{bmatrix}^T$ is the premise variable vector; A_i, A_{di}, B_i, L_i and $L_{\omega i}$ are known constant matrices with appropriate dimensions; $\phi(k)$ denotes the initial condition.

For every p ($p = 1, 2, \ldots, n$), the pth sensor node is given as follows:

$$y_p(k) = \beta_p(k) C_{pi} x(k) + D_{pi}\omega(k),$$

where $y_p(k) \in \mathbf{R}^{n_y}$ is the measured output received by the pth sensor node from the plant, C_{pi} and D_{pi} are known constant matrices with appropriate dimensions, and the stochastic variable $\beta_p(k)$ is Bernoulli-distributed white noise sequences specified by the following distribution laws:

$$\mathrm{Prob}\{\beta_p(k) = 1\} = \bar{\beta}_p, \quad \mathrm{Prob}\{\beta_p(k) = 0\} = 1 - \bar{\beta}_p,$$

where $\bar{\beta} \in [0, 1]$ is a known constant. Obviously, for the stochastic variable $\beta_p(k)$, it has $\bar{\alpha}_p^2 \triangleq \mathbf{E}\left\{[\beta_p(k) - \bar{\beta}_p]^2\right\} = \bar{\beta}_p(1 - \bar{\beta}_p)$.

A more compact presentation of the sensor network in the pth sensor node can be given by

$$\begin{aligned} x(k+1) &= A(k)x(k) + A_d(k)x(k - d(k)) + B(k)\omega(k), &(5.1a)\\ y_p(k) &= \beta_p(k) C_p(k) x(k) + D_p(k)\omega(k), &(5.1b)\\ z(k) &= L(k)x(k) + L_\omega(k)\omega(k), &(5.1c)\end{aligned}$$

where

$$\begin{aligned} A(k) &\triangleq \sum_{i=1}^{r} h_i(\theta) A_i, \quad A_d(k) \triangleq \sum_{i=1}^{r} h_i(\theta) A_{di}, \quad D_p(k) \triangleq \sum_{i=1}^{r} h_i(\theta) D_{pi},\\ B(k) &\triangleq \sum_{i=1}^{r} h_i(\theta) B_i, \quad C_p(k) \triangleq \sum_{i=1}^{r} h_i(\theta) C_{pi}, \quad L_\omega(k) \triangleq \sum_{i=1}^{r} h_i(\theta) L_{\omega i},\\ L(k) &\triangleq \sum_{i=1}^{r} h_i(\theta) L_i,\end{aligned}$$

with $h_i(\theta)$, $i = 1, 2, \ldots, r$ are the normalized membership functions, which are defined as that of (1.2) in Chapter 1.

Here, we shall design the fuzzy filter of the following structure on the pth sensor node. Based on the PDC, the fuzzy-rule-dependent filter is designed to share the same IF-THEN parts with the following structure:

♦ **Distributed Fuzzy Filter of the pth Sensor Node:**

Filter Rule i: IF $\theta_1(k)$ is $\mathcal{M}_{i1}$ and $\theta_2(k)$ is $\mathcal{M}_{i2}$ and … and $\theta_p(k)$ is $\mathcal{M}_{ip}$, THEN

$$\begin{aligned} \hat{x}_p(k+1) &= \sum_{q \in \mathcal{N}_p} l_{pq} H_{pqi}\left[y_q(k) - \bar{\beta}_q C_q(k)\hat{x}_q(k)\right] + \sum_{q \in \mathcal{N}_p} l_{pq} K_{pqi}\hat{x}_q(k), &(5.2a)\\ \hat{z}_p(k) &= L_i \hat{x}_p(k) + L_{\omega i}\omega(k), &(5.2b)\end{aligned}$$

where $\hat{x}_p(k) \in \mathbf{R}^{n_x}$ is the state estimate of the pth sensor node and $\hat{z}_p(k) \in \mathbf{R}^{n_z}$ is the estimate for $z(k)$ from the fuzzy filter on the pth sensor node. Here, matrices K_{pqi} and H_{pqi} ($q \in \mathcal{N}_p$) in (5.2) are parameters of the fuzzy filter for the pth sensor node which are to be determined. Moreover, the initial values of fuzzy filters are assumed to be $\hat{x}_p(0) = 0$ for $k = -d_2, -d_2+1, \ldots, 0$. Thus, the fuzzy filter for the pth sensor node can be represented by the following form:

$$\begin{aligned} \hat{x}_p(k+1) &= \sum_{q \in \mathcal{N}_p} l_{pq} H_{pq}(k) \left[y_q(k) - \bar{\beta}_q C_q(k) \hat{x}_q(k) \right] \\ &\quad + \sum_{q \in \mathcal{N}_p} l_{pq} K_{pq}(k) \hat{x}_q(k), \qquad (5.3a) \\ \hat{z}_p(k) &= L(k)\hat{x}(k) + L_\omega(k)\omega(k), \qquad (5.3b) \end{aligned}$$

where

$$K_{pq}(k) \triangleq \sum_{i=1}^{r} h_i(\theta) K_{pqi}, \quad H_{pq}(k) \triangleq \sum_{i=1}^{r} h_i(\theta) H_{pqi}.$$

Letting $e_p \triangleq x(k) - \hat{x}_p(k)$ and $\tilde{z}_p(k) \triangleq z(k) - \hat{z}_p(k)$, we can obtain the following system that governs the fuzzy filtering error dynamics for the sensor network:

$$\begin{aligned} e_p(k+1) &= \sum_{q \in \mathcal{N}_p} l_{pq} K_{pq}(k) e_q(k) - \sum_{q \in \mathcal{N}_p} \bar{\beta}_q l_{pq} H_{pq}(k) C_q(k) e_q(k) \\ &\quad + \left[A(k) - \sum_{q \in \mathcal{N}_p} (\beta_q(k) - \bar{\beta}_q) l_{pq} H_{pq}(k) C_q(k) - \sum_{q \in \mathcal{N}_p} l_{pq} K_{pq}(k) \right] x(k) \\ &\quad + A_d(k) x(k - d(k)) + \left[B(k) - \sum_{q \in \mathcal{N}_p} l_{pq} H_{pq}(k) D_q(k) \right] \omega(k), \qquad (5.4a) \\ \tilde{z}_p(k) &= L(k) e_p(k), \quad p = 1, 2, \ldots, n. \qquad (5.4b) \end{aligned}$$

We denote

$$\begin{aligned} &\bar{A}(k) \triangleq \mathrm{diag}_n\{A(k)\}, && \bar{x}(k) \triangleq \mathrm{col}_n\{x(k)\}, && e(k) \triangleq \mathrm{col}_n\{e_p(k)\}, \\ &\bar{B}(k) \triangleq \mathrm{col}_n\{B(k)\}, && \tilde{z}(k) \triangleq \mathrm{col}_n\{\tilde{z}_p(k)\}, && \bar{D}(k) \triangleq \mathrm{col}_n\{D_p(k)\}, \\ &\bar{L}(k) \triangleq \mathrm{diag}_n\{L(k)\}, && \bar{G}_{\bar{\beta}}(k) \triangleq \mathrm{diag}_n\{\bar{\beta}_p C_p(k)\}, && \\ &\bar{A}_d(k) \triangleq \mathrm{diag}_n\{A_d(k)\}, && \bar{E}_n^p(k) \triangleq \mathrm{diag}_n^p\{C_p(k)\}. && \end{aligned}$$

Then, based on system (5.1) with n sensors whose topology are determined by the given graph $\mathscr{G} = (\mathcal{V}, \mathcal{E}, \mathcal{L})$, the error dynamics governed by (5.4) can be rewritten in the following compact form

$$
\begin{aligned}
e(k+1) = & \left[\bar{A}(k) - \sum_{q=1}^{n}(\beta_q(k) - \bar{\beta}_q)\bar{H}(k)\bar{E}_n^q(k) - \bar{K}(k)\right]\bar{x}(k) \\
& + \left[\bar{B}(k) - \bar{H}(k)\bar{D}(k)\right]\omega(k) + \bar{A}_d(k)\bar{x}(k - d(k)) \\
& + \left(\bar{K}(k) - \bar{H}(k)\bar{G}_{\bar{\beta}}(k)\right)e(k), \qquad (5.5a)\\
\tilde{z}(k) = & \bar{L}(k)e(k), \qquad (5.5b)
\end{aligned}
$$

where

$$
\begin{cases}
\bar{K}(k) = \left[\mathscr{O}_{pq}(k)\right]_{n\times n}, & \text{with } \mathscr{O}_{pq}(k) = l_{pq}K_{pq}(k), \\
\bar{H}(k) = \left[\bar{\mathscr{O}}_{pq}(k)\right]_{n\times n}, & \text{with } \bar{\mathscr{O}}_{pq}(k) = l_{pq}H_{pq}(k).
\end{cases} \qquad (5.6)
$$

Augmenting the original model (5.1) to include the fuzzy filter error dynamics (5.5), we obtain the following system:

$$
\begin{aligned}
\zeta(k+1) = & \left[\bar{\mathscr{A}}(k) + \sum_{q=1}^{n}(\beta_q(k) - \bar{\beta}_q)\bar{\mathscr{F}}_q(k)\right]\zeta(k) \\
& + \bar{\mathscr{A}}_d(k)\zeta(k - d(k)) + \bar{\mathscr{B}}(k)\omega(k), \qquad (5.7a)\\
\tilde{z}(k) = & \bar{\mathscr{L}}(k)\zeta(k), \qquad (5.7b)
\end{aligned}
$$

where $\zeta(k) \triangleq \begin{bmatrix} \bar{x}(k) \\ e(k) \end{bmatrix}$ and

$$
\bar{\mathscr{A}}(k) \triangleq \begin{bmatrix} \bar{A}(k) & 0 \\ \bar{A}(k) - \bar{K}(k) & \bar{K}(k) - \bar{H}(k)\bar{G}_{\bar{\beta}}(k) \end{bmatrix}, \quad \bar{\mathscr{F}}_q(k) \triangleq \begin{bmatrix} 0 & 0 \\ -\bar{H}(k)\bar{E}_n^q(k) & 0 \end{bmatrix},
$$

$$
\bar{\mathscr{B}}(k) \triangleq \begin{bmatrix} \bar{B}(k) \\ \bar{B}(k) - \bar{H}(k)\bar{D}(k) \end{bmatrix}, \quad \bar{\mathscr{L}}(k) \triangleq \begin{bmatrix} 0 \\ \bar{L}^T(k) \end{bmatrix}, \quad \bar{\mathscr{A}}_d(k) \triangleq \begin{bmatrix} \bar{A}_d(k) & 0 \\ \bar{A}_d(k) & 0 \end{bmatrix}.
$$

Definition 5.1. The fuzzy filtering error system in (5.7) is said to be mean-square asymptotically stable if under $\omega(k) = 0$,

$$
\lim_{k\to\infty} \mathbf{E}\{\|\zeta(k)\|\} = 0.
$$

Definition 5.2. Given a scalar $\gamma > 0$, the fuzzy filtering error system in (5.7) is said to be mean-square asymptotically stable with an average $\mathcal{H}_\infty$ performance level γ if it is asymptotically stable under $\omega(k) = 0$, and under zero initial condition and for $\omega(k) \neq 0$, it holds that

$$
\frac{1}{n}\|\tilde{z}(k)\|_{\mathbf{E}_2}^2 < \gamma^2\|\omega(k)\|_2^2, \qquad (5.8)
$$

where

$$
\|\tilde{z}(k)\|_{\mathbf{E}_2} = \mathbf{E}\left\{\sqrt{\sum_{k=0}^{\infty}\tilde{z}^T(k)\tilde{z}(k)}\right\}, \quad \|\omega(k)\|_2 = \sqrt{\sum_{k=0}^{\infty}\omega^T(k)\omega(k)}.
$$

Define $\hat{\omega}(k) \triangleq \sqrt{n}\gamma\omega(k)$, and consider an auxiliary system:

$$\zeta(k+1) = \left[\bar{\mathscr{A}}(k) + \sum_{q=1}^{n}(\beta_q(k) - \bar{\beta}_q)\bar{\mathscr{F}}_q(k)\right]\zeta(k)$$
$$+\bar{\mathscr{A}}_d(k)\zeta(k-d(k)) + (\sqrt{n}\gamma)^{-1}\bar{\mathscr{B}}(k)\hat{\omega}(k), \tag{5.9a}$$
$$\tilde{z}(k) = \bar{\mathscr{L}}(k)\zeta(k), \tag{5.9b}$$

It is clear that the average $\mathcal{H}_\infty$ performance in (5.8) is equivalent to

$$\|\tilde{z}(k)\|_{\mathbf{E}_2} < \|\hat{\omega}(k)\|_2, \quad \forall\, 0 \neq \hat{\omega}(k) \in \ell_2[0,\infty). \tag{5.10}$$

Our objective is to determine the distributed fuzzy filter matrices (K_{pqi}, H_{pqi}) in (5.3) such that the filtering error system in (5.10) is mean-square asymptotically stable with a guaranteed average $\mathcal{H}_\infty$ performance.

5.3 Main Results

Considering the fuzzy filtering error system in (5.9), we now estimate the time-varying $\zeta(k-d(k))$ using its lower bound d_1 and upper bound d_2. The two-term approximation $\frac{\zeta(k-d_1)+\zeta(k-d_2)}{2}$ results in the estimation error:

$$\sigma(k) = \frac{2}{d}\left\{\zeta(k-d(k)) - \frac{1}{2}\left[\zeta(k-d_1) + \zeta(k-d_2)\right]\right\}$$
$$= \frac{1}{d}\left(\sum_{i=k-d_2}^{k-d_1-1}\beta(i)\varsigma(i)\right),$$

where $d = d_2 - d_1$, $\varsigma(i) \triangleq \zeta(i+1) - \zeta(i)$ and

$$\beta(i) \triangleq \begin{cases} 1, & \text{when } i \leq k-d(k)-1, \\ -1, & \text{when } i > k-d(k)-1. \end{cases}$$

To employ the input-output approach, the following auxiliary system is introduced to replace system in (5.9):

$$\zeta(k+1) = \left[\bar{\mathscr{A}}(k) + \sum_{q=1}^{n}(\beta_q(k) - \bar{\beta}_q)\bar{\mathscr{F}}_q(k)\right]\zeta(k)$$
$$+\frac{d}{2}\bar{\mathscr{A}}_d(k)\sigma(k) + (\sqrt{n}\gamma)^{-1}\bar{\mathscr{B}}(k)\hat{\omega}(k)$$
$$+\frac{1}{2}\bar{\mathscr{A}}_d(k)\left(\zeta(k-d_1) + \zeta(k-d_2)\right), \tag{5.11a}$$
$$\tilde{z}(k) = \bar{\mathscr{L}}(k)\zeta(k). \tag{5.11b}$$

The following model can formulate system (5.11) in the interconnection frame shown in Fig. 1.1:

$$(\mathcal{S}_1): \quad \begin{bmatrix} \zeta(k+1) \\ \varsigma(k) \\ \tilde{z}(k) \end{bmatrix} = \begin{bmatrix} \Sigma_1(k) & \frac{d}{2}\bar{\mathscr{A}}_d(k) & (\sqrt{n}\gamma)^{-1}\bar{\mathscr{B}}(k) \\ \Sigma_2(k) & \frac{d}{2}\bar{\mathscr{A}}_d(k) & (\sqrt{n}\gamma)^{-1}\bar{B}(k) \\ \Sigma_3(k) & 0 & 0 \end{bmatrix} \begin{bmatrix} \bar{\zeta}(k) \\ \sigma(k) \\ \hat{\omega}(k) \end{bmatrix}, \tag{5.12a}$$

$$(\mathcal{S}_2): \quad \sigma(k) = \mathcal{K}\varsigma(k), \tag{5.12b}$$

where $\bar{\zeta}(k) \triangleq \begin{bmatrix} \zeta(k) \\ \zeta(k-d_1) \\ \zeta(k-d_2) \end{bmatrix}$ and

$$\Sigma_1(k) \triangleq \left[\left(\bar{\mathscr{A}}(k) + \sum_{q=1}^{n} (\beta_q(k) - \bar{\beta}_q)\bar{\mathscr{F}}_q(k) \right) \; \frac{1}{2}\bar{\mathscr{A}}_d(k) \; \frac{1}{2}\bar{\mathscr{A}}_d(k) \right],$$

$$\Sigma_2(k) \triangleq \left[\left(\bar{\mathscr{A}}(k) + \sum_{q=1}^{n} (\beta_q(k) - \bar{\beta}_q)\bar{\mathscr{F}}_q(k) - I \right) \; \frac{1}{2}\bar{\mathscr{A}}_d(k) \; \frac{1}{2}\bar{\mathscr{A}}_d(k) \right],$$

$$\Sigma_3(k) \triangleq \begin{bmatrix} \bar{\mathscr{L}}(k) & 0 & 0 \end{bmatrix}.$$

For brevity, let us use the following operator:

$$(\mathcal{K}): \quad \varsigma(k) \to \sigma(k) = \frac{1}{d}\left(\sum_{i=k-d_2}^{k-d_1-1} \beta(i)\varsigma(i) \right), \tag{5.13}$$

to denote the relation $(\mathcal{S}_2)$ from $\varsigma(k)$ to $\sigma(k)$ in Fig. 1.1. By Lemma 3.4, we can obtain $\|\mathcal{K}\|_\infty \leq 1$ in (5.13).

We can see that the average $\mathcal{H}_\infty$ performance of $(\mathcal{S}_2)$ in (5.12) from input to output is bounded by one. Then based on Lemma 1.33, we focus on researching the scaled small gain of $(\mathcal{S}_1)$ for the interconnection frame (5.12).

Lemma 5.3. *Assume $(\mathcal{S}_1)$ in (5.12a) is internally stable, the closed-loop system of interconnection system described by (5.12) is mean-square asymptotically stable and has an average $\mathcal{H}_\infty$ performance level γ for $(\mathcal{K})$ if there exists a matrix $\hat{\mathscr{X}} \triangleq \mathrm{diag}\{\mathscr{X}, I\} > 0$ such that*

$$\|\hat{\mathscr{X}} \circ \mathcal{G} \circ \hat{\mathscr{X}}^{-1}\|_\infty < 1,$$

where

$$\mathcal{G} \triangleq \begin{bmatrix} \Sigma_1(k) & \frac{d}{2}\bar{\mathscr{A}}_d(k) & (\sqrt{n}\gamma)^{-1}\bar{\mathscr{B}}(k) \\ \Sigma_2(k) & \frac{d}{2}\bar{\mathscr{A}}_d(k) & (\sqrt{n}\gamma)^{-1}\bar{\mathscr{B}}(k) \\ \bar{\mathscr{L}}(k) & 0 & 0 \end{bmatrix}.$$

Remark 5.4. By Lemma 5.3, assume $(\mathcal{S}_1)$ in (5.12a) is internally stable, the closed-loop system of interconnection system described by (5.12) is mean-square asymptotically stable with an average $\mathcal{H}_\infty$ performance level γ for $\mathscr{K}_d(k)$ if there exists a matrix $\mathscr{X} \triangleq \bar{\mathscr{X}}^T \bar{\mathscr{X}}$ such that

$$\mathcal{J} \triangleq \mathbf{E}\left\{\sum_{k=0}^{\infty}\left[\varsigma^T(k)\mathscr{X}\varsigma(k) - \sigma^T(k)\mathscr{X}\sigma(k) + \tilde{z}^T(k)\tilde{z}(k) - \hat{\omega}^T(k)\hat{\omega}(k)\right]\right\} < 0. \quad \blacklozenge$$

Since $\|\mathcal{K}\|_\infty \leq 1$, it follows that the $\mathcal{H}_\infty$ norm of $(\mathcal{S}_2)$ in (5.12) from input to output is bounded by one. In the following, by applying the input-output approach, we will derive a LMI formulation of an average $\mathcal{H}_\infty$ performance for system (5.12). Firstly, let $d = d_2 - d_1$ and

$$\begin{cases}
\bar{\mathscr{T}}_1(k) \triangleq \sum_{i=1}^{r} h_i(\theta)\mathscr{T}_{1i}, & \bar{\mathscr{T}}_2(k) \triangleq \sum_{i=1}^{r} h_i(\theta)\mathscr{T}_{2i}, \\
\bar{\mathscr{S}}_1(k) \triangleq \sum_{i=1}^{r} h_i(\theta)\mathscr{S}_{1i}, & \bar{\mathscr{S}}_2(k) \triangleq \sum_{i=1}^{r} h_i(\theta)\mathscr{S}_{2i}, \\
\bar{\mathscr{N}}_1(k) \triangleq \sum_{i=1}^{r} h_i(\theta)\mathscr{N}_{1i}, & \bar{\mathscr{N}}_2(k) \triangleq \sum_{i=1}^{r} h_i(\theta)\mathscr{N}_{2i}, \\
\bar{\mathscr{M}}_1(k) \triangleq \sum_{i=1}^{r} h_i(\theta)\mathscr{M}_{1i}, & \bar{\mathscr{M}}_2(k) \triangleq \sum_{i=1}^{r} h_i(\theta)\mathscr{M}_{2i}, \\
\bar{\mathscr{Q}}_1(k) \triangleq \sum_{i=1}^{r} h_i(\theta)\mathscr{Q}_{1i}, & \bar{\mathscr{Q}}_2(k) \triangleq \sum_{i=1}^{r} h_i(\theta)\mathscr{Q}_{2i}, \\
\bar{\mathscr{T}}_3(k) \triangleq \sum_{i=1}^{r} h_i(\theta)\mathscr{T}_{3i}, & \bar{\mathscr{S}}_3(k) \triangleq \sum_{i=1}^{r} h_i(\theta)\mathscr{S}_{3i}, \\
\bar{\mathscr{N}}_3(k) \triangleq \sum_{i=1}^{r} h_i(\theta)\mathscr{N}_{3i}, & \bar{\mathscr{M}}_3(k) \triangleq \sum_{i=1}^{r} h_i(\theta)\mathscr{M}_{3i},
\end{cases} \tag{5.14}$$

where $\mathscr{Q}_{1i} > 0$, $\mathscr{Q}_{2i} > 0$, $\mathscr{S}_{1i} > 0$, $\mathscr{S}_{2i} > 0$, $\mathscr{S}_{3i} > 0$, $\mathscr{T}_{1i} > 0$, $\mathscr{T}_{2i} > 0$, $\mathscr{T}_{3i} > 0$, $i = 1, 2, \ldots, r$, are all $(2nn_x) \times (2nn_x)$ matrices.

Thus, we construct the following LKF:

$$V(k) \triangleq \sum_{i=1}^{3} V_i(k), \tag{5.15}$$

where

$$\begin{aligned}
V_1(k) &\triangleq \zeta^T(k)\mathscr{P}\zeta(k), \\
V_2(k) &\triangleq \sum_{i=k-d_1}^{k-1} \zeta^T(i)\bar{\mathscr{Q}}_1(i)\zeta(i) + \sum_{i=k-d_2}^{k-1} \zeta^T(i)\bar{\mathscr{Q}}_2(i)\zeta(i), \\
V_3(k) &\triangleq \sum_{i=-d_1}^{-1}\sum_{j=k+i}^{k-1} \varsigma^T(j)\mathscr{Z}_1\varsigma(j) + \sum_{i=-d_2}^{-1}\sum_{j=k+i}^{k-1} \varsigma^T(j)\mathscr{Z}_2\varsigma(j).
\end{aligned}$$

Then, based on (5.15), we can obtain the following result.

Theorem 5.5. *The filtering error system in (5.12) is mean-square asymptotically stable with an average $\mathcal{H}_\infty$ performance level γ if there exist matrices $\mathscr{P} > 0$, $\mathscr{X} > 0$, $\mathscr{Z}_1 > 0$, $\mathscr{Z}_2 > 0$, $\mathscr{Q}_{1i} > 0$, $\mathscr{Q}_{2i} > 0$, $\mathscr{S}_{1i} > 0$, $\mathscr{S}_{2i} > 0$, $\mathscr{S}_{3i} > 0$, $\mathscr{T}_{1i} > 0$, $\mathscr{T}_{2i} > 0$, $\mathscr{T}_{3i} > 0$, $\mathscr{M}_{1i}$, $\mathscr{M}_{2i}$, $\mathscr{M}_{3i}$, $\mathscr{N}_{1i}$, $\mathscr{N}_{2i}$, and $\mathscr{N}_{3i}$, which are defined in (5.14), such that for $i, j, l, s = 1, \ldots, r$,*

$$\left.\begin{aligned} \frac{1}{r-1}\Omega_{iils} + \frac{1}{2}\left(\Omega_{ijls} + \Omega_{jils}\right) < 0, \quad 0 \le i \ne j \le r, \\ \Omega_{iils} < 0, \end{aligned}\right\} \tag{5.16a}$$

$$\Lambda_{1i} \triangleq \begin{bmatrix} \mathscr{S}_{1i} & 0 & 0 & \mathscr{M}_{1i} \\ \star & \mathscr{S}_{2i} & 0 & \mathscr{M}_{2i} \\ \star & \star & \mathscr{S}_{3i} & \mathscr{M}_{3i} \\ \star & \star & \star & \mathscr{Z}_1 \end{bmatrix} \ge 0, \tag{5.16b}$$

$$\Lambda_{2i} \triangleq \begin{bmatrix} \mathscr{T}_{1i} & 0 & 0 & \mathscr{N}_{1i} \\ \star & \mathscr{T}_{2i} & 0 & \mathscr{N}_{2i} \\ \star & \star & \mathscr{T}_{3i} & \mathscr{N}_{3i} \\ \star & \star & \star & \mathscr{Z}_2 \end{bmatrix} \ge 0, \tag{5.16c}$$

where

$$\Omega_{ijls} \triangleq \begin{bmatrix} \Omega_{11} & \Omega_{12ij} & \Omega_{13i} & \Omega_{13i} & d\Omega_{13i} & \Omega_{14ij} \\ \star & \Omega_{22i} & \Omega_{23i} & \Omega_{24i} & 0 & 0 \\ \star & \star & \Omega_{33il} & \Omega_{34i} & 0 & 0 \\ \star & \star & \star & \Omega_{44is} & 0 & 0 \\ \star & \star & \star & \star & -\mathscr{X} & 0 \\ \star & \star & \star & \star & \star & -n\gamma^2 I \end{bmatrix},$$

with

$$\begin{aligned}
\Omega_{11} &\triangleq \operatorname{diag}\left\{-\mathscr{P}, -d_1^{-1}\mathscr{Z}_1, -d_2^{-1}\mathscr{Z}_2, -\mathscr{X}, -I, -\mathscr{P}, -d_1^{-1}\mathscr{Z}_1, -d_2^{-1}\mathscr{Z}_2, -\mathscr{X}\right\}, \\
\Omega_{12ij} &\triangleq \operatorname{col}\left\{\mathscr{P}\bar{\mathscr{A}}_{ij}, \mathscr{Z}_1(\bar{\mathscr{A}}_{ij}-I), \mathscr{Z}_2(\bar{\mathscr{A}}_{ij}-I), \mathscr{X}(\bar{\mathscr{A}}_{ij}-I), \bar{\mathscr{L}}_i, \mathscr{P}\left(\sum_{q=1}^{n}\bar{\alpha}_p\bar{\mathscr{F}}_{qij}\right),\right. \\
&\qquad \left.\mathscr{Z}_1\left(\sum_{q=1}^{n}\bar{\alpha}_p\bar{\mathscr{F}}_{qij}\right), \mathscr{Z}_2\left(\sum_{q=1}^{n}\bar{\alpha}_p\bar{\mathscr{F}}_{qij}\right), \mathscr{X}\left(\sum_{q=1}^{n}\bar{\alpha}_p\bar{\mathscr{F}}_{qij}\right)\right\}, \\
\Omega_{13i} &\triangleq \operatorname{col}\left\{\frac{1}{2}\mathscr{P}\bar{\mathscr{A}}_{di}, \frac{1}{2}\mathscr{Z}_1\bar{\mathscr{A}}_{di}, \frac{1}{2}\mathscr{Z}_2\bar{\mathscr{A}}_{di}, \frac{1}{2}\mathscr{X}\bar{\mathscr{A}}_{di}, 0, 0, 0, 0, 0\right\}, \\
\Omega_{22i} &\triangleq -\mathscr{P} + \mathscr{Q}_{1i} + \mathscr{Q}_{2i} + \mathscr{M}_{1i} + \mathscr{M}_{1i}^T + \mathscr{N}_{1i} + \mathscr{N}_{1i}^T + d_1\mathscr{S}_{1i} + d_2\mathscr{T}_{1i}, \\
\Omega_{23i} &\triangleq -\mathscr{M}_{1i}^T + \mathscr{M}_{2i} + \mathscr{N}_{2i}, \quad \Omega_{33il} \triangleq -\mathscr{Q}_{1l} - \mathscr{M}_{2i} - \mathscr{M}_{2i}^T + d_1\mathscr{S}_{2i} + d_2\mathscr{T}_{2i}, \\
\Omega_{14i} &\triangleq \operatorname{col}\left\{\mathscr{P}\bar{\mathscr{B}}_{ij}, \mathscr{Z}_1\bar{\mathscr{B}}_{ij}, \mathscr{Z}_2\bar{\mathscr{B}}_{ij}, \mathscr{X}\bar{\mathscr{B}}_{ij}, 0, 0, 0, 0, 0\right\}, \Omega_{24i} \triangleq \mathscr{M}_{3i} + \mathscr{N}_{3i} - \mathscr{N}_{1i}^T, \\
\Omega_{34i} &\triangleq -\mathscr{M}_{3i} - \mathscr{N}_{2i}^T, \quad \Omega_{44is} \triangleq -\mathscr{Q}_{2s} - \mathscr{N}_{3i} - \mathscr{N}_{3i}^T + d_1\mathscr{S}_{3i} + d_2\mathscr{T}_{3i}.
\end{aligned}$$

Proof. Based on the fuzzy basis functions, from (5.16a)–(5.16c), we obtain

$$\sum_{s=1}^{r}\sum_{l=1}^{r}\sum_{i=1}^{r}\sum_{j=1}^{r} h_s(\theta(k-d_1))h_l(\theta(k-d_2))h_i(\theta)h_j(\theta)\Omega_{ijls} < 0, \quad (5.17)$$

$$\sum_{i=1}^{r} h_i(\theta)\Lambda_{1i} \geq 0, \quad (5.18)$$

$$\sum_{i=1}^{r} h_i(\theta)\Lambda_{2i} \geq 0. \quad (5.19)$$

By Schur complement, inequality (5.17) implies

$$\hat{\Omega}(k) \triangleq \begin{bmatrix} \hat{\Omega}_{11}(k) & \hat{\Omega}_{12}(k) & \hat{\Omega}_{13}(k) & d\hat{\Omega}_{14}(k) & \hat{\Omega}_{15}(k) \\ \star & \hat{\Omega}_{22}(k) & \hat{\Omega}_{23}(k) & d\hat{\Omega}_{24}(k) & \hat{\Omega}_{25}(k) \\ \star & \star & \hat{\Omega}_{33}(k) & d\hat{\Omega}_{24}(k) & \hat{\Omega}_{25}(k) \\ \star & \star & \star & d^2\hat{\Omega}_{44}(k) & d\hat{\Omega}_{45}(k) \\ \star & \star & \star & \star & \hat{\Omega}_{55}(k) \end{bmatrix} < 0, \quad (5.20)$$

where

$$\begin{aligned}
\hat{\Omega}_{11}(k) &\triangleq \bar{\mathscr{A}}^T(k)\mathscr{P}\bar{\mathscr{A}}(k) + \sum_{q=1}^{n}\bar{\alpha}_p\bar{\mathscr{F}}_q^T(k)\mathscr{P}\bar{\mathscr{F}}_q(k) + \sum_{q=1}^{n}\bar{\alpha}_p\bar{\mathscr{F}}_q^T(k)\mathscr{Z}\bar{\mathscr{F}}_q(k) \\
&\quad + \bar{\mathscr{L}}^T(k)\bar{\mathscr{L}}(k) + \left[\bar{\mathscr{A}}(k)-I\right]^T\mathscr{Z}\left[\bar{\mathscr{A}}(k)-I\right] + \mathscr{Q}_1(k) + \mathscr{Q}_2(k) \\
&\quad - \mathscr{P} + \mathscr{M}_1(k) + \mathscr{M}_1^T(k) + \mathscr{N}_1(k) + \mathscr{N}_1^T(k) + d_1\mathscr{S}_1(k) + d_2\mathscr{T}_1(k), \\
\hat{\Omega}_{12}(k) &\triangleq \hat{\Omega}_{14}(k) - \mathscr{M}_1^T(k) + \mathscr{M}_2(k) + \mathscr{N}_2(k), \\
\hat{\Omega}_{13}(k) &\triangleq \hat{\Omega}_{14}(k) + \mathscr{M}_3^T(k) + \mathscr{N}_3(k) - \mathscr{N}_1^T(k), \\
\hat{\Omega}_{14}(k) &\triangleq \frac{1}{2}\bar{\mathscr{A}}^T(k)\mathscr{P}\bar{\mathscr{A}}_d(k) + \frac{1}{2}\left[\bar{\mathscr{A}}(k)-I\right]^T\mathscr{Z}\bar{\mathscr{A}}_d(k), \\
\hat{\Omega}_{15}(k) &\triangleq (\sqrt{n}\gamma)^{-1}\left\{\bar{\mathscr{A}}^T(k)\mathscr{P}\bar{\mathscr{B}}(k) + \left[\bar{\mathscr{A}}(k)-I\right]^T\mathscr{Z}\bar{\mathscr{B}}(k)\right\}, \\
\hat{\Omega}_{22}(k) &\triangleq \hat{\Omega}_{24}(k) - \mathscr{Q}_1(k-d_1) - \mathscr{M}_2(k) - \mathscr{M}_2^T(k) + d_1\mathscr{S}_2(k) + d_2\mathscr{T}_2(k), \\
\hat{\Omega}_{23}(k) &\triangleq \hat{\Omega}_{24}(k) - \mathscr{M}_3(k) - \mathscr{N}_2^T(k), \\
\hat{\Omega}_{24}(k) &\triangleq \frac{1}{4}\bar{\mathscr{A}}_d^T(k)\left(\mathscr{Z}+\mathscr{P}\right)\bar{\mathscr{A}}_d(k), \\
\hat{\Omega}_{25}(k) &\triangleq \frac{1}{2}(\sqrt{n}\gamma)^{-1}\bar{\mathscr{A}}_d^T(k)\left(\mathscr{Z}+\mathscr{P}\right)\bar{\mathscr{B}}(k), \\
\hat{\Omega}_{33}(k) &\triangleq \hat{\Omega}_{24}(k) + d_1\mathscr{S}_3(k) + d_2\mathscr{T}_3(k) - \mathscr{Q}_2(k-d_2) - \mathscr{N}_3(k) - \mathscr{N}_3^T(k), \\
\hat{\Omega}_{44}(k) &\triangleq \hat{\Omega}_{24}(k) - \mathscr{X}, \quad \mathscr{Z} \triangleq d_1\mathscr{Z}_1 + d_2\mathscr{Z}_2 + \mathscr{X}, \\
\hat{\Omega}_{55}(k) &\triangleq (\sqrt{n}\gamma)^{-2}\bar{\mathscr{B}}^T(k)\left(\mathscr{Z}+\mathscr{P}\right)\bar{\mathscr{B}}(k) - I.
\end{aligned}$$

Along the trajectories system (5.12), and considering the mathematical expectation and the difference of the fuzzy LKF in (5.15), we have

$$\mathbf{E}\{\Delta V(k)\} \triangleq \mathbf{E}\left\{V\left(k+1|k\right)\right\} - V(k), \tag{5.21}$$

where

$$\begin{aligned}
\mathbf{E}\left\{\Delta V_1(k)\right\} &= \mathbf{E}\left\{\zeta^T(k+1)\mathscr{P}\zeta(k+1)\right\} - \mathbf{E}\left\{\zeta^T(k)\mathscr{P}\zeta(k)\right\}, \\
\mathbf{E}\left\{\Delta V_2(k)\right\} &= -\mathbf{E}\left\{\zeta^T(k-d_1)\bar{\mathscr{Q}}_1(k-d_1)\zeta(k-d_1)\right\} \\
&\quad +\mathbf{E}\left\{\zeta^T(k)\left(\bar{\mathscr{Q}}_1(k)+\bar{\mathscr{Q}}_2(k)\right)\zeta(k)\right\} \\
&\quad -\mathbf{E}\left\{\zeta^T(k-d_2)\bar{\mathscr{Q}}_2(k-d_2)\zeta(k-d_2)\right\}, \\
\mathbf{E}\left\{\Delta V_3(k)\right\} &= \mathbf{E}\left\{\varsigma^T(k)\left(d_1\mathscr{Z}_1+d_2\mathscr{Z}_2\right)\varsigma(k)\right\} \\
&\quad -\mathbf{E}\left\{\sum_{i=k-d_1}^{k-1}\varsigma^T(i)\mathscr{Z}_1\varsigma(i)\right\} - \mathbf{E}\left\{\sum_{i=k-d_2}^{k-1}\varsigma^T(i)\mathscr{Z}_2\varsigma(i)\right\}.
\end{aligned}$$

Moreover, according to the definition of $\varsigma(k)$, for any matrices $\bar{\mathscr{M}}(k) \triangleq \left[\bar{\mathscr{M}}_1(k) \;\; \bar{\mathscr{M}}_2(k) \;\; \bar{\mathscr{M}}_3(k)\right]$ and $\bar{\mathscr{N}}(k) \triangleq \left[\bar{\mathscr{N}}_1(k) \;\; \bar{\mathscr{N}}_2(k) \;\; \bar{\mathscr{N}}_3(k)\right]$, the following equations always hold:

$$\begin{aligned}
2\bar{\zeta}^T(k)\bar{\mathscr{M}}^T(k)\left[\zeta(k)-\zeta(k-d_1)-\sum_{s=k-d_1}^{k-1}\varsigma(s)\right] &= 0, \\
2\bar{\zeta}^T(k)\bar{\mathscr{N}}^T(k)\left[\zeta(k)-\zeta(k-d_2)-\sum_{s=k-d_2}^{k-1}\varsigma(s)\right] &= 0,
\end{aligned}$$

where $\bar{\zeta}(k)$ is defined in (5.12).

Moreover, for any appropriately dimensioned matrices $\bar{\mathscr{S}}(k) \triangleq \operatorname{diag}\{\bar{\mathscr{S}}_1(k), \bar{\mathscr{S}}_2(k), \bar{\mathscr{S}}_3(k)\} > 0$ and $\bar{\mathscr{T}}(k) \triangleq \operatorname{diag}\{\bar{\mathscr{T}}_1(k), \bar{\mathscr{T}}_2(k), \bar{\mathscr{T}}_3(k)\} > 0$, we have

$$\begin{aligned}
d_1\bar{\zeta}^T(k)\bar{\mathscr{S}}(k)\bar{\zeta}(k) - \sum_{s=k-d_1}^{k-1}\bar{\zeta}^T(k)\bar{\mathscr{S}}(k)\bar{\zeta}(k) &= 0, \\
d_2\bar{\zeta}^T(k)\bar{\mathscr{T}}(k)\bar{\zeta}(k) - \sum_{s=k-d_2}^{k-1}\bar{\zeta}^T(k)\bar{\mathscr{T}}(k)\bar{\zeta}(k) &= 0.
\end{aligned}$$

Therefore, from (5.21) and the above analysis, under zero inputs, that is, $\sigma(k) = 0$ and $\hat{\omega}(k) = 0$, we can see that $\mathbf{E}\{\Delta V(k)\} < 0$, thus the system $(\mathcal{S}_1)$ is mean-square asymptotically stable.

Let $\mathscr{X} > 0$ and consider the following index:

$$\mathcal{J} \triangleq \mathbf{E}\left\{\sum_{k=0}^{\infty}\left[\varsigma^T(k)\mathscr{X}\varsigma(k) - \sigma^T(k)\mathscr{X}\sigma(k) + \tilde{z}^T(k)\tilde{z}(k) - \hat{\omega}^T(k)\hat{\omega}(k)\right]\right\}.$$

Considering zero initial condition, we have that $V(k)|_{k=0} = 0$ and

$$\mathcal{J} \leq \sum_{k=0}^{\infty} \begin{bmatrix} \bar{\zeta}(k) \\ \sigma(k) \\ \hat{\omega}(k) \end{bmatrix}^T \hat{\Omega}(k) \begin{bmatrix} \bar{\zeta}(k) \\ \sigma(k) \\ \hat{\omega}(k) \end{bmatrix} - \sum_{k=0}^{\infty} \sum_{s=k-d_1}^{k-1} \begin{bmatrix} \bar{\zeta}(k) \\ \varsigma(s) \end{bmatrix}^T \bar{\Lambda}_1(k) \begin{bmatrix} \bar{\zeta}(k) \\ \varsigma(s) \end{bmatrix}$$
$$- \sum_{k=0}^{\infty} \sum_{s=k-d_2}^{k-1} \begin{bmatrix} \bar{\zeta}(k) \\ \varsigma(s) \end{bmatrix}^T \bar{\Lambda}_2(k) \begin{bmatrix} \bar{\zeta}(k) \\ \varsigma(s) \end{bmatrix}.$$

Therefore, considering (5.16a)–(5.16c), for any nonzero $\hat{\omega}(k) \in \ell_2[0, \infty)$, we have $\mathcal{J} < 0$, which means $\|\tilde{z}(k)\|_{\mathbf{E}_2} < \|\hat{\omega}(k)\|_2$. The proof is completed. ■

We present a solution to the $\mathcal{H}_\infty$ distributed fuzzy filter design.

Theorem 5.6. *The filtering error system in (5.12) is mean-square asymptotically stable with an average $\mathcal{H}_\infty$ performance level γ if there exist matrices $\mathcal{P} > 0$, $\mathcal{X} > 0$, $\mathcal{Z}_1 > 0$, $\mathcal{Z}_2 > 0$, $\mathscr{P} > 0$, $\mathscr{X} > 0$, $\mathscr{Z}_1 > 0$, $\mathscr{Z}_2 > 0$, $\mathscr{Q}_{1i} > 0$, $\mathscr{Q}_{2i} > 0$, $\mathscr{S}_{1i} > 0$, $\mathscr{S}_{2i} > 0$, $\mathscr{S}_{3i} > 0$, $\mathscr{T}_{1i} > 0$, $\mathscr{T}_{2i} > 0$, $\mathscr{T}_{3i} > 0$, $\mathscr{M}_{1i}$, $\mathscr{M}_{2i}$, $\mathscr{M}_{3i}$, $\mathscr{N}_{1i}$, $\mathscr{N}_{2i}$, $\mathscr{N}_{3i}$ and $\mathscr{G}_i$, such that (5.16b)–(5.16c) and the following conditions hold for $i, j, l, s = 1, 2, \ldots, r$,*

$$\left.\begin{aligned} \frac{1}{r-1}\bar{\Omega}_{iils} + \frac{1}{2}\left(\bar{\Omega}_{ijls} + \bar{\Omega}_{jils}\right) < 0, \quad 0 \leq i \neq j \leq r, \\ \bar{\Omega}_{iils} < 0, \end{aligned}\right\} \tag{5.22a}$$

$$\mathcal{P}\mathscr{P} = I, \quad \mathcal{Z}_1\mathscr{Z}_1 = I, \quad \mathcal{Z}_2\mathscr{Z}_2 = I, \quad \mathcal{X}\mathscr{X} = I, \tag{5.22b}$$

where

$$\bar{\Omega}_{ijls} \triangleq \begin{bmatrix} \bar{\Omega}_{11} & \bar{\Omega}_{12ij} & \frac{1}{2}\bar{\Omega}_{13i} & \frac{1}{2}\bar{\Omega}_{13i} & \frac{d}{2}\bar{\Omega}_{13i} & \bar{\Omega}_{14ij} \\ \star & \Omega_{22i} & \Omega_{23i} & \Omega_{24i} & 0 & 0 \\ \star & \star & \Omega_{33il} & \Omega_{34i} & 0 & 0 \\ \star & \star & \star & \Omega_{44is} & 0 & 0 \\ \star & \star & \star & \star & -\mathscr{X} & 0 \\ \star & \star & \star & \star & \star & -n\gamma^2 I \end{bmatrix},$$

with

$$\bar{\Omega}_{11} \triangleq -\mathrm{diag}\left\{\mathcal{P}, d_1^{-1}\mathcal{Z}_1, d_2^{-1}\mathcal{Z}_2, \mathcal{X}, I, \mathcal{P}, d_1^{-1}\mathcal{Z}_1, d_2^{-1}\mathcal{Z}_2, \mathcal{X}\right\},$$

$$\begin{aligned} \bar{\Omega}_{12ij} \triangleq \mathrm{col}\Big\{ & \left(\bar{\mathscr{A}}_{0i} + \mathcal{I}\mathscr{G}_j\mathcal{G}_{\bar{\beta}i}\right), \left(\bar{\mathscr{A}}_{0i} + \mathcal{I}\mathscr{G}_j\mathcal{G}_{\bar{\beta}i} - I\right), \\ & \left(\bar{\mathscr{A}}_{0i} + \mathcal{I}\mathscr{G}_j\mathcal{G}_{\bar{\beta}i} - I\right), \left(\bar{\mathscr{A}}_{0i} + \mathcal{I}\mathscr{G}_j\mathcal{G}_{\bar{\beta}i} - I\right), \bar{\mathscr{L}}_i, \\ & \sum_{q=1}^{n}\bar{\alpha}_p\mathcal{I}\mathscr{G}_j\mathcal{F}_{qi}, \sum_{q=1}^{n}\bar{\alpha}_p\mathcal{I}\mathscr{G}_j\mathcal{F}_{qi}, \sum_{q=1}^{n}\bar{\alpha}_p\mathcal{I}\mathscr{G}_j\mathcal{F}_{qi}, \sum_{q=1}^{n}\bar{\alpha}_p\mathcal{I}\mathscr{G}_j\mathcal{F}_{qi}\Big\}, \end{aligned}$$

$$\bar{\Omega}_{13i} \triangleq \left[\mathscr{A}_{di}^T \ \mathscr{A}_{di}^T \ \mathscr{A}_{di}^T \ \mathscr{A}_{di}^T \ 0 \ 0 \ 0 \ 0 \ 0\right]^T,$$

$$\bar{\Omega}_{14ij} \triangleq \left[\bar{\mathscr{B}}_{ij}^T \ \bar{\mathscr{B}}_{ij}^T \ \bar{\mathscr{B}}_{ij}^T \ \bar{\mathscr{B}}_{ij}^T \ 0 \ 0 \ 0 \ 0 \ 0\right]^T, \quad \mathscr{G}_i \triangleq \left[\bar{K}_i \ \ \bar{H}_i\right],$$

$$\bar{\mathscr{A}}_{0i} \triangleq \begin{bmatrix} \bar{A}_i & 0 \\ \bar{A}_i & 0 \end{bmatrix}, \ \bar{\mathscr{B}}_{ij} \triangleq \begin{bmatrix} \bar{B}_i \\ \bar{B}_i - \bar{H}_j \bar{D}_i \end{bmatrix}, \ \bar{\mathscr{B}}_{0i} \triangleq \begin{bmatrix} \bar{B}_i \\ \bar{B}_i \end{bmatrix}, \ \mathcal{I} \triangleq \begin{bmatrix} 0 \\ I \end{bmatrix},$$

$$\mathcal{D}_i \triangleq \begin{bmatrix} 0 \\ -\bar{D}_i \end{bmatrix}, \ \mathcal{F}_{qi} \triangleq \begin{bmatrix} 0 & 0 \\ -\bar{E}^q_{ni} & 0 \end{bmatrix}, \ \mathcal{G}_{\bar{\beta}i} \triangleq \begin{bmatrix} -I & I \\ 0 & -\bar{G}_{\bar{\beta}i} \end{bmatrix}, \tag{5.23}$$

and Ω_{22i}, Ω_{23i}, Ω_{24i}, Ω_{33il}, Ω_{34i} and Ω_{44is} are defined in Theorem 5.5. Moreover, if the above conditions have a set of feasible solutions, then the desired filter parameters K_{pqi} and H_{pqi} can be computed by (5.6).

Proof. With (5.23), we rewrite $\bar{\mathscr{A}}_{ij}$, $\bar{\mathscr{B}}_{ij}$ and $\mathscr{F}_{qij}$ in Theorem 5.5 as $\bar{\mathscr{A}}_{ij} \triangleq \bar{\mathscr{A}}_{0i}+\mathcal{I}\mathscr{G}_j\mathcal{G}_{\bar{\beta}i}$, $\bar{\mathscr{B}}_{ij} \triangleq \bar{\mathscr{B}}_{0i}+\mathcal{I}\mathscr{G}_j\mathcal{D}_i$ and $\mathscr{F}_{qij} \triangleq \mathcal{I}\mathscr{G}_j\mathcal{F}_{qi}$, respectively. Then, performing a congruence transformation to (5.16a) by diag$\{\mathscr{P}^{-1}, \mathscr{Z}_1^{-1}, \mathscr{Z}_2^{-1}, \mathscr{X}^{-1}, I, \mathscr{P}^{-1}, \mathscr{Z}_1^{-1}, \mathscr{Z}_2^{-1}, \mathscr{X}^{-1}, I, I, I, I, I\}$, we have

$$\frac{1}{r-1}\hat{\Omega}_{iils} + \frac{1}{2}\left(\hat{\Omega}_{ijls} + \hat{\Omega}_{jils}\right) < 0, \quad 0 \le i \ne j \le r,$$
$$\hat{\Omega}_{iils} < 0,$$

where

$$\hat{\Omega}_{ijls} \triangleq \begin{bmatrix} \hat{\Omega}_{11} & \bar{\Omega}_{12ij} & \frac{1}{2}\bar{\Omega}_{13i} & \frac{1}{2}\bar{\Omega}_{13i} & \frac{d}{2}\bar{\Omega}_{13i} & \bar{\Omega}_{14ij} \\ \star & \Omega_{22i} & \Omega_{23i} & \Omega_{24i} & 0 & 0 \\ \star & \star & \Omega_{33il} & \Omega_{34i} & 0 & 0 \\ \star & \star & \star & \Omega_{44is} & 0 & 0 \\ \star & \star & \star & \star & -\mathscr{X} & 0 \\ \star & \star & \star & \star & \star & -n\gamma^2 I \end{bmatrix},$$

$$\hat{\Omega}_{11} \triangleq -\text{diag}\{\mathscr{P}^{-1}, d_1^{-1}\mathscr{Z}_1^{-1}, d_2^{-1}\mathscr{Z}_2^{-1}, \mathscr{X}^{-1}, I, \mathscr{P}^{-1}, d_1^{-1}\mathscr{Z}_1^{-1}, d_2^{-1}\mathscr{Z}_2^{-1}, \mathscr{X}^{-1}\},$$

with $\bar{\Omega}_{12ij}$, $\bar{\Omega}_{13i}$, $\bar{\Omega}_{14ij}$, Ω_{22i}, Ω_{23i}, Ω_{24i}, Ω_{33il}, Ω_{34i} and Ω_{44is} defined in (5.23) and Theorem 5.5. Thus, the desired result can be worked out by noting (5.22b). This completes the proof. ■

Remark 5.7. Note that the conditions in Theorem 5.6 are not all in LMI form due to the matrix equalities (5.22b). We suggest the following minimization problem (which can be solved by the CCL algorithm [39]) involving LMI conditions instead of the original nonconvex problem in Theorem 5.6. ♦

Problem $\mathcal{H}_\infty$ Distributed Fuzzy Filter Design ($\mathcal{H}_\infty$-*DFFD*):

$$\begin{aligned} &\min \ \text{trace}\left(\mathcal{P}\mathscr{P} + \mathcal{Z}_1\mathscr{Z}_1 + \mathcal{Z}_2\mathscr{Z}_2 + \mathcal{X}\mathscr{X}\right) \\ &\text{subject to (5.16b)–(5.22a) and} \\ &\begin{bmatrix} \mathcal{P} & I \\ I & \mathscr{P} \end{bmatrix} \ge 0, \ \begin{bmatrix} \mathcal{Z}_1 & I \\ I & \mathscr{Z}_1 \end{bmatrix} \ge 0, \ \begin{bmatrix} \mathcal{Z}_2 & I \\ I & \mathscr{Z}_2 \end{bmatrix} \ge 0, \ \begin{bmatrix} \mathcal{X} & I \\ I & \mathscr{X} \end{bmatrix} \ge 0. \end{aligned}$$

5.4 Illustrative Example

Example 5.8. Consider a sensor network described by the following modified Henon mapping system with time-varying delay:

$$x_1(k+1) = -\left[\mathcal{C}x_1(k) + (1-\mathcal{C})x_1(k-d(k))\right]^2 + 0.3x_2(k) + \omega(k),$$
$$x_2(k+1) = \mathcal{C}x_1(k) + (1-\mathcal{C})x_1(k-d(k)),$$

where $\omega(k)$ is the disturbance input. The constant $\mathcal{C} \in [0,1]$ is the retarded coefficient.

The topology of this sensor network (with $n = 2$ nodes) is represented by a directed graph $\mathscr{G} = (\mathcal{V}, \mathcal{E}, \mathcal{L})$ with the set of nodes $\mathcal{V} = \{1, 2\}$, set of edges $\mathcal{E} = \{(1,1), (1,2), (2,2)\}$ and the adjacency matrix $\mathcal{L} = [l_{pq}]_{2\times 2}$, where adjacency elements $l_{pq} = 1$ when $(p,q) \in \mathcal{E}$; otherwise, $l_{pq} = 0$. The adjacency matrix $\mathcal{L}$ is given as $\mathcal{L} = \begin{bmatrix} 1 & 1 \\ 0 & 1 \end{bmatrix}$. For each p $(p = 1, 2)$, the pth sensor node is described as

$$y_p(k) = \beta_p(k)\mathcal{C}_p x(k) + \mathcal{D}_p \omega(k).$$

Let $\theta = \mathcal{C}x_1(k) + (1-\mathcal{C})x_1(k-d)$. Assume that $\theta \in [-\mathcal{M}, \mathcal{M}]$, $\mathcal{M} > 0$. By using the same procedure as in [203], the nonlinear term θ^2 can be exactly represented as $\theta^2 = h_1(\theta)(-\mathcal{M})\theta + h_2(\theta)\mathcal{M}\theta$, where $h_1(\theta), h_2(\theta) \in [0,1]$, and $h_1(\theta) + h_2(\theta) = 1$. Thus, the membership functions $h_1(\theta)$ and $h_2(\theta)$ can be set as $h_1(\theta) = \frac{1}{2}\left(1 - \frac{\theta}{\mathcal{M}}\right)$ and $h_2(\theta) = \frac{1}{2}\left(1 + \frac{\theta}{\mathcal{M}}\right)$. It can be seen that $h_1(\theta) = 1$ and $h_2(\theta) = 0$ when θ is $-\mathcal{M}$ and that $h_1(\theta) = 0$ and $h_2(\theta) = 1$ when θ is $\mathcal{M}$. Then, the above nonlinear system in the pth sensor node with the output $z(k)$ can be approximately represented by the following T-S fuzzy model:

♦ **Plant Form:**

Rule 1: IF θ is $-\mathcal{M}$, THEN

$$\begin{cases} x(k+1) = A_1 x(k) + A_{d1} x(k-d(k)) + B_1 \omega(k), \\ y_p(k) = \beta_p(k)\mathcal{C}_{p1} x(k) + \mathcal{D}_{p1}\omega(k), \\ z(k) = L_1 x(k) + L_{w1}\omega(k), \end{cases}$$

Rule 2: IF θ is $\mathcal{M}$, THEN

$$\begin{cases} x(k+1) = A_2 x(k) + A_{d2} x(k-d(k)) + B_2 \omega(k), \\ y_p(k) = \beta_p(k)\mathcal{C}_{p2} x(k) + \mathcal{D}_{p2}\omega(k), \\ z(k) = L_2 x(k) + L_{w2}\omega(k), \end{cases}$$

where

$$A_1 = \begin{bmatrix} \mathcal{C}\mathcal{M} & 0.3 \\ \mathcal{C} & 0 \end{bmatrix}, \quad A_{d1} = \begin{bmatrix} (1-\mathcal{C})\mathcal{M} & 0 \\ 1-\mathcal{C} & 0 \end{bmatrix}, \quad B_1 = B_2 = \begin{bmatrix} 1 \\ 0 \end{bmatrix},$$

$$A_2 = \begin{bmatrix} -\mathcal{CM} & 0.3 \\ \mathcal{C} & 0 \end{bmatrix}, \ A_{d2} = \begin{bmatrix} -(1-\mathcal{C})\mathcal{M} & 0 \\ 1-\mathcal{C} & 0 \end{bmatrix}, \ \mathcal{D}_{11} = \mathcal{D}_{21} = 1,$$
$$\mathcal{C}_{11} = \begin{bmatrix} 0.8 & 0 \end{bmatrix}, \quad \mathcal{C}_{21} = \begin{bmatrix} 0.9 & 0 \end{bmatrix}, \quad \mathcal{D}_{12} = 0.45, \quad L_{w1} = 2,$$
$$\mathcal{C}_{12} = \begin{bmatrix} 0.6 & 0 \end{bmatrix}, \quad \mathcal{C}_{22} = \begin{bmatrix} 0.7 & 0 \end{bmatrix}, \quad \mathcal{D}_{22} = 0.55, \quad L_{w2} = 3,$$
$$L_1 = \begin{bmatrix} 1.7 & -0.8 \end{bmatrix}, \quad L_2 = \begin{bmatrix} 0.2 & 2.6 \end{bmatrix}.$$

In the example, $x(k) = \begin{bmatrix} x_1(k) \\ x_2(k) \end{bmatrix}$, $\mathcal{C} = 0.8$, $\mathcal{M} = 0.2$ and $1 \leq d(k) \leq 3$. The probabilities are taken as $\bar{\beta}_1 = 0.95$ and $\bar{\beta}_2 = 0.93$, and the disturbance attenuation level is given as $\gamma = 4.140$. Then, by solving the conditions in Theorem 5.6, the parameters of the desired distributed filter are as follows:

$$K_{111} = \begin{bmatrix} 0.3487 & 0.1471 \\ 0.4949 & 0.1002 \end{bmatrix}, \quad K_{121} = \begin{bmatrix} 0.1387 & -0.1100 \\ 0.0787 & -0.0244 \end{bmatrix}, \ H_{111} = \begin{bmatrix} 0.1377 \\ 0.5861 \end{bmatrix},$$
$$K_{221} = \begin{bmatrix} 0.3998 & 0.1194 \\ 0.7468 & 0.0645 \end{bmatrix}, \quad K_{112} = \begin{bmatrix} -0.0984 & 0.2214 \\ 0.6905 & 0.0101 \end{bmatrix}, \ H_{122} = \begin{bmatrix} 0.0539 \\ 0.1665 \end{bmatrix},$$
$$K_{122} = \begin{bmatrix} 0.0124 & -0.0192 \\ 0.0901 & -0.0208 \end{bmatrix}, \quad H_{112} = \begin{bmatrix} -0.2687 \\ 0.6212 \end{bmatrix}, \quad H_{121} = \begin{bmatrix} 0.1852 \\ 0.1816 \end{bmatrix},$$
$$K_{222} = \begin{bmatrix} -0.0943 & 0.1794 \\ 0.9298 & -0.0172 \end{bmatrix}, \ H_{222} = \begin{bmatrix} -0.3486 \\ 1.2794 \end{bmatrix}, \quad H_{221} = \begin{bmatrix} 0.2145 \\ 0.8952 \end{bmatrix}.$$

Let the initial conditions be zero, and suppose the disturbance input $\omega(k)$ be $\omega(k) = 0.1e^{-0.25k}\sin(0.29k)$. The simulation results are shown in Figs. 5.2–5.4. Among them, Fig. 5.2 shows the time-varying delay $d(k)$ which changes

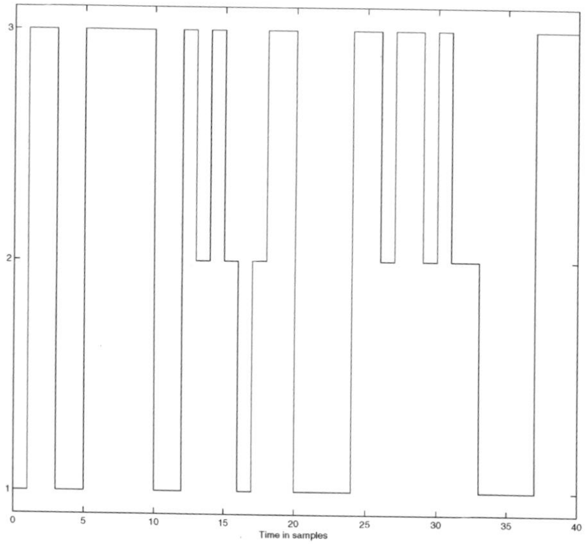

Fig. 5.2. Time-varying delay $d(k)$

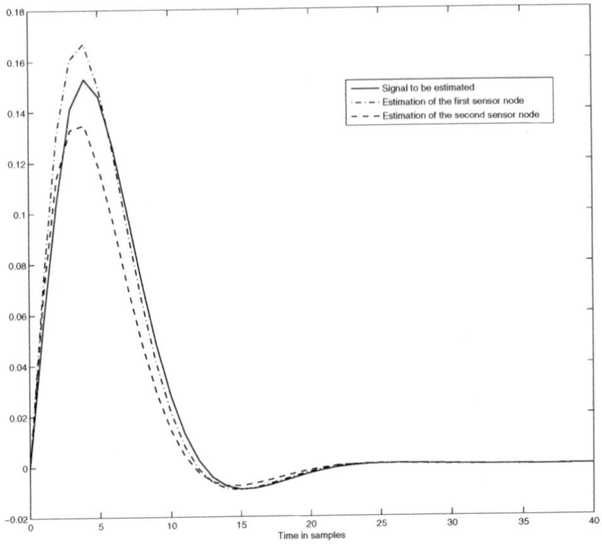

Fig. 5.3. Signal $z(k)$ and its estimation $\hat{z}_p(k)$ of two sensor nodes

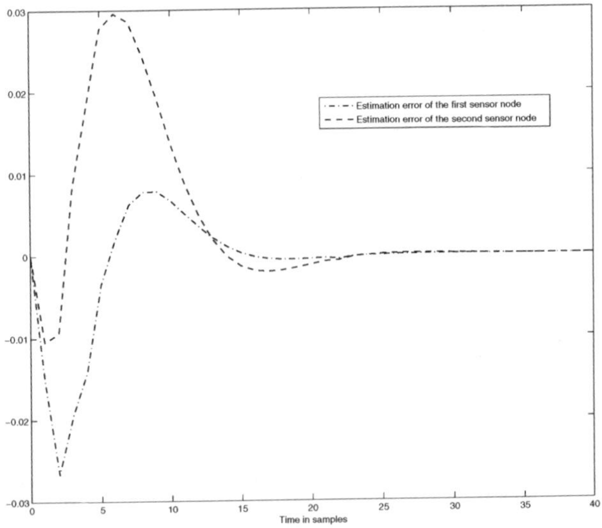

Fig. 5.4. Estimation error $e_p(k)$ of two sensor nodes

randomly between $\underline{d} = 1$ and $\bar{d} = 3$. Fig. 5.3 plots the signal $z(k)$, and its estimations $\hat{z}_p(k)$ with the pth distributed fuzzy filter. The corresponding estimation errors $e_p(k)$ are shown in Fig. 5.4.

5.5 Conclusion

In this chapter, the distributed fuzzy filtering problem has been investigated for a class of sensor networks described by discrete-time T-S fuzzy systems with time-varying delays under multiple probabilistic packet losses. The distributed fuzzy filters have been designed such that the filtering error dynamic system is mean-square stable with an average $\mathcal{H}_\infty$ performance. Based on the input-output approach and the two-term approximation idea, the application of the small scale gain theorem has been proved to be an effective technique in dealing with the time-varying delay system.

Chapter 6
Model Approximation of Discrete-Time T-S Fuzzy Time-Delay Systems

6.1 Introduction

In this chapter, the $\mathcal{H}_\infty$ model approximation problem is studied for discrete-time T-S fuzzy time-delay systems. For a given high-order T-S fuzzy system, our attention is focused on the construction of reduced-order models, which approximate the original system well in an $\mathcal{H}_\infty$ sense. To reduced the conservativeness caused by the time-delay, the delay partitioning approach is advanced to derive a delay-dependent stability condition with an $\mathcal{H}_\infty$ performance for the approximation error system. Based on which, the $\mathcal{H}_\infty$ model approximation problem is settled by applying the projection approach, which casts the model approximation into a sequential minimization problem subject to LMI constraints by employing a CCL algorithm. In addition, the $\mathcal{H}_\infty$ model approximation problems with special structures, delay-free model and zero-order model, are also obtained.

6.2 System Description and Preliminaries

Consider the following T-S fuzzy time-delay system:

♦ **Plant Form:**

Rule i: IF $\theta_1(k)$ is $\mathcal{M}_{i1}$ and $\theta_2(k)$ is $\mathcal{M}_{i2}$ and $\cdots$ and $\theta_p(k)$ is $\mathcal{M}_{ip}$, THEN

$$\begin{aligned} x(k+1) &= A_i x(k) + A_{di} x(k-d) + B_i u(k), \\ y(k) &= C_i x(k) + C_{di} x(k-d) + D_i u(k), \end{aligned}$$

where $i = 1, 2, \ldots, r$, and r is the number of IF-THEN rules; $\mathcal{M}_{ij}(i = 1, 2, \ldots, r; j = 1, 2, \ldots, p)$ are the fuzzy sets; $\theta(k) = \begin{bmatrix} \theta_1(k) \ \theta_2(k) \ \cdots \ \theta_p(k) \end{bmatrix}^T$ is the premise variable vector. $x(k) \in \mathbf{R}^n$ is the state vector; $u(k) \in \mathbf{R}^p$ is the control input; $y(k) \in \mathbf{R}^l$ is the output; and the time-delay d is positive integer and assumed to be constant in the whole dynamic process. The delay d can

L. Wu et al., *Fuzzy Control Systems with Time-Delay and Stochastic Perturbation*, Studies in Systems, Decision and Control 12, DOI: 10.1007/978-3-319-11316-6_6

always be described by $d = \tau m$, where τ and m are integers. A_i, A_{di}, B_i, C_i, C_{di} and D_i are known real constant matrices with appropriate dimensions.

Given a pair of $(x(k), u(k))$, a more compact presentation of the discrete-time T-S fuzzy time-delay model can be given by

$$x(k+1) = \sum_{i=1}^{r} h_i(\theta)\left[A_i x(k) + A_{di} x(k-d) + B_i u(k)\right], \tag{6.1a}$$

$$y(k) = \sum_{i=1}^{r} h_i(\theta)\left[C_i x(k) + C_{di} x(k-d) + D_i u(k)\right], \tag{6.1b}$$

where $h_i(\theta)$, $i = 1, 2, \ldots, r$ are the normalized membership functions, which are defined as that of (1.2) in Chapter 1.

We approximate system (6.1) by a reduced-order model described by

$$\hat{x}(k+1) = \hat{A}\hat{x}(k) + \hat{A}_d\hat{x}(k-d) + \hat{B}u(k), \tag{6.2a}$$

$$\hat{y}(k) = \hat{C}\hat{x}(k) + \hat{C}_d\hat{x}(k-d) + \hat{D}u(k), \tag{6.2b}$$

where $\hat{x}(k) \in \mathbf{R}^k$ is the state vector of the reduced-order model with $k < n$, and $\hat{A}$, $\hat{A}_d$, $\hat{B}$, $\hat{C}$, $\hat{C}_d$ and $\hat{D}$ are real matrices to be determined.

Augmenting the model of (6.1) to include the states of system (6.2), we obtain the approximation error system as

$$\xi(k+1) = \sum_{i=1}^{r} h_i(\theta)\left[\bar{A}_i \xi(k) + \bar{A}_{di}\xi(k-d) + \bar{B}_i u(k)\right], \tag{6.3a}$$

$$e(k) = \sum_{i=1}^{r} h_i(\theta)\left[\bar{C}_i \xi(k) + \bar{C}_{di}\xi(k-d) + \bar{D}_i u(k)\right], \tag{6.3b}$$

where $\xi(k) \triangleq \begin{bmatrix} x(k) \\ \hat{x}(k) \end{bmatrix}$, $e(k) \triangleq y(k) - \hat{y}(k)$ and

$$\bar{A}_i \triangleq \begin{bmatrix} A_i & 0 \\ 0 & \hat{A} \end{bmatrix}, \quad \bar{A}_{di} \triangleq \begin{bmatrix} A_{di} & 0 \\ 0 & \hat{A}_d \end{bmatrix}, \quad \bar{B}_i \triangleq \begin{bmatrix} B_i \\ \hat{B} \end{bmatrix},$$
$$\bar{C}_i \triangleq \begin{bmatrix} C_i & -\hat{C} \end{bmatrix}, \quad \bar{C}_{di} \triangleq \begin{bmatrix} C_{di} & -\hat{C}_d \end{bmatrix}, \quad \bar{D}_i \triangleq D_i - \hat{D}.$$

Moreover, we define

$$\bar{A}(k) \triangleq \sum_{i=1}^{r} h_i(\theta)\bar{A}_i, \quad \bar{A}_d(k) \triangleq \sum_{i=1}^{r} h_i(\theta)\bar{A}_{di}, \quad \bar{B}(k) \triangleq \sum_{i=1}^{r} h_i(\theta)\bar{B}_i,$$
$$\bar{C}(k) \triangleq \sum_{i=1}^{r} h_i(\theta)\bar{C}_i, \quad \bar{C}_d(k) \triangleq \sum_{i=1}^{r} h_i(\theta)\bar{C}_{di}, \quad \bar{D}(k) \triangleq \sum_{i=1}^{r} h_i(\theta)\bar{D}_i.$$

Definition 6.1. The approximation error system in (6.3) is said to be asymptotically stable if under $u(k) = 0$,

$$\lim_{k\to\infty} \|\xi(k)\| = 0.$$

Definition 6.2. Given a scalar $\gamma > 0$, the approximation error system in (6.3) is said to be asymptotically stable with an $\mathcal{H}_\infty$ performance level γ if it is asymptotically stable and $\|e(k)\|_2 < \gamma \|u(k)\|_2$ for all nonzero $u(k) \in \ell_2[0,\infty)$, where

$$\|e(k)\|_2 \triangleq \sqrt{\sum_{k=0}^{\infty} e^T(k)e(k)}.$$

Our objective in this work is to determine the matrices $\hat{A}$, $\hat{A}_d$, $\hat{B}$, $\hat{C}$, $\hat{C}_d$ and $\hat{D}$ in reduced-order model (6.2) such that the approximation error system in (6.3) is asymptotically stable with a guaranteed $\mathcal{H}_\infty$ performance level γ.

Assumption 6.1 *System (6.1a) with $u(k) = 0$ is asymptotically stable.*

When system (6.1a) with $u(k) = 0$ is asymptotically stable, it follows that $e(k) \in \ell_2[0,\infty)$ when $u(k) \in \ell_2[0,\infty)$. Assumption 6.1 is made based on the fact that there is no control in system (6.1). Therefore, the original system to be approximated has to be asymptotically stable, which is a prerequisite for the approximation error system in (6.3) to be asymptotically stable.

Lemma 6.3. [62] *Let $W = W^T \in \mathbf{R}^{n\times n}$, $U \in \mathbf{R}^{n\times m}$ and $V \in \mathbf{R}^{k\times n}$ be given matrices, and suppose that* rank$(U) < n$ *and* rank$(V) < n$. *Consider the problem of finding some matrix $\mathcal{G}$ satisfying*

$$W + U\mathcal{G}V + (U\mathcal{G}V)^T < 0. \tag{6.4}$$

Then, (6.4) is solvable for $\mathcal{G}$ if and only if

$$U^{\perp}WU^{\perp T} < 0, \quad V^{T\perp}WV^{T\perp T} < 0. \tag{6.5}$$

Furthermore, if (6.5) holds, then all the solutions of $\mathcal{G}$ are given by

$$\mathcal{G} = U_R^{+}\Psi V_L^{+} + \Phi - U_R^{+}U_R\Phi V_L V_L^{+},$$

with

$$\begin{cases} \Psi = \Pi^{-1}\Xi^{1/2}L(V_R\Lambda V_R^T)^{-1/2} - \Pi^{-1}U_L^T\Lambda V_R^T(V_R\Lambda V_R^T)^{-1}, \\ \Lambda = (U_L\Pi^{-1}U_L^T - W)^{-1} > 0, \\ \Xi = \Pi - U_L^T(\Lambda - \Lambda V_R^T(V_R\Lambda V_R^T)^{-1}V_R\Lambda)U_L > 0, \end{cases}$$

where Φ, Π and L are any appropriately dimensioned matrices satisfying $\Pi > 0$ and $\|L\| < 1$.

6.3 Main Results

6.3.1 Performance Analysis

In this section, by applying the delay partitioning approach, we derive an LMI formulation of $\mathcal{H}_\infty$ performance for the approximation error system in (6.3). To this end, we set $d = \tau m$ and construct the following fuzzy LKF:

$$V(k) \triangleq \xi^T(k)P\xi(k) + \sum_{\alpha=0}^{m-1}\sum_{l=k-\tau}^{k-1} \xi^T(l-\alpha\tau)Q_\alpha(l)\xi(l-\alpha\tau) + \sum_{s=-\tau}^{-1}\sum_{l=k+s}^{k-1} \delta^T(l)Z^{-1}(l)\delta(l), \tag{6.6}$$

where $\delta(l) \triangleq \xi(l+1) - \xi(l)$ and

$$P > 0, \quad Z(k) \triangleq \sum_{i=1}^{r} h_i(\theta)Z_i, \quad Q_\alpha(k) \triangleq \sum_{i=1}^{r} h_i(\theta)Q_{\alpha i}.$$

Theorem 6.4. *Given positive integers τ and m, the approximation error system in (6.3) is asymptotically stable and has an $\mathcal{H}_\infty$ performance, if there exist matrices $P > 0$, $Z_i > 0$, $Q_{\alpha i} > 0$, $X_{vi} > 0$, $R_i > 0$, and Y_{vi}, $i = 1, 2, \ldots, r$, $\alpha = 0, 1, \ldots, m-1$, $v = 0, 1, \ldots, m$, such that for $i, j = 1, 2, \ldots, r$,*

$$\begin{bmatrix} \Pi_{11i} & \Pi_{12i} & 0 & 0 & \cdots & 0 & \Pi_{16i} & \Pi_{17i} \\ \star & \Pi_{22i} & \Pi_{23i} & Y_{2i} & \cdots & Y_{(m-1)i} & Y_{mi} & 0 \\ \star & \star & \Pi_{33ij} & -Y_{2i} & \cdots & -Y_{(m-1)i} & -Y_{mi} & 0 \\ \star & \star & \star & \Pi_{44ij} & \cdots & 0 & 0 & 0 \\ \star & \star & \star & \star & \ddots & \vdots & \vdots & \vdots \\ \star & \star & \star & \star & \star & \Pi_{55ij} & 0 & 0 \\ \star & \star & \star & \star & \star & \star & \Pi_{66ij} & 0 \\ \star & \star & \star & \star & \star & \star & \star & -\gamma^2 I \end{bmatrix} < 0, \tag{6.7a}$$

$$\begin{bmatrix} X_{0i} & 0 & \cdots & 0 & Y_{0i} \\ \star & X_{1i} & \cdots & 0 & Y_{1i} \\ \star & \star & \ddots & \vdots & \vdots \\ \star & \star & \star & X_{mi} & Y_{mi} \\ \star & \star & \star & \star & R_i \end{bmatrix} \geq 0, \tag{6.7b}$$

$$Z_j - R_i < 0, \tag{6.7c}$$

where

$$\begin{aligned}
&\Pi_{11i} \triangleq \mathrm{diag}\{-P^{-1}, -\tau^{-1}Z_i, -I\}, && \Pi_{16i} \triangleq \mathrm{col}\left\{\bar{A}_{di}, \bar{A}_{di}, \bar{C}_{di}\right\}, \\
&\Pi_{12i} \triangleq \mathrm{col}\left\{\bar{A}_i, \bar{A}_i - I, \bar{C}_i\right\}, && \Pi_{17i} \triangleq \mathrm{col}\left\{\bar{B}_i, \bar{B}_i, \bar{D}_i^T\right\}, \\
&\Pi_{22i} \triangleq -P + Q_{0i} + Y_{0i} + Y_{0i}^T + \tau X_{0i}, && \Pi_{23i} \triangleq Y_{1i} - Y_{0i}^T, \\
&\Pi_{33ij} \triangleq Q_{1i} - Q_{0j} - Y_{1i} - Y_{1i}^T + \tau X_{1i}, && \Pi_{44ij} \triangleq Q_{2i} - Q_{1j} + \tau X_{2i}, \\
&\Pi_{55ij} \triangleq Q_{(m-1)i} - Q_{(m-2)j} + \tau X_{(m-1)i}, && \Pi_{66ij} \triangleq -Q_{(m-1)j} + \tau X_{mi}.
\end{aligned}$$

Proof. Based on the fuzzy basis functions and from (6.7a)–(6.7c), we have

$$\sum_{i=1}^{r}\sum_{j=1}^{r} h_i(\theta(k))h_j(\theta(k-\tau))\times$$

$$\begin{bmatrix}
\Pi_{11i} & \Pi_{12i} & 0 & 0 & \cdots & 0 & \Pi_{16i} & \Pi_{17i} \\
\star & \Pi_{22i} & \Pi_{23i} & Y_{2i} & \cdots & Y_{(m-1)i} & Y_{mi} & 0 \\
\star & \star & \Pi_{33ij} & -Y_{2i} & \cdots & -Y_{(m-1)i} & -Y_{mi} & 0 \\
\star & \star & \star & \Pi_{44ij} & \cdots & 0 & 0 & 0 \\
\star & \star & \star & \star & \ddots & \vdots & \vdots & \vdots \\
\star & \star & \star & \star & \star & \Pi_{55ij} & 0 & 0 \\
\star & \star & \star & \star & \star & \star & \Pi_{66ij} & 0 \\
\star & \star & \star & \star & \star & \star & \star & -\gamma^2 I
\end{bmatrix} < 0,$$

$$\sum_{i=1}^{r} h_i(\theta(k)) \begin{bmatrix}
X_{0i} & 0 & \cdots & 0 & Y_{0i} \\
\star & X_{1i} & \cdots & 0 & Y_{1i} \\
\star & \star & \ddots & \vdots & \vdots \\
\star & \star & \star & X_{mi} & Y_{mi} \\
\star & \star & \star & \star & R_i
\end{bmatrix} \geq 0,$$

$$\sum_{i=1}^{r} h_i(\theta(s))Z_i - \sum_{i=1}^{r} h_i(\theta(k))R_i < 0.$$

A more compact presentation of the above inequalities is given by

$$\Pi(k) < 0, \tag{6.8}$$

$$\begin{bmatrix}
X_0(k) & 0 & \cdots & 0 & Y_0(k) \\
\star & X_1(k) & \cdots & 0 & Y_1(k) \\
\star & \star & \ddots & \vdots & \vdots \\
\star & \star & \star & X_m(k) & Y_m(k) \\
\star & \star & \star & \star & R(k)
\end{bmatrix} \geq 0, \tag{6.9}$$

$$Z(s) - R(k) < 0, \tag{6.10}$$

where

$$\Pi(k) \triangleq \begin{bmatrix} \Pi_{11}(k) & \Pi_{12}(k) & 0 & 0 & \cdots & 0 & \Pi_{16}(k) & \Pi_{17}(k) \\ \star & \Pi_{22}(k) & \Pi_{23}(k) & Y_2(k) & \cdots & Y_{m-1}(k) & Y_m(k) & 0 \\ \star & \star & \Pi_{33}(k) & -Y_2(k) & \cdots & -Y_{m-1}(k) & -Y_m(k) & 0 \\ \star & \star & \star & \Pi_{44}(k) & \cdots & 0 & 0 & 0 \\ \star & \star & \star & \star & \ddots & \vdots & \vdots & \vdots \\ \star & \star & \star & \star & \star & \Pi_{55}(k) & 0 & 0 \\ \star & \star & \star & \star & \star & \star & \Pi_{66}(k) & 0 \\ \star & \star & \star & \star & \star & \star & \star & -\gamma^2 I \end{bmatrix},$$

with

$$\begin{aligned} \Pi_{11}(k) &\triangleq \operatorname{diag}\{-P^{-1}, -\tau^{-1}Z(k), -I\}, \\ \Pi_{12}(k) &\triangleq \operatorname{col}\left\{\bar{A}(k), \bar{A}(k) - I, \bar{C}(k)\right\}, \\ \Pi_{16}(k) &\triangleq \operatorname{col}\left\{\bar{A}_d(k), \bar{A}_d(k), \bar{C}_d(k)\right\}, \\ \Pi_{17}(k) &\triangleq \operatorname{col}\left\{\bar{B}(k), \bar{B}(k), \bar{D}(k)\right\}, \\ \Pi_{22}(k) &\triangleq -P + Q_0(k) + Y_0(k) + Y_0^T(k) + \tau X_0(k), \\ \Pi_{23}(k) &\triangleq Y_1(k) - Y_0^T(k), \\ \Pi_{33}(k) &\triangleq Q_1(k) - Q_0(k-\tau) - Y_1(k) - Y_1^T(k) + \tau X_1(k), \\ \Pi_{44}(k) &\triangleq Q_2(k) - Q_1(k-\tau) + \tau X_2(k), \\ \Pi_{55}(k) &\triangleq Q_{m-1}(k) - Q_{m-2}(k-\tau) + \tau X_{m-1}(k), \\ \Pi_{66}(k) &\triangleq -Q_{m-1}(k-\tau) + \tau X_m(k). \end{aligned}$$

By Schur complement, inequality (6.8) implies

$$\hat{\Pi}(k) \triangleq \begin{bmatrix} \hat{\Pi}_{22}(k) & \Pi_{23}(k) & Y_2(k) & \cdots & Y_{m-1}(k) & \hat{\Pi}_{26}(k) & \hat{\Pi}_{27}(k) \\ \star & \Pi_{33}(k) & -Y_2(k) & \cdots & -Y_{m-1}(k) & -Y_m(k) & 0 \\ \star & \star & \Pi_{44}(k) & \cdots & 0 & 0 & 0 \\ \star & \star & \star & \ddots & \vdots & \vdots & \vdots \\ \star & \star & \star & \star & \Pi_{55}(k) & 0 & 0 \\ \star & \star & \star & \star & \star & \hat{\Pi}_{66}(k) & \hat{\Pi}_{67}(k) \\ \star & \star & \star & \star & \star & \star & \hat{\Pi}_{77}(k) \end{bmatrix} < 0, \quad (6.11)$$

where

$$\begin{aligned} \hat{\Pi}_{22}(k) &\triangleq \bar{A}^T(k)P\bar{A}(k) + \tau\left[\bar{A}(k) - I\right]^T Z^{-1}(k)\left[\bar{A}(k) - I\right] + \bar{C}^T(k)\bar{C}(k) + \Pi_{22}(k), \\ \hat{\Pi}_{26}(k) &\triangleq \bar{A}^T(k)P\bar{A}_d(k) + \tau\left[\bar{A}(k) - I\right]^T Z^{-1}(k)\bar{A}_d(k) + \bar{C}^T(k)\bar{C}_d(k) + Y_m(k), \\ \hat{\Pi}_{27}(k) &\triangleq \bar{A}^T(k)P\bar{B}(k) + \tau[\bar{A}(k) - I]^T Z^{-1}(k)\bar{B}(k) + \bar{C}^T(k)\bar{D}(k), \\ \hat{\Pi}_{66}(k) &\triangleq \bar{A}_d^T(k)P\bar{A}_d(k) + \tau\bar{A}_d^T(k)Z^{-1}(k)\bar{A}_d(k) + \bar{C}_d^T(k)\bar{C}_d(k) + \Pi_{66}(k), \\ \hat{\Pi}_{67}(k) &\triangleq \bar{A}_d^T(k)P\bar{B}(k) + \tau\bar{A}_d^T(k)Z^{-1}(k)\bar{B}(k) + \bar{C}_d^T(k)\bar{D}(k), \\ \hat{\Pi}_{77}(k) &\triangleq \bar{B}^T(k)P\bar{B}(k) + \tau\bar{B}^T(k)Z^{-1}(k)\bar{B}(k) + \bar{D}^T(k)\bar{D}(k) - \gamma^2 I. \end{aligned}$$

Consider the LKF in (6.6), and then along the trajectories of the approximation error system in (6.3), we have

$$\begin{aligned}\Delta V(k) &\triangleq V(k+1)-V(k),\\ &\leq \xi^T(k+1)P\xi(k+1)-\xi^T(k)P\xi(k)+\sum_{\alpha=0}^{m-1}\xi^T(k-\alpha\tau)Q_\alpha(k)\xi(k-\alpha\tau)\\ &\quad-\sum_{\alpha=0}^{m-1}\xi^T(k-(\alpha+1)\tau)Q_N(k-\tau)\xi(k-(\alpha+1)\tau)\\ &\quad+\tau\delta^T(k)Z^{-1}(k)\delta(k)-\sum_{s=k-\tau}^{k-1}\delta^T(s)R^{-1}(k)\delta(s),\end{aligned} \tag{6.12}$$

where

$$R(k) \triangleq \sum_{i=1}^{r} h_i(\theta)R_i, \quad R_i > 0, \quad i=1,2,\ldots,r.$$

Moreover, according to the definition of $\delta(l)$ and for any matrix $Y(k)$, the following equation holds:

$$2\eta^T(k)Y(k)\left[\xi(k)-\xi(k-\tau)-\sum_{s=k-\tau}^{k-1}\delta(s)\right]=0, \tag{6.13}$$

where

$$\eta(k) \triangleq \begin{bmatrix}\xi(k)\\ \xi(k-\tau)\\ \vdots\\ \xi(k-(m-1)\tau)\\ \xi(k-d)\end{bmatrix}, \quad Y(k) \triangleq \begin{bmatrix}Y_0^T(k)\\ Y_1^T(k)\\ \vdots\\ Y_m^T(k)\end{bmatrix}, \quad \begin{array}{l}Y_v(k) \triangleq \sum_{i=1}^{r} h_i(\theta)Y_{vi},\\ v=0,1,\ldots,m.\end{array}$$

Moreover, for any appropriately dimensioned matrices

$$\begin{aligned}X(k) &\triangleq \operatorname{diag}\{X_0(k),X_1(k),\ldots,X_m(k)\}>0,\\ X_v(k) &\triangleq \sum_{i=1}^{r} h_i(\theta)X_{vi}, \quad X_{vi}>0,\ v=0,1,\ldots,m,\end{aligned}$$

the following holds:

$$\tau\eta^T(k)X(k)\eta(k)-\sum_{s=k-\tau}^{k-1}\eta^T(k)X(k)\eta(k)=0. \tag{6.14}$$

Therefore, from (6.12)–(6.14) we have

$$\Delta V(k) \leq -\sum_{s=k-\tau}^{k-1} \begin{bmatrix} \eta(k) \\ \delta(s) \end{bmatrix}^T \Psi(k) \begin{bmatrix} \eta(k) \\ \delta(s) \end{bmatrix} + \begin{bmatrix} \eta(k) \\ u(k) \end{bmatrix}^T \tilde{\Pi}(k) \begin{bmatrix} \eta(k) \\ u(k) \end{bmatrix}, \quad (6.15)$$

where

$$\tilde{\Pi}(k) \triangleq \begin{bmatrix} \tilde{\Pi}_{22}(k) & \Pi_{23}(k) & Y_2(k) & \cdots & Y_{m-1}(k) & \tilde{\Pi}_{26}(k) & \tilde{\Pi}_{27}(k) \\ \star & \Pi_{33}(k) & -Y_2(k) & \cdots & -Y_{m-1}(k) & -Y_m(k) & 0 \\ \star & \star & \Pi_{44}(k) & \cdots & 0 & 0 & 0 \\ \star & \star & \star & \ddots & \vdots & \vdots & \vdots \\ \star & \star & \star & \star & \Pi_{55}(k) & 0 & 0 \\ \star & \star & \star & \star & \star & \tilde{\Pi}_{66}(k) & \tilde{\Pi}_{67}(k) \\ \star & \star & \star & \star & \star & \star & \tilde{\Pi}_{77}(k) \end{bmatrix},$$

with

$$\begin{aligned} \tilde{\Pi}_{22}(k) &\triangleq \bar{A}^T(k)P\bar{A}(k) + \tau[\bar{A}(k) - I]^T Z^{-1}(k)[\bar{A}(k) - I] + \bar{\Pi}_{22}(k), \\ \tilde{\Pi}_{26}(k) &\triangleq \bar{A}^T(k)P\bar{A}_d(k) + \tau\left[\bar{A}(k) - I\right]^T Z^{-1}(k)\bar{A}_d(k) + Y_m(k), \\ \tilde{\Pi}_{27}(k) &\triangleq \bar{A}^T(k)P\bar{B}(k) + \tau\left[\bar{A}(k) - I\right]^T Z^{-1}(k)\bar{B}(k), \\ \tilde{\Pi}_{66}(k) &\triangleq \bar{A}_d^T(k)P\bar{A}_d(k) + \tau\bar{A}_d^T(k)Z^{-1}(k)\bar{A}_d(k) + \bar{\Pi}_{66}(k), \\ \tilde{\Pi}_{67}(k) &\triangleq \bar{A}_d^T(k)P\bar{B}(k) + \tau\bar{A}_d^T(k)Z^{-1}(k)\bar{B}(k), \\ \tilde{\Pi}_{77}(k) &\triangleq \bar{B}^T(k)P\bar{B}(k) + \tau\bar{B}^T(k)Z^{-1}(k)\bar{B}(k). \end{aligned}$$

Based on (6.9) and (6.11), we can conclude that $\Delta V(k) < 0$ in (6.15).

To establish the stability of system (6.3), assume $u(k) = 0$, then we have

$$\Delta V(k) \leq -\sum_{s=k-\tau}^{k-1} \begin{bmatrix} \eta(k) \\ \delta(s) \end{bmatrix}^T \Psi(k) \begin{bmatrix} \eta(k) \\ \delta(s) \end{bmatrix} + \eta^T(k)\breve{\Pi}(k)\eta(k), \quad (6.16)$$

where

$$\breve{\Pi}(k) \triangleq \begin{bmatrix} \tilde{\Pi}_{22}(k) & \Pi_{23}(k) & Y_2(k) & \cdots & Y_{m-1}(k) & \tilde{\Pi}_{26}(k) \\ \star & \Pi_{33}(k) & -Y_2(k) & \cdots & -Y_{m-1}(k) & -Y_m(k) \\ \star & \star & \Pi_{44}(k) & \cdots & 0 & 0 \\ \star & \star & \star & \ddots & \vdots & \vdots \\ \star & \star & \star & \star & \Pi_{55}(k) & 0 \\ \star & \star & \star & \star & \star & \tilde{\Pi}_{66}(k) \end{bmatrix}.$$

Obviously, (6.9) and (6.11) assure $\Delta V(k) < 0$ in (6.16), thus the approximation error system in (6.3) is asymptotically stable.

Now, to establish the $\mathcal{H}_\infty$ performance, we consider the following index:

$$\mathcal{J} \triangleq \sum_{k=0}^{\infty} \left[e^T(k)e(k) - \gamma^2 u^T(k)u(k)\right].$$

Under zero initial condition, $V(k)|_{k=0} = 0$. Then, considering $\Delta V(k) < 0$ in (6.15), we have

$$\begin{aligned}
\mathcal{J} &\leqslant \sum_{k=0}^{\infty} \left[e^T(k)e(k) - \gamma^2 u^T(k)u(k) \right] + V(k)|_{k=\infty} - V(k)|_{k=0}, \\
&= \sum_{k=0}^{\infty} \left[e^T(k)e(k) - \gamma^2 u^T(k)u(k) + \Delta V(k) \right], \\
&= \sum_{k=0}^{\infty} \begin{bmatrix} \eta(k) \\ u(k) \end{bmatrix}^T \hat{\Pi}(k) \begin{bmatrix} \eta(k) \\ u(k) \end{bmatrix} - \sum_{s=k-\tau}^{k-1} \begin{bmatrix} \eta(k) \\ \delta(s) \end{bmatrix}^T \Psi(k) \begin{bmatrix} \eta(k) \\ \delta(s) \end{bmatrix}.
\end{aligned}$$

Therefore, for all nonzero $u(k) \in \ell_2[0, \infty)$, we have $\mathcal{J} < 0$, which implies $\|e(k)\|_2 < \|u(k)\|_2$, thus the proof is completed. ■

6.3.2 Model Approximation

Now, we present a solution to the $\mathcal{H}_\infty$ model approximation problem.

Theorem 6.5. *Given positive integers τ and m, the approximation error system in (6.3) is asymptotically stable and has an $\mathcal{H}_\infty$ performance, if there exist matrices $P > 0$, $\mathscr{P} > 0$, $Z_i > 0$, $Q_{\alpha i} > 0$, $R_i > 0$, $X_{\upsilon i}$ and $Y_{\upsilon i}$, $i = 1, 2, \ldots, r$, $\alpha = 0, 1, \ldots, m-1$, $\upsilon = 0, 1, \ldots, m$, such that (6.7b)–(6.7c) and the followings hold for $i = 1, 2, \ldots, r$,*

$$\begin{bmatrix}
\bar{\Pi}_{11i} & \bar{\Pi}_{12i} & 0 & 0 & \cdots & 0 & \bar{\Pi}_{16i} & \bar{\Pi}_{17i} & \bar{\Pi}_{18i} \\
\star & \Pi_{22i} & \Pi_{23i} & Y_{2i} & \cdots & Y_{(m-1)i} & Y_{mi} & 0 & J^T \\
\star & \star & \Pi_{33ij} & -Y_{2i} & \cdots & -Y_{(m-1)i} & -Y_{mi} & 0 & 0 \\
\star & \star & \star & \Pi_{44ij} & \cdots & 0 & 0 & 0 & 0 \\
\star & \star & \star & \star & \ddots & \vdots & \vdots & \vdots & \vdots \\
\star & \star & \star & \star & \star & \Pi_{55ij} & 0 & 0 & 0 \\
\star & \star & \star & \star & \star & \star & \Pi_{66ij} & 0 & 0 \\
\star & \star & \star & \star & \star & \star & \star & -\gamma^2 I & 0 \\
\star & \star & \star & \star & \star & \star & \star & \star & \bar{\Pi}_{88i}
\end{bmatrix} < 0, \quad (6.17a)$$

$$\begin{bmatrix}
\hat{\Pi}_{11i} & \hat{\Pi}_{12i} & 0 & 0 & \cdots & 0 & \hat{\Pi}_{16i} \\
\star & H\Pi_{22i}H^T & H\Pi_{23i} & HY_{2i} & \cdots & HY_{(m-1)i} & HY_{mi}H^T \\
\star & \star & \Pi_{33ij} & -Y_{2i} & \cdots & -Y_{(m-1)i} & -Y_{mi}H^T \\
\star & \star & \star & \Pi_{44ij} & \cdots & 0 & 0 \\
\star & \star & \star & \star & \ddots & \vdots & \vdots \\
\star & \star & \star & \star & \star & \Pi_{55ij} & 0 \\
\star & \star & \star & \star & \star & \star & H\Pi_{66ij}H^T
\end{bmatrix} < 0, \quad (6.17b)$$

$$P\mathscr{P} = I, \quad (6.17c)$$

where

$$\begin{aligned}
\bar{\Pi}_{11i} &\triangleq \mathrm{diag}\{-H\mathscr{P}H^T, -\tau^{-1}HZ_iH^T\},\\
\bar{\Pi}_{12i} &\triangleq \mathrm{col}\left\{H\bar{A}_{i0}, H\bar{A}_{i0} - H\right\},\\
\bar{\Pi}_{16i} &\triangleq \mathrm{col}\left\{H\bar{A}_{di0}, H\bar{A}_{di0}\right\},\\
\bar{\Pi}_{17i} &\triangleq \mathrm{col}\left\{H\bar{B}_{i0}, H\bar{B}_{i0}\right\},\\
\bar{\Pi}_{18i} &\triangleq \mathrm{col}\left\{-H\mathscr{P}J^T, \tau^{-1}HZ_iJ^T\right\},\\
\bar{\Pi}_{88i} &\triangleq -J\left(\mathscr{P} + \tau^{-1}Z_i\right)J^T,\\
\hat{\Pi}_{11i} &\triangleq \mathrm{diag}\{-\mathscr{P}, -\tau^{-1}Z_i, -I\},\\
\hat{\Pi}_{12i} &\triangleq \mathrm{col}\left\{\bar{A}_{i0}H^T, \bar{A}_{i0}H^T - H^T, \bar{C}_{i0}H^T\right\},\\
\hat{\Pi}_{16i} &\triangleq \mathrm{col}\left\{\bar{A}_{di0}H^T, \bar{A}_{di0}H^T, \bar{C}_{di0}H^T\right\},
\end{aligned}$$

with

$$\begin{aligned}
&\bar{A}_{i0} \triangleq \begin{bmatrix} A_i & 0_{n\times k} \\ 0_{k\times n} & 0_{k\times k} \end{bmatrix}, \bar{A}_{di0} \triangleq \begin{bmatrix} A_{di} & 0_{n\times k} \\ 0_{k\times n} & 0_{k\times k} \end{bmatrix}, \bar{B}_{i0} \triangleq \begin{bmatrix} B_i \\ 0_{k\times p} \end{bmatrix},\\
&\bar{C}_{i0} \triangleq \begin{bmatrix} C_i & 0_{l\times k} \end{bmatrix}, \quad \bar{C}_{di0} \triangleq \begin{bmatrix} C_{di} & 0_{l\times k} \end{bmatrix}, \quad \bar{D}_{i0} \triangleq D_i.
\end{aligned} \tag{6.18}$$

Moveover, the system matrices of an admissible $\mathcal{H}_\infty$ reduced-order model in the form of (6.2) are given by

$$\mathcal{G} \triangleq \begin{bmatrix} \hat{A} & \hat{A}_d & \hat{B} \\ \hat{C} & \hat{C}_d & \hat{D} \end{bmatrix}, \tag{6.19}$$

where

$$\begin{cases}
\mathcal{G} = -\Pi^{-1}U_L^T\Lambda V_R^T(V_R\Lambda V_R^T)^{-1} + \Pi^{-1}\Xi^{1/2}L(V_R\Lambda V_R^T)^{-1/2},\\
\Lambda = (U_L\Pi^{-1}U_L^T - W)^{-1} > 0,\\
\Xi = \Pi - U_L^T(\Lambda - \Lambda V_R^T(V_R\Lambda V_R^T)^{-1}V_R\Lambda)U_L,
\end{cases}$$

where Π and L are any appropriately dimensioned matrices satisfying $\Pi > 0$ and $\|L\| < 1$ and

$$\begin{aligned}
&H \triangleq \begin{bmatrix} I_{n\times n} & 0_{n\times k} \end{bmatrix}, \; J \triangleq \begin{bmatrix} 0_{k\times n} & I_{k\times k} \end{bmatrix}, \; F \triangleq \begin{bmatrix} 0_{l\times k} & -I_{l\times l} \end{bmatrix},\\
&R \triangleq \begin{bmatrix} 0_{k\times n} & I_{k\times k} \\ 0_{k\times n} & 0_{k\times k} \\ 0_{p\times n} & 0_{p\times k} \end{bmatrix}, S \triangleq \begin{bmatrix} 0_{k\times n} & 0_{k\times k} \\ 0_{k\times n} & I_{k\times k} \\ 0_{p\times n} & 0_{p\times k} \end{bmatrix}, T \triangleq \begin{bmatrix} 0_{k\times p} \\ 0_{k\times p} \\ I_{p\times p} \end{bmatrix}, E \triangleq \begin{bmatrix} 0_{n\times k} & 0_{n\times l} \\ I_{k\times k} & 0_{k\times l} \end{bmatrix},
\end{aligned}$$

$$W \triangleq \begin{bmatrix} \Pi_{11i} & \breve{\Pi}_{12i} & 0 & 0 & \cdots & 0 & \breve{\Pi}_{16i} & \breve{\Pi}_{17i} \\ \star & \Pi_{22i} & \Pi_{23i} & Y_{2i} & \cdots & Y_{(m-1)i} & Y_{mi} & 0 \\ \star & \star & \Pi_{33ij} & -Y_{2i} & \cdots & -Y_{(m-1)i} & -Y_{mi} & 0 \\ \star & \star & \star & \Pi_{44ij} & \cdots & 0 & 0 & 0 \\ \star & \star & \star & \star & \ddots & \vdots & \vdots & \vdots \\ \star & \star & \star & \star & \star & \Pi_{55ij} & 0 & 0 \\ \star & \star & \star & \star & \star & \star & \Pi_{66ij} & 0 \\ \star & \star & \star & \star & \star & \star & \star & -\gamma^2 I \end{bmatrix},$$

$$U \triangleq \begin{bmatrix} E \\ E \\ F \\ 0_{(n+k)\times(l+k)} \\ 0_{(n+k)\times(l+k)} \\ 0_{(n+k)\times(l+k)} \\ \vdots \\ 0_{(n+k)\times(l+k)} \\ 0_{(n+k)\times(l+k)} \\ 0_{p\times(l+k)} \end{bmatrix}, V \triangleq \begin{bmatrix} 0_{(n+k)\times(2k+p)} \\ 0_{(n+k)\times(2k+p)} \\ 0_{l\times(2k+p)} \\ R^T \\ 0_{(n+k)\times(2k+p)} \\ 0_{(n+k)\times(2k+p)} \\ \vdots \\ 0_{(n+k)\times(2k+p)} \\ S^T \\ T^T \end{bmatrix}^T, \quad \breve{\Pi}_{12i} \triangleq \begin{bmatrix} \bar{A}_{i0} \\ \bar{A}_{i0} - I \\ \bar{C}_{i0} \end{bmatrix}, \quad \breve{\Pi}_{16i} \triangleq \begin{bmatrix} \bar{A}_{di0} \\ \bar{A}_{di0} - I \\ \bar{C}_{di0} \end{bmatrix}, \quad \breve{\Pi}_{17i} \triangleq \begin{bmatrix} \bar{B}_{i0} \\ \bar{B}_{i0} \\ \bar{D}_{i0} \end{bmatrix}. \tag{6.20}$$

Proof. Rewrite $\bar{A}_i$, $\bar{A}_{di}$, $\bar{B}_i$, $\bar{C}_i$, $\bar{C}_{di}$ and $\bar{D}_i$ in the form of

$$\begin{cases} \bar{A}_i \triangleq \bar{A}_{i0} + E\mathcal{G}R, & \bar{A}_{di} \triangleq \bar{A}_{di0} + E\mathcal{G}S, \\ \bar{C}_i \triangleq \bar{C}_{i0} + F\mathcal{G}R, & \bar{C}_{di} \triangleq \bar{C}_{di0} + F\mathcal{G}S, \\ \bar{B}_i \triangleq \bar{B}_{i0} + E\mathcal{G}T, & \bar{D}_i \triangleq \bar{D}_{i0} + F\mathcal{G}T, \end{cases} \tag{6.21}$$

where $\bar{A}_{i0}$, $\bar{A}_{di0}$, $\bar{B}_{i0}$, $\bar{C}_{i0}$, $\bar{C}_{di0}$, $\bar{D}_{i0}$, E, F, R, S and T are defined in (6.18) and (6.20). By considering (6.21), thus (6.7a) can be rewritten as

$$W + U\mathcal{G}V + (U\mathcal{G}V)^T < 0, \tag{6.22}$$

where W, U and V are defined in (6.20). We choose

$$U^{\perp} \triangleq \begin{bmatrix} H & 0 & 0 & 0 & 0 & 0 & \cdots & 0 & 0 & 0 \\ 0 & H & 0 & 0 & 0 & 0 & \cdots & 0 & 0 & 0 \\ 0 & 0 & 0 & I & 0 & 0 & \cdots & 0 & 0 & 0 \\ 0 & 0 & 0 & 0 & I & 0 & \cdots & 0 & 0 & 0 \\ 0 & 0 & 0 & 0 & 0 & I & \cdots & 0 & 0 & 0 \\ 0 & 0 & 0 & 0 & 0 & 0 & \cdots & 0 & 0 & 0 \\ \vdots & \vdots & \vdots & \vdots & \vdots & \vdots & \ddots & \vdots & \vdots & \vdots \\ 0 & 0 & 0 & 0 & 0 & 0 & \cdots & 0 & I & 0 \\ 0 & 0 & 0 & 0 & 0 & 0 & \cdots & 0 & 0 & I \\ J & -J & 0 & 0 & 0 & 0 & \cdots & 0 & 0 & 0 \end{bmatrix}, \quad V^{T\perp} \triangleq \begin{bmatrix} I & 0 & 0 & 0 & 0 & 0 & \cdots & 0 & 0 & 0 \\ 0 & I & 0 & 0 & 0 & 0 & \cdots & 0 & 0 & 0 \\ 0 & 0 & I & 0 & 0 & 0 & \cdots & 0 & 0 & 0 \\ 0 & 0 & 0 & H & 0 & 0 & \cdots & 0 & 0 & 0 \\ 0 & 0 & 0 & 0 & I & 0 & \cdots & 0 & 0 & 0 \\ 0 & 0 & 0 & 0 & 0 & I & \cdots & 0 & 0 & 0 \\ \vdots & \vdots & \vdots & \vdots & \vdots & \vdots & \ddots & \vdots & \vdots & \vdots \\ 0 & 0 & 0 & 0 & 0 & 0 & \cdots & I & 0 & 0 \\ 0 & 0 & 0 & 0 & 0 & 0 & \cdots & 0 & H & 0 \end{bmatrix}, \tag{6.23}$$

where H and J are defined in (6.20). By Lemma 6.3, inequality (6.22) is solvable for $\mathcal{G}$ if and only if

$$\Gamma \triangleq U^{\perp} W U^{\perp T} < 0, \quad \Upsilon \triangleq V^{T\perp} W V^{T\perp T} < 0, \tag{6.24}$$

where

$$\Gamma \triangleq \begin{bmatrix} \breve{\Pi}_{11i} & \bar{\Pi}_{12i} & 0 & 0 & \cdots & 0 & \bar{\Pi}_{16i} & \bar{\Pi}_{17i} & \bar{\Pi}_{18i} \\ \star & \Pi_{22i} & \Pi_{23i} & Y_{2i} & \cdots & Y_{(m-1)i} & Y_{mi} & 0 & J^T \\ \star & \star & \Pi_{33ij} & -Y_{2i} & \cdots & -Y_{(m-1)i} & -Y_{mi} & 0 & 0 \\ \star & \star & \star & \Pi_{44ij} & \cdots & 0 & 0 & 0 & 0 \\ \star & \star & \star & \star & \ddots & \vdots & \vdots & \vdots & \vdots \\ \star & \star & \star & \star & \star & \Pi_{55ij} & 0 & 0 & 0 \\ \star & \star & \star & \star & \star & \star & \Pi_{66ij} & 0 & 0 \\ \star & \star & \star & \star & \star & \star & \star & -\gamma^2 I & 0 \\ \star & \star & \star & \star & \star & \star & \star & \star & \bar{\Pi}_{88i} \end{bmatrix},$$

$$\Upsilon \triangleq \begin{bmatrix} \hat{\Pi}_{11i} & \hat{\Pi}_{12i} & 0 & 0 & \cdots & 0 & \hat{\Pi}_{16i} \\ \star & H\Pi_{22i}H^T & H\Pi_{23i} & HY_{2i} & \cdots & HY_{(m-1)i} & HY_{mi}H^T \\ \star & \star & \Pi_{33ij} & -Y_{2i} & \cdots & -Y_{(m-1)i} & -Y_{mi}H^T \\ \star & \star & \star & \Pi_{44ij} & \cdots & 0 & 0 \\ \star & \star & \star & \star & \ddots & \vdots & \vdots \\ \star & \star & \star & \star & \star & \Pi_{55ij} & 0 \\ \star & \star & \star & \star & \star & \star & H\Pi_{66ij}H^T \end{bmatrix},$$

with $\breve{\Pi}_{11i} \triangleq \text{diag}\left\{-HP^{-1}H^T, -\tau^{-1}HZ_iH^T\right\}$.

By noting $\mathscr{P} \triangleq P^{-1}$, it follows that (6.24) implies (6.17a) and (6.17b). The second part of the theorem is immediate by applying Lemma 6.3, and the proof is completed. ■

Notice that the conditions in Theorem 6.5 are not all in LMI form due to the matrix equality (6.17c). We suggest the following minimization problem involving LMI conditions instead of the original nonconvex feasible problem formulated in Theorem 6.5.

Problem $\mathcal{H}_\infty$-MRTSFS ($\mathcal{H}_\infty$ Model Approximation for T-S Fuzzy Systems):

$$\min \ \text{trace}(P\mathscr{P}) \ \text{subject to (6.17a)–(6.17b) and}$$
$$\begin{bmatrix} P & I \\ I & \mathscr{P} \end{bmatrix} \geq 0. \tag{6.25}$$

Suggest the following algorithm to solve the above minimization problem.

Algorithm $\mathcal{H}_\infty$-MRTSFS

Step 1. Find a feasible set $\left(P^{(0)}, \mathscr{P}^{(0)}\right)$ satisfying (6.17a)–(6.17b) and (6.25). Set $\kappa = 0$.

Step 2. Solve the following optimization problem:

$$\begin{aligned} \min \quad & \text{trace}\left(P^{(\kappa)}\mathscr{P} + P\mathscr{P}^{(\kappa)}\right) \\ \text{subject to} \quad & \text{(6.17a)–(6.17b) and (6.25)} \end{aligned}$$

and denote f^* to be the optimized value.

Step 3. Substitute the obtained matrix variables $(P, \mathscr{P})$ into (6.24). If (6.24) is satisfied, with

$$|f^* - 2(n+k)| < \delta,$$

for a sufficiently small scalar $\delta > 0$, then output the feasible solutions $(P, \mathscr{P})$. EXIT.

Step 4. If $\kappa > \mathbb{N}$ where $\mathbb{N}$ is the maximum number of iterations allowed, EXIT.

Step 5. Set $\kappa = \kappa + 1$, $(P^{\kappa}, \mathscr{P}^{\kappa}) = (P, \mathscr{P})$, and go to Step 2.

6.4 Special Cases

6.4.1 Delay-Free Model Case

We will now further extend the results obtained in the above sections, to consider the problem of $\mathcal{H}_\infty$ model approximation by delay-free reduced-order model, that is, we use the following reduced-order model to approximate system (6.1):

$$\hat{x}(k+1) = \hat{A}\hat{x}(k) + \hat{B}u(k), \tag{6.26a}$$
$$\hat{y}(k) = \hat{C}\hat{x}(k) + \hat{D}u(k). \tag{6.26b}$$

Then, the approximation error system is also given by (6.3) with

$$\begin{cases} \bar{A}_i \triangleq \begin{bmatrix} A_i & 0 \\ 0 & \hat{A} \end{bmatrix}, \quad \bar{A}_{di} \triangleq \begin{bmatrix} A_{di} & 0 \\ 0 & 0 \end{bmatrix}, \quad \bar{B}_i \triangleq \begin{bmatrix} B_i \\ \hat{B} \end{bmatrix}, \\ \bar{C}_i \triangleq \begin{bmatrix} C_i & -\hat{C} \end{bmatrix}, \quad \bar{C}_{di} \triangleq \begin{bmatrix} C_{di} & 0 \end{bmatrix}, \quad \bar{D}_i \triangleq D_i - \hat{D}. \end{cases} \tag{6.27}$$

In the following, we will present the result for the delay-free model case.

Theorem 6.6. *Consider the approximation error system in (6.3). An admissible delay-free $\mathcal{H}_\infty$ reduced-order model in the form of (6.26) exists if there exist matrices $P > 0$, $\mathscr{P} > 0$, $Z_i > 0$, $Q_{\alpha i} > 0$, $X_{vi} > 0$, $R_i > 0$, and Y_{vi} ($i = 1, 2, \ldots, r$; $\alpha = 0, 1, \ldots, m-1$; $v = 0, 1, \ldots, m$), satisfying (6.7b)–(6.7c), (6.17a), (6.17c) and*

$$\begin{bmatrix} \hat{\Pi}_{11i} & \hat{\Pi}_{12i} & 0 & 0 & \cdots & 0 & \breve{\Pi}_{16i} \\ \star & H\Pi_{22i}H^T & H\Pi_{23i} & HY_{2i} & \cdots & HY_{(m-1)i} & HY_{mi} \\ \star & \star & \Pi_{33ij} & -Y_{2i} & \cdots & -Y_{(m-1)i} & -Y_{mi}H^T \\ \star & \star & \star & \Pi_{44ij} & \cdots & 0 & 0 \\ \star & \star & \star & \star & \ddots & \vdots & \vdots \\ \star & \star & \star & \star & \star & \Pi_{55ij} & 0 \\ \star & \star & \star & \star & \star & \star & \Pi_{66ij} \end{bmatrix} < 0. \quad (6.28)$$

Moreover, if the above conditions have a feasible solution then the parameter matrices of an admissible delay-free $\mathcal{H}_\infty$ reduced-order model in the form of (6.26) can be solved by

$$\mathcal{G}_f \triangleq \begin{bmatrix} \hat{A} & \hat{B} \\ \hat{C} & \hat{D} \end{bmatrix},$$

where

$$\begin{cases} \mathcal{G}_f = -\Pi^{-1}U_L^T\Lambda V_R^T(V_R\Lambda V_R^T)^{-1} + \Pi^{-1}\Xi^{1/2}L(V_R\Lambda V_R^T)^{-1/2}, \\ \Lambda = (U_L\Pi^{-1}U_L^T - W)^{-1} > 0, \\ \Xi = \Pi - U_L^T(\Lambda - \Lambda V_R^T(V_R\Lambda V_R^T)^{-1}V_R\Lambda)U_L, \end{cases}$$

where $\bar{A}_{i0}$, $\bar{A}_{di0}$, $\bar{B}_{i0}$, $\bar{C}_{i0}$, $\bar{C}_{di0}$, $\bar{D}_{i0}$ and H are given in (6.18) and (6.20), Π and L are any appropriately dimensioned matrices satisfying $\Pi > 0$ and $\|L\| < 1$. W, U, V, E and F are defined in (6.20) with matrices given by

$$R \triangleq \begin{bmatrix} 0_{k\times n} & I_{k\times k} \\ 0_{p\times n} & 0_{p\times k} \end{bmatrix}, \quad S \triangleq \begin{bmatrix} 0_{k\times n} & 0_{k\times k} \\ 0_{p\times n} & 0_{p\times k} \end{bmatrix}, \quad T \triangleq \begin{bmatrix} 0_{k\times p} \\ I_{p\times p} \end{bmatrix}. \quad (6.29)$$

Proof. The desired result can be obtained by following similar lines as in the proof of Theorem 6.5. Rewrite (6.27) in the form of

$$\begin{cases} \bar{A}_i \triangleq \bar{A}_{0i} + E\mathcal{G}_fR, & \bar{A}_{di} \triangleq \bar{A}_{d0i} + E\mathcal{G}_fS, \\ \bar{C}_i \triangleq \bar{C}_{0i} + F\mathcal{G}_fR, & \bar{C}_{di} \triangleq \bar{C}_{d0i} + F\mathcal{G}_fS, \\ \bar{B}_i \triangleq \bar{B}_{0i} + E\mathcal{G}_fT, & \bar{D}_i \triangleq \bar{D}_{0i} + F\mathcal{G}_fT, \end{cases} \quad (6.30)$$

where $\bar{A}_{0i}$, $\bar{A}_{d0i}$, $\bar{B}_{0i}$, $\bar{C}_{0i}$, $\bar{C}_{d0i}$, $\bar{D}_{0i}$, E, F, R, S and T are defined in (6.19), (6.20) and (6.29). By considering (6.30), thus (6.7a) can be rewritten as

$$W + U\mathcal{G}_fV + (U\mathcal{G}_fV)^T < 0, \quad (6.31)$$

where W, U and V are defined in (6.20). In addition, $U^\perp$ is given in (6.23) and

$$V^{T\perp} \triangleq \begin{bmatrix} I\,0\,0 & 0 & 0\,0 & \cdots & 0\,0\,0 \\ 0\,I\,0 & 0 & 0\,0 & \cdots & 0\,0\,0 \\ 0\,0\,I & 0 & 0\,0 & \cdots & 0\,0\,0 \\ 0\,0\,0 & H & 0\,0 & \cdots & 0\,0\,0 \\ 0\,0\,0 & 0 & I\,0 & \cdots & 0\,0\,0 \\ 0\,0\,0 & 0 & 0\,I & \cdots & 0\,0\,0 \\ \vdots & \vdots & \vdots & \ddots & \vdots \\ 0\,0\,0 & 0 & 0\,0 & \cdots & I\,0\,0 \\ 0\,0\,0 & 0 & 0\,0 & \cdots & 0\,I\,0 \end{bmatrix}, \tag{6.32}$$

where H is defined in (6.20). Then, by Lemma 6.3, inequality (6.31) is solvable for $\mathcal{G}_f$ if and only if $\Gamma < 0$ (given in (6.24)) and

$$\begin{bmatrix} \Pi_{11i} & \hat{\Pi}_{12i} & 0 & 0 & \cdots & 0 & \breve{\Pi}_{16i} \\ \star & H\Pi_{22i}H^T & H\Pi_{23i} & HY_{2i} & \cdots & HY_{(m-1)i} & HY_{mi} \\ \star & \star & \Pi_{33ij} & -Y_{2i} & \cdots & -Y_{(m-1)i} & -Y_{mi}H^T \\ \star & \star & \star & \Pi_{44ij} & \cdots & 0 & 0 \\ \star & \star & \star & \star & \ddots & \vdots & \vdots \\ \star & \star & \star & \star & \star & \Pi_{55ij} & 0 \\ \star & \star & \star & \star & \star & \star & \Pi_{66ij} \end{bmatrix} < 0. \tag{6.33}$$

By defining $\mathscr{P} \triangleq P^{-1}$ in (6.24) and (6.33), we readily obtain (6.17a), (6.17c) and (6.28). The second part of the theorem is true by applying Lemma 6.3, and the proof is completed. ∎

6.4.2 Zero-Order Model Case

The problem of finding a zero-order model to approximate the original system has also been considered in some prior works, see for example, [64, 107]. More specially, the zero-order model approximation problem is to approximate system (6.1) by a zero-order system described by

$$\hat{y}(k) = \hat{D}u(k). \tag{6.34}$$

In this case, the corresponding approximation error system is given by

$$x(k+1) = \sum_{i=1}^{r} h_i(\theta)\left[A_i x(k) + A_{di}x(k-d) + B_i u(k)\right], \tag{6.35a}$$

$$e(k) = \sum_{i=1}^{r} h_i(\theta)\left[C_i x(k) + C_{di}x(k-d) + (D_i - \hat{D})u(k)\right]. \tag{6.35b}$$

The following theorem provides a solution to the zero-order model design.

Theorem 6.7. *Consider system (6.1). An admissible zero-order $\mathcal{H}_\infty$ reduced-order model in the form of (6.34) exists if there exist matrices $P > 0$, $Z_i > 0$, $Q_{\alpha i} > 0$, $X_{vi} > 0$, $R_i > 0$, Y_{vi} and $\mathcal{N}$ satisfying (6.7b)–(6.7c) and*

$$\begin{bmatrix} -P & 0 & 0 & PA_i & 0 & 0 & \cdots & 0 & PA_{di} & PB_i \\ \star & -\tau^{-1}Z_i & 0 & A_i - I & 0 & 0 & \cdots & 0 & A_{di} & B_i \\ \star & \star & -I & C_i & 0 & 0 & \cdots & 0 & C_{di} & D_i - \mathcal{N} \\ \star & \star & \star & \Pi_{22i} & \Pi_{23i} & Y_{2i} & \cdots & Y_{(m-1)i} & Y_{mi} & 0 \\ \star & \star & \star & \star & \Pi_{33ij} & -Y_{2i} & \cdots & -Y_{(m-1)i} & -Y_{mi} & 0 \\ \star & \star & \star & \star & \star & \Pi_{44ij} & \cdots & 0 & 0 & 0 \\ & & & & & & & \vdots & \vdots & \vdots \\ \star & \star & \star & \star & \star & \star & \ddots & & & \\ \star & \star & \star & \star & \star & \star & \star & \Pi_{55ij} & 0 & 0 \\ \star & \star & \star & \star & \star & \star & \star & \star & \Pi_{66ij} & 0 \\ \star & \star & \star & \star & \star & \star & \star & \star & \star & -\gamma^2 I \end{bmatrix} < 0.$$

Moreover, if the above condition has a feasible solution then the parameter matrices of an admissible zero-order $\mathcal{H}_\infty$ reduced-order model in the form of (6.34) is given by $\hat{D} = \mathcal{N}$.

With the previous results, the proof of Theorem 6.7 can be carried out in a straightforward way, thus the detailed proof is omitted.

6.5 Illustrative Example

Example 6.8. Consider system (6.1) with the parameters given as follows:

$$A_1 = \begin{bmatrix} 0.1612 & 0.0574 & -0.0144 & 0.1846 \\ 0.0434 & -0.3638 & 0.5258 & -0.0357 \\ -0.0747 & -0.3146 & -0.0487 & -0.1043 \\ -0.1664 & 0.4031 & 0.0347 & 0.2864 \end{bmatrix}, \quad B_1 = \begin{bmatrix} 0.2023 \\ -0.2313 \\ -0.1137 \\ 0.1279 \end{bmatrix},$$

$$A_2 = \begin{bmatrix} 0.1312 & 0.0474 & -0.0044 & 0.1546 \\ 0.0234 & -0.3018 & 0.4258 & -0.0357 \\ -0.0554 & -0.2421 & -0.0367 & -0.0843 \\ -0.1551 & 0.3031 & 0.0247 & 0.1864 \end{bmatrix}, \quad B_2 = \begin{bmatrix} 0.0123 \\ -0.1313 \\ -0.1138 \\ 0.1179 \end{bmatrix},$$

$$A_{d1} = \begin{bmatrix} 0.2 & 0.1 & 0 & 0 \\ 0 & 0.2 & 0 & 0 \\ 0 & 0 & 0.2 & 0.1 \\ 0 & 0 & 0 & 0.2 \end{bmatrix}, \quad A_{d2} = \begin{bmatrix} 0.1 & 0.05 & 0 & 0 \\ 0 & 0.1 & 0 & 0 \\ 0 & 0 & 0.1 & 0.05 \\ 0 & 0 & 0 & 0.1 \end{bmatrix},$$

$$C_1 = \begin{bmatrix} 1.4419 & 0.6720 & 0.1387 & -0.8595 \end{bmatrix}, \quad D_1 = 1,$$

$$C_2 = \begin{bmatrix} 1.3329 & 0.6720 & 0.1387 & -0.8478 \end{bmatrix}, \quad D_2 = 0.5,$$

$$C_{d1} = \begin{bmatrix} 0.2 & 0.5 & 0.1 & 0.9 \end{bmatrix}, \quad C_{d2} = \begin{bmatrix} 0.1 & 0.25 & 0.05 & 0.45 \end{bmatrix}.$$

It is easy to verify by Theorem 6.4 that this system is asymptotically stable when the delay size $d = 3$. Here we are interested in finding one-order, two-order and three-order systems in the form of (6.2) to approximate the above system. By solving the non-convex feasibility problem in Theorem 6.5 with the application of Algorithm $\mathcal{H}_\infty$-MRTSFS, the obtained results for different cases are as follows.

Case 1. (with $k = 3$, in this case we obtain $\gamma_{\min} = 0.3114$):

$$\begin{aligned}
\hat{A} &= \begin{bmatrix} -0.0034 & -0.3343 & 0.1511 \\ -0.4011 & -0.0432 & 0.1776 \\ 0.2308 & 0.2251 & 0.2405 \end{bmatrix}, \quad \hat{B} = \begin{bmatrix} -0.0705 \\ -0.0626 \\ 0.0245 \end{bmatrix}, \\
\hat{A}_d &= \begin{bmatrix} 0.1768 & 0.1598 & -0.0634 \\ 0.1535 & 0.1398 & -0.0556 \\ -0.0579 & -0.0536 & 0.0215 \end{bmatrix}, \\
\hat{C} &= \begin{bmatrix} 0.3527 & 0.4341 & -0.2672 \end{bmatrix}, \\
\hat{C}_d &= \begin{bmatrix} -0.0977 & -0.1056 & 0.0484 \end{bmatrix}, \quad \hat{D} = 0.9872.
\end{aligned} \tag{6.36}$$

Case 2. (with $k = 2$, in this case we obtain $\gamma_{\min} = 0.3224$):

$$\begin{aligned}
\hat{A} &= \begin{bmatrix} -0.3647 & 0.3637 \\ 0.3671 & 0.1567 \end{bmatrix}, \quad \hat{A}_d = \begin{bmatrix} 0.2280 & -0.1159 \\ -0.1157 & 0.0588 \end{bmatrix}, \\
\hat{B} &= \begin{bmatrix} 0.0809 \\ -0.0415 \end{bmatrix}, \quad \hat{D} = 0.9857, \\
\hat{C} &= \begin{bmatrix} -0.5367 & 0.2773 \end{bmatrix}, \quad \hat{C}_d = \begin{bmatrix} 0.1346 & -0.0617 \end{bmatrix}.
\end{aligned} \tag{6.37}$$

Case 3. (with $k = 1$, in this case we obtain $\gamma_{\min} = 0.3391$):

$$\begin{aligned}
\hat{A} &= -0.5433, \quad \hat{A}_d = 0.2255, \quad \hat{B} = -0.0853, \\
\hat{C} &= 0.6211, \quad \hat{C}_d = -0.0855, \quad \hat{D} = 0.9900.
\end{aligned} \tag{6.38}$$

To further illustrate the effectiveness of the obtained reduced-order models, in the following, we will show the approximation performance of the obtained reduced-order systems. To this end, let the initial condition be zero ,that is, $x(0) = 0$ ($\hat{x}(0) = 0$), the membership functions are

$$h_1(x_1(k)) = \frac{1 - \sin(x_1(k))}{2}, \quad h_2(x_1(k)) = \frac{1 + \sin(x_1(k))}{2},$$

and the input $u(k)$ is

$$u(k) = \begin{cases} e^{(-0.2k+1)} \sin(0.2k), & 10 \le k \le 50, \\ 0.3\cos(0.25k), & 80 \le k \le 120, \\ 0, & \text{otherwise}. \end{cases}$$

Fig. 6.1 depicts the output trajectories of the original system (solid line), the three-order reduced model (6.36) (dashed line), the two-order reduced model (6.37) (dash-dot line) and the one-order reduced model (6.38) (dotted line) due to the above input signal. The corresponding output errors between the original system and the reduced models are shown in Fig. 6.2.

The output error energy, denoted by $\mathbf{E}(k)$, is the extraction of the summation of $e^T(k)e(k)$, that is, $\mathbf{E}(k) \triangleq \sqrt{\sum_{s=0}^{k} e^T(s)e(s)}$. The input energy, denoted by $\mathbf{U}(k)$, is the extraction of the summation of $u^T(k)u(k)$, that is, $\mathbf{U}(k) \triangleq \sqrt{\sum_{s=0}^{k} u^T(s)u(s)}$. Moveover, we use $\mathbf{Y}(k)$ to denote the ratio between the output error energy $\mathbf{E}(k)$ and the input energy $\mathbf{U}(k)$, that is, $\mathbf{Y}(k) = \mathbf{E}(k)/\mathbf{U}(k)$. Fig. 6.3 shows the values of $\mathbf{Y}(k)$ of the approximation error system with the different reduced-order models.

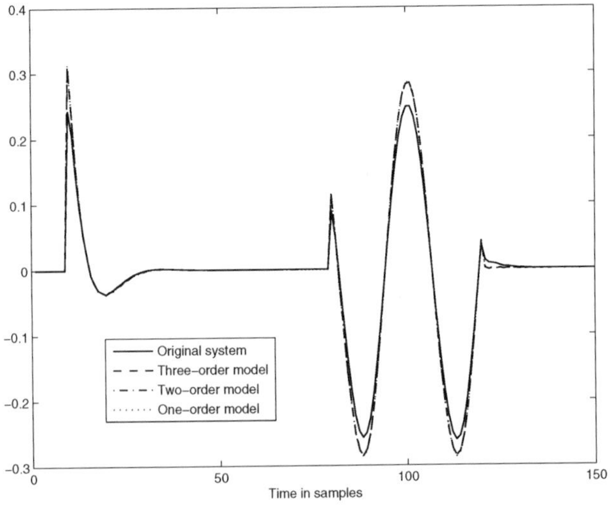

Fig. 6.1. Outputs of the original system and the reduced-order models

Example 6.9. Consider the following Henon system with time-delay:

$$\begin{aligned} x_1(k+1) &= -\left[\mathcal{C}x_1(k) + (1-\mathcal{C})x_1(k-d)\right]^2 + 0.3x_2(k) + u(k), \\ x_2(k+1) &= \mathcal{C}x_1(k) + (1-\mathcal{C})x_1(k-d), \\ y(k) &= \mathcal{C}x_1(k) + (1-\mathcal{C})x_1(k-d) + u(k), \end{aligned}$$

where $u(k)$ is the control input, and the constant $\mathcal{C} \in [0, 1]$ is the retarded coefficient.

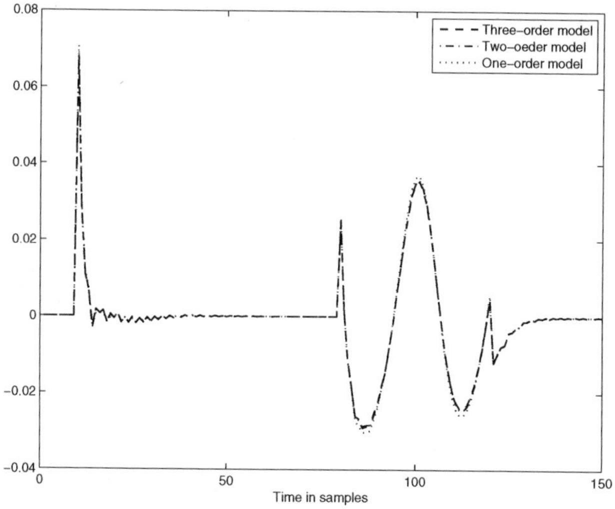

Fig. 6.2. Output errors between the original system and the reduced-order models

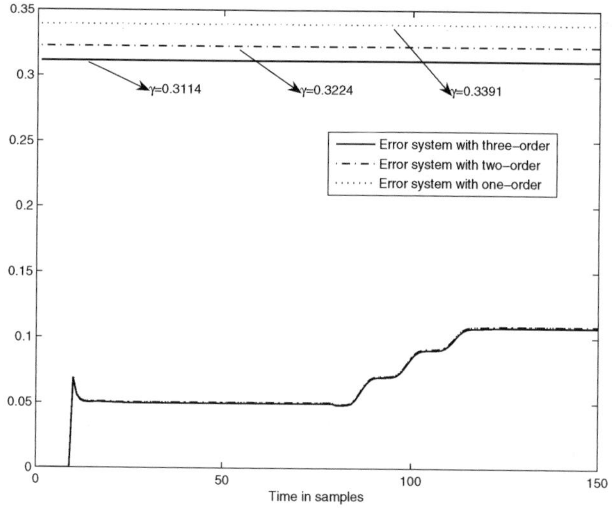

Fig. 6.3. Ratios of the error energy to the input energy for the error systems with different reduced-order models

Let $\theta(k) = \mathcal{C}x_1(k)+(1-\mathcal{C})x_1(k-d)$, and assume $\theta(k) \in [-\mathcal{M}, \mathcal{M}]$, $\mathcal{M} > 0$. By the same procedure as in [194], the nonlinear term $\theta^2(k)$ can be exactly represented as

$$\theta^2(k) = h_1(\theta(k))(-\mathcal{M})\theta(k) + h_2(\theta(k))\mathcal{M}\theta(k),$$

where $h_1(\theta(k)), h_2(\theta(k)) \in [0, 1]$, and $h_1(\theta(k)) + h_2(\theta(k)) = 1$. By solving the equations, the membership functions $h_1(\theta(k))$ and $h_2(\theta(k))$ are obtained as

$$h_1(\theta(k)) = \frac{1}{2}\left(1 - \frac{\theta(k)}{\mathcal{M}}\right), \quad h_2(\theta(k)) = \frac{1}{2}\left(1 + \frac{\theta(k)}{\mathcal{M}}\right).$$

It can be seen from the aforementioned expressions that when $\theta(k)$ is $-\mathcal{M}$ we know that $h_1(\theta(k)) = 1$ and $h_2(\theta(k)) = 0$, and when $\theta(k)$ is $\mathcal{M}$ we have that $h_1(\theta(k)) = 0$ and $h_2(\theta(k)) = 1$. Then, the above nonlinear system can be approximately represented by the following T-S fuzzy model:

♦ **Plant Form:**

Rule 1: IF $\theta(k)$ is $-\mathcal{M}$, THEN

$$\begin{aligned} x(k+1) &= A_1x(k) + A_{d1}x(k-d) + B_1u(k), \\ y(k) &= C_1x(k) + C_{d1}(x-d) + D_1u(k), \end{aligned}$$

Rule 2: IF $\theta(k)$ is $\mathcal{M}$, THEN

$$\begin{aligned} x(k+1) &= A_2x(k) + A_{d2}x(k-d(k)) + B_2u(k), \\ y(k) &= C_2x(k) + C_{d2}(x-d) + D_2u(k), \end{aligned}$$

where

$$\begin{aligned} A_1 &= \begin{bmatrix} \mathcal{C}\mathcal{M} & 0.3 \\ \mathcal{C} & 0 \end{bmatrix}, \quad A_{d1} = \begin{bmatrix} (1-\mathcal{C})\mathcal{M} & 0 \\ 1-\mathcal{C} & 0 \end{bmatrix}, \quad B_1 = \begin{bmatrix} 1 \\ 0 \end{bmatrix}, \\ A_2 &= \begin{bmatrix} -\mathcal{C}\mathcal{M} & 0.3 \\ \mathcal{C} & 0 \end{bmatrix}, \quad A_{d2} = \begin{bmatrix} -(1-\mathcal{C})\mathcal{M} & 0 \\ 1-\mathcal{C} & 0 \end{bmatrix}, \quad B_2 = \begin{bmatrix} 1 \\ 0 \end{bmatrix}, \\ C_1 &= \begin{bmatrix} \mathcal{C} & 0 \end{bmatrix}, \quad C_{d1} = \begin{bmatrix} 1-\mathcal{C} & 0 \end{bmatrix}, \quad D_1 = 1, \\ C_2 &= \begin{bmatrix} \mathcal{C} & 0 \end{bmatrix}, \quad C_{d2} = \begin{bmatrix} 1-\mathcal{C} & 0 \end{bmatrix}, \quad D_2 = 1. \end{aligned}$$

In this example, $x(k) = \begin{bmatrix} x_1(k) \\ x_2(k) \end{bmatrix}$, $\mathcal{C} = 0.8$, $\mathcal{M} = 0.2$ and $d = 3$ represents a time state delay, then by solving the conditions in Theorem 6.7, we obtain the minimum $\mathcal{H}_\infty$ level $\gamma_{\min} = 0.4638$ with the corresponding feedthrough matrix given by $\hat{D} = 0.9975$.

In the following, we will present the simulation result to show the special model approximation performance. Let the initial condition be zero, that is, $x(0) = 0$ and $\hat{x}(0) = 0$, the input $u(k)$ be

$$u(k) = \begin{cases} e^{(-0.2k+1)}\sin(0.8k), & 10 \le k \le 50, \\ 0.5\sin(0.81k), & 80 \le k \le 120, \\ 0, & \text{otherwise.} \end{cases}$$

Fig. 6.4 depicts the output trajectories of the original nonlinear system (solid line) and the zero-order model (dashed line) due to the above input signal.

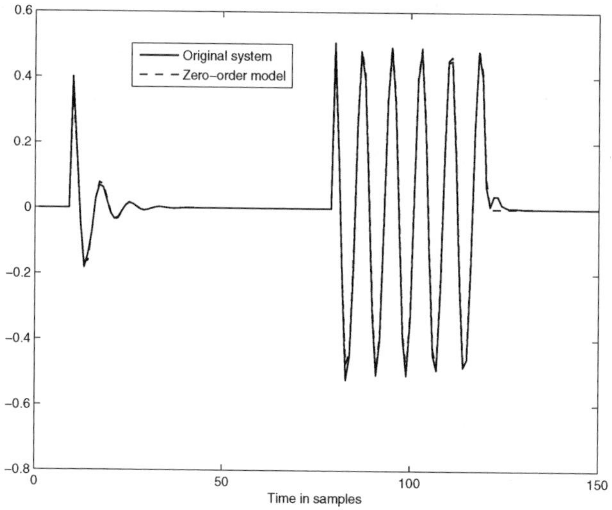

Fig. 6.4. Outputs of the original system and the zero-order model

6.6 Conclusion

In this chapter, the $\mathcal{H}_\infty$ model approximation problem has been investigated for discrete-time T-S fuzzy time-delay system. By employing the delay partitioning approach, a delay-dependent sufficient condition has been proposed in terms of LMIs for the asymptotic stability with an $\mathcal{H}_\infty$ error performance of the approximation error system. The solvability condition for the reduced-order model has been established by using the projection approach, which casts the model approximation into a sequential minimization problem subject to LMI constraints by employing the CCL algorithm. Moreover, we have further extended the results to two special structures, that is, the delay-free model case and the zero-order model case. Two numerical examples have been provided to demonstrate the effectiveness of the proposed methods.

Part II
Analysis and Synthesis of T-S Fuzzy Stochastic Systems

Chapter 7
Stability and Stabilization of Discrete-Time T-S Fuzzy Stochastic Systems

7.1 Introduction

In this chapter, we intend to investigate the stability and stabilization problems for discrete-time T-S fuzzy stochastic systems with time-varying delay. For a given T-S fuzzy stochastic system, our attention is focused on obtaining the sufficient conditions assuring its mean-square asymptotic stability, and designing the stabilization controller for those unstable systems. Employing the novel idea of delay-partitioning technique, we first get the parameterized LMIs by constructing basis-dependent LKF, and then transform those parameterized LMIs into strict LMIs which can be directly solved by computer software. Finally, those stability conditions are extended to settle stabilization problem via the non-PDC scheme, and the proposed theory will be applied to stabilize an inverted pendulum system.

7.2 System Description and Preliminaries

Consider the following discrete-time T-S fuzzy stochastic time-delay system:

♦ **Plant Form:**

Rule i: IF $\theta_1(k)$ is $\mathcal{M}_{i1}$ and $\theta_2(k)$ is $\mathcal{M}_{i2}$ and … and $\theta_p(k)$ is $\mathcal{M}_{ip}$, THEN

$$\begin{aligned} x(k+1) &= A_i x(k) + A_{di} x(k-d(k)) + B_{1i} u(k) \\ &\quad + \left[E_i x(k) + E_{di} x(k-d(k)) + B_{2i} u(k)\right] \varpi(k), \qquad (7.1\text{a}) \\ x(k) &= \phi(k), \quad k = -d_2, -d_2+1, \ldots, 0, \qquad (7.1\text{b}) \end{aligned}$$

where $i = 1, 2, \ldots, r$, and r is the number of IF-THEN rules; $\mathcal{M}_{ij} (i = 1, 2, \ldots, r; j = 1, 2, \ldots, p)$ are the fuzzy sets; $\theta(k) = \left[\theta_1(k)\ \theta_2(k)\ \cdots\ \theta_p(k)\right]^T$ is the premise variable vector. where $x(k) \in \mathbf{R}^n$ is the state vector; $u(k) \in \mathbf{R}^p$ is the control input; $\varpi(k)$ is a scalar Brownian motion, which is independent and satisfies $\mathbf{E}\{\varpi(k)\} = 0$ and $\mathbf{E}\{\varpi^2(k)\} = 1$; $d(k)$ is the time-delay, which is

L. Wu et al., *Fuzzy Control Systems with Time-Delay and Stochastic Perturbation*, Studies in Systems, Decision and Control 12, DOI: 10.1007/978-3-319-11316-6_7

a positive integer and satisfies $1 \leq d_1 \leq d(k) \leq d_2$, where d_1 and d_2 are positive constants representing the lower and upper bounds, respectively. Clearly, $d_1 = d_2$ means that $d(k)$ is time-invariant. $\phi(k)$ is the initial condition sequence. A_i, A_{di}, B_{1i}, E_i, E_{di} and B_{2i} are known real constant matrices with appropriate dimensions.

It is assumed that the premise variables do not depend on input variable $u(k)$ explicitly. Then the defuzzified output of the T-S fuzzy system (7.1) can be represented as

$$\begin{aligned} x(k+1) &= \sum_{i=1}^{r} h_i(\theta) \left[A_i x(k) + A_{di} x(k - d(k)) + B_{1i} u(k)\right] \\ &\quad + \sum_{i=1}^{r} h_i(\theta) \left[E_i x(k) + E_{di} x(k - d(k)) + B_{2i} u(k)\right] \varpi(k), \end{aligned} \tag{7.2}$$

where $h_i(\theta)$, $i = 1, 2, \ldots, r$ are the normalized membership functions, which are defined as that of (1.2) in Chapter 1.

The open-loop system of (7.2) in a compact form is presented as

$$\begin{aligned} x(k+1) &= \bar{A}(k)x(k) + \bar{A}_d(k)x(k - d(k)) \\ &\quad + \left[\bar{E}(k)x(k) + \bar{E}_d(k)x(k - d(k))\right] \varpi(k), \end{aligned} \tag{7.3}$$

where

$$\begin{aligned} \bar{A}(k) &\triangleq \sum_{i=1}^{r} h_i(\theta) A_i, & \bar{A}_d(k) &\triangleq \sum_{i=1}^{r} h_i(\theta) A_{di}, \\ \bar{E}(k) &\triangleq \sum_{i=1}^{r} h_i(\theta) E_i, & \bar{E}_d(k) &\triangleq \sum_{i=1}^{r} h_i(\theta) E_{di}. \end{aligned}$$

Now, consider the following fuzzy control:

♦ **Controller Form:**

Rule i: IF $\theta_1(k)$ is $\mathcal{M}_{i1}$ and $\theta_2(k)$ is $\mathcal{M}_{i2}$ and ... and $\theta_p(k)$ is $\mathcal{M}_{ip}$, THEN

$$u(k) = K_i x(k), \quad i = 1, 2, \ldots, r,$$

where K_i is the gain matrix of the state feedback controller in each rule, and the state feedback controller in (7.2) is given by

$$u(k) = \sum_{i=1}^{r} h_i(\theta) K_i x(k). \tag{7.4}$$

Under control of (7.4), the closed-loop system is obtained as

$$x(k+1) = \sum_{i=1}^{r}\sum_{j=1}^{r} h_i(\theta)h_j(\theta)A_{ij}x(k) + \sum_{i=1}^{r} h_i(\theta)A_{di}x(k-d(k))$$
$$+ \left[\sum_{i=1}^{r}\sum_{j=1}^{r} h_i(\theta)h_j(\theta)E_{ij}x(k) + \sum_{i=1}^{r} h_i(\theta)E_{di}x(k-d(k))\right]\varpi(k),$$

where $A_{ij} \triangleq A_i + B_{1i}K_j$ and $E_{ij} \triangleq E_i + B_{2i}K_j$.

The compact form of the above closed-loop system can be formulated by

$$x(k+1) = \hat{A}(k)x(k) + \bar{A}_d(k)x(k-d(k))$$
$$+ \left[\hat{E}(k)x(k) + \bar{E}_d(k)x(k-d(k))\right]\varpi(k), \tag{7.5}$$

where

$$\hat{A}(k) \triangleq \sum_{i=1}^{r}\sum_{j=1}^{r} h_i(\theta)h_j(\theta)A_{ij}, \quad \hat{E}(k) \triangleq \sum_{i=1}^{r}\sum_{j=1}^{r} h_i(\theta)h_j(\theta)E_{ij}.$$

We introduce the following definition for system (7.2), which is essential for the main results in the sequel.

Definition 7.1. The T-S fuzzy stochastic system in (7.2) with $u(k) = 0$ is said to be stochastically stable if there exists a scalar $\sigma > 0$ such that

$$\mathbf{E}\left\{\sum_{k=0}^{\infty} \|x(k)\|^2\right\} \le \sigma \mathbf{E}\left\{\|\phi(0)\|_a^2\right\},$$

where $\varsigma(l) \triangleq \phi(l+1) - \phi(l)$ and

$$\|\phi(0)\|_a^2 \triangleq \max_{l=-d_2,-d_2+1\ldots,-1}\left\{\|\phi(0)\|^2, \|\phi(l)\|^2, \|\varsigma(l)\|^2\right\}.$$

7.3 Main Results

7.3.1 Stability Analysis

This section is concerned with the stability analysis problem. First we assume that, the lower bound of the delay d_1 can always be described by $d_1 = \tau m$, where τ and m are integers, we represent the time-delay $d(k)$ in two parts: constant part τm and time-varying part $h(k)$, that is $d(k) = \tau m + h(k)$. To facilitate the analysis, we make the following definitions:

$$W_1 \triangleq \begin{bmatrix} I_n & 0_{n\times mn} & 0_n & 0_n \end{bmatrix}, \; W_2 \triangleq \begin{bmatrix} 0_{n\times mn} & I_n & 0_n & 0_n \end{bmatrix},$$
$$W_3 \triangleq \begin{bmatrix} 0_{n\times mn} & 0_n & I_n & 0_n \end{bmatrix}, \; W_4 \triangleq \begin{bmatrix} 0_{n\times mn} & 0_n & 0_n & I_n \end{bmatrix},$$
$$W_5 \triangleq \begin{bmatrix} I_{mn} & 0_{mn\times 3n} \end{bmatrix}, \qquad W_6 \triangleq \begin{bmatrix} 0_{mn\times n} & I_{mn} & 0_{mn\times 2n} \end{bmatrix}.$$

Then we employ a novel idea of delay partitioning to obtain a less conservative stability condition. To this end, the lower delay bound $d_1 = \tau m$ is divided into m parts, and by constructing a LKF, we have the following result.

Theorem 7.2. *Given positive integers τ, m, and d_2, the fuzzy stochastic system in (7.3) is stochastically stable if there exist matrices $P_i > 0$, $Q_{1i} > 0$, $Q_{2i} > 0$, $R_i > 0$, $S_{1i} > 0$, $S_{2i} > 0$, $R_{1i} > 0$, $R_{2i} > 0$, M_i, N_i, X_i, Y_i, Z_i, G $(i = 1, 2, \ldots, r)$, and a scalar $\varepsilon > 0$, such that for any integers k and s,*

$$\begin{bmatrix} \Phi(k) + \Psi(k) + \Psi^T(k) & \Xi(k) \\ \star & -\Lambda(k) \end{bmatrix} < 0, \tag{7.6a}$$

$$\begin{bmatrix} \bar{M}(k) & \bar{X}(k) \\ \star & \varepsilon G + \varepsilon G^T - \bar{R}_1(k) \end{bmatrix} \geq 0, \tag{7.6b}$$

$$\begin{bmatrix} \bar{N}(k) & \bar{Y}(k) \\ \star & \varepsilon G + \varepsilon G^T - \bar{R}_2(k) \end{bmatrix} \geq 0, \tag{7.6c}$$

$$\begin{bmatrix} \bar{N}(k) & \bar{Z}(k) \\ \star & \varepsilon G + \varepsilon G^T - \bar{R}_2(k) \end{bmatrix} \geq 0, \tag{7.6d}$$

$$\bar{S}_1(s) - \bar{R}_1(k) < 0, \tag{7.6e}$$

$$\bar{S}_2(s) - \bar{R}_2(k) < 0, \tag{7.6f}$$

where

$$\begin{aligned}
\Phi(k) &\triangleq \varepsilon^{-2} W_1^T \bar{Q}_2(k) W_1 + \varepsilon^{-2}(d_2 - \tau m + 1) W_1^T \bar{R}(k) W_1 - W_1^T \bar{P}(k) W_1 \\
&\quad - W_4^T \bar{Q}_2(k - d_2) W_4 + \tau \mathcal{T}_\varepsilon^{-1} \bar{M}(k) \mathcal{T}_\varepsilon^{-1} - W_3^T \bar{R}(k - d(k)) W_3 \\
&\quad + \mathcal{T}_\varepsilon^{-1} W_5^T \bar{Q}_1(k) W_5 \mathcal{T}_\varepsilon^{-1} - W_6^T \bar{Q}_1(k - \tau) W_6 + (d_2 - \tau m) \mathcal{T}_\varepsilon^{-1} \bar{N}(k) \mathcal{T}_\varepsilon^{-1}, \\
\Psi(k) &\triangleq \varepsilon \mathcal{T}_\varepsilon^{-1} \begin{bmatrix} \bar{X}(k) & \bar{Y}(k) & \bar{Z}(k) \end{bmatrix} \begin{bmatrix} \varepsilon^{-1} I_n & -I_n & 0_{n \times mn} & 0_n \\ 0_{n \times mn} & I_n & -I_n & 0_n \\ 0_{n \times mn} & 0_n & I_n & -I_n \end{bmatrix}, \\
\Lambda(k) &\triangleq \operatorname{diag}\left\{ \left(G + G^T - \bar{P}(k+1)\right), \left(G + G^T - \bar{P}(k+1)\right), \left(\tau^{-1} \bar{S}_1(k)\right), \right. \\
&\quad \left. (d_2 - \tau m)^{-1} \bar{S}_2(k), \tau^{-1} \bar{S}_1(k), (d_2 - \tau m)^{-1} \bar{S}_2(k) \right\}, \\
\Xi(k) &\triangleq \begin{bmatrix} \bar{\Xi}_1^T(k) & \bar{\Xi}_2^T(k) & \bar{\Xi}_2^T(k) & \bar{\Xi}_2^T(k) & \bar{\Xi}_3^T(k) & \bar{\Xi}_3^T(k) \end{bmatrix}, \\
\bar{\Xi}_1(k) &\triangleq \begin{bmatrix} \bar{A}(k) G & 0_{n \times mn} & \varepsilon \bar{A}_d(k) G & 0_n \end{bmatrix}, \\
\bar{\Xi}_2(k) &\triangleq \begin{bmatrix} \bar{E}(k) G & 0_{n \times mn} & \varepsilon \bar{E}_d(k) G & 0_n \end{bmatrix}, \\
\bar{\Xi}_3(k) &\triangleq \begin{bmatrix} \bar{A}(k) G - G & 0_{n \times mn} & \varepsilon \bar{A}_d(k) G & 0_n \end{bmatrix}, \\
\mathcal{T}_\varepsilon &\triangleq \operatorname{diag}\{\varepsilon I_n, \underbrace{I_n, \ldots, I_n}_{m+2}\},
\end{aligned}$$

with

$$\bar{P}(k) \triangleq \sum_{i=1}^{r} h_i(\theta)P_i, \quad \bar{R}(k) \triangleq \sum_{i=1}^{r} h_i(\theta)R_i, \quad \bar{X}(k) \triangleq \sum_{i=1}^{r} h_i(\theta)X_i,$$
$$\bar{Y}(k) \triangleq \sum_{i=1}^{r} h_i(\theta)Y_i, \quad \bar{Z}(k) \triangleq \sum_{i=1}^{r} h_i(\theta)Z_i, \quad \bar{M}(k) \triangleq \sum_{i=1}^{r} h_i(\theta)M_i,$$
$$\bar{N}(k) \triangleq \sum_{i=1}^{r} h_i(\theta)N_i, \quad \bar{Q}_1(k) \triangleq \sum_{i=1}^{r} h_i(\theta)Q_{1i}, \quad \bar{Q}_2(k) \triangleq \sum_{i=1}^{r} h_i(\theta)Q_{2i},$$
$$\bar{S}_1(k) \triangleq \sum_{i=1}^{r} h_i(\theta)S_{1i}, \quad \bar{S}_2(k) \triangleq \sum_{i=1}^{r} h_i(\theta)S_{2i},$$
$$\bar{R}_1(k) \triangleq \sum_{i=1}^{r} h_i(\theta)R_{1i}, \quad \bar{R}_2(k) \triangleq \sum_{i=1}^{r} h_i(\theta)R_{2i}.$$

Proof. Defining $F \triangleq \varepsilon G$ and considering

$$\begin{aligned}
\left(\bar{P}(k+1) - G\right)^T \bar{P}^{-1}(k+1)\left(\bar{P}(k+1) - G\right) &\geq 0,\\
\left(\bar{R}_1(k) - F\right)^T \bar{R}_1^{-1}(k)\left(\bar{R}_1(k) - F\right) &\geq 0,\\
\left(\bar{R}_2(k) - F\right)^T \bar{R}_2^{-1}(k)\left(\bar{R}_2(k) - F\right) &\geq 0,
\end{aligned}$$

we have

$$\begin{aligned}
-G^T\bar{P}^{-1}(k+1)G &\leq -G - G^T + \bar{P}(k+1),\\
F^T\bar{R}_1^{-1}(k)F &\geq \varepsilon G + \varepsilon G^T - \bar{R}_1(k),\\
F^T\bar{R}_2^{-1}(k)F &\geq \varepsilon G + \varepsilon G^T - \bar{R}_2(k).
\end{aligned}$$

Then, from (7.6a)–(7.6d), we have

$$\begin{bmatrix} \Phi(k) + \Psi(k) + \Psi^T(k) & \Xi(k) \\ \star & -\bar{\Lambda}(k) \end{bmatrix} < 0, \tag{7.7a}$$

$$\begin{bmatrix} \bar{M}(k) & \bar{X}(k) \\ \star & F^T\bar{R}_1^{-1}(k)F \end{bmatrix} \geq 0, \tag{7.7b}$$

$$\begin{bmatrix} \bar{N}(k) & \bar{Y}(k) \\ \star & F^T\bar{R}_2^{-1}(k)F \end{bmatrix} \geq 0, \tag{7.7c}$$

$$\begin{bmatrix} \bar{N}(k) & \bar{Z}(k) \\ \star & F^T\bar{R}_2^{-1}(k)F \end{bmatrix} \geq 0, \tag{7.7d}$$

where

$$\begin{aligned}
\bar{\Lambda}(k) \triangleq \operatorname{diag}\{&G^T\bar{P}^{-1}(k+1)G, G^T\bar{P}^{-1}(k+1)G, \tau^{-1}\bar{S}_1(k),\\
&(d_2 - \tau m)^{-1}\bar{S}_2(k), \tau^{-1}\bar{S}_1(k), (d_2 - \tau m)^{-1}\bar{S}_2(k)\}.
\end{aligned}$$

Define the following matrices:

$$\begin{cases} \mathcal{T}_1 \triangleq \mathrm{diag}\{G, \underbrace{F, \ldots, F}_{m+2}, I_n, I_n, I_n, I_n, I_n, I_n\}, \\ \mathcal{T}_2 \triangleq \mathrm{diag}\{\underbrace{G, \ldots, G}_{m+3}, F\}, \\ \mathcal{T}_3 \triangleq \mathrm{diag}\{\underbrace{G, \ldots, G}_{m+3}\}, \\ \mathcal{T}_4 \triangleq \mathrm{diag}\{\underbrace{F, \ldots, F}_{m}\}. \end{cases} \tag{7.8}$$

It's clear from (7.6b) that $\varepsilon G + \varepsilon G^T - \bar{R}_1(k) \geq 0$. Since $\bar{R}_1(k) > 0$ and $\varepsilon > 0$, we have $G + G^T > 0$, which implies that G is nonsingular. Considering (7.8) and $F \triangleq \varepsilon G$, we know that matrices $\mathcal{T}_1$, $\mathcal{T}_2$, $\mathcal{T}_3$ and $\mathcal{T}_4$ are all nonsingular. Performing congruence transformations to (7.7a)–(7.7d) by $\mathcal{T}_1^{-1}$, $\mathcal{T}_2^{-1}$, $\mathcal{T}_2^{-1}$ and $\mathcal{T}_2^{-1}$, respectively, we have

$$\begin{bmatrix} \hat{\Phi}(k) + \hat{\Psi}(k) + \hat{\Psi}^T(k) & \hat{\Xi}(k) \\ \star & -\hat{\Lambda}(k) \end{bmatrix} < 0, \tag{7.9a}$$

$$\begin{bmatrix} \hat{M}(k) & \hat{X}(k) \\ \star & \hat{R}_1(k) \end{bmatrix} \geq 0, \tag{7.9b}$$

$$\begin{bmatrix} \hat{N}(k) & \hat{Y}(k) \\ \star & \hat{R}_2(k) \end{bmatrix} \geq 0, \tag{7.9c}$$

$$\begin{bmatrix} \hat{N}(k) & \hat{Z}(k) \\ \star & \hat{R}_2(k) \end{bmatrix} \geq 0, \tag{7.9d}$$

where

$$\begin{aligned} \hat{\Phi}(k) \triangleq & -W_1^T \hat{P}(k) W_1 + W_1^T \hat{Q}_2(k) W_1 + (d_2 - \tau m + 1) W_1^T \hat{R}(k) W_1 \\ & - W_3^T \hat{R}(k - d(k)) W_3 - W_4^T \hat{Q}_2(k - d_2) W_4 - W_6^T \hat{Q}_1(k - \tau) W_6 \\ & + W_5^T \hat{Q}_1(k) W_5 + \tau \hat{M}(k) + (d_2 - \tau m) \hat{N}(k), \\ \hat{\Psi}(k) \triangleq & \begin{bmatrix} \hat{X}(k) & \hat{Y}(k) & \hat{Z}(k) \end{bmatrix} \begin{bmatrix} I_n & -I_n & 0_{n\times mn} & 0_n \\ 0_{n\times mn} & I_n & -I_n & 0_n \\ 0_{n\times mn} & 0_n & I_n & -I_n \end{bmatrix}, \\ \hat{\Lambda}(k) \triangleq & \ \mathrm{diag}\Big\{\hat{P}^{-1}(k+1), \hat{P}^{-1}(k+1), \tau^{-1}\hat{S}_1^{-1}(k), (d_2 - \tau m)^{-1}\hat{S}_2^{-1}(k), \\ & \tau^{-1}\hat{S}_1^{-1}(k), (d_2 - \tau m)^{-1}\hat{S}_2^{-1}(k)\Big\}, \\ \hat{\Xi}(k) \triangleq & \begin{bmatrix} \hat{\Xi}_1^T(k) & \hat{\Xi}_2^T(k) & \hat{\Xi}_2^T(k) & \hat{\Xi}_2^T(k) & \hat{\Xi}_3^T(k) & \hat{\Xi}_3^T(k) \end{bmatrix}, \\ \hat{\Xi}_1(k) \triangleq & \begin{bmatrix} \bar{A}(k) & 0_{n\times mn} & \bar{A}_d(k) & 0_n \end{bmatrix}, \\ \hat{\Xi}_2(k) \triangleq & \begin{bmatrix} \bar{E}(k) & 0_{n\times mn} & \bar{E}_d(k) & 0_n \end{bmatrix}, \\ \hat{\Xi}_3(k) \triangleq & \begin{bmatrix} \bar{A}(k) - I_n & 0_{n\times mn} & \bar{A}_d(k) & 0_n \end{bmatrix}, \end{aligned}$$

with $\hat{S}_1(k) \triangleq \bar{S}_1^{-1}(k)$, $\hat{S}_2(k) \triangleq \bar{S}_2^{-1}(k)$, $\hat{R}_1(k) \triangleq \bar{R}_1^{-1}(k)$, $\hat{R}_2(k) \triangleq \bar{R}_2^{-1}(k)$ and

$$\hat{P}(k) \triangleq G^{-T}\bar{P}(k)G^{-1}, \quad \hat{Q}_1(k) \triangleq \mathcal{T}_4^{-T}\bar{Q}_1(k)\mathcal{T}_4^{-1}, \quad \hat{Q}_2(k) \triangleq F^{-T}\bar{Q}_2(k)F^{-1},$$
$$\hat{R}(k) \triangleq F^{-T}\bar{R}(k)F^{-1}, \quad \hat{M}(k) \triangleq \mathcal{T}_3^{-T}\bar{M}(k)\mathcal{T}_3^{-1}, \quad \hat{N}(k) \triangleq \mathcal{T}_3^{-T}\bar{N}(k)\mathcal{T}_3^{-1},$$
$$\hat{X}(k) \triangleq \mathcal{T}_3^{-T}\bar{X}(k)F^{-1}, \quad \hat{Y}(k) \triangleq \mathcal{T}_3^{-T}\bar{Y}(k)F^{-1}, \quad \hat{Z}(k) \triangleq \mathcal{T}_3^{-T}\bar{Z}(k)F^{-1}.$$

By Schur complement, (7.9a) can be transformed to

$$\Omega(k) < 0, \tag{7.10}$$

where

$$\begin{aligned}\Omega(k) \triangleq\ & \hat{\Phi}(k) + \hat{\Psi}(k) + \hat{\Psi}^T(k) + \hat{\Xi}_1^T(k)\hat{P}(k+1)\hat{\Xi}_1(k) \\ & + \hat{\Xi}_3^T(k)\left[\tau\hat{S}_1(k) + (d_2 - \tau m)\hat{S}_2(k)\right]\hat{\Xi}_3(k) \\ & + \hat{\Xi}_2^T(k)\left[\tau\hat{S}_1(k) + (d_2 - \tau m)\hat{S}_2(k)\right]\hat{\Xi}_2(k) + \hat{\Xi}_2^T(k)\hat{P}(k+1)\hat{\Xi}_2(k).\end{aligned}$$

Moreover, it is shown from (7.10) that there exists a scalar $\rho > 0$ such that

$$\Omega(k) < \mathrm{diag}\{-\rho I_n, \underbrace{0_n, \ldots, 0_n}_{m+2}\}. \tag{7.11}$$

Define an LKF as

$$V(k) \triangleq V_1(k) + V_2(k) + V_3(k) + V_4(k),$$

with

$$\begin{cases} V_1(k) \triangleq x^T(k)\hat{P}(k)x(k), \\ V_2(k) \triangleq \sum_{i=k-\tau}^{k-1} \Upsilon^T(i)\hat{Q}_1(i)\Upsilon(i) + \sum_{i=k-d_2}^{k-1} x^T(i)\hat{Q}_2(i)x(i), \\ V_3(k) \triangleq \sum_{j=-d_2+1}^{-\tau m+1}\sum_{i=k+j-1}^{k-1} x^T(i)\hat{R}(i)x(i), \\ V_4(k) \triangleq \sum_{i=-\tau}^{-1}\sum_{j=k+i}^{k-1} \delta^T(j)\hat{S}_1(j)\delta(j) + \sum_{i=-d_2}^{-\tau m-1}\sum_{j=k+i}^{k-1} \delta^T(j)\hat{S}_2(j)\delta(j), \\ \Upsilon(k) \triangleq \begin{bmatrix} x(k) \\ x(k-\tau) \\ \vdots \\ x(k-\tau m+\tau) \end{bmatrix}, \\ \delta(j) \triangleq x(j+1) - x(j). \end{cases}$$

By the former definition, we have

$$\begin{aligned} x(k+1) &= \hat{\Xi}_1(k)\eta(k) + \hat{\Xi}_2(k)\eta(k)\varpi(k), \\ \delta(k) &= \hat{\Xi}_3(k)\eta(k) + \hat{\Xi}_2(k)\eta(k)\varpi(k), \end{aligned}$$

where

$$\eta(k) \triangleq \begin{bmatrix} \Upsilon(k) \\ x(k-\tau m) \\ x(k-d(k)) \\ x(k-d_2) \end{bmatrix}.$$

By calculating the difference of $V(k)$ along the trajectory of system (7.3) and taking expectation, we have

$$\mathbf{E}\{\Delta V(k)\} = \sum_{i=1}^{4} \mathbf{E}\{\Delta V_i(k)\} = \sum_{i=1}^{4} \mathbf{E}\left\{V_i(k+1) - V_i(k)\right\},$$

where

$$\begin{aligned} \mathbf{E}\{\Delta V_1(k)\} &= -x^T(k)\hat{P}(k)x(k) \\ &\quad +\eta^T(k)\left[\hat{\Xi}_1^T(k)\hat{P}(k+1)\hat{\Xi}_1(k) + \hat{\Xi}_2^T(k)\hat{P}(k+1)\hat{\Xi}_2(k)\right]\eta(k) \\ &= \eta^T(k)\left[\hat{\Xi}_1^T(k)\hat{P}(k+1)\hat{\Xi}_1(k) - W_1^T\hat{P}(k)W_1 \right. \\ &\quad \left. + \hat{\Xi}_2^T(k)\hat{P}(k+1)\hat{\Xi}_2(k)\right]\eta(k), \end{aligned} \tag{7.12}$$

$$\begin{aligned} \mathbf{E}\left\{\Delta V_2(k)\right\} &= x^T(k)\hat{Q}_2(k)x(k) - x^T(k-d_2)\hat{Q}_2(k-d_2)x(k-d_2) \\ &\quad +\Upsilon^T(k)\hat{Q}_1(k)\Upsilon(k) - \Upsilon^T(k-\tau)\hat{Q}_1(k-\tau)\Upsilon(k-\tau) \\ &= \eta^T(k)\left[W_1^T\hat{Q}_2(k)W_1 - W_4^T\hat{Q}_2(k-d_2)W_4 \right. \\ &\quad \left. + W_5^T\hat{Q}_1(k)W_5 - W_6^T\hat{Q}_1(k-\tau)W_6\right]\eta(k), \end{aligned} \tag{7.13}$$

$$\begin{aligned} \mathbf{E}\left\{\Delta V_3(k)\right\} &= (d_2-\tau m+1)x^T(k)\hat{R}(k)x(k) - \sum_{i=k-d_2}^{k-\tau m} x^T(i)\hat{R}(i)x(i) \\ &\le (d_2-\tau m+1)x^T(k)\hat{R}(k)x(k) \\ &\quad -x^T(k-d(k))\hat{R}(k-d(k))x(k-d(k)) \\ &= \eta^T(k)\left[(d_2-\tau m+1)W_1^T\hat{R}(k)W_1\right]\eta(k) \\ &\quad -\eta^T(k)\left[W_3^T\hat{R}(k-d(k))W_3\right]\eta(k). \end{aligned} \tag{7.14}$$

Considering (7.6e) and (7.6f), we have $\hat{S}_1(s) > \hat{R}_1(k)$ and $\hat{S}_2(s) > \hat{R}_2(k)$, $\forall\ s, k$, then

$$\begin{aligned}
\mathbf{E}\{\Delta V_4(k)\} &= \mathbf{E}\Big\{\tau\delta^T(k)\hat{S}_1(k)\delta(k) + (d_2 - \tau m)\delta^T(k)\hat{S}_2(k)\delta(k) \\
&\quad - \sum_{j=k-\tau}^{k-1}\delta^T(j)\hat{S}_1(j)\delta(j) - \sum_{j=k-d_2}^{k-\tau m-1}\delta^T(j)\hat{S}_2(j)\delta(j)\Big\} \\
&\leq \eta^T(k)\Big\{\hat{\Xi}_3^T(k)\left[\tau\hat{S}_1(k) + (d_2 - \tau m)\hat{S}_2(k)\right]\hat{\Xi}_3^T(k) \\
&\quad + \hat{\Xi}_2^T(k)\left[\tau\hat{S}_1(k) + (d_2 - \tau m)\hat{S}_2(k)\right]\hat{\Xi}_2^T(k)\eta(k)\Big\} \\
&\quad - \mathbf{E}\Big\{\sum_{j=k-\tau}^{k-1}\delta^T(j)\hat{R}_1(k)\delta(j) + \sum_{j=k-d(k)}^{k-\tau m-1}\delta^T(j)\hat{R}_2(k)\delta(j) \\
&\quad + \sum_{j=k-d_2}^{k-d(k)-1}\delta^T(j)\hat{R}_2(k)\delta(j)\Big\}. \qquad (7.15)
\end{aligned}$$

Summing up (7.12)–(7.15), it follows that

$$\begin{aligned}
\mathbf{E}\{\Delta V(k)\} &\leq \eta^T(k)\hat{\Omega}(k)\eta(k) - \mathbf{E}\Big\{\sum_{j=k-\tau}^{k-1}\delta^T(j)\hat{R}_1(k)\delta(j) \\
&\quad + \sum_{j=k-d(k)}^{k-\tau m-1}\delta^T(j)\hat{R}_2(k)\delta(j) + \sum_{j=k-d_2}^{k-d(k)-1}\delta^T(j)\hat{R}_2(k)\delta(j)\Big\} \\
&= \eta^T(k)\hat{\Omega}(k)\eta(k) - \mathbf{E}\Big\{\sum_{j=k-\tau}^{k-1}\xi^T(k,j)\begin{bmatrix}0 & 0 \\ 0 & \hat{R}_1(k)\end{bmatrix}\xi(k,j) \\
&\quad + \sum_{j=k-d(k)}^{k-\tau m-1}\xi^T(k,j)\begin{bmatrix}0 & 0 \\ 0 & \hat{R}_2(k)\end{bmatrix}\xi(k,j) \\
&\quad + \sum_{j=k-d_2}^{k-d(k)-1}\xi^T(k,j)\begin{bmatrix}0 & 0 \\ 0 & \hat{R}_2(k)\end{bmatrix}\xi(k,j)\Big\}, \qquad (7.16)
\end{aligned}$$

where $\xi(k,j) \triangleq \begin{bmatrix}\eta(k) \\ \delta(j)\end{bmatrix}$ and

$$\begin{aligned}
\hat{\Omega}(k) \triangleq\ & \hat{\Xi}_1^T(k)\hat{P}(k+1)\hat{\Xi}_1(k) + \hat{\Xi}_2^T(k)\hat{P}(k+1)\hat{\Xi}_2(k) - W_1^T\hat{P}(k)W_1 \\
& + W_1^T\hat{Q}_2(k)W_1 + W_5^T\hat{Q}_1(k)W_5 + (d_2 - \tau m + 1)W_1^T\hat{R}(k)W_1 \\
& - W_3^T\hat{R}(k - d(k))W_3 - W_4^T\hat{Q}_2(k - d_2)W_4 - W_6^T\hat{Q}_1(k - \tau)W_6 \\
& + \hat{\Xi}_2^T(k)\left[\tau\hat{S}_1(k) + (d_2 - \tau m)\hat{S}_2(k)\right]\hat{\Xi}_2(k) \\
& + \hat{\Xi}_3^T(k)\left[\tau\hat{S}_1(k) + (d_2 - \tau m)\hat{S}_2(k)\right]\hat{\Xi}_3(k).
\end{aligned}$$

Next, we will introduce several slack matrices to further reduce the conservatism. According to the definition of $\delta(j)$, for any matrices $\hat{X}(k)$, $\hat{Y}(k)$ and $\hat{Z}(k)$, we have

$$\left.\begin{aligned} 2\eta^T(k)\hat{X}(k)\left[x(k)-x(k-\tau)-\sum_{j=k-\tau}^{k-1}\delta(j)\right]&=0,\\ 2\eta^T(k)\hat{Y}(k)\left[x(k-\tau m)-x(k-d(k))-\sum_{j=k-d(k)}^{k-\tau m-1}\delta(j)\right]&=0,\\ 2\eta^T(k)\hat{Z}(k)\left[x(k-d(k))-x(k-d_2)-\sum_{j=k-d_2}^{k-d(k)-1}\delta(j)\right]&=0.\end{aligned}\right\}\quad(7.17)$$

Combining (7.17) yields

$$2\eta^T(k)\begin{bmatrix}\hat{X}(k) & \hat{Y}(k) & \hat{Z}(k)\end{bmatrix}\begin{bmatrix} I_n & -I_n & 0_{n\times mn} & 0_n\\ 0_{n\times mn} & I_n & -I_n & 0_n\\ 0_{n\times mn} & 0_n & I_n & -I_n\end{bmatrix}\eta(k)$$
$$=\sum_{j=k-\tau}^{k-1}2\eta^T(k)\hat{X}(k)\delta(j)+\sum_{j=k-d(k)}^{k-\tau m-1}2\eta^T(k)\hat{Y}(k)\delta(j)+\sum_{j=k-d_2}^{k-d(k)-1}2\eta^T(k)\hat{Z}(k)\delta(j).$$

Considering former definitions, we have

$$\begin{aligned}\eta^T(k)\left(\hat{\Psi}(k)+\hat{\Psi}^T(k)\right)\eta(k)=&\sum_{j=k-\tau}^{k-1}\xi^T(k,j)\begin{bmatrix}0 & \hat{X}(k)\\ \hat{X}^T(k) & 0\end{bmatrix}\xi(k,j)\\ &+\sum_{j=k-d(k)}^{k-\tau m-1}\xi^T(k,j)\begin{bmatrix}0 & \hat{Y}(k)\\ \hat{Y}^T(k) & 0\end{bmatrix}\xi(k,j)\\ &+\sum_{j=k-d_2}^{k-d(k)-1}\xi^T(k,j)\begin{bmatrix}0 & \hat{Z}(k)\\ \hat{Z}^T(k) & 0\end{bmatrix}\xi(k,j)\end{aligned}\quad(7.18)$$

Furthermore, for any matrices $\hat{M}(k)$ and $\hat{N}(k)$, we have

$$\begin{aligned}0=&\ \tau\eta^T(k)\hat{M}(k)\eta(k)-\sum_{j=k-\tau}^{k-1}\eta^T(k)\hat{M}(k)\eta(k),\\ 0=&\ (d_2-\tau m)\eta^T(k)\hat{N}(k)\eta(k)\\ &-\sum_{j=k-d(k)}^{k-\tau m-1}\eta^T(k)\hat{N}(k)\eta(k)-\sum_{j=k-d_2}^{k-d(k)-1}\eta^T(k)\hat{N}(k)\eta(k),\end{aligned}$$

that is,

$$\tau\eta^T(k)\hat{M}(k)\eta(k) = \sum_{j=k-\tau}^{k-1} \xi^T(k,j) \begin{bmatrix} \hat{M}(k) & 0 \\ 0 & 0 \end{bmatrix} \xi(k,j), \qquad (7.19)$$

$$\begin{aligned}(d_2-\tau m)\eta^T(k)\hat{N}(k)\eta(k) = &\sum_{j=k-d(k)}^{k-\tau m-1} \xi^T(k,j) \begin{bmatrix} \hat{N}(k) & 0 \\ 0 & 0 \end{bmatrix} \xi(k,j) \\ &+ \sum_{j=k-d_2}^{k-d(k)-1} \xi^T(k,j) \begin{bmatrix} \hat{N}(k) & 0 \\ 0 & 0 \end{bmatrix} \xi(k,j). \qquad (7.20)\end{aligned}$$

Summing up the expectations of (7.16), (7.18), (7.19) and (7.20), we have

$$\begin{aligned}\mathbf{E}\{\Delta V(k)\} \leq -\mathbf{E}\Bigg\{ &\sum_{j=k-\tau}^{k-1} \xi^T(k,j) \begin{bmatrix} \hat{M}(k) & \hat{X}(k) \\ \star & \hat{R}_1(k) \end{bmatrix} \xi(k,j) \\ &+ \sum_{j=k-d(k)}^{k-\tau m-1} \xi^T(k,j) \begin{bmatrix} \hat{N}(k) & \hat{Y}(k) \\ \star & \hat{R}_2(k) \end{bmatrix} \xi(k,j) \\ &+ \sum_{j=k-d_2}^{k-d(k)-1} \xi^T(k,j) \begin{bmatrix} \hat{N}(k) & \hat{Z}(k) \\ \star & \hat{R}_2(k) \end{bmatrix} \xi(k,j) \Bigg\} + \eta^T(k)\Omega(k)\eta(k).\end{aligned}$$

Considering (7.9b)–(7.9d) and (7.11), for all $\eta(k) \neq 0$, we have

$$\mathbf{E}\{\Delta V(k)\} \leq \eta^T(k)\Omega(k)\eta(k) < -\rho\|x(k)\|^2,$$

that is,

$$\mathbf{E}\{V(k+1)\} - \mathbf{E}\{V(k)\} < -\rho\mathbf{E}\left\{\|x(k)\|^2\right\}. \qquad (7.21)$$

Therefore, for any integer $N > 1$, summing up both sides of (7.21) from $k = 0$ to $k = N$ will results in

$$\mathbf{E}\{V(N+1)\} - \mathbf{E}\{V(0)\} < -\rho\mathbf{E}\left\{\sum_{k=0}^{N} \|x(k)\|^2\right\},$$

which means that

$$\begin{aligned}\mathbf{E}\left\{\sum_{k=0}^{N} \|x(k)\|^2\right\} &< \frac{1}{\rho}\left(\mathbf{E}\{V(0)\} - \mathbf{E}\{V(N+1)\}\right) \\ &\leq \mathbf{E}\{V(0)\}. \qquad (7.22)\end{aligned}$$

Reconsidering the LKF defined previously, we have

$$
\begin{aligned}
V_1(0) &= x^T(0)\hat{P}(0)x(0) \\
&\le \sum_{i=1}^{r} h_i(\theta(0))\lambda_{\max}(P_i)x^T(0)G^{-T}G^{-1}x(0) \\
&\le \max_{i=1,2,\ldots,r}\left\{\frac{\lambda_{\max}(P_i)}{\lambda_{\min}(G^TG)}\right\}\|\phi(0)\|^2, \\
V_2(0) &= \sum_{i=-\tau}^{-1}\Upsilon^T(i)\hat{Q}_1(i)\Upsilon(i) + \sum_{i=-d_2}^{-1}x^T(i)\hat{Q}_2(i)x(i) \\
&\le \max_{i=1,2,\ldots,r}\{\lambda_{\max}(Q_{1i})\}\sum_{j=-\tau}^{-1}\Upsilon^T(j)\mathcal{T}_4^{-T}\mathcal{T}_4^{-1}\Upsilon(j) \\
&\quad + \max_{i=1,2,\ldots,r}\{\lambda_{\max}(Q_{2i})\}\sum_{j=-d_2}^{-1}x^T(j)G^{-T}G^{-1}x(j) \\
&\le \left(\max_{i=1,2,\ldots,r}\left\{\frac{\tau m\lambda_{\max}(Q_{1i})}{\varepsilon\lambda_{\min}(G^TG)}\right\} + \max_{i=1,2,\ldots,r}\left\{\frac{d_2\lambda_{\max}(Q_{2i})}{\varepsilon\lambda_{\min}(G^TG)}\right\}\right) \\
&\quad \times \max_{-d_2\le l\le 1}\|\phi(l)\|^2, \\
V_3(0) &= \sum_{j=-d_2+1}^{-\tau m+1}\sum_{i=j-1}^{-1}x^T(i)\hat{R}(i)x(i) \\
&\le \max_{i=1,2,\ldots,r}\left\{\frac{(d_2+\tau m)(d_2-\tau m+1)\lambda_{\max}(R_i)}{2\varepsilon\lambda_{\min}(G^TG)}\right\}\max_{-d_2\le l\le 1}\|\phi(l)\|^2, \\
V_4(0) &= \sum_{i=-\tau}^{-1}\sum_{j=i}^{-1}\delta^T(j)\hat{S}_1(j)\delta(j) + \sum_{i=-d_2}^{-\tau m-1}\sum_{j=i}^{-1}\delta^T(j)\hat{S}_2(j)\delta(j) \\
&\le \sum_{i=-\tau}^{-1}\sum_{j=i}^{-1}\frac{\delta^T(j)\delta(j)}{\lambda_{\min}(\bar{S}_1(j))} + \sum_{i=-d_2}^{-\tau m-1}\sum_{j=i}^{-1}\frac{\delta^T(j)\delta(j)}{\lambda_{\min}(\bar{S}_2(j))} \\
&\le \max_{i=1,2,\ldots,r}\left\{\frac{(\tau m+d_2+1)(d_2-\tau m)}{2\lambda_{\min}(S_{2i})}\right\}\max_{-d_2\le l\le 1}\|\varsigma(l)\|^2 \\
&\quad + \max_{i=1,2,\ldots,r}\left\{\frac{\tau^2+1}{2\lambda_{\min}(S_{1i})}\right\}\max_{-d_2\le l\le 1}\|\varsigma(l)\|^2.
\end{aligned}
$$

Summing up the above inequalities, we have

$$
V(0) = V_1(0) + V_2(0) + V_3(0) + V_4(0) \le \kappa\|\phi(0)\|_a^2, \tag{7.23}
$$

where

$$
\kappa \triangleq \max_{i=1,2,\ldots,r}\left\{\frac{(\tau m+d_2+1)(d_2-\tau m)}{2\lambda_{\min}(S_{2i})}\right\} + \max_{i=1,2,\ldots,r}\left\{\frac{d_2\lambda_{\max}(Q_{2i})}{\varepsilon\lambda_{\min}(G^TG)}\right\}
$$

$$+ \max_{i=1,2,\ldots,r} \left\{ \frac{\lambda_{\max}(P_i)}{\lambda_{\min}(G^T G)} \right\} + \max_{i=1,2,\ldots,r} \left\{ \frac{\tau m \lambda_{\max}(Q_{1i})}{\varepsilon \lambda_{\min}(G^T G)} \right\}$$
$$+ \max_{i=1,2,\ldots,r} \left\{ \frac{(d_2 + \tau m)(d_2 - \tau m + 1)\lambda_{\max}(R_i)}{2\varepsilon \lambda_{\min}(G^T G)} \right\}$$
$$+ \max_{i=1,2,\ldots,r} \left\{ \frac{\tau^2 + 1}{2\lambda_{\min}(S_{1i})} \right\}. \tag{7.24}$$

Considering (7.22) and (7.23), we have

$$\mathbf{E}\left\{ \sum_{k=0}^{N} \|x(k)\|^2 \right\} < \frac{\kappa}{\rho} \mathbf{E}\left\{ \|\phi(0)\|_a^2 \right\} = \sigma \mathbf{E}\left\{ \|\phi(0)\|_a^2 \right\},$$

where $\sigma \triangleq \frac{\kappa}{\rho}$. It is obvious that

$$\lim_{N\to\infty} \mathbf{E}\left\{ \sum_{k=0}^{N} \|x(k)\|^2 \right\} \leq \sigma \mathbf{E}\left\{ \|\phi(0)\|_a^2 \right\}.$$

Then, by Definition 7.1, we can conclude that the open-loop fuzzy stochastic system in (7.3) is stochastically stable, thus the proof is completed. ■

Note that the conditions obtained in Theorem 7.2 contain time-varying parameters. Those parameters are merely available online, so it will be impossible for us to check the feasibility of those inequalities. We need to transform those parameterized LMIs [203] into strict LMIs, and then check their feasibility by computer software.

Theorem 7.3. *Given positive integers τ, m, and d_2, system (7.3) is stochastically stable if there exist matrices $P_i > 0$, $Q_{1i} > 0$, $Q_{2i} > 0$, $R_i > 0$, $S_{1i} > 0$, $S_{2i} > 0$, $R_{1i} > 0$, $R_{2i} > 0$, M_i, N_i, X_i, Y_i, Z_i, G $(i = 1, 2, \ldots, r)$, and a scalar $\varepsilon > 0$, such that for any $o, s, t, l, i = 1, 2, \ldots, r$, the following inequalities hold:*

$$\Pi_{ostli} \triangleq \begin{bmatrix} \Phi_{osli} + \Psi_i + \Psi_i^T & \Xi_i \\ \star & -\Lambda_{ti} \end{bmatrix} < 0, \tag{7.25a}$$
$$\begin{bmatrix} M_i & X_i \\ \star & \varepsilon G + \varepsilon G^T - R_{1i} \end{bmatrix} > 0, \tag{7.25b}$$
$$\begin{bmatrix} N_i & Y_i \\ \star & \varepsilon G + \varepsilon G^T - R_{2i} \end{bmatrix} > 0, \tag{7.25c}$$
$$\begin{bmatrix} N_i & Z_i \\ \star & \varepsilon G + \varepsilon G^T - R_{2i} \end{bmatrix} > 0, \tag{7.25d}$$
$$S_{1i} - R_{1j} < 0, \tag{7.25e}$$
$$S_{2i} - R_{2j} < 0, \tag{7.25f}$$

where

$$\begin{aligned}
\Phi_{osli} &\triangleq (d_2 - \tau m)\mathcal{T}_\varepsilon^{-1} N_i \mathcal{T}_\varepsilon^{-1} + \varepsilon^{-2}(d_2 - \tau m + 1)W_1^T R_i W_1 \\
&\quad + \varepsilon^{-2} W_1^T Q_{2i} W_1 - W_1^T P_i W_1 - W_3^T R_l W_3 - W_4^T Q_{2o} W_4 \\
&\quad + \mathcal{T}_\varepsilon^{-1} W_5^T Q_{1i} W_5 \mathcal{T}_\varepsilon^{-1} - W_6^T Q_{1s} W_6 + \tau \mathcal{T}_\varepsilon^{-1} M_i \mathcal{T}_\varepsilon^{-1}, \\
\Psi_i &\triangleq \varepsilon \mathcal{T}_\varepsilon^{-1} \begin{bmatrix} X_i & Y_i & Z_i \end{bmatrix} \begin{bmatrix} \varepsilon^{-1} I_n & -I_n & 0_{n\times mn} & 0_n \\ 0_{n\times mn} & I_n & -I_n & 0_n \\ 0_{n\times mn} & 0_n & I_n & -I_n \end{bmatrix}, \\
\Lambda_{ti} &\triangleq \operatorname{diag}\left\{ \left(G + G^T - P_t\right), \left(G + G^T - P_t\right), \tau^{-1} S_{1i}, \right. \\
&\qquad \left. (d_2 - \tau m)^{-1} S_{2i}, \tau^{-1} S_{1i}, (d_2 - \tau m)^{-1} S_{2i} \right\}, \\
\Xi_i &\triangleq \begin{bmatrix} \Xi_{1i}^T & \Xi_{2i}^T & \Xi_{2i}^T & \Xi_{2i}^T & \Xi_{3i}^T & \Xi_{3i}^T \end{bmatrix}, \\
\Xi_{1i} &\triangleq \begin{bmatrix} A_i G & 0_{n\times mn} & \varepsilon A_{di} G & 0_n \end{bmatrix}, \\
\Xi_{2i} &\triangleq \begin{bmatrix} E_i G & 0_{n\times mn} & \varepsilon E_{di} G & 0_n \end{bmatrix}, \\
\Xi_{3i} &\triangleq \begin{bmatrix} A_i G - G & 0_{n\times mn} & \varepsilon A_{di} G & 0_n \end{bmatrix}.
\end{aligned}$$

Proof. Note that the matrices in inequality (7.6a) of Theorem 7.2 can be unfolded as

$$\begin{aligned}
\begin{bmatrix} \Phi(k) + \Psi(k) + \Psi^T(k) & \Xi(k) \\ \star & -\Lambda(k) \end{bmatrix} &= \sum_{l=1}^{r}\sum_{o=1}^{r}\sum_{s=1}^{r}\sum_{t=1}^{r}\sum_{i=1}^{r} h_l(\theta(k - d(k))) \\
&\quad \times h_o(\theta(k - d_2)) h_s(\theta(k - \tau)) h_t(\theta(k+1)) h_i(\theta(k)) \Pi_{ostli}.
\end{aligned}$$

Obviously, condition (7.25a) implies (7.6a). Moreover, we have

$$\begin{aligned}
\begin{bmatrix} \bar{M}(k) & \bar{X}(k) \\ \star & \varepsilon G + \varepsilon G^T - \bar{R}_1(k) \end{bmatrix} &= \sum_{i=1}^{r} h_i(\theta) \begin{bmatrix} M_i & X_i \\ \star & \varepsilon G + \varepsilon G^T - R_{1i} \end{bmatrix} > 0, \\
\begin{bmatrix} \bar{N}(k) & \bar{Y}(k) \\ \star & \varepsilon G + \varepsilon G^T - \bar{R}_2(k) \end{bmatrix} &= \sum_{i=1}^{r} h_i(\theta) \begin{bmatrix} N_i & Y_i \\ \star & \varepsilon G + \varepsilon G^T - R_{2i} \end{bmatrix} > 0, \\
\begin{bmatrix} \bar{N}(k) & \bar{Z}(k) \\ \star & \varepsilon G + \varepsilon G^T - \bar{R}_2(k) \end{bmatrix} &= \sum_{i=1}^{r} h_i(\theta) \begin{bmatrix} N_i & Z_i \\ \star & \varepsilon G + \varepsilon G^T - R_{2i} \end{bmatrix} > 0,
\end{aligned}$$

and

$$\begin{aligned}
\bar{S}_1(s) - \bar{R}_1(k) &= \sum_{i=1}^{r}\sum_{j=1}^{r} h_i(\theta(s)) h_j(\theta(k)) \left(S_{1i} - R_{1j}\right) < 0, \\
\bar{S}_2(s) - \bar{R}_2(k) &= \sum_{i=1}^{r}\sum_{j=1}^{r} h_i(\theta(s)) h_j(\theta(k)) \left(S_{2i} - R_{2j}\right) < 0.
\end{aligned}$$

Then the condition in Theorem 7.2 is fulfilled if the LMI conditions in Theorem 7.3 hold, and it follows from the analysis in Theorem 7.2 that the open-loop system in (7.3) is stochastically stable if conditions of Theorem 7.3 are satisfied. The proof is completed. ■

Remark 7.4. By relaxing inequalities (7.6a)–(7.6f), we have converted the stability conditions into strict LMI form. This may introduce some conservatism, but the LMI-based conditions can be easily checked. ♦

For a special case, that is, the time-delay $d(k)$ is constant, the corresponding result becomes more simple. Replacing time-delay $d(k)$ by $d = d_2 = d_1 = \tau m$, the open-loop in system (7.3) is then rewritten as

$$\begin{aligned} x(k+1) &= \bar{A}(k)x(k) + \bar{A}_d(k)x(k-d) \\ &\quad + \left[\bar{E}(k)x(k) + \bar{E}_d(k)x(k-d)\right]\varpi(k), \end{aligned} \tag{7.26}$$

and the LKF can be chosen as

$$\begin{aligned} V(k) &\triangleq x^T(k)\hat{P}(k)x(k) + \sum_{i=k-\tau}^{k-1} \Upsilon^T(i)\hat{Q}_1(i)\Upsilon(i) + \sum_{i=k-d}^{k-1} x^T(i)\hat{Q}_2(i)x(i) \\ &\quad + \sum_{i=-\tau}^{-1}\sum_{j=k+i}^{k-1} \delta^T(j)\hat{S}_1(j)\delta(j). \end{aligned}$$

We give the new result in the following Corollary.

Corollary 7.5. *Given positive integers τ and m, the system in (7.26) is stochastically stable if there exist matrices $P_i > 0$, $Q_{1i} > 0$, $Q_{2i} > 0$, $S_{1i} > 0$, $R_{1i} > 0$, M_i, X_i, G ($i = 1, 2, \ldots, r$), and a scalar $\varepsilon > 0$, such that for any $o, s, t, i = 1, 2, \ldots, r$, the following inequalities hold:*

$$\begin{bmatrix} \check{\Phi}_{osi} + \check{\Psi}_i + \check{\Psi}_i^T & \check{\Xi}_i \\ \star & -\check{\Lambda}_{ti} \end{bmatrix} < 0,$$

$$\begin{bmatrix} M_i & X_i \\ \star & \varepsilon G + \varepsilon G^T - R_{1i} \end{bmatrix} > 0,$$

$$S_{1i} - R_{1j} < 0,$$

where

$$\begin{aligned} \check{\Phi}_{osi} &\triangleq -\check{W}_1^T P_i \check{W}_1 + \varepsilon^{-2}\check{W}_1^T Q_{2i}\check{W}_1 - \check{W}_2^T Q_{2o}\check{W}_2 - \check{W}_4^T Q_{1s}\check{W}_4 \\ &\quad + \check{\mathcal{T}}_\varepsilon^{-1}\check{W}_3^T Q_{1i}\check{W}_3\check{\mathcal{T}}_\varepsilon^{-1} + \tau\check{\mathcal{T}}_\varepsilon^{-1}M_i\check{\mathcal{T}}_\varepsilon^{-1}, \\ \check{\Psi}_i &\triangleq \varepsilon\check{\mathcal{T}}_\varepsilon^{-1}X_i\left[\varepsilon^{-1}I_n \;\; -I_n \;\; 0_{n\times(m-1)n}\right], \\ \check{\Lambda}_{ti} &\triangleq \mathrm{diag}\left\{\left(G + G^T - P_t\right), \left(G + G^T - P_t\right), \tau^{-1}S_{1i}, \tau^{-1}S_{1i}\right\}, \\ \check{\Xi}_i &\triangleq \left[\check{\Xi}_{1i}^T \;\; \check{\Xi}_{2i}^T \;\; \check{\Xi}_{2i}^T \;\; \check{\Xi}_{3i}^T\right], \\ \check{\Xi}_{1i} &\triangleq \left[A_i G \;\; 0_{n\times(m-1)n} \;\; \varepsilon A_{di}G\right], \\ \check{\Xi}_{2i} &\triangleq \left[E_i G \;\; 0_{n\times(m-1)n} \;\; \varepsilon E_{di}G\right], \end{aligned}$$

$$\begin{aligned}
\check{\Xi}_{3i} &\triangleq \begin{bmatrix} A_iG - G & 0_{n\times(m-1)n} & \varepsilon A_{di}G \end{bmatrix},\\
\check{W}_1 &\triangleq \begin{bmatrix} I_n & 0_{n\times mn} \end{bmatrix}, \quad \check{W}_2 \triangleq \begin{bmatrix} 0_{n\times mn} & I_n \end{bmatrix},\\
\check{W}_3 &\triangleq \begin{bmatrix} I_{mn} & 0_{mn\times n} \end{bmatrix}, \quad \check{W}_4 \triangleq \begin{bmatrix} 0_{mn\times n} & I_{mn} \end{bmatrix},\\
\check{\mathcal{T}}_\varepsilon &\triangleq \operatorname{diag}\{\varepsilon I_n, \underbrace{I_n, \ldots, I_n}_{m}\},
\end{aligned}$$

Proof. The result in this corollary is a special case of Theorem 7.3, so we omit the proof here. ■

What's more, for a system without stochastic noise, the stability condition will be further simplified. We give the T-S fuzzy system in the following form,

$$x(k+1) = \bar{A}(k)x(k) + \bar{A}_d(k)x(k-d). \tag{7.27}$$

Corresponding result is given in the following corollary without proof.

Corollary 7.6. *Given positive integers τ and m, system (7.27) is stable if there exist matrices $P_i > 0$, $Q_{1i} > 0$, $Q_{2i} > 0$, $S_{1i} > 0$, $R_{1i} > 0$, M_i, X_i, G $(i = 1, 2, \ldots, r)$, and a scalar $\varepsilon > 0$, such that for any $o, s, t, i = 1, 2, \ldots, r$, the following inequalities hold:*

$$\begin{bmatrix} \check{\Phi}_{osi} + \check{\Psi}_i + \check{\Psi}_i^T & \check{\Xi}_{1i}^T & \check{\Xi}_{1i}^T \\ \star & -G - G^T + P_t & 0 \\ \star & \star & -\tau^{-1}S_{1i} \end{bmatrix} < 0,$$

$$\begin{bmatrix} M_i & X_i \\ \star & \varepsilon G + \varepsilon G^T - R_{1i} \end{bmatrix} > 0,$$

$$S_{1i} - R_{1j} < 0.$$

7.3.2 Stabilization

Having analyzed the stability conditions of open-loop system (7.3), we will extended the former results to design a fuzzy state-feedback controller for the closed-loop system (7.5).

Theorem 7.7. *Given positive integers τ, m, and d_2, system (7.5) is stochastically stable if there exist matrices $P_i > 0$, $Q_{1i} > 0$, $Q_{2i} > 0$, $R_i > 0$, $S_{1i} > 0$, $S_{2i} > 0$, $R_{1i} > 0$, $R_{2i} > 0$, M_i, N_i, X_i, Y_i, Z_i, G, H_i $(i = 1, 2, \ldots, r)$, and a scalar $\varepsilon > 0$, such that for any $o, s, t, l, i, j = 1, 2, \ldots, r$, the following inequalities hold:*

$$\tilde{\Pi}_{ostlij} \triangleq \begin{bmatrix} \Phi_{osli} + \Psi_i + \Psi_i^T & \tilde{\Xi}_{ij} \\ \star & -\Lambda_{ti} \end{bmatrix} < 0, \tag{7.28a}$$

$$\frac{1}{r-1}\tilde{\Pi}_{ostlii} + \frac{1}{2}(\tilde{\Pi}_{ostlij} + \tilde{\Pi}_{ostlji}) < 0, \ \ i \neq j, \tag{7.28b}$$

$$\begin{bmatrix} M_i & X_i \\ \star & \varepsilon G + \varepsilon G^T - R_{1i} \end{bmatrix} > 0, \tag{7.28c}$$

$$\begin{bmatrix} N_i & Y_i \\ \star & \varepsilon G + \varepsilon G^T - R_{2i} \end{bmatrix} > 0, \tag{7.28d}$$

$$\begin{bmatrix} N_i & Z_i \\ \star & \varepsilon G + \varepsilon G^T - R_{2i} \end{bmatrix} > 0, \tag{7.28e}$$

$$S_{1i} - R_{1j} < 0, \tag{7.28f}$$

$$S_{2i} - R_{2j} < 0, \tag{7.28g}$$

where

$$\begin{aligned}
\tilde{\Xi}_{ij} &\triangleq \begin{bmatrix} \tilde{\Xi}_{1ij}^T & \tilde{\Xi}_{2ij}^T & \tilde{\Xi}_{2ij}^T & \tilde{\Xi}_{2ij}^T & \tilde{\Xi}_{3ij}^T & \tilde{\Xi}_{3ij}^T \end{bmatrix}, \\
\tilde{\Xi}_{1ij} &\triangleq \begin{bmatrix} A_i G + B_{1i} H_j & 0_{n\times mn} & \varepsilon A_{di} G & 0_n \end{bmatrix}, \\
\tilde{\Xi}_{2ij} &\triangleq \begin{bmatrix} E_i G + B_{2i} H_j & 0_{n\times mn} & \varepsilon E_{di} G & 0_n \end{bmatrix}, \\
\tilde{\Xi}_{3ij} &\triangleq \begin{bmatrix} A_i G + B_{1i} H_j - G_j & 0_{n\times mn} & \varepsilon A_{di} G & 0_n \end{bmatrix},
\end{aligned}$$

and the fuzzy controller is given as

$$u(k) = \bar{K}(k)x(k) = \bar{H}(k)G^{-1}x(k), \tag{7.29}$$

with

$$\bar{K}(k) \triangleq \sum_{i=1}^{r} h_i(\theta)K_i, \quad \bar{H}(k) \triangleq \sum_{i=1}^{r} h_i(\theta)H_i.$$

Proof. To stabilize system (7.5), we just need to replace $\bar{A}(k)$ and $\bar{A}_d(k)$ in (7.6a) by $\hat{A}(k)$ and $\hat{A}_d(k)$. Firstly, we have the following expressions:

$$\hat{A}(k)G = \bar{A}(k)G + \bar{B}_1(k)\bar{K}(k)G, \tag{7.30}$$

$$\hat{E}(k)G = \bar{E}(k)G + \bar{B}_2(k)\bar{K}(k)G, \tag{7.31}$$

where G and K_i $(i = 1, 2, \ldots, r)$, are coupled with each other, making it impossible for us to solve them through LMI technique. To decouple them, define new variables of $H_i \triangleq K_i G$ $(i = 1, 2, \ldots, r)$, thus we have $\bar{H}(k) = \bar{K}(k)G$. Obviously, $\bar{K}(k)$ can be obtained by $\bar{K}(k) = \bar{H}(k)G^{-1}$, which is also called the non-PDC [34]. So (7.30) and (7.31) can be rewritten as

$$\hat{A}(k)G = \bar{A}(k)G + \bar{B}_1(k)\bar{H}(k) = \sum_{i=1}^{r}\sum_{j=1}^{r} h_i(\theta)h_j(\theta)\left(A_i G + B_{1i}H_j\right),$$

$$\hat{E}(k)G = \bar{E}(k)G + \bar{B}_2(k)\bar{H}(k) = \sum_{i=1}^{r}\sum_{j=1}^{r} h_i(\theta)h_j(\theta)\left(E_i G + B_{2i}H_j\right).$$

Accordingly, inequality (7.6a) will be replaced by

$$\begin{bmatrix} \Phi(k)+\Psi(k)+\Psi^T(k) & \tilde{\Xi}(k) \\ \star & -\Lambda(k) \end{bmatrix} < 0, \tag{7.32}$$

where

$$\begin{aligned}
\tilde{\Xi}(k) &\triangleq \left[\tilde{\Xi}_1^T(k) \ \ \tilde{\Xi}_2^T(k) \ \ \tilde{\Xi}_2^T(k) \ \ \tilde{\Xi}_2^T(k) \ \ \tilde{\Xi}_3^T(k) \ \ \tilde{\Xi}_3^T(k) \right], \\
\tilde{\Xi}_1(k) &\triangleq \left[\bar{A}(k)G+\bar{B}_1(k)\bar{H}(k) \ \ 0_{n\times mn} \ \ \varepsilon\bar{A}_d(k)G \ \ 0_n \right], \\
\tilde{\Xi}_2(k) &\triangleq \left[\bar{E}(k)G+\bar{B}_2(k)\bar{H}(k) \ \ 0_{n\times mn} \ \ \varepsilon\bar{E}_d(k)G \ \ 0_n \right], \\
\tilde{\Xi}_3(k) &\triangleq \left[\bar{A}(k)G+\bar{B}_1(k)\bar{H}(k)-G \ \ 0_{n\times mn} \ \ \varepsilon\bar{A}_d(k)G \ \ 0_n \right].
\end{aligned}$$

Then,

$$\begin{aligned}
\begin{bmatrix} \Phi(k)+\Psi(k)+\Psi^T(k) & \tilde{\Xi}(k) \\ \star & -\Lambda(k) \end{bmatrix} &= \sum_{l=1}^{r}\sum_{o=1}^{r}\sum_{s=1}^{r}\sum_{t=1}^{r}\sum_{1\le i<j\le r} h_l(\theta(k-d(k))) \\
&\quad \times h_o(\theta(k-d_2))h_s(\theta(k-\tau))h_t(\theta(k+1)) \\
&\quad \times \left[\frac{1}{r-1}\left(h_i^2(\theta(k))\tilde{\Pi}_{ostlii}+h_j^2(\theta(k))\tilde{\Pi}_{ostljj} \right) \right. \\
&\quad \left. + \ h_i(\theta(k))h_j(\theta(k))(\tilde{\Pi}_{ostlij}+\tilde{\Pi}_{ostlji}) \right].
\end{aligned}$$

The following part is analyzed in two cases:

Case 1: supposing $\tilde{\Pi}_{ostlij}+\tilde{\Pi}_{ostlji}<0$, then we have

$$\begin{aligned}
\begin{bmatrix} \Phi(k)+\Psi(k)+\Psi^T(k) & \tilde{\Xi}(k) \\ \star & -\Lambda(k) \end{bmatrix} &\le \sum_{l=1}^{r}\sum_{o=1}^{r}\sum_{s=1}^{r}\sum_{t=1}^{r}\sum_{1\le i<j\le r} h_l(\theta(k-d(k))) \\
&\quad \times h_o(\theta(k-d_2))h_s(\theta(k-\tau))h_t(\theta(k+1)) \\
&\quad \times \frac{1}{r-1}\left[h_i^2(\theta(k))\tilde{\Pi}_{ostlii}+h_j^2(\theta(k))\tilde{\Pi}_{ostljj} \right].
\end{aligned}$$

A sufficient condition for this inequality is

$$\tilde{\Pi}_{ostlii}<0, \quad o,s,t,l,i=1,2,\ldots,r.$$

Case 2: supposing $\tilde{\Pi}_{ostlij}+\tilde{\Pi}_{ostlji}\ge 0$, then we have

$$\begin{aligned}
\begin{bmatrix} \Phi(k)+\Psi(k)+\Psi^T(k) & \tilde{\Xi}(k) \\ \star & -\Lambda(k) \end{bmatrix} &\le \sum_{l=1}^{r}\sum_{o=1}^{r}\sum_{s=1}^{r}\sum_{t=1}^{r}\sum_{1\le i<j\le r} h_l(\theta(k-d(k))) \\
&\quad \times h_o(\theta(k-d_2))h_s(\theta(k-\tau))h_t(\theta(k+1)) \\
&\quad \times \left[\frac{1}{r-1}\left(h_i^2(\theta(k))\tilde{\Pi}_{ostlii}+h_j^2(\theta(k))\tilde{\Pi}_{ostljj} \right. \right. \\
&\quad \left. + \ \frac{1}{2}\left(h_i^2(\theta(k))+h_j^2(\theta(k))\right)\left(\tilde{\Pi}_{ostlij}+\tilde{\Pi}_{ostlji}\right) \right]
\end{aligned}$$

$$= \sum_{l=1}^{r}\sum_{o=1}^{r}\sum_{s=1}^{r}\sum_{t=1}^{r}\sum_{1\le i<j\le r} h_l(\theta(k-d(k)))$$
$$\times\, h_o(\theta(k-d_2))h_s(\theta(k-\tau))h_t(\theta(k+1))$$
$$\times \left\{ h_i^2(\theta(k))\left[\frac{1}{r-1}\tilde{\Pi}_{ostlii}+\frac{1}{2}(\tilde{\Pi}_{ostlij}+\tilde{\Pi}_{ostlij})\right]\right.$$
$$\left. +\, h_j^2(\theta(k))\left[\frac{1}{r-1}\tilde{\Pi}_{ostljj}+\frac{1}{2}(\tilde{\Pi}_{ostlij}+\tilde{\Pi}_{ostlij})\right]\right\}.$$

A sufficient condition for the above inequality is

$$\frac{1}{r-1}\tilde{\Pi}_{ostlii} + \frac{1}{2}\left(\tilde{\Pi}_{ostlij} + \tilde{\Pi}_{ostlji}\right) < 0, \quad o,s,t,l,i,j = 1,2,\ldots,r;\ i \neq j.$$

Based on the above analysis, we know that the inequality (7.32) is guaranteed by (7.28a) and (7.28b). Then it follows from the analysis of Theorems 7.2 and 7.3 that the closed-loop system in (7.5) is stochastically stable. ■

Remark 7.8. The unfolding methods of parameterized LMIs was investigated in [203]. Comparing with [66], our method has less conservativeness, since condition (7.28b) may still hold in case of $\tilde{\Pi}_{ostlij} + \tilde{\Pi}_{ostlji} \geq 0$. ♦

7.4 Illustrative Example

Three examples will now be provided. Example 7.9 is given to validate the effectiveness of delay-partitioning method; Example 7.10 presents the comparison results with some existing work; and Example 7.11 shows the effectiveness of the fuzzy controller proposed in Theorem 7.7 through an inverted pendulum system.

Example 7.9. Consider the following discrete-time T-S fuzzy stochastic system with time-varying delay:

$$x(k+1) = \sum_{i=1}^{2} h_i(\theta)\left[A_i x(k) + A_{di}x(k-d(k))\right] + \sum_{i=1}^{2} h_i(\theta)\left[E_i x(k) + E_{di}x(k-d(k))\right]\varpi(k),$$

with

$$A_1 = \begin{bmatrix} 0.4 & 0 \\ 0.01 & -0.3 \end{bmatrix},\ A_{d1} = \begin{bmatrix} 0.2 & 0 \\ 0 & -0.2 \end{bmatrix},\ E_1 = \begin{bmatrix} 0.1 & 0 \\ 0 & 0.1 \end{bmatrix},\ E_{d1} = \begin{bmatrix} 0.1 & 0 \\ 0 & 0.1 \end{bmatrix},$$
$$A_2 = \begin{bmatrix} -0.4 & 0 \\ 0.02 & 0.2 \end{bmatrix},\ A_{d2} = \begin{bmatrix} -0.2 & 0 \\ 0 & 0.2 \end{bmatrix},\ E_2 = \begin{bmatrix} 0.1 & 0 \\ 0 & 0.1 \end{bmatrix},\ E_{d2} = \begin{bmatrix} 0.1 & 0 \\ 0 & 0.1 \end{bmatrix}.$$

By this numerical example, we will investigate the important role of delay partitioning method in reducing conservatism of stability analysis. Set $m = 1, 2, 3$ respectively, then for each m, obtain the allowable upper bound of d_2 on the basis of different d_1 (d_1 can be divided by m with no remainder). The obtained results of this procedure are plotted in Fig. 7.1.

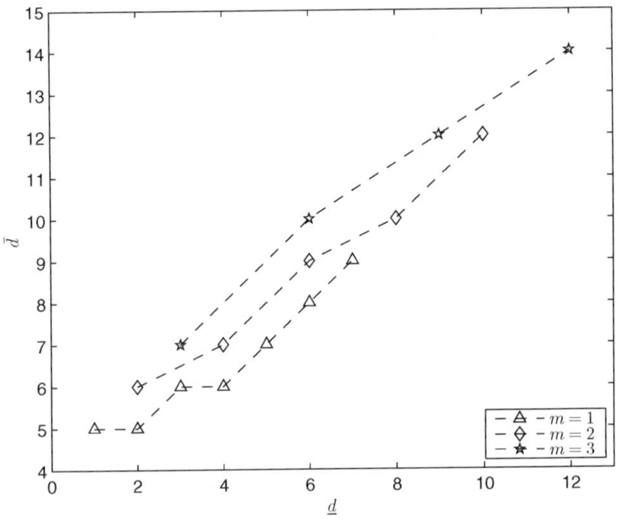

Fig. 7.1. Simulation results with different m

It is obvious that, for the same d_1, the allowable upper bound of d_2 is enhanced by increasing the partition number m, which means that conservatism is reduced as the fractions become thinner. On the other hand, with the increase of m, the number of variables also increases, which adds inevitable computational burden. A trade-off between conservatism and computational burden can be made by choosing a proper partition number m.

Example 7.10. In this example, we will compare the result in Corollary 7.6 with some existing ones. Consider the following open-loop discrete-time T-S fuzzy system:

$$x(k+1) = \sum_{i=1}^{2} h_i(\theta) \left[A_i x(k) + A_{di} x(k-d)\right],$$

with

$$A_1 = \begin{bmatrix} -0.29 & 1 \\ 0 & 0.95 \end{bmatrix}, \quad A_{d1} = \begin{bmatrix} 0.012 & 0.014 \\ 0 & 0.015 \end{bmatrix},$$
$$A_2 = \begin{bmatrix} -0.1 & 0 \\ 0.98 & -0.2 \end{bmatrix}, \quad A_{d2} = \begin{bmatrix} 0.01 & 0 \\ 0.01 & 0.015 \end{bmatrix}.$$

The achieved upper bounds of time-delay by different methods are given in Table 7.1. Getting a better view of this table, we can find the achieved upper bounds of the first two methods are obviously bigger than that of the last one. This is because Corollary 7.6 in this chapter and Corollary 1 in [223] have employed the idea of delay partitioning method, and conservatism can be reduced by increasing the partitioning number m. Based on this fact, we can conclude that the method in this chapter has an advantage over those without delay-partitioning technique. What's more, for the same partitioning number m, conservatism of Corollary 7.6 in this chapter is relatively smaller than that of Corollary 1 in [223], though both of them have implied delay-partitioning technique. The reason is that, matrices Q_{1i} and M_i $(i = 1, 2, \ldots, r)$ in Corollary 1 have general structures, but corresponding matrices in Corollary 2 of [223] are of block-diagonal forms. Those block-diagonal matrices are special cases of the general ones, so the method presented in this chapter has lower conservatism.

Table 7.1. Achieved upper bounds of d

m	Corollary 2	Corollary 1 of [223]	Theorem 1 of [217]
2	6	4	3
3	9	6	3
4	12	8	3

Example 7.11. The inverted pendulum is a typical nonlinear system whose dynamics has been investigated in [128] and [242]. Obviously, the upward equilibrium of the pendulum is an unstable state of this system. Due to its nonlinear characteristic, this system is difficult to stabilize by traditional linear methods.

In this example, we consider the inverted pendulum on a cart with delayed resonator. Schematic diagram of this pendulum system is shown in Fig. 7.2. To illustrate our approach, firstly we will establish the nonlinear dynamic equations of this system, then obtain its T-S fuzzy model by some approximation procedure, and finally stabilize the pendulum system by our controller design method. The parameters of the pendulum system are presented as

M	mass of the cart
m	mass of the block on the pendulum
l	length of the pendulum
g	acceleration due to gravity
g_r	coefficient of the delayed resonator
c_r	coefficient of the damper
$\theta(t)$	angle the pendulum makes with vertical
$y(t)$	displacement of the cart
$d(t)$	time-varying delay
$u(t)$	applied force

For simplicity, in some places the notation "(t)" of the variables will be elided (for example, d stands for $d(t)$). By Newton's law, we know, the motion of the system can be described by

$$\begin{cases} M\dfrac{d^2y}{dt^2} + m\dfrac{d^2}{dt^2}(y + l\sin\theta) = F_\omega - F_r + u, \\ m\dfrac{d^2}{dt^2}(y + l\sin\theta)\cdot l\cos\theta = mgl\sin\theta. \end{cases}$$

where $F_r(t) = g_r\dot{y}(t-d) + c_r\dot{y}(t)$ is the resultant force of the damper and delayed resonator; $F_\omega(t)$ is the position-dependent stochastic perturbation which is caused by the rough road. With the choice of the state variables $x_1 = y$, $x_2 = \theta$, $x_3 = \dot{y}$, $x_4 = \dot{\theta}$ the state-space equations of the system are

$$\begin{cases} \dot{x}_1 = x_3, \\ \dot{x}_2 = x_4, \\ \dot{x}_3 = \dfrac{-mg\sin x_2}{M\cos x_2} - \dfrac{(c_r x_3 + g_r x_{3d} - F_\omega - u)}{M}, \\ \dot{x}_4 = \dfrac{(M+m)g\sin x_2}{Ml\cos^2 x_2} + \dfrac{x_4^2\sin x_2}{\cos x_2} + \dfrac{c_r x_3 + g_r x_{3d} - F_\omega - u}{Ml\cos x_2}, \end{cases}$$

where x_{3d} denotes $x_3(t-d)$. To applying our controller design method, firstly we need to describe the original nonlinear system by a T-S fuzzy model. And we will obtain this model by the following approximation method.

i) When x_2 is near zero, the nonlinear equations can be simplified as

$$\begin{cases} \dot{x}_1 = x_3, \\ \dot{x}_2 = x_4, \\ \dot{x}_3 = \dfrac{-mgx_2}{M} - \dfrac{c_r x_3 + g_r x_{3d} - F_\omega - u}{M}, \\ \dot{x}_4 = \dfrac{(M+m)gx_2}{Ml} + \dfrac{c_r x_3 + g_r x_{3d} - F_\omega - u}{Ml}. \end{cases}$$

ii) When x_2 is near γ $(0 < |\gamma| < 90°)$, the nonlinear equations can be simplified as

$$\begin{cases} \dot{x}_1 = x_3, \\ \dot{x}_2 = x_4, \\ \dot{x}_3 = \dfrac{-mg\beta x_2}{M\alpha} - \dfrac{c_r x_3 + g_r x_{3d} - F_\omega - u}{M} \\ \dot{x}_4 = \dfrac{(M+m)g\beta x_2}{Ml\alpha^2} + \dfrac{c_r x_3 + g_r x_{3d} - F_\omega - u}{Ml\alpha}. \end{cases}$$

where $\alpha = \cos\gamma$ and $\beta = (\sin\gamma)/\gamma$. Firstly using the local approximation in fuzzy partition spaces [194], we will obtain the corresponding two-rules T-S fuzzy model. Then employing the Euler first-order approximation, we get the following discrete-time T-S fuzzy model:

♦ **Plant Form:**

Rule 1: IF $x_2(k)$ is $\mathcal{M}_1$, THEN

$$x(k+1) = A_1 x(k) + A_{d1} x(k - d(k)) + B_{11} u(k) + E_1 x(k) \varpi(k),$$

Rule 2: IF $x_2(k)$ is $\mathcal{M}_2$, THEN

$$x(k+1) = A_2 x(k) + A_{d2} x(k - d(k)) + B_{12} u(k) + E_2 x(k) \varpi(k),$$

where $\mathcal{M}_1 = 0$, $\mathcal{M}_2 = \gamma$, $\varpi(k)$ is the stochastic perturbation introduced by $F_\omega(t)$, and the system matrices are expressed as

$$A_1 = \begin{bmatrix} 1 & 0 & T & 0 \\ 0 & 1 & 0 & T \\ 0 & -\dfrac{Tmg}{M} & 1 - \dfrac{Tc_r}{M} & 0 \\ 0 & \dfrac{T(M+m)g}{Ml} & \dfrac{Tc_r}{Ml} & 1 \end{bmatrix}, \quad B_{11} = \begin{bmatrix} 0 \\ 0 \\ \dfrac{T}{M} \\ \dfrac{-T}{Ml} \end{bmatrix},$$

$$A_2 = \begin{bmatrix} 1 & 0 & T & 0 \\ 0 & 1 & 0 & T \\ 0 & -\dfrac{Tmg\beta}{M\alpha} & 1 - \dfrac{Tc_r}{M} & 0 \\ 0 & \dfrac{T(M+m)g\beta}{Ml\alpha^2} & \dfrac{Tc_r}{Ml\alpha} & 1 \end{bmatrix}, \quad B_{12} = \begin{bmatrix} 0 \\ 0 \\ \dfrac{T}{M} \\ \dfrac{-T}{Ml\alpha} \end{bmatrix},$$

$$A_{d1} = \begin{bmatrix} 0\,0 & 0 & 0 \\ 0\,0 & 0 & 0 \\ 0\,0 & \dfrac{-Tg_r}{M} & 0 \\ 0\,0 & \dfrac{Tg_r}{Ml} & 0 \end{bmatrix}, \quad E_1 = \begin{bmatrix} 0 & 0 & 0\,0 \\ 0 & 0 & 0\,0 \\ 0 & \dfrac{Tg_\omega}{M} & 0\,0 \\ 0 & \dfrac{-Tg_\omega}{Ml} & 0\,0 \end{bmatrix},$$

$$A_{d2} = \begin{bmatrix} 0\,0 & 0 & 0 \\ 0\,0 & 0 & 0 \\ 0\,0 & \frac{-Tg_r}{M} & 0 \\ 0\,0 & \frac{Tg_r}{Ml\alpha} & 0 \end{bmatrix}, \quad E_2 = \begin{bmatrix} 0 & 0 & 0\,0 \\ 0 & 0 & 0\,0 \\ 0 & \frac{Tg_\omega}{M} & 0\,0 \\ 0 & \frac{-Tg_\omega}{Ml\alpha} & 0\,0 \end{bmatrix},$$

with T the sampling time, g_ω the coefficient of the stochastic perturbation. For convenience, the membership functions $\mathcal{M}_1(x_2(k))$ and $\mathcal{M}_2(x_2(k))$ are simply chosen as triangular ones, which are shown in Fig. 7.3. Under the assumption that $|x_2(k)|$ should be smaller then $|\gamma|$, we can represent $\mathcal{M}_1(x_2(k))$ and $\mathcal{M}_2(x_2(k))$ by the following expressions:

$$\mathcal{M}_1(x_2(k)) = 1 - \frac{|x_2(k)|}{|\gamma|}, \quad \mathcal{M}_2(x_2(k)) = \frac{|x_2(k)|}{|\gamma|}.$$

We further get the following fuzzy basis functions:

$$h_1(x_2(k)) = 1 - \frac{|x_2(k)|}{|\gamma|}, \quad h_2(x_2(k)) = \frac{|x_2(k)|}{|\gamma|}.$$

To simulate this model, we set $M = 1.378$ Kg, $m = 0.051$ Kg, $l = 0.325$ m, $g = 9.8$ m/s^2, $c_r = 5.98$ Kg/s, $g_r = 0.7$ Kg/s, $g_\omega = 1.2$ Kg/(ms), $T = 0.025$ s and $\gamma = 30°$. Then by simple calculation, we have

$$A_1 = \begin{bmatrix} 1 & 0 & 0.0250 & 0 \\ 0 & 1.0000 & 0 & 0.0250 \\ 0 & -0.0091 & 0.8915 & 0 \\ 0 & 0.7817 & 0.3338 & 1.0000 \end{bmatrix}, \quad B_{11} = \begin{bmatrix} 0 \\ 0 \\ 0.0181 \\ -0.0558 \end{bmatrix},$$

$$A_2 = \begin{bmatrix} 1 & 0 & 0.0250 & 0 \\ 0 & 1.0000 & 0 & 0.0250 \\ 0 & -0.0100 & 0.8915 & 0 \\ 0 & 0.9954 & 0.3855 & 1.0000 \end{bmatrix}, \quad B_{12} = \begin{bmatrix} 0 \\ 0 \\ 0.0181 \\ -0.0645 \end{bmatrix},$$

$$A_{d1} = \begin{bmatrix} 0 & 0 & 0 & 0 \\ 0 & 0 & 0 & 0 \\ 0 & 0 & -0.0127 & 0 \\ 0 & 0 & 0.0391 & 0 \end{bmatrix}, \quad E_1 = \begin{bmatrix} 0 & 0 & 0 & 0 \\ 0 & 0 & 0 & 0 \\ 0 & 0.0581 & 0 & 0 \\ 0 & -0.1786 & 0 & 0 \end{bmatrix},$$

$$A_{d2} = \begin{bmatrix} 0 & 0 & 0 & 0 \\ 0 & 0 & 0 & 0 \\ 0 & 0 & -0.0127 & 0 \\ 0 & 0 & 0.0451 & 0 \end{bmatrix}, \quad E_2 = \begin{bmatrix} 0 & 0 & 0 & 0 \\ 0 & 0 & 0 & 0 \\ 0 & 0.0581 & 0 & 0 \\ 0 & -0.2063 & 0 & 0 \end{bmatrix},$$

and

$$h_1(x_2(k)) = 1 - 1.91|x_2(k)|, \quad h_2(x_2(k)) = 1.91|x_2(k)|.$$

Perturbation $\varpi(k)$ is assumed to be the standard Gaussian white noise sequence with $\mathbf{E}\{\varpi(k)\} = 0$ and $\mathbf{E}\{\varpi^2(k)\} = 1$. The initial condition is set to be $\phi(k) = \begin{bmatrix} 0.3\ 0.3\ 0.3\ 0.3 \end{bmatrix}^T$, $k = -d_2, -d_2 + 1, \ldots, 0$. The time-varying delay $d(k)$, which is shown in Fig. 7.4, randomly change between $d_1 = 1$ and $d_2 = 3$. The motion of the original system is shown in Fig. 7.5, from which we can see that the inverted pendulum is unstable. By LMI Toolbox in Matlab, we find the following solutions which satisfy the conditions in Theorem 7.7.

$$
\begin{aligned}
H_1 &= \begin{bmatrix} -21.0430\ 3.2988\ -36.6337\ 184.3749 \end{bmatrix}, \\
H_2 &= \begin{bmatrix} -21.5827\ 4.5925\ -26.7420\ 150.2708 \end{bmatrix}, \\
G &= \begin{bmatrix} 5.8925 & -0.2659 & -2.4554 & -0.6674 \\ -0.2637 & 0.7426 & -1.3622 & -0.7044 \\ -2.9839 & -1.0400 & 8.1850 & -1.2593 \\ 0.8206 & -1.8082 & -2.6845 & 11.4835 \end{bmatrix}.
\end{aligned}
$$

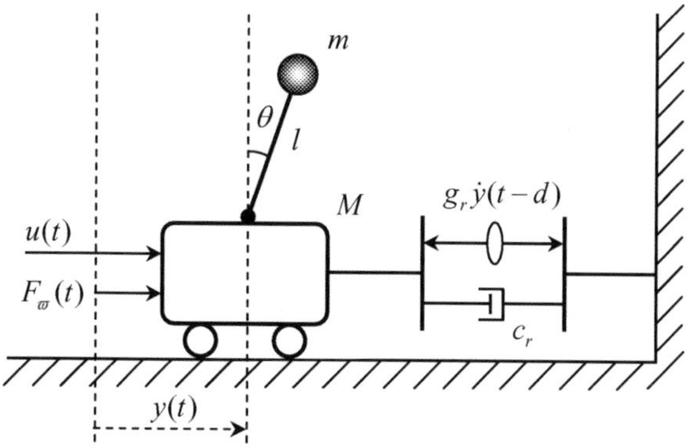

Fig. 7.2. Inverted pendulum on a cart with delayed resonator

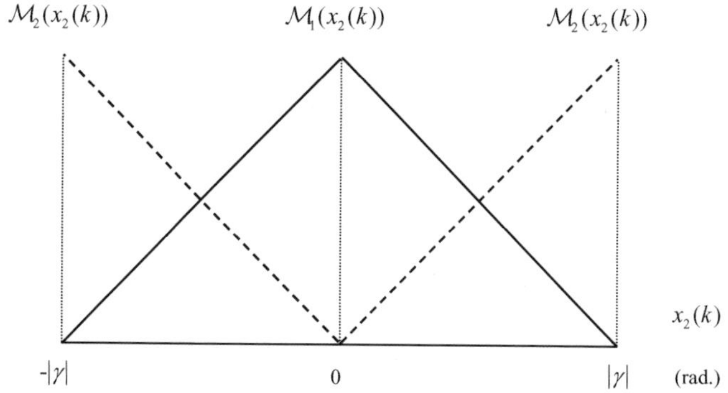

Fig. 7.3. Membership functions

With the controller constructed by (7.29), we can get the experimental results of closed-loop system. Fig. 7.6 shows the control result of the fuzzy model, and Fig. 7.7 shows the control result of the original system. Figs. 7.6 and 7.7 manifest that both models can be stabilized by the constructed fuzzy controller.

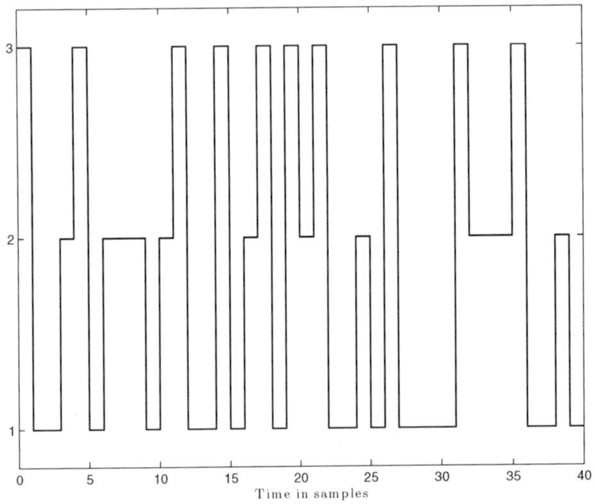

Fig. 7.4. Random time-varying delay

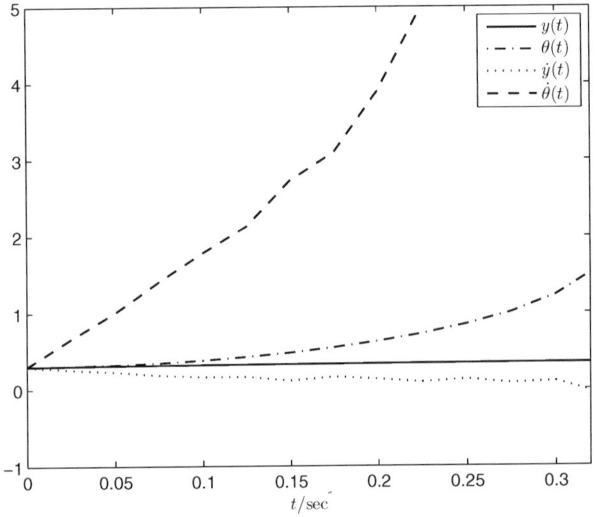

Fig. 7.5. Original system without control

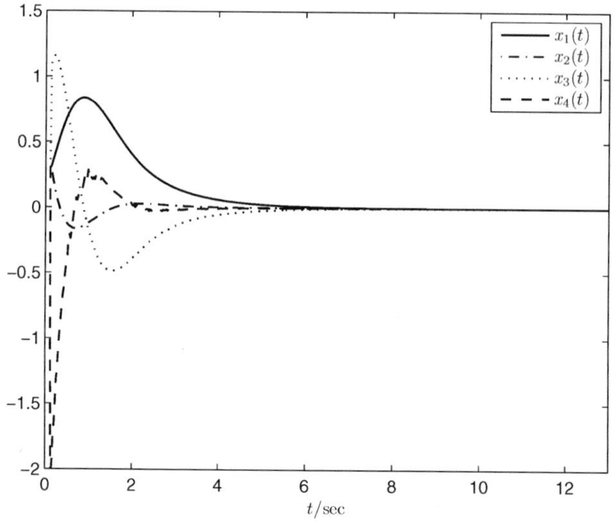

Fig. 7.6. Control result of the fuzzy model

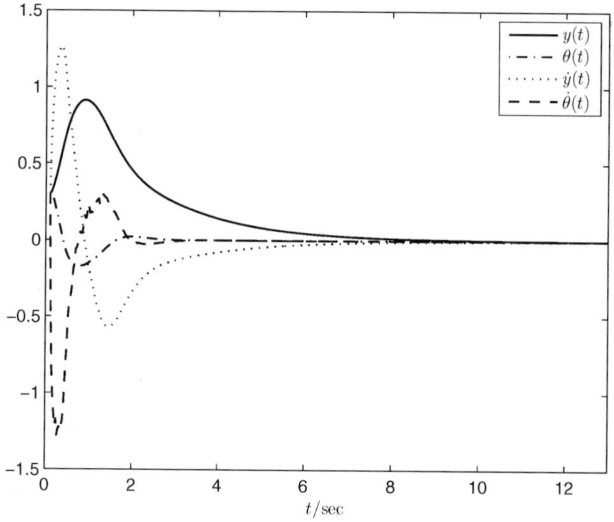

Fig. 7.7. Control result of the original model

7.5 Conclusion

The problems of stability and stabilization of discrete-time T-S fuzzy stochastic systems with time-varying delay have been investigated in this chapter. By constructing a basis-dependent LKF, delay partitioning method is employed to reduce the conservatism of the results. Firstly, delay-dependent stability conditions in terms of LMI have been derived. Secondly, a non-PDC control law is obtained based on the former analyzed stability conditions, where the cross products is replaced by the newly defined matrices. Finally, three illustrative examples have been provided to show the effectiveness of the proposed methods.

Chapter 8
Dissipativity Analysis and Synthesis of Discrete-Time T-S Fuzzy Stochastic Systems

8.1 Introduction

Dissipativity theory has played a critical part in the analysis and control design of linear and nonlinear systems, especially for high-order systems, since from the practical application point of view, many systems need to be dissipative for achieving effective noise attenuation [51]. It has been recognized that for more abstract systems one can still associate them with an energy-like function (called the storage function) and an input-power-like function (called the supply rate). Dissipativity is then characterized by storage functions and supply rates, which represent the energy stored inside the system and energy supplied from outside the system, respectively. Generally speaking, dissipative systems are those for which the increase in stored energy is never larger than the amount of energy supplied by the environment, i.e., dissipative systems can only dissipate but not generate energy. The dissipative systems theory is closely related to the dynamic properties of a process and, in particular, to its stability properties.

In this chapter, we investigate the problems of dissipativity analysis and synthesis for discrete-time T-S fuzzy stochastic systems with time-varying delay. A new model transformation method is first introduced to pull time-delay uncertainty out of the original system. Consequently, the uncertainty is confined to a subsystem and the approximated main system contains only constant delays. A sufficient condition of dissipativity is established in terms of LMIs by using LKF approach. Then, a fuzzy controller, which guarantees that the closed-loop system is dissipative, is designed based on the dissipativity conditions. The main contributions of this work can be summarized as: 1) Proposing a novel two-term approximation method in the process of model transformation; 2) A basis-dependent LKF is employed to reduce the conservativeness; 3) The obtained parameterized LMIs are replaced by the less conservative LMIs.

L. Wu et al., *Fuzzy Control Systems with Time-Delay and Stochastic Perturbation*,
Studies in Systems, Decision and Control 12, DOI: 10.1007/978-3-319-11316-6_8

8.2 System Description and Preliminaries

Consider the following T-S fuzzy stochastic system with time-varying delay:

♦ **Plant Form:**

Rule i: IF $\theta_1(k)$ is $\mathcal{M}_{i1}$ and $\theta_2(k)$ is $\mathcal{M}_{i2}$ and … and $\theta_p(k)$ is $\mathcal{M}_{ip}$, THEN

$$\begin{aligned} x(k+1) &= A_i x(k) + A_{di} x(k-d(k)) + B_{1i} u(k) + B_{\omega i} \omega(k) \\ &\quad + \left[E_i x(k) + E_{di} x(k-d(k)) + B_{2i} u(k)\right] \varpi(k), && \text{(8.1a)} \\ z(k) &= C_i x(k) + C_{di} x(k-d(k)) + D_{1i} u(k) + D_{\omega i} \omega(k) \\ &\quad + \left[F_i x(k) + F_{di} x(k-d(k)) + D_{2i} u(k)\right] \varpi(k), && \text{(8.1b)} \\ x(k) &= \phi(k), \quad k = -d_2, -d_2+1, \ldots, 0, && \text{(8.1c)} \end{aligned}$$

where $i = 1, 2, \ldots, r$, and r is the number of IF-THEN rules; $\mathcal{M}_{ij} (i = 1, 2, \ldots, r; j = 1, 2, \ldots, p)$ are the fuzzy sets; $\theta(k) = \left[\theta_1(k)\ \theta_2(k)\ \cdots\ \theta_p(k)\right]^T$ is the premise variable vector. where $x(k) \in \mathbf{R}^n$ is the state vector; $u(k) \in \mathbf{R}^p$ is the control input; $\omega(k) \in \mathbf{R}^q$ is a deterministic exogenous input; $\varpi(k)$ is a one-dimensional, zero mean Gaussian white noise sequence on a probability space $(\Omega, \mathcal{F}, \mathcal{P})$ with

$$\mathbf{E}\{\varpi(k)\} = 0, \quad \mathbf{E}\{\varpi^2(k)\} = 1, \quad \mathbf{E}\{\varpi(l)\varpi(k)\} = 0, \quad l \neq k, \qquad (8.2)$$

where $z(k) \in \mathbf{R}^q$ is the measurement output vector; $d(k)$ is the time-delay, which is a positive integer and satisfies $1 \leq d_1 \leq d(k) \leq d_2$, where d_1 and d_2 are positive constants representing the lower and upper bounds of the time-varying delay, respectively. Clearly, $d_1 = d_2$ means that the time-delay $d(k)$ is time-invariant. $\phi(k)$ is the initial condition sequence.

It is assumed that the premise variables do not depend on input variable $u(k)$ explicitly. Then the defuzzified dynamics of the T-S fuzzy stochastic system in (8.1) can be represented as

$$\begin{aligned} x(k+1) &= \sum_{i=1}^{r} h_i(\theta) \left[A_i x(k) + A_{di} x(k-d(k)) + B_{1i} u(k) + B_{\omega i} \omega(k)\right] \\ &\quad + \sum_{i=1}^{r} h_i(\theta) \left[E_i x(k) + E_{di} x(k-d(k)) + B_{2i} u(k)\right] \varpi(k), && \text{(8.3a)} \\ z(k) &= \sum_{i=1}^{r} h_i(\theta) \left[C_i x(k) + C_{di} x(k-d(k)) + D_{1i} u(k) + D_{\omega i} \omega(k)\right] \\ &\quad + \sum_{i=1}^{r} h_i(\theta) \left[F_i x(k) + F_{di} x(k-d(k)) + D_{2i} u(k)\right] \varpi(k). && \text{(8.3b)} \end{aligned}$$

where $h_i(\theta)$, $i = 1, 2, \ldots, r$ are the normalized membership functions, which are defined as that of (1.2) in Chapter 1.

We present the open-loop system of (8.3) in a compact form as

$$\begin{aligned} x(k+1) &= \bar{A}(k)x(k) + \bar{A}_d(k)x(k-d(k)) + \bar{B}_\omega(k)\omega(k) \\ &\quad + \left[\bar{E}(k)x(k) + \bar{E}_d(k)x(k-d(k))\right]\varpi(k), \qquad (8.4a)\\ z(k) &= \bar{C}(k)x(k) + \bar{C}_d(k)x(k-d(k)) + \bar{D}_\omega(k)\omega(k) \\ &\quad + \left[\bar{F}(k)x(k) + \bar{F}_d(k)x(k-d(k))\right]\varpi(k), \qquad (8.4b) \end{aligned}$$

where

$$\begin{aligned} &\bar{A}(k) \triangleq \sum_{i=1}^{r} h_i(\theta)A_i, \quad \bar{A}_d(k) \triangleq \sum_{i=1}^{r} h_i(\theta)A_{di}, \quad \bar{B}_\omega(k) \triangleq \sum_{i=1}^{r} h_i(\theta)B_{\omega i}, \\ &\bar{C}(k) \triangleq \sum_{i=1}^{r} h_i(\theta)C_i, \quad \bar{C}_d(k) \triangleq \sum_{i=1}^{r} h_i(\theta)C_{di}, \quad \bar{D}_\omega(k) \triangleq \sum_{i=1}^{r} h_i(\theta)D_{\omega i}, \\ &\bar{E}(k) \triangleq \sum_{i=1}^{r} h_i(\theta)E_i, \quad \bar{E}_d(k) \triangleq \sum_{i=1}^{r} h_i(\theta)E_{di}, \\ &\bar{F}(k) \triangleq \sum_{i=1}^{r} h_i(\theta)F_i, \quad \bar{F}_d(k) \triangleq \sum_{i=1}^{r} h_i(\theta)F_{di}. \end{aligned}$$

Consider the following fuzzy controller:

♦ **Controller Form:**

Rule i: IF $\theta_1(k)$ is $\mathcal{M}_{i1}$ and $\theta_2(k)$ is $\mathcal{M}_{i2}$ and … and $\theta_p(k)$ is $\mathcal{M}_{ip}$, THEN

$$u(k) = K_i x(k), \quad i = 1, 2, \ldots, r,$$

where K_i is the gain matrix of the state feedback controller in each rule, and the state-feedback controller in (8.3) is given by

$$u(k) = \sum_{i=1}^{r} h_i K_i x(k). \qquad (8.5)$$

Under the control of (8.5), the closed-loop system is obtained as

$$\begin{aligned} x(k+1) &= \sum_{i=1}^{r}\sum_{j=1}^{r} h_i(\theta)h_j(\theta)A_{ij}x(k) + \sum_{i=1}^{r} h_i(\theta)A_{di}x(k-d(k)) \\ &\quad + \sum_{i=1}^{r} h_i(\theta)B_{\omega i}\omega(k) + \left[\sum_{i=1}^{r}\sum_{j=1}^{r} h_i(\theta)h_j(\theta)E_{ij}x(k)\right. \\ &\quad \left. + \sum_{i=1}^{r} h_i(\theta)E_{di}x(k-d(k))\right]\varpi(k), \qquad (8.6a)\\ z(k) &= \sum_{i=1}^{r}\sum_{j=1}^{r} h_i(\theta)h_j(\theta)C_{ij}x(k) + \sum_{i=1}^{r} h_i(\theta)C_{di}x(k-d(k)) \end{aligned}$$

$$+\sum_{i=1}^{r} h_i(\theta) D_{\omega i}\omega(k) + \left[\sum_{i=1}^{r}\sum_{j=1}^{r} h_i(\theta)h_j(\theta)F_{ij}x(k)\right.$$
$$\left. + \sum_{i=1}^{r} h_i(\theta)F_{di}x(k-d(k))\right]\varpi(k), \tag{8.6b}$$

where

$$A_{ij} \triangleq A_i + B_{1i}K_j, \quad E_{ij} \triangleq E_i + B_{2i}K_j,$$
$$C_{ij} \triangleq C_i + D_{1i}K_j, \quad F_{ij} \triangleq F_i + D_{2i}K_j.$$

The compact form of the closed-loop system (8.6) can be given as

$$x(k+1) = \hat{A}(k)x(k) + \bar{A}_d(k)x(k-d(k)) + \bar{B}_\omega(k)\omega(k)$$
$$+ \left[\hat{E}(k)x(k) + \bar{E}_d(k)x(k-d(k))\right]\varpi(k), \tag{8.7a}$$
$$z(k) = \hat{C}(k)x(k) + \bar{C}_d(k)x(k-d(k)) + \bar{D}_\omega(k)\omega(k)$$
$$+ \left[\hat{F}(k)x(k) + \bar{F}_d(k)x(k-d(k))\right]\varpi(k), \tag{8.7b}$$

where

$$\hat{A}(k) \triangleq \sum_{i=1}^{r}\sum_{j=1}^{r} h_i(\theta)h_j(\theta)A_{ij}, \quad \hat{E}(k) \triangleq \sum_{i=1}^{r}\sum_{j=1}^{r} h_i(\theta)h_j(\theta)E_{ij},$$
$$\hat{C}(k) \triangleq \sum_{i=1}^{r}\sum_{j=1}^{r} h_i(\theta)h_j(\theta)C_{ij}, \quad \hat{F}(k) \triangleq \sum_{i=1}^{r}\sum_{j=1}^{r} h_i(\theta)h_j(\theta)F_{ij}.$$

Before presenting the main results of this chapter, we first introduce the following definitions for system (8.4), which will be essential for the main results of this chapter.

The energy supply function of system (8.4) is defined as

$$J(\omega, z, T) = \langle z, \mathcal{X}z\rangle_T + 2\langle z, \mathcal{Y}\omega\rangle_T + \langle \omega, \mathcal{Z}\omega\rangle_T, \quad \forall\, T \geq 0,$$

where $\mathcal{X}$, $\mathcal{Y}$ and $\mathcal{Z}$ are real matrices of appropriate dimensions, with $\mathcal{X}$ and $\mathcal{Z}$ being symmetric matrices, and

$$\langle u, v\rangle_T \triangleq \sum_{k=0}^{T} u^T(k)v(k).$$

Definition 8.1. Under zero initial condition, system (8.4) is said to be stochastically $(\mathcal{X}, \mathcal{Y}, \mathcal{Z})$-dissipative if for all $\omega(k) \in \ell_2[0, \infty)$, the energy supply function satisfies

$$\mathbf{E}\left\{J(\omega, z, T)\right\} \geq 0, \quad \forall\, T > 0.$$

Furthermore, system (8.4) is called strictly stochastically $(\mathcal{X}, \mathcal{Y}, \mathcal{Z})$-$\alpha$-dissipative if for a sufficiently small scalar $\alpha > 0$, the energy supply function satisfies

$$\mathbf{E}\{J(\omega, z, T)\} \geq \alpha \langle \omega, \omega \rangle_T, \quad \forall\, T > 0. \tag{8.8}$$

Remark 8.2. The study of dissipative system was initiated by Willems [215] to tie together ideas common to network theory and feedback control theory. Original definition of dissipativity is defined for deterministic systems. In view of the stochastic characteristics of the T-S fuzzy system in this chapter, we expended the original definition into stochastic dissipativity by taking expectation. Stochastic stability can also be defined in this way. ♦

Remark 8.3. In most cases, matrices $\mathcal{X}$, $\mathcal{Y}$ and $\mathcal{Z}$ are given with $\mathcal{Z}$ real symmetric and $\mathcal{X} < 0$. A protruding feature of the dissipativity theorem is that it generalizes many system theorems. Those theorems include the bounded real theorem, passivity theorem, circle criterion and sector bounded nonlinearity. We can get these theorems by setting the $\mathcal{X}$, $\mathcal{Y}$, $\mathcal{Z}$ parameters [226]. Some special cases are listed as follows:

i) $\mathcal{H}_\infty$ performance: $\mathcal{Z} = \gamma^2 I$, $\gamma > 0$, $\mathcal{Y} = 0$, and $\mathcal{X} = -I$;
ii) Positive real performance: $\mathcal{Z} = 0$, $\mathcal{Y} = I$, and $\mathcal{X} = 0$;
iii) Mixed performance: $\mathcal{Z} = \theta\gamma^2 I$, $\gamma > 0$, $\mathcal{Y} = (1-\theta)$, $\theta \in [0, 1]$, and $\mathcal{X} = -\theta I$;
iv) Sector bounded performance: $\mathcal{Z} = -\frac{1}{2}\left(\mathcal{K}_1^T\mathcal{K}_2 + \mathcal{K}_2^T\mathcal{K}_1\right)$, $\gamma > 0$, $\mathcal{Y} = \frac{1}{2}\left(\mathcal{K}_1 + \mathcal{K}_2\right)^T$, and $\mathcal{X} = -I$, for some constant matrices $\mathcal{K}_1$ and $\mathcal{K}_2$. ♦

The problems concerned in this chapter are listed as follows:

Problem 8.4. (Model transformation) Consider the open-loop system in (8.4), by pulling out the uncertainty in $x(k - d(k))$, transform the original system into two subsystems: certain part and uncertain part.

Problem 8.5. (Dissipativity analysis) Consider the open-loop system in (8.4), with the given system matrices, determine under what condition system (8.4) is stochastically dissipative.

Problem 8.6. (Dissipative control) Consider the closed-loop system in (8.6), with the given system matrices, find a fuzzy controller $u(k)$ in the form of (8.5), such that the closed-loop system in (8.6) is stochastically dissipative.

8.3 Main Results

8.3.1 Model Transformation

The time-varying time-delay $d(k)$ is an undesired uncertain part in the system, which perplexes the system analysis and brings conservatism to the delay-dependent conditions. To avoid the disadvantages of time-varying $d(k)$, we employ a novel model transformation method to pull out the uncertainty in $d(k)$, then the delayed state $x(k-d(k))$ can be expressed as

$$x(k-d(k)) = \frac{1}{2}\left[x(k-d_1)+x(k-d_2)\right] + \frac{\tau}{2}\omega_d(k), \tag{8.9}$$

where $\tau \triangleq d_2 - d_1$.

Remark 8.7. By this manipulation, $x(k-d(k))$ is divided into two parts: certain part $\frac{1}{2}\left[x(k-d_1)+x(k-d_2)\right]$ and uncertain part $\frac{\tau}{2}\omega_d(k)$. The certain part can be regarded as the approximation of $x(k-d(k))$ and $\frac{\tau}{2}\omega_d(k)$ is undoubtedly the approximation error. This approximation method has been adopted in [137] to analyze the stability problems of uncertain time-delay systems. Obviously, there are also some other approximation methods, such as $x(k-d(k)) \approx x(k-d_1)$ in [106], and $x(k-d(k)) \approx x(k-(d_1+d_2)/2)$ in [61] (if $(d_1+d_2)/2$ is not an integer, it can be replaced by the minimum integer which is more than or equal to $(d_1+d_2)/2$). Comparison of the three kinds of methods can be found in Remark 8.8 and Example 8.21. ♦

By defining $\delta(k) \triangleq x(k+1) - x(k)$ and simple calculation, we can find

$$\begin{aligned}
\omega_d(k) &= \frac{2}{\tau}\left\{x(k-d(k)) - \frac{1}{2}\left[x(k-d_1)+x(k-d_2)\right]\right\} \\
&= \frac{1}{\tau}\left[\sum_{i=k-d_2}^{k-d(k)-1}\delta(i) - \sum_{i=k-d(k)}^{k-d_1-1}\delta(i)\right] \\
&= \frac{1}{\tau}\left[\sum_{i=k-d_2}^{k-d_1-1}\psi(i)\delta(i)\right],
\end{aligned} \tag{8.10}$$

with

$$\psi(i) \triangleq \begin{cases} 1, & \text{when } i \le k-d(k)-1, \\ -1, & \text{when } i > k-d(k)-1. \end{cases}$$

Define

$$\begin{aligned}
\bar{\Psi}_1(k) &\triangleq \left[\bar{A}(k)\ \ \frac{1}{2}\bar{A}_d(k)\ \ \frac{1}{2}\bar{A}_d(k)\ \ \frac{\tau}{2}\bar{A}_d(k)\ \ \bar{B}_\omega(k)\right], \\
\bar{\Psi}_2(k) &\triangleq \left[\bar{E}(k)\ \ \frac{1}{2}\bar{E}_d(k)\ \ \frac{1}{2}\bar{E}_d(k)\ \ \frac{\tau}{2}\bar{E}_d(k)\ \ 0_n\right], \\
\bar{\Psi}_3(k) &\triangleq \left[\bar{A}(k)-I_n\ \ \frac{1}{2}\bar{A}_d(k)\ \ \frac{1}{2}\bar{A}_d(k)\ \ \frac{\tau}{2}\bar{A}_d(k)\ \ \bar{B}_\omega(k)\right], \\
\bar{\Psi}_4(k) &\triangleq \left[\bar{C}(k)\ \ \frac{1}{2}\bar{C}_d(k)\ \ \frac{1}{2}\bar{C}_d(k)\ \ \frac{\tau}{2}\bar{C}_d(k)\ \ \bar{D}_\omega(k)\right], \\
\bar{\Psi}_5(k) &\triangleq \left[\bar{F}(k)\ \ \frac{1}{2}\bar{F}_d(k)\ \ \frac{1}{2}\bar{F}_d(k)\ \ \frac{\tau}{2}\bar{F}_d(k)\ \ 0_n\right].
\end{aligned}$$

Replacing $x(k-d(k))$ by (8.9), then the original open-loop system (8.4) can be transformed to the following two interconnected subsystems,

$$\begin{cases} x(k+1) = \bar{\Psi}_1(k)\eta(k) + \bar{\Psi}_2(k)\eta(k)\varpi(k), \\ \delta(k) = \bar{\Psi}_3(k)\eta(k) + \bar{\Psi}_2(k)\eta(k)\varpi(k), \\ z(k) = \bar{\Psi}_4(k)\eta(k) + \bar{\Psi}_5(k)\eta(k)\varpi(k), \end{cases} \tag{8.11a}$$

$$\omega_d(k) = \Delta_d(\delta(k)), \tag{8.11b}$$

where

$$\eta(k) \triangleq \begin{bmatrix} x(k) \\ x(k-d_1) \\ x(k-d_2) \\ \omega_d(k) \\ \omega(k) \end{bmatrix},$$

and the mapping $\Delta_d(\cdot) : \delta(k) \to \omega_d(k)$ is an operator used to denote the relationship of $\omega_d(k)$ and $\delta_d(k)$ in (8.10). In light of model transformation, the resulting subsystem (8.11a) (main system) has only two known constant delays, and the original uncertainty in $d(k)$ has been moved into subsystem (8.11b).

Remark 8.8. Let $\varrho(k)$ denote the approximation error (obviously, $\varrho(k) = \frac{\tau}{2}\omega_d(k)$ in this chapter), then, for the method in this chapter, we have $\frac{\|\varrho\|_2}{\|\delta\|_2} < \frac{\tau}{2}$ which can be derived from Lemma 2 in [137]. For the method proposed in [61], the bound of its approximation error in continuous-time cases has been given in [80], that is $\frac{\|\varrho\|_2}{\|\delta\|_2} < \frac{\tau}{\sqrt{2}}$ for $\dot{d}(t) = \infty$ and $\frac{\|\varrho\|_2}{\|\delta\|_2} < \frac{\tau}{2}$ for $\dot{d}(t) < 1$, where $\dot{d}(t)$ is the first derivative of $d(t)$ with respect to t and $d(t)$ is the continuous-time case of $d(k)$. So, if the changing rate of $d(k)$ is not very high, the approximation error in [61] is same as that of (8.9). But, when $d(k)$ changes very fast, the approximation error of (8.9) will be relatively smaller than that of [61]. Moreover, it has been given in Lemma 2 of [106] that, $\varrho(k) = x(k-d(k)) - x(k-d_1)$ is bounded by $\frac{\|\varrho\|_2}{\|\delta\|_2} < \tau$. So the approximation error of [106] is much bigger than that of [61] and (8.9). ♦

Proposition 8.9. *Suppose $V_s(k)$ is an LKF of the subsystem (8.11a), and $S > 0$ is a matrix of appropriate dimension, then an LKF of the interconnected system (8.4) can be constructed as*

$$V(k) = V_s(k) + \frac{1}{\tau} \sum_{i=-d_2}^{-d_1-1} \sum_{j=k+i}^{k-1} \delta^T(j)S\delta(j). \tag{8.12}$$

Moreover, if $V_s(k)$ and S satisfy

$$\mathbf{E}\{\Delta V_s(k)\} + \mathbf{E}\{\delta^T(k)S\delta(k)\} - \mathbf{E}\{\omega_d^T(k)S\omega_d(k)\} < 0, \tag{8.13}$$

then $V(k)$ will directly demonstrate that system (8.4) is asymptotically stochastically stable.

Proof. Obviously, the specially constructed functional $V(k)$ always has the property $V(k) \geq 0$. And $V(k) = 0$ if and only if $V_s(k) = 0$ and $\delta(k) = 0$. In addition, taking the forward difference of $V(k)$ along the trajectory of system (8.4) and taking expectation, we have

$$\begin{aligned}
\mathbf{E}\{\Delta V(k)\} &= \mathbf{E}\{\Delta V_s(k)\} + \mathbf{E}\left\{\frac{1}{\tau}\sum_{i=-d_2}^{-d_1-1}\left[\delta^T(k)S\delta(k) - \delta^T(k+i)S\delta(k+i)\right]\right\} \\
&= \mathbf{E}\{\Delta V_s(k)\} + \mathbf{E}\left\{\delta^T(k)S\delta(k)\right\} \\
&\quad - \mathbf{E}\left\{\frac{1}{\tau}\sum_{i=-d_2}^{-d_1-1}\left[\delta^T(k+i)S\delta(k+i)\right]\right\} \\
&= \mathbf{E}\{\Delta V_s(k)\} + \mathbf{E}\left\{\delta^T(k)S\delta(k)\right\} \\
&\quad - \mathbf{E}\left\{\frac{1}{\tau}\sum_{i=k-d_2}^{k-d_1-1}\left[\psi(i)\delta^T(i)S\delta(i)\psi(i)\right]\right\}.
\end{aligned}$$

Applying the Jensen inequality in Lemma 1.22 of Chapter 1, and considering (8.13), we have

$$\begin{aligned}
\mathbf{E}\{\Delta V(k)\} &\leq \mathbf{E}\{\Delta V_s(k)\} + \mathbf{E}\left\{\delta^T(k)S\delta(k)\right\} \\
&\quad -\mathbf{E}\left\{\frac{1}{\tau^2}\left[\sum_{i=k-d_2}^{k-d_1-1}\psi(i)\delta(i)\right]^T S\left[\sum_{i=k-d_2}^{k-d_1-1}\psi(i)\delta(i)\right]\right\} \\
&= \mathbf{E}\{\Delta V_s(k)\} + \mathbf{E}\left\{\delta^T(k)S\delta(k)\right\} - \mathbf{E}\left\{\omega_d^T(k)S\omega_d(k)\right\} \\
&< 0.
\end{aligned} \tag{8.14}$$

Based on the Lyapunov stability theory, (8.14) means that system (8.4) is asymptotically stochastically stable. The proof is completed. ■

Remark 8.10. By a simple example, we will illustrate the important role of this proposition. For simplicity, consider a simple interconnected system (8.15), which is composed of the following two subsystems,

$$x(k+1) = Ax(k) + A_s\omega_d(k), \tag{8.15a}$$

$$\omega_d(k) = \Delta_d(\delta(k)). \tag{8.15b}$$

Suppose the LKF $V_s(k)$ of subsystem (8.15a) is chosen as

$$V_s(k) \triangleq x^T(k)Px(k).$$

Calculating the difference of $V_s(k)$ along the trajectories of system (8.15), we have $\Delta V_s(k) = \xi^T(k)\Upsilon_s\xi(k)$, where

$$\xi(k) \triangleq \begin{bmatrix} x(k) \\ \omega_d(k) \end{bmatrix}, \quad \Upsilon_s \triangleq \begin{bmatrix} A^T PA - P & A^T PA_s \\ \star & A_s^T PA_s \end{bmatrix}.$$

It is obvious that $V_s(k)$ cannot be regarded as an LKF of (8.15), because the term $A_s^T PA_s$ ($P > 0$) of Υ_s can never be smaller than zero. Moreover, with the help of Proposition 8.9, we construct a new LKF for (8.15) as

$$V(k) = V_s(k) + \frac{1}{\tau} \sum_{i=-d_2}^{-d_1-1} \sum_{j=k+i}^{k-1} \delta^T(j) S \delta(j),$$

then $\Delta V(k) \leq \xi^T(k) \Upsilon \xi(k)$ with

$$\Upsilon \triangleq \begin{bmatrix} A^T PA + (A - I_n)^T S(A - I_n) - P & A^T PA_s + (A - I_n)^T SA_s \\ \star & A_s^T PA_s + A_s^T SA_s - S \end{bmatrix}.$$

For appropriate A, A_s, P and S, matrix Υ will be able to satisfy $\Upsilon < 0$. Thus $V(k)$ can be regarded as an LKF of system (8.15). ♦

Remark 8.11. Proposition 8.9 gives us an effective method to construct an LKF for system (8.4) based on the LKF of subsystem (8.11a). Inequality (8.13) is a commonly used sufficient condition to ensure $\frac{\|\delta(k)\|_2}{\|\omega_d(k)\|_2} < 1$ for (8.11a). So accompanied with $\|\Delta_d\|_\infty < 1$, the proof of stability can also be completed by the scaled small gain theorem. It should be pointed out that, (8.13) is a sufficient condition of $\mathbf{E}\{\Delta V(k)\} < 0$. Therefore, the stability condition obtained by $\mathbf{E}\{\Delta V(k)\} < 0$ will be less conservative than the commonly used sufficient condition induced by scaled small gain theorem. ♦

8.3.2 Dissipativity Analysis

The dissipativity problem is minutely analyzed in this section. Based on the transformed system and LKF method, we are determined to find a sufficient condition of dissipativity for the given system (8.4). To make analysis simpler, we make the following definitions:

$$W_1 \triangleq \begin{bmatrix} I_n & 0_{n\times 3n} & 0_{n\times n} \end{bmatrix}, \quad W_2 \triangleq \begin{bmatrix} 0_{n\times n} & I_n & 0_{n\times 3n} \end{bmatrix},$$
$$W_3 \triangleq \begin{bmatrix} 0_{n\times 2n} & I_n & 0_{n\times 2n} \end{bmatrix}, W_4 \triangleq \begin{bmatrix} 0_{n\times 3n} & I_n & 0_{n\times n} \end{bmatrix}, W_5 \triangleq \begin{bmatrix} 0_{n\times 3n} & 0_{n\times n} & I_n \end{bmatrix},$$

and

$$\bar{P}(k) \triangleq \sum_{i=1}^{r} h_i(\theta) P_i, \quad \bar{Q}_1(k) \triangleq \sum_{i=1}^{r} h_i(\theta) Q_{1i}, \quad \bar{Q}_2(k) \triangleq \sum_{i=1}^{r} h_i(\theta) Q_{2i}.$$

Improved sufficient condition of dissipativity is given in the following theorem.

Theorem 8.12. *Given positive integers d_1, d_2 and a scalar $\alpha > 0$, system (8.4) is strictly stochastically $(\mathcal{X}, \mathcal{Y}, \mathcal{Z})$-$\alpha$-dissipative, if there exist matrices*

$P_i > 0$, $Q_{1i} > 0$, $Q_{2i} > 0$, $R_1 > 0$, $R_2 > 0$, $S > 0$ $(i = 1, 2, \ldots, r)$, *such that for any integer* k,

$$\begin{bmatrix} \Phi(k) & \Psi(k) \\ \star & -\Xi(k) \end{bmatrix} < 0, \tag{8.16}$$

where

$$\begin{aligned}
\Phi(k) &\triangleq -W_1^T\bar{P}(k)W_1 + W_1^T\bar{Q}_1(k)W_1 - W_2^T\bar{Q}_1(k-d_1)W_2 + W_1^T\bar{Q}_2(k)W_1 \\
&\quad - (W_1 - W_2)^T R_1 (W_1 - W_2) - W_3^T\bar{Q}_2(k-d_2)W_3 - W_4^T S W_4 \\
&\quad - (W_1 - W_3)^T R_2 (W_1 - W_3) - 2W_5^T\mathcal{Y}\bar{\Psi}_4(k) - W_5^T(\mathcal{Z} - \alpha I)W_5, \\
\Psi(k) &\triangleq \big[\bar{\Psi}_1^T(k)\bar{P}(k+1) \ \ \bar{\Psi}_2^T(k)\bar{P}(k+1) \ \ d_1\bar{\Psi}_3^T(k)R_1 \ \ d_2\bar{\Psi}_3^T(k)R_2 \\
&\quad\ d_1\bar{\Psi}_2^T(k)R_1 \ \ d_2\bar{\Psi}_2^T(k)R_2 \ \ \bar{\Psi}_3^T(k)S \ \ \bar{\Psi}_2^T(k)S \ \ \bar{\Psi}_4^T(k) \ \ \bar{\Psi}_5^T(k)\big], \\
\Xi(k) &\triangleq \mathrm{diag}\left\{\bar{P}(k+1), \bar{P}(k+1), R_1, R_2, R_1, R_2, S, S, -\mathcal{X}^{-1}, -\mathcal{X}^{-1}\right\}.
\end{aligned}$$

Proof. Choose a basis-dependent LKF for (8.11a) as $V_s(k) \triangleq \sum_{i=1}^3 V_i(k)$ with

$$\begin{aligned}
V_1(k) &\triangleq x^T(k)\bar{P}(k)x(k), \\
V_2(k) &\triangleq \sum_{i=k-d_1}^{k-1} x^T(i)\bar{Q}_1(i)x(i) + \sum_{i=k-d_2}^{k-1} x^T(i)\bar{Q}_2(i)x(i), \\
V_3(k) &\triangleq \sum_{i=-d_1}^{-1}\sum_{j=k+i}^{k-1} d_1\delta^T(j)R_1\delta(j) + \sum_{i=-d_2}^{-1}\sum_{j=k+i}^{k-1} d_2\delta^T(j)R_2\delta(j).
\end{aligned}$$

Calculating the increment of $V_s(k)$ along the trajectory of system (8.4) and taking expectation, we have $\mathbf{E}\{\Delta V_s(k)\} = \sum_{i=1}^3 \mathbf{E}\{\Delta V_i(k)\}$, where

$$\begin{aligned}
\mathbf{E}\{\Delta V_1(k)\} &= \mathbf{E}\left\{V_1(k+1) - V_1(k)\right\} \\
&= \mathbf{E}\left\{x^T(k+1)\bar{P}(k+1)x(k+1) - x^T(k)\bar{P}(k)x(k)\right\} \\
&= \eta^T(k)\left[\bar{\Psi}_1^T(k)\bar{P}(k+1)\bar{\Psi}_1(k) + \bar{\Psi}_2^T(k)\bar{P}(k+1)\bar{\Psi}_2(k)\right]\eta(k) \\
&\quad - \mathbf{E}\left\{x^T(k)\bar{P}(k)x(k)\right\} \\
&= \eta^T(k)\big[\bar{\Psi}_1^T(k)\bar{P}(k+1)\bar{\Psi}_1(k) + \bar{\Psi}_2^T(k)\bar{P}(k+1)\bar{\Psi}_2(k) \\
&\quad - W_1^T\bar{P}(k)W_1\big]\eta(k), \\
\mathbf{E}\left\{\Delta V_2(k)\right\} &= \mathbf{E}\left\{V_2(k+1) - V_2(k)\right\} \\
&= x^T(k)\bar{Q}_1(k)x(k) - x^T(k-d_1)\bar{Q}_1(k-d_1)x(k-d_1) \\
&\quad + x^T(k)\bar{Q}_2(k)x(k) - x^T(k-d_2)\bar{Q}_2(k-d_2)x(k-d_2) \\
&= \eta^T(k)\big[W_1^T\bar{Q}_1(k)W_1 - W_2^T\bar{Q}_1(k-d_1)W_2 + W_1^T\bar{Q}_2(k)W_1 \\
&\quad - W_3^T\bar{Q}_2(k-d_2)W_3\big]\eta(k).
\end{aligned}$$

By virtue of Jensen inequality again, it follows that

$$\begin{aligned}
&\mathbf{E}\{\Delta V_3(k)\} = \mathbf{E}\{V_3(k+1) - V_3(k)\}\\
&= \mathbf{E}\left\{d_1^2\delta^T(k)R_1\delta(k) - \sum_{i=k-d_1}^{k-1} d_1\delta^T(i)R_1\delta(i)\right.\\
&\qquad \left. +d_2^2\delta^T(k)R_2\delta(k) - \sum_{i=k-d_2}^{k-1} d_2\delta^T(i)R_2\delta(i)\right\}\\
&\leq \eta^T(k)\left[\bar{\Psi}_3^T(k)\left(d_1^2R_1+d_2^2R_2\right)\bar{\Psi}_3(k)+\bar{\Psi}_2^T(k)\left(d_1^2R_1+d_2^2R_2\right)\bar{\Psi}_2(k)\right]\eta(k)\\
&-\eta^T(k)\left[(W_1-W_2)^T R_1(W_1-W_2)+(W_1-W_3)^T R_2(W_1-W_3)\right]\eta(k).
\end{aligned}$$

Based on Proposition 8.9, an LKF for system (8.4) can be constructed as

$$V(k) = V_s(k) + \frac{1}{\tau}\sum_{i=-d_2}^{-d_1-1}\sum_{j=k+i}^{k-1}\delta^T(j)S\delta(j).$$

Accordingly,

$$\begin{aligned}
\mathbf{E}\{\Delta V(k)\} &\leq \mathbf{E}\{\Delta V_s(k)\}\\
&\quad +\eta^T(k)\left[\bar{\Psi}_3^T(k)S\bar{\Psi}_3(k)+\bar{\Psi}_2^T(k)S\bar{\Psi}_2(k)-W_4^TSW_4\right]\eta(k)\\
&\leq \eta^T(k)\Omega(k)\eta(k),
\end{aligned}$$

where

$$\begin{aligned}
\Omega(k) \triangleq\ & \bar{\Psi}_1^T(k)\bar{P}(k+1)\bar{\Psi}_1(k)+\bar{\Psi}_2^T(k)\bar{P}(k+1)\bar{\Psi}_2(k)+\bar{\Psi}_2^T(k)S\bar{\Psi}_2(k)\\
&+\bar{\Psi}_3^T(k)S\bar{\Psi}_3(k)+\bar{\Psi}_2^T(k)\left(d_1^2R_1+d_2^2R_2\right)\bar{\Psi}_2(k)\\
&+\bar{\Psi}_3^T(k)\left(d_1^2R_1+d_2^2R_2\right)\bar{\Psi}_3(k)-W_1^T\bar{P}(k)W_1+W_1^T\bar{Q}_1(k)W_1\\
&+W_1^T\bar{Q}_2(k)W_1-W_2^T\bar{Q}_1(k-d_1)W_2-W_3^T\bar{Q}_2(k-d_2)W_3-W_4^TSW_4\\
&-(W_1-W_2)^T\bar{R}_1(W_1-W_2)-(W_1-W_3)^T\bar{R}_2(W_1-W_3).
\end{aligned}$$

Furthermore, the above inequality implies

$$\begin{aligned}
&\mathbf{E}\{\Delta V(k)\} - \mathbf{E}\left\{z^T(k)\mathcal{X}z(k)+2\omega^T(k)\mathcal{Y}z(k)+\omega^T(k)\mathcal{Z}\omega(k)\right\}\\
&\quad +\alpha\mathbf{E}\left\{\omega^T(k)\omega(k)\right\}\\
&\quad \leq \eta^T(k)\Omega(k)\eta(k)-\eta^T(k)\left[\bar{\Psi}_4^T(k)\mathcal{X}\bar{\Psi}_4(k)+\bar{\Psi}_5^T(k)\mathcal{X}\bar{\Psi}_5(k)\right]\eta(k)\\
&\qquad -2\eta^T(k)W_5^T\mathcal{Y}\bar{\Psi}_4(k)\eta(k)-\eta^T(k)W_5^T(\mathcal{Z}-\alpha I)W_5\eta(k)\\
&\quad = \eta^T(k)\bar{\Omega}(k)\eta(k),
\end{aligned} \tag{8.17}$$

where

$$\begin{aligned}\bar{\Omega}(k) \triangleq\ & \bar{\Psi}_1^T(k)\bar{P}(k+1)\bar{\Psi}_1(k) + \bar{\Psi}_2^T(k)\bar{P}(k+1)\bar{\Psi}_2(k) + \bar{\Psi}_2^T(k)S\bar{\Psi}_2(k) \\ & +\bar{\Psi}_3^T(k)S\bar{\Psi}_3(k) - \bar{\Psi}_4^T(k)\mathcal{X}\bar{\Psi}_4(k) - \bar{\Psi}_5^T(k)\mathcal{X}\bar{\Psi}_5(k) \\ & +\bar{\Psi}_2^T(k)\left(d_1^2R_1 + d_2^2R_2\right)\bar{\Psi}_2(k) + \bar{\Psi}_3^T(k)\left(d_1^2R_1 + d_2^2R_2\right)\bar{\Psi}_3(k) \\ & -W_1^T\bar{P}(k)W_1 + W_1^T\bar{Q}_1(k)W_1 + W_1^T\bar{Q}_2(k)W_1 - W_2^T\bar{Q}_1(k-d_1)W_2 \\ & -W_3^T\bar{Q}_2(k-d_2)W_3 - W_4^TSW_4 - 2W_5^T\mathcal{Y}\bar{\Psi}_4(k) - W_5^T(\mathcal{Z} - \alpha I)W_5 \\ & -\left(W_1 - W_2\right)^T R_1\left(W_1 - W_2\right) - \left(W_1 - W_3\right)^T R_2\left(W_1 - W_3\right).\end{aligned}$$

By Schur complement, (8.16) gives rise to

$$\bar{\Omega}(k) < 0.$$

Considering (8.17), we have

$$\begin{aligned}& \mathbf{E}\left\{z^T(k)\mathcal{X}z(k) + 2\omega^T(k)\mathcal{Y}z(k) + \omega^T(k)\mathcal{Z}\omega(k)\right\} \\ & -\alpha\mathbf{E}\left\{\omega^T(k)\omega(k)\right\} - \mathbf{E}\left\{V(k+1)\right\} + \mathbf{E}\left\{V(k)\right\} > 0.\end{aligned} \tag{8.18}$$

Therefore, for any integer $T > 0$, summing up both sides of (8.18) from $k = 0$ to $k = T - 1$ will result in

$$\mathbf{E}\left\{V(0)\right\} + \mathbf{E}\left\{J(\omega, z, T)\right\} > \alpha\mathbf{E}\left\{\langle\omega, \omega\rangle_T\right\} + \mathbf{E}\left\{V(T)\right\}. \tag{8.19}$$

Under zero initial condition, (8.19) means that

$$\mathbf{E}\left\{J(\omega, z, T)\right\} > \alpha\left\langle\omega, \omega\right\rangle_T.$$

Then, by Definition 8.1, the open-loop fuzzy stochastic system in (8.4) is strictly stochastically $(\mathcal{X}, \mathcal{Y}, \mathcal{Z})$-$\alpha$-dissipative. ■

Remark 8.13. If $V(k)$ is regarded as a storage function of system (8.4), inequality (8.19) will well satisfy the dissipativity definition in [215]. So the definition in this chapter is same as that in [215]. Moreover, it should be noted that, the definition in the form of (8.8) is less general but more standard than that in [215], for (8.8) is much easier to manipulate and there is no need to worry about the existence of a storage function. ♦

Remark 8.14. Note that $V_1(k)$ and $V_2(k)$ in the proof are basis-dependent. Compared with the basis-independent ones in [148], they are capable of adapting to the linear systems in each rule, which can further reduce the conservatism. ♦

Note that the obtained dissipativity condition cannot be used directly to analyze a fuzzy system since the inequality in Theorem 8.12 contains time-varying parameters, which are merely available online. To solve this problem, we need to relax this parameterized LMI [203] into strict LMIs.

Theorem 8.15. *Given positive integers d_1, d_2 and a scalar $\alpha > 0$, system (8.4) is strictly stochastically $(\mathcal{X}, \mathcal{Y}, \mathcal{Z})$-$\alpha$-dissipative, if there exist matrices $P_i > 0$, $Q_{1i} > 0$, $Q_{2i} > 0$, $R_1 > 0$, $R_2 > 0$ and $S > 0$ $(i = 1, 2, \ldots, r)$, such that the following inequalities hold,*

$$\begin{bmatrix} \Phi_{oli} & \Psi_{it} \\ \star & -\Xi_t \end{bmatrix} < 0, \quad o, l, t, i = 1, 2, \ldots, r, \tag{8.20}$$

where

$$\begin{aligned}
\Phi_{oli} &\triangleq -W_1^T P_i W_1 + W_1^T Q_{1i} W_1 + W_1^T Q_{2i} W_1 - W_2^T Q_{1l} W_2 \\
&\quad -W_3^T Q_{2o} W_3 - W_4^T S W_4 - 2W_5^T \mathcal{Y} \Psi_{4i} - W_5^T (\mathcal{Z} - \alpha I) W_5 \\
&\quad - (W_1 - W_2)^T R_1 (W_1 - W_2) - (W_1 - W_3)^T R_2 (W_1 - W_3), \\
\Psi_{it} &\triangleq \big[\Psi_{1i}^T P_t \;\; \Psi_{2i}^T P_t \;\; d_1 \Psi_{3i}^T R_1 \;\; d_2 \Psi_{3i}^T R_2 \;\; d_1 \Psi_{2i}^T R_1 \;\; d_2 \Psi_{2i}^T R_2 \\
&\quad \Psi_{3i}^T S \;\; \Psi_{2i}^T S \;\; \Psi_{4i}^T \;\; \Psi_{5i}^T \big], \\
\Psi_{1i} &\triangleq \begin{bmatrix} A_i & \frac{1}{2} A_{di} & \frac{1}{2} A_{di} & \frac{\tau}{2} A_{di} & B_{\omega i} \end{bmatrix}, \\
\Psi_{2i} &\triangleq \begin{bmatrix} E_i & \frac{1}{2} E_{di} & \frac{1}{2} E_{di} & \frac{\tau}{2} E_{di} & 0_n \end{bmatrix}, \\
\Psi_{3i} &\triangleq \begin{bmatrix} A_i - I_n & \frac{1}{2} A_{di} & \frac{1}{2} A_{di} & \frac{\tau}{2} A_{di} & B_{\omega i} \end{bmatrix}, \\
\Psi_{4i} &\triangleq \begin{bmatrix} C_i & \frac{1}{2} C_{di} & \frac{1}{2} C_{di} & \frac{\tau}{2} C_{di} & D_{\omega i} \end{bmatrix}, \\
\Psi_{5i} &\triangleq \begin{bmatrix} F_i & \frac{1}{2} F_{di} & \frac{1}{2} C_{di} & \frac{\tau}{2} C_{di} & 0_n \end{bmatrix}, \\
\Xi_t &\triangleq \operatorname{diag} \left\{ P_t, P_t, R_1, R_2, R_1, R_2, S, S, -\mathcal{X}^{-1}, -\mathcal{X}^{-1} \right\}.
\end{aligned}$$

Proof. Note that the matrices in (8.16) of Theorem 8.12 can be unfolded as

$$\begin{aligned}
\begin{bmatrix} \Phi(k) & \Psi(k) \\ \star & -\Xi(k) \end{bmatrix} &= \sum_{o=1}^{r} \sum_{l=1}^{r} \sum_{t=1}^{r} \sum_{i=1}^{r} h_o(\theta(k - d_2)) h_l(\theta(k - d_1)) \\
&\quad \times h_t(\theta(k+1)) h_i(\theta(k)) \begin{bmatrix} \Phi_{oli} & \Psi_{it} \\ \star & -\Xi_t \end{bmatrix}.
\end{aligned}$$

Then the condition in Theorem 8.12 is fulfilled if the LMIs in Theorem 8.15 hold, and it follows from the analysis in Theorem 8.12 that the open-loop system in (8.4) is strictly stochastically $(\mathcal{X}, \mathcal{Y}, \mathcal{Z})$-$\alpha$-dissipative. Then the proof is completed. ■

Remark 8.16. By Theorem 8.15, the parameterized LMI in Theorem 8.12 has been replaced by a set of strict LMIs. This means that, regardless of $\theta(k)$, a system satisfying Theorem 8.15 will always fulfil the required dissipativity criterion. ♦

For a system without stochastic noise, the result will be simplified to a large extent. System (8.4) will be rewritten as

$$x(k+1) = \bar{A}(k)x(k) + \bar{A}_d(k)x(k-d(k)) + \bar{B}_\omega(k)\omega(k), \quad (8.21a)$$
$$z(k) = \bar{C}(k)x(k) + \bar{C}_d(k)x(k-d(k)) + \bar{D}_\omega(k)\omega(k). \quad (8.21b)$$

Accordingly, the required criterion is dissipativity, but not stochastic dissipativity. The result is given in the following corollary.

Corollary 8.17. *Given positive integers d_1, d_2 and a scalar $\alpha > 0$, system (8.21) is strictly $(\mathcal{X}, \mathcal{Y}, \mathcal{Z})$-$\alpha$-dissipative, if there exist matrices $P_i > 0$, $Q_{1i} > 0$, $Q_{2i} > 0$, $(i = 1, 2, \ldots, r)$, $R_1 > 0$, $R_2 > 0$ and S, such that the following inequalities hold:*

$$\begin{bmatrix} \Phi_{oli} & \Gamma_{it} \\ \star & -\Theta_t \end{bmatrix} < 0, \quad o, l, t, i = 1, 2, \ldots, r, \quad (8.22)$$

where

$$\Gamma_{it} \triangleq \begin{bmatrix} \Psi_{1i}^T P_t & d_1 \Psi_{3i}^T R_1 & d_2 \Psi_{3i}^T R_2 & \Psi_{3i}^T S & \Psi_{4i}^T \end{bmatrix},$$
$$\Theta_t \triangleq \text{diag}\left\{P_t, R_1, R_2, S, -\mathcal{X}^{-1}\right\}.$$

Proof. This corollary is a special case of Theorem 8.15, thus it can be easily proved following the same lines as those in the proofs of Theorems 8.12 and 8.15. Here, we omit the details of the proof. ■

8.3.3 Dissipative Controller Design

In what follows, the dissipativity condition obtained in the previous section will be used to design a fuzzy controller such that the closed-loop system in (8.7) satisfies a required dissipativity criterion.

Theorem 8.18. *Given positive integers d_1, d_2 and a scalar $\alpha > 0$, the closed-loop system in (8.7) is strictly stochastically $(\mathcal{X}, \mathcal{Y}, \mathcal{Z})$-$\alpha$-dissipative, if there exist matrices $P_i > 0$, $Q_{1i} > 0$, $Q_{2i} > 0$, $R_1 > 0$, $R_2 > 0$, S, G, H_i $(i = 1, 2, \ldots, r)$, and a scalar $\varepsilon > 0$, such that for any $o, l, t, i, j = 1, 2, \ldots, r$,*

$$\Pi_{oltii} < 0, \quad (8.23)$$
$$\frac{1}{r-1}\Pi_{oltii} + \frac{1}{2}(\Pi_{oltij} + \Pi_{oltji}) < 0, \quad i \neq j, \quad (8.24)$$

where

$$\Pi_{oltij} \triangleq \begin{bmatrix} \tilde{\Phi}_{oli} & \tilde{\Psi}_{ij} \\ \star & \tilde{\Xi}_t \end{bmatrix},$$
$$\tilde{\Phi}_{oli} \triangleq -W_1^T P_i W_1 + \varepsilon^{-2} W_1^T Q_{1i} W_1 + \varepsilon^{-2} W_1^T Q_{2i} W_1 - W_2^T Q_{1l} W_2$$

$$
\begin{aligned}
&-W_3^T Q_{2o} W_3 - \varepsilon^{-2} W_4^T S W_4 - 2W_5^T \mathcal{Y}\Psi_{4i} - W_5^T(\mathcal{Z} - \alpha I)W_5 \\
&- (W_1 - \varepsilon W_3)^T R_2 (W_1 - \varepsilon W_3) - (W_1 - \varepsilon W_2)^T R_1 (W_1 - \varepsilon W_2),
\end{aligned}
$$

$$
\begin{aligned}
\tilde{\Psi}_{ij} &\triangleq \begin{bmatrix} \Psi_{1ij}^T & \Psi_{2ij}^T & d_1\Psi_{3ij}^T & d_2\Psi_{3ij}^T & d_1\Psi_{2ij}^T & d_2\Psi_{2ij}^T & \Psi_{3ij}^T & \Psi_{2ij}^T & \Psi_{4ij}^T & \Psi_{5ij}^T \end{bmatrix}, \\
\Psi_{1ij} &\triangleq \begin{bmatrix} A_i G + B_{1i} H_j & \frac{\varepsilon}{2} A_{di} G & \frac{\varepsilon}{2} A_{di} G & \frac{\varepsilon\tau}{2} A_{di} G & B_{\omega i} \end{bmatrix}, \\
\Psi_{2ij} &\triangleq \begin{bmatrix} E_i G + B_{2i} H_j & \frac{\varepsilon}{2} E_{di} G & \frac{\varepsilon}{2} E_{di} G & \frac{\varepsilon\tau}{2} E_{di} G & 0_n \end{bmatrix}, \\
\Psi_{3ij} &\triangleq \begin{bmatrix} A_i G + B_{1i} H_j - G & \frac{\varepsilon}{2} A_{di} G & \frac{\varepsilon}{2} A_{di} G & \frac{\varepsilon\tau}{2} A_{di} G & B_{\omega i} \end{bmatrix}, \\
\Psi_{4ij} &\triangleq \begin{bmatrix} C_i G + D_{1i} H_j & \frac{\varepsilon}{2} C_{di} G & \frac{\varepsilon}{2} C_{di} G & \frac{\varepsilon\tau}{2} C_{di} G & D_{\omega i} \end{bmatrix}, \\
\Psi_{5ij} &\triangleq \begin{bmatrix} F_i G + D_{2i} H_j & \frac{\varepsilon}{2} F_{di} G & \frac{\varepsilon}{2} C_{di} G & \frac{\varepsilon\tau}{2} C_{di} G & 0_n \end{bmatrix}, \\
\tilde{\Xi}_t &\triangleq \mathrm{diag}\left\{ \left(P_t - G - G^T\right), \left(P_t - G - G^T\right), \left(R_1 - G - G^T\right), \right. \\
&\quad \left(R_2 - G - G^T\right), \left(R_1 - G - G^T\right), \left(R_2 - G - G^T\right), \\
&\quad \left. (S - \varepsilon G - \varepsilon G), (S - \varepsilon G - \varepsilon G), -\mathcal{X}^{-1}, -\mathcal{X}^{-1} \right\},
\end{aligned}
$$

and the fuzzy controller is given as

$$
u(k) = \bar{K}(k)x(k) = \bar{H}(k)G^{-1}x(k), \tag{8.25}
$$

with

$$
\bar{K}(k) \triangleq \sum_{i=1}^{r} h_i(\theta)K_i, \quad \bar{H}(k) \triangleq \sum_{i=1}^{r} h_i(\theta)H_i.
$$

Proof. Define

$$
\begin{aligned}
\tilde{\Phi}(k) &\triangleq -W_1^T \bar{P}(k) W_1 - W_2^T \bar{Q}_1(k-d_1) W_2 + \varepsilon^{-2}\left(W_1^T \bar{Q}_1(k) W_1 + W_1^T \bar{Q}_2(k) W_1\right) \\
&\quad - W_3^T \bar{Q}_2(k-d_2) W_3 - W_4^T S W_4 - 2W_5^T \mathcal{Y}\bar{\Psi}_4(k) - W_5^T(\mathcal{Z} - \alpha I)W_5 \\
&\quad - (W_1 - \varepsilon W_2)^T R_1 (W_1 - \varepsilon W_2) - (W_1 - \varepsilon W_3)^T R_2 (W_1 - \varepsilon W_3), \\
\tilde{\Psi}(k) &\triangleq \begin{bmatrix} \tilde{\Psi}_1^T(k) & \tilde{\Psi}_2^T(k) & d_1\tilde{\Psi}_3^T(k) & d_2\tilde{\Psi}_3^T(k) & d_1\tilde{\Psi}_2^T(k) & d_2\tilde{\Psi}_2^T(k) \end{bmatrix. \\
&\quad \begin{matrix} \tilde{\Psi}_3^T(k) & \tilde{\Psi}_2^T(k) & \tilde{\Psi}_4^T(k) & \tilde{\Psi}_5^T(k) \end{matrix} \Big], \\
\tilde{\Psi}_1(k) &\triangleq \begin{bmatrix} \bar{A}(k)G + \bar{B}_1(k)\bar{K}(k)G & \frac{\varepsilon}{2}\bar{A}_d(k)G & \frac{\varepsilon}{2}\bar{A}_d(k)G & \frac{\varepsilon\tau}{2}\bar{A}_d(k)G & \bar{B}_\omega(k) \end{bmatrix}, \\
\tilde{\Psi}_2(k) &\triangleq \begin{bmatrix} \bar{E}(k)G + \bar{B}_2(k)\bar{K}(k)G & \frac{\varepsilon}{2}\bar{E}_d(k)G & \frac{\varepsilon}{2}\bar{E}_d(k)G & \frac{\varepsilon\tau}{2}\bar{E}_d(k)G & 0_n \end{bmatrix}, \\
\tilde{\Psi}_3(k) &\triangleq \begin{bmatrix} \bar{A}(k)G + \bar{B}_1(k)\bar{K}(k)G - G & \frac{\varepsilon}{2}\bar{A}_d(k)G & \frac{\varepsilon}{2}\bar{A}_d(k)G & \frac{\varepsilon\tau}{2}\bar{A}_d(k)G & \bar{B}_\omega(k) \end{bmatrix}, \\
\tilde{\Psi}_4(k) &\triangleq \begin{bmatrix} \bar{C}(k)G + \bar{D}_1(k)\bar{K}(k)G & \frac{\varepsilon}{2}\bar{C}_d(k)G & \frac{\varepsilon}{2}\bar{C}_d(k)G & \frac{\varepsilon\tau}{2}\bar{C}_d(k)G & \bar{D}_\omega(k) \end{bmatrix}, \\
\tilde{\Psi}_5(k) &\triangleq \begin{bmatrix} \bar{F}(k)G + \bar{D}_2(k)\bar{K}(k)G & \frac{\varepsilon}{2}\bar{F}_d(k)G & \frac{\varepsilon}{2}\bar{F}_d(k)G & \frac{\varepsilon\tau}{2}\bar{F}_d(k)G & 0_n \end{bmatrix}, \\
\tilde{\Xi}(k) &\triangleq \mathrm{diag}\left\{ \left(\bar{P}(k+1) - G - G^T\right), \left(\bar{P}(k+1) - G - G^T\right), \right.
\end{aligned}
$$

$$\left(R_1 - G - G^T\right), \left(R_2 - G - G^T\right), \left(R_1 - G - G^T\right),$$
$$\left(R_2 - G - G^T\right), (S - \varepsilon G - \varepsilon G), (S - \varepsilon G - \varepsilon G), -\mathcal{X}^{-1}, -\mathcal{X}^{-1}\Big\}.$$

Realizing $H_i = K_i G$ $(i = 1, 2, \ldots, r)$, we proceed to express

$$\begin{bmatrix} \tilde{\Phi}(k) & \tilde{\Psi}(k) \\ \star & \tilde{\Xi}(k) \end{bmatrix} = \sum_{o=1}^{r}\sum_{l=1}^{r}\sum_{t=1}^{r}\sum_{i=1}^{r}\sum_{j=1}^{r} h_o(\theta(k-d_2))h_l(\theta(k-d_1))h_t(\theta(k+1)) \times h_i(\theta(k))h_j(\theta(k))\Pi_{oltij} < 0.$$

By the same method as [203], it leads to

$$\sum_{i=1}^{r}\sum_{j=1}^{r} h_i(\theta)h_j(\theta)\Pi_{oltij} = \sum_{1\le i<j\le r}\left[\frac{1}{r-1}\left(h_i^2(\theta)\Pi_{otlii} + h_j^2(\theta)\Pi_{otljj}\right) + h_i(\theta)h_j(\theta(k))(\Pi_{otlij} + \Pi_{otlji})\right].$$

Then we analyze the above expression in two cases.

Case 1: When $\Pi_{ostlij} + \Pi_{ostlji} \le 0$, we have

$$\sum_{i=1}^{r}\sum_{j=1}^{r} h_i(\theta)h_j(\theta)\Pi_{oltij} \le \sum_{1\le i<j\le r}\left[\frac{1}{r-1}h_i^2(\theta)\Pi_{oltii} + \frac{1}{r-1}h_j^2(\theta)\Pi_{oltjj}\right],$$

then (8.23) implies that

$$\sum_{i=1}^{r}\sum_{j=1}^{r} h_i(\theta)h_j(\theta)\Pi_{oltij} < 0.$$

Case 2: When $\Pi_{ostlij} + \Pi_{ostlji} > 0$, we have

$$\begin{aligned}\sum_{i=1}^{r}\sum_{j=1}^{r} h_i(\theta)h_j(\theta)\Pi_{oltij} &\le \sum_{1\le i<j\le r}\left[\frac{1}{r-1}\left(h_i^2(\theta)\Pi_{oltii} + h_j^2(\theta)\Pi_{oltjj}\right) + \frac{1}{2}\left(h_i^2(\theta) + h_j^2(\theta)\right)(\Pi_{oltij} + \Pi_{oltji})\right] \\ &= \sum_{1\le i<j\le r}\left\{h_i^2(\theta)\left[\frac{1}{r-1}\Pi_{oltii} + \frac{1}{2}(\Pi_{oltij} + \Pi_{oltij})\right] + h_j^2(\theta)\left[\frac{1}{r-1}\Pi_{oltjj} + \frac{1}{2}(\Pi_{oltij} + \Pi_{oltij})\right]\right\},\end{aligned}$$

then (8.24) gives rise to

$$\sum_{i=1}^{r}\sum_{j=1}^{r} h_i(\theta)h_j(\theta)\Pi_{oltij} < 0.$$

Obviously, (8.23) and (8.24) imply

$$\begin{bmatrix} \tilde{\Phi}(k) & \tilde{\Psi}(k) \\ \star & \tilde{\Xi}(k) \end{bmatrix} < 0. \tag{8.26}$$

Define

$$L \triangleq \varepsilon G, \quad \mathcal{T} \triangleq \mathrm{diag}\{G, L, L, L, \underbrace{I_n, \ldots, I_n}_{11}\}. \tag{8.27}$$

From the facts

$$\begin{aligned} \left(\bar{P}(k+1) - G\right)^T \bar{P}^{-1}(k+1)\left(\bar{P}(k+1) - G\right) &\geq 0, \\ (R_1 - G)^T R_1^{-1}(R_1 - G) &\geq 0, \\ (R_2 - G)^T R_2^{-1}(R_2 - G) &\geq 0, \\ (S - L)^T S^{-1}(S - L) &\geq 0, \end{aligned}$$

it yields that

$$\begin{aligned} -G^T\bar{P}^{-1}(k+1)G &\leq \bar{P}(k+1) - G - G^T, \\ -G^T R_1^{-1}G &\leq R_1 - G - G^T, \\ -G^T R_2^{-1}G &\leq R_2 - G - G^T, \\ -L^T S^{-1}L &\leq S - L - L^T. \end{aligned}$$

Therefore, (8.26) implies

$$\begin{bmatrix} \tilde{\Phi}(k) & \tilde{\Psi}(k) \\ \star & -\hat{\Xi}(k) \end{bmatrix} < 0, \tag{8.28}$$

where

$$\hat{\Xi}(k) \triangleq \mathrm{diag}\left\{\hat{P}^{-1}(k+1), \hat{P}^{-1}(k+1), \hat{R}_1^{-1}, \hat{R}_2^{-1}, \hat{R}_1^{-1}, \hat{R}_2^{-1}, \hat{S}^{-1}, \hat{S}^{-1}, -\mathcal{X}^{-1}, -\mathcal{X}^{-1}\right\}.$$

From (8.23) we know $G + G^T - R_1 \geq 0$. Since $R_1 > 0$, we have $G + G^T > 0$, which implies that G is nonsingular, and then $\mathcal{T}$ is also nonsingular from (8.27). Performing a congruence transformation to (8.28) by $\mathcal{T}^{-1}$, we have

$$\begin{bmatrix} \hat{\Phi}(k) & \hat{\Psi}(k) \\ \star & -\hat{\Xi}(k) \end{bmatrix} < 0, \tag{8.29}$$

where

$$\begin{aligned}\hat{\Phi}(k) &\triangleq -W_1^T\hat{P}(k)W_1 + W_1^T\hat{Q}_1(k)W_1 - W_2^T\hat{Q}_1(k-d_1)W_2 + W_1^T\hat{Q}_2(k)W_1\\ &\quad -2W_5^T\mathcal{Y}\hat{\Psi}_4(k) - W_5^T(\mathcal{Z}-\alpha I)W_5 - W_3^T\hat{Q}_2(k-d_2)W_3 - W_4^T\hat{S}W_4\\ &\quad -(W_1-W_2)^T\hat{R}_1(W_1-W_2) - (W_1-W_3)^T\hat{R}_2(W_1-W_3),\\ \hat{\Psi}(k) &\triangleq \left[\hat{\Psi}_1^T(k)\ \hat{\Psi}_2^T(k)\ d_1\hat{\Psi}_3^T(k)\ d_2\hat{\Psi}_3^T(k)\ d_1\hat{\Psi}_2^T(k)\ d_2\hat{\Psi}_2^T(k)\right.\\ &\quad \left.\hat{\Psi}_3^T(k)\ \hat{\Psi}_2^T(k)\ \hat{\Psi}_4^T(k)\ \hat{\Psi}_5^T(k)\right],\\ \hat{\Psi}_1(k) &\triangleq \left[\hat{A}(k)\ \ \frac{\varepsilon}{2}\bar{A}_d(k)\ \ \frac{\varepsilon}{2}\bar{A}_d(k)\ \ \frac{\varepsilon\tau}{2}\bar{A}_d(k)\ \ \bar{B}_\omega(k)\right],\\ \hat{\Psi}_2(k) &\triangleq \left[\hat{E}(k)\ \ \frac{\varepsilon}{2}\bar{E}_d(k)\ \ \frac{\varepsilon}{2}\bar{E}_d(k)\ \ \frac{\varepsilon\tau}{2}\bar{E}_d(k)\ \ 0_n\right],\\ \hat{\Psi}_3(k) &\triangleq \left[\hat{A}(k)-I_n\ \ \frac{\varepsilon}{2}\bar{A}_d(k)\ \ \frac{\varepsilon}{2}\bar{A}_d(k)\ \ \frac{\varepsilon\tau}{2}\bar{A}_d(k)\ \ \bar{B}_\omega(k)\right],\\ \hat{\Psi}_4(k) &\triangleq \left[\hat{C}(k)\ \ \frac{\varepsilon}{2}\bar{C}_d(k)\ \ \frac{\varepsilon}{2}\bar{C}_d(k)\ \ \frac{\varepsilon\tau}{2}\bar{C}_d(k)\ \ \bar{D}_\omega(k)\right],\\ \hat{\Psi}_5(k) &\triangleq \left[\hat{F}(k)\ \ \frac{\varepsilon}{2}\bar{F}_d(k)\ \ \frac{\varepsilon}{2}\bar{F}_d(k)\ \ \frac{\varepsilon\tau}{2}\bar{F}_d(k)\ \ 0_n\right].\end{aligned}$$

Choose an LKF as

$$\begin{aligned}V(k) &\triangleq x^T(k)\hat{P}(k)x(k) + \sum_{i=k-d_1}^{k-1} x^T(i)\hat{Q}_1(i)x(i) + \sum_{i=k-d_2}^{k-1} x^T(i)\hat{Q}_2(i)x(i),\\ &\quad + \sum_{i=-d_1}^{-1}\sum_{j=k+i}^{k-1} d_1\delta^T(j)\hat{R}_1\delta(j) + \sum_{i=-d_2}^{-1}\sum_{j=k+i}^{k-1} d_2\delta^T(j)\hat{R}_2\delta(j),\\ &\quad +\frac{1}{\tau}\sum_{i=-d_2}^{-d_1-1}\sum_{j=k+i}^{k-1}\delta^T(j)\hat{S}\delta(j),\end{aligned}$$

where

$$\begin{aligned}\hat{P}(k) &\triangleq G^{-T}\bar{P}(k)G^{-1}, & \hat{S} &\triangleq L^{-T}SL^{-1},\\ \hat{Q}_1(k) &\triangleq L^{-T}\bar{Q}_1(k)L^{-1}, & \hat{R}_1 &\triangleq G^{-T}R_1G^{-1},\\ \hat{Q}_2(k) &\triangleq L^{-T}\bar{Q}_2(k)L^{-1}, & \hat{R}_2 &\triangleq G^{-T}R_2G^{-1}.\end{aligned}$$

Following the same lines as those in the proof of Theorem 8.12, we have

$$\mathbf{E}\{\Delta V(k)\} \le \eta^T(k)\tilde{\Omega}(k)\eta(k),$$

where

$$\begin{aligned}\tilde{\Omega}(k) &\triangleq \hat{\Psi}_1^T(k)\hat{P}(k+1)\hat{\Psi}_1(k) + \hat{\Psi}_2^T(k)\hat{P}(k+1)\hat{\Psi}_2(k) + \hat{\Psi}_2^T(k)\hat{S}\hat{\Psi}_2(k)\\ &\quad +\hat{\Psi}_3^T(k)\hat{S}\hat{\Psi}_3(k) + W_1^T\hat{Q}_1(k)W_1 + W_1^T\hat{Q}_2(k)W_1 - W_1^T\hat{P}(k)W_1\\ &\quad -W_2^T\hat{Q}_1(k-d_1)W_2 - W_3^T\hat{Q}_2(k-d_2)W_3 - W_4^T\hat{S}W_4\\ &\quad +\hat{\Psi}_3^T(k)\left(d_1^2\hat{R}_1 + d_2^2\hat{R}_2\right)\hat{\Psi}_3(k) + \hat{\Psi}_2^T(k)\left(d_1^2\hat{R}_1 + d_2^2\hat{R}_2\right)\hat{\Psi}_2(k)\\ &\quad -(W_1-W_2)^T\hat{R}_1(W_1-W_2) - (W_1-W_3)^T\hat{R}_2(W_1-W_3).\end{aligned}$$

Accordingly,

$$\mathbf{E}\left\{\Delta V(k)\right\} - \mathbf{E}\left\{z^T(k)\mathcal{X}z(k) + 2\omega^T(k)\mathcal{Y}z(k) + \omega^T(k)\mathcal{Z}\omega(k)\right\} \\ + \alpha\mathbf{E}\left\{\omega^T(k)\omega(k)\right\} \leq \eta^T(k)\hat{\Omega}(k)\eta(k), \tag{8.30}$$

where

$$\begin{aligned}\hat{\Omega}(k) \triangleq\ & \hat{\Psi}_1^T(k)\hat{P}(k+1)\hat{\Psi}_1(k) + \hat{\Psi}_2^T(k)\hat{P}(k+1)\hat{\Psi}_2(k) + \hat{\Psi}_2^T(k)\hat{S}\hat{\Psi}_2(k) \\ & +\hat{\Psi}_3^T(k)\hat{S}\hat{\Psi}_3(k) - \hat{\Psi}_4^T(k)\mathcal{X}\hat{\Psi}_4(k) - \hat{\Psi}_5^T(k)\mathcal{X}\hat{\Psi}_5(k) - 2W_5^T\mathcal{Y}\hat{\Psi}_4(k) \\ & -W_1^T\hat{P}(k)W_1 + W_1^T\hat{Q}_1(k)W_1 + W_1^T\hat{Q}_2(k)W_1 - W_4^T\hat{S}W_4 \\ & -W_2^T\hat{Q}_1(k-d_1)W_2 - W_3^T\hat{Q}_2(k-d_2)W_3 - W_5^T(\mathcal{Z}-\alpha I)W_5 \\ & +\hat{\Psi}_3^T(k)\left(d_1^2\hat{R}_1 + d_2^2\hat{R}_2\right)\hat{\Psi}_3(k) + \hat{\Psi}_2^T(k)\left(d_1^2\hat{R}_1 + d_2^2\hat{R}_2\right)\hat{\Psi}_2(k) \\ & -\left(W_1 - W_2\right)^T\hat{R}_1\left(W_1 - W_2\right) - \left(W_1 - W_3\right)^T\hat{R}_2\left(W_1 - W_3\right).\end{aligned}$$

By Schur complement, inequality (8.29) yields $\hat{\Omega}(k) < 0$. Considering (8.30) and using the same method as those in the proof of Theorem 8.12, we can obtain $\mathbf{E}\left\{J(\omega, z, T)\right\} > \alpha\left\langle\omega, \omega\right\rangle_T$. Then the proof is completed. ■

Remark 8.19. In the proof of Theorem 8.18, instead of directly requiring $\Pi_{oltij} < 0$ $(o, l, t, i, j = 1, 2, \ldots, r)$, we substitute it by two less conservative conditions (8.23) and (8.24). Other condition, such as $\Pi_{oltii} < 0$ and $\Pi_{oltij} + \Pi_{oltji} < 0$ for all $o, l, t, i, j = 1, 2, \ldots, r$, in [66], is just the Case 1 in this proof. Therefore, the method in this theorem is less conservative. ♦

8.4 Illustrative Example

In this section, three examples are provided to illustrate the effectiveness of the previously developed methods.

Example 8.20. Consider the following T-S fuzzy stochastic system:

$$\begin{aligned} x(k+1) &= \sum_{i=1}^{2} h_i(\theta)\left[A_i x(k) + A_{di}x(k-d(k)) + B_{\omega i}\omega(k)\right] \\ &\quad + \sum_{i=1}^{2} h_i(\theta)\left[E_i x(k) + E_{di}x(k-d(k))\right]\varpi(k), \\ z(k) &= \sum_{i=1}^{2} h_i(\theta)\left[C_i x(k) + C_{di}x(k-d(k)) + D_{\omega i}\omega(k)\right], \end{aligned}$$

where

$$A_1 = \begin{bmatrix} -0.21 & 0 \\ 0.1 & -0.13 \end{bmatrix}, A_{d1} = \begin{bmatrix} 0.047 & -0.01 \\ 0 & 0.012 \end{bmatrix}, \quad B_{\omega 1} = \begin{bmatrix} -0.02 & -0.1 \\ 0.1 & -0.2 \end{bmatrix},$$

$$A_2 = \begin{bmatrix} -0.23 & 0 \\ 0.05 & -0.15 \end{bmatrix}, A_{d2} = \begin{bmatrix} -0.033 & 0.01 \\ -0.01 & 0.062 \end{bmatrix}, B_{\omega 2} = \begin{bmatrix} -0.2 & -0.1 \\ 0.18 & -0.23 \end{bmatrix},$$

$$C_1 = \begin{bmatrix} -0.17 & 0.13 \\ 0.012 & -0.36 \end{bmatrix}, C_{d1} = \begin{bmatrix} 0.14 & 0.07 \\ 0.09 & -0.04 \end{bmatrix}, \quad D_{\omega 1} = \begin{bmatrix} -0.12 & 0.1 \\ 0 & 0.23 \end{bmatrix},$$

$$C_2 = \begin{bmatrix} -0.25 & 0.16 \\ 0.05 & -0.28 \end{bmatrix}, C_{d2} = \begin{bmatrix} 0.10 & -0.011 \\ 0.017 & -0.04 \end{bmatrix}, D_{\omega 2} = \begin{bmatrix} -0.14 & 0 \\ 0.11 & 0.32 \end{bmatrix},$$

$$E_1 = \begin{bmatrix} -0.1 & 0 \\ 0.13 & 0.01 \end{bmatrix}, \quad E_{d1} = \begin{bmatrix} 0.01 & -0.06 \\ 0.09 & 0 \end{bmatrix},$$

$$E_2 = \begin{bmatrix} 0.11 & 0.13 \\ 0 & -0.01 \end{bmatrix}, \quad E_{d2} = \begin{bmatrix} 0.03 & 0 \\ 0 & -0.07 \end{bmatrix}.$$

Firstly, assume that $E_i = E_{di} = 0_{n \times n}$ $(i = 1, 2, \ldots, r)$, and analyze the given numerical example by Corollary 8.17. Set

$$\mathcal{X} = \begin{bmatrix} -0.5 & 0 \\ 0 & -1 \end{bmatrix}, \quad \mathcal{Y} = \mathcal{Z} = \begin{bmatrix} 1 & 0 \\ 0 & 1 \end{bmatrix}.$$

To find the relations of dissipative margin α with d_1 and d_2, we primarily set $d_2 = 7$ and $d_2 = 10$, respectively, then change the value of d_1 between 1 and d_2, recording the maximum of α for each d_1. The resulting data is printed in Fig. 8.1. And conversely, we fix d_1 at 1 and 4, respectively, then change the value of d_2 between d_1 and 10, recording the maximum of α for each d_2. Accordingly, the resulting data is plotted in Fig. 8.2. From the resulting data, we can find, margin α monotonously increases with the increasing of d_1, and monotonously decreases with the increasing of d_2. The reason for this phenomenon is that, both the decrease of d_1 and the increase of d_2 will result in the accretion of τ. From (8.9), we know, the delayed state is separated into certain and uncertain terms. The uncertain term $\frac{\tau}{2}\omega_d(k)$, which is also called approximation error, will deteriorate the system performance. So the increase of τ will undoubtedly decrease the dissipative margin α.

Furthermore, consider the system whose state function contains stochastic noise. Applying the method in Theorem 8.15 and using the same simulation method, we can get the maximum of α for each pair of (d_1, d_2). After plotting the resulting data in Figs. 8.1 and 8.2, we will find two facts from the curves. First of all, the curves have the same trends as those without stochastic noise. This fact also illustrates that, the increase of delay uncertainty will deprave the dissipative performance. In the second place, for each pair of (d_1, d_2), the maximum of α we get in this situation is smaller than that without stochastic noise. This fact indicates that, stochastic noise also deteriorates the system performance.

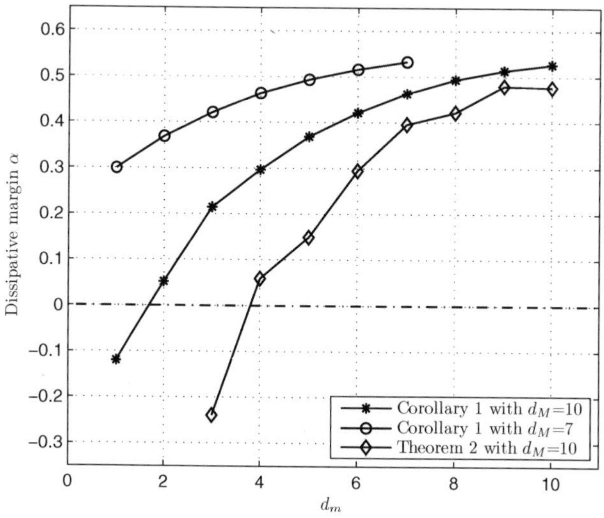

Fig. 8.1. α versus d_1 with fixed d_2

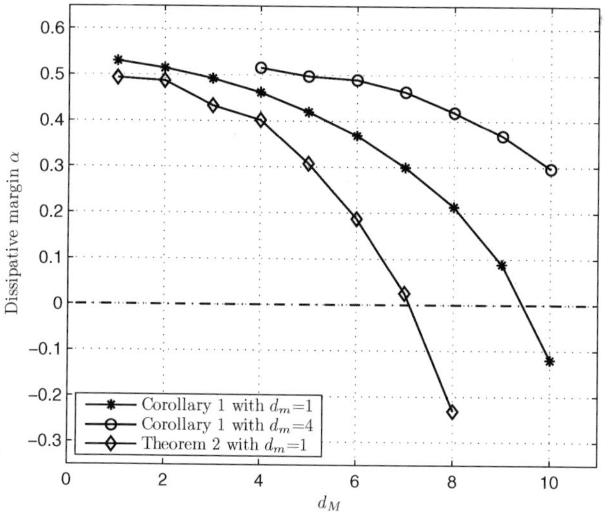

Fig. 8.2. α versus d_2 with fixed d_1

Example 8.21. As it is mentioned earlier, we have adopted a two-terms approximation method in the model transformation. Here we would like to compare it with the one-term approximation methods of [61] and [106]. Firstly, the system state is set to be $x(k) = \exp(-0.004k)$ for $k \geq 0$, and $x(k) = 1$ for $k < 0$. The time-varying delay is assumed to be $d(k) = 9(1 - \exp(-\beta k)) + 1$ for $k \geq 0$, and $d(k) = 1$ for $k < 0$, then we have $d_1 = 1$, $d_2 = 10$ and $\tau = 9$ for all $k \geq 0$. For convenience, we directly define: $\varrho_1(k) \triangleq x(k - d(k)) - x(k - d_1)$ for [106], $\varrho_2(k) \triangleq x(k - d(k)) - x(k - \frac{d_1+d_2}{2})$ for [61] and $\varrho_3(k) \triangleq x(k - d(k)) - \frac{x(k-d_1)+x(k-d_2)}{2}$ for this work. To compare the three kinds of approximation, we will calculate the ℓ_2-norms of their approximation errors. The numerical results are presented in Table 8.1.

Table 8.1. Comparison of different approximation methods

β	$\lVert\varrho_1(k)\rVert_2$	$\lVert\varrho_2(k)\rVert_2$	$\lVert\varrho_3(k)\rVert_2$	$\lVert\varrho_3(k)\rVert_2 - \lVert\varrho_3(k)\rVert_2$	$\lVert\delta(k)\rVert_2$
0.01	2.5825	1.2591	1.2560	0.0032	0.4599
0.02	3.1681	1.5097	1.4990	0.0107	0.4625
0.04	3.5775	1.7417	1.7271	0.0145	0.4645
0.08	3.8264	1.8987	1.8823	0.0164	0.4657

From Table 8.1, we can notice that, $\|\varrho_1(k)\|_2$ is much bigger than $\|\varrho_2(k)\|_2$ and $\|\varrho_3(k)\|_2$. Thus, the approximations of [61] and this chapter are much better than that of [106]. On the other hand, we will also find, $\|\varrho_3(k)\|_2$ is slightly smaller than $\|\varrho_2(k)\|_2$, and with the increase of β, $\frac{\|\varrho_3\|_2 - \|\varrho_2\|_2}{\|\delta\|_2}$ also increases. So the approximation in this chapter is relatively better than that of [61], especially when the changing rate of $d(k)$ is very high. In addition, the numerical results also satisfy $\frac{\|\varrho_1\|_2}{\|\delta\|_2} < \tau$, $\frac{\|\varrho_2\|_2}{\|\delta\|_2} < \frac{\tau}{\sqrt{2}}$ and $\frac{\|\varrho_3\|_2}{\|\delta\|_2} < \frac{\tau}{2}$, which exactly verifies our conclusion in Remark 8.8.

Noting that, if we directly remove the lines related with $\mathcal{X}$, $\mathcal{Y}$ and $\mathcal{Z}$, the result in Corollary 8.17 can be also applied to analyze the stability property of T-S fuzzy systems. Further, if $r = 1$, the result will be applicable for linear systems. Thus, we will further compare our result of Corollary 8.17 with the alternative approaches which are derived by direct Lyapunov methods [65] and [245]. Consider the following system which has been used in [65],

$$A_1 = \begin{bmatrix} 0.80 & 0.00 \\ 0.05 & 0.09 \end{bmatrix}, \quad A_{d1} = \begin{bmatrix} -0.10 & 0.00 \\ -0.20 & -0.10 \end{bmatrix}, \tag{8.31}$$

For a given d_1, let's calculate the upper bound of d_2 by different methods. Corresponding result is presented in Table 8.2, where NoV is the abbreviated form of "Number of Variables".

Table 8.2. Calculated upper bound of d_2 for different d_1

d_1	3	5	7	9	11	13	15	NoV
[65]	13	13	14	15	16	17	18	51
[245]	13	14	15	16	17	19	20	42
Corollary 8.17	17	18	18	19	20	22	23	18

Then, the obtained result of d_2 demonstrates the lower conservatism of our method. And it is obvious that, the NoV for Corollary 8.17 is much smaller than that of other alternative approaches, which indicates the lower computational burden of our approach.

Example 8.22. Consider the model car shown in Fig. 8.3. Its ideal system model, which has been used in [20], is given as

$$\begin{cases} x_1(k+1) = x_1(k) + \dfrac{vT}{l} \tan(u(k)), \\ x_2(k+1) = x_2(k) + vT \sin(x_1(k)), \\ x_3(k+1) = x_3(k) + vT \cos(x_1(k)), \end{cases}$$

where $x_1(k)$ is the angle of the car; $x_2(k)$ is the vertical position of the car; $x_3(k)$ is the horizontal position of the car; $u(k)$ is the steering angle; l is the length of this car; T is the sampling time, and v is the constant speed. The parameters are set as $v = 1.0(\text{m/s})$, $l = 2.8(\text{m})$, and $T = 1.0(\text{s})$.

In practical situation, this system will be more complex. Suppose that, this car is moving on the gravel road and it also sustains strong wind, then the influence of disturbances should not be neglected. The corresponding model in practical situation can be described as

$$\begin{cases} x_1(k+1) = x_1(k) + \mu_d x_1(k - d(k)) + \dfrac{vT}{l} \tan(u(k)), \\ x_2(k+1) = x_2(k) + vT \sin(x_1(k)) + \mu_\omega \omega(k) + \mu_\varpi x_1(k)\varpi(k), \\ x_3(k+1) = x_3(k) + vT \cos(x_1(k)), \end{cases}$$

where $d(k)$ is the inner operation state delay of this car; $\omega(k)$ is the exogenous disturbance input caused by the strong wind; $\varpi(k)$ is a Gaussian white noise representing the affection of the uneven road. This stochastic noise undoubtedly has relation with the states of the car, for simplicity, we just multiply $x_1(k)$ in front of it. To simulate the trajectories of the car, we choose $\mu_d = 0.12$, $\mu_\omega = -0.05$ and $\mu_\varpi = 0.01$.

The purpose of this simulation is to steer the car along a desired trajectories, in other words, to regulate $x_1(k)$, $x_2(k)$ and $x_3(k)$ by manipulating the steering angle $u(k)$. We choose $x_2(k) = 0$ as the desired trajectories and

$z(k) = 0.1x_2(k)$ as the output to be controlled of this system. In this case of trajectory control, $x_3(k)$ is not necessary for us, so we omit it in the following analysis to reduce the system dimension. Next, we need to approximate this nonlinear system by a fuzzy model. Considering the property of $\sin(x_1(k))$, we may find that, when $x_1(k)$ is very small, $\sin(x_2(k))$ can be approximated by $x_1(k)$. And $\sin(x_1(k)) \to 0$ (rad) when $x_1(k) \to \pi$ (rad) or $x_1(k) \to -\pi$ (rad). Realizing that the system will be uncontrollable if we set $\sin(x_1(k)) = 0$, so we choose $\mu_s x_1(k)$ to replace $\sin(x_1(k))$ when $x_1(k)$ is about π (rad) or $-\pi$ (rad), where $\mu_s = 0.03$ (m/rad) is a small scalar. The other nonlinear part $\tan(u(k))$ can be simply replaced by $u(k)$ since the value of $u(k)$ is small. Based on above analysis, the fuzzy model of the car can be expressed as

♦ **Plant Form:**

Rule 1: IF $x_1(k)$ is $\mathcal{M}_1$, THEN

$$\begin{aligned} x(k+1) &= A_1 x(k) + A_{d1} x(k-d(k)) + B_{11} u(k) + B_{\omega 1} \omega(k) \\ &\quad + \left[E_1 x(k) + E_{d1} x(k-d(k)) + B_{21} u(k)\right] \varpi(k), \end{aligned}$$

Rule 2: IF $x_1(k)$ is $\mathcal{M}_2$, THEN

$$\begin{aligned} x(k+1) &= A_2 x(k) + A_{d2} x(k-d(k)) + B_{12} u(k) + B_{\omega 2} \omega(k) \\ &\quad + \left[E_2 x(k) + E_{d2} x(k-d(k)) + B_{22} u(k)\right] \varpi(k), \end{aligned}$$

where $\mathcal{M}_1 \triangleq 0$ (rad) and $\mathcal{M}_2 \triangleq \pi$ (rad) or $\mathcal{M}_2 \triangleq -\pi$ (rad), and

$$\begin{aligned} A_1 &= \begin{bmatrix} 1 & 0 \\ 1 & 1 \end{bmatrix}, \quad A_{d1} = A_{d2} = \begin{bmatrix} 0.12 & 0 \\ 0 & 0 \end{bmatrix}, \quad B_{11} = B_{12} = \begin{bmatrix} 0.357 \\ 0 \end{bmatrix}, \\ A_2 &= \begin{bmatrix} 1 & 0 \\ 0.03 & 1 \end{bmatrix}, \quad B_{\omega 1} = B_{\omega 2} = \begin{bmatrix} 0 \\ -0.05 \end{bmatrix}, \quad B_{21} = B_{22} = \begin{bmatrix} 0 \\ 0 \end{bmatrix}, \\ E_1 &= E_2 = \begin{bmatrix} 0 & 0 \\ 0.01 & 0 \end{bmatrix}, \quad E_{d1} = E_{d2} = \begin{bmatrix} 0 & 0 \\ 0 & 0 \end{bmatrix}. \end{aligned}$$

The output $z(k)$ in each rule is uniformly chosen as $z(k) = 0.1x_2(k)$, that is, $C_1 = C_2 = \begin{bmatrix} 0 & 0.1 \end{bmatrix}$, and the rest of matrices in (8.3b) are all zero ones.

Membership functions $\mathcal{M}_1(x_1(k))$ and $\mathcal{M}_2(x_1(k))$ are shown in Fig. 8.4, and they can be expressed by the following equations.

$$\mathcal{M}_1(\theta) = 1 - \frac{|x_1(k)|}{\pi}, \quad \mathcal{M}_2(\theta) = \frac{|x_1(k)|}{\pi}.$$

Accordingly, the fuzzy basis functions are

$$h_1(\theta) = 1 - \frac{|x_1(k)|}{\pi}, \quad h_2(\theta) = \frac{|x_1(k)|}{\pi}.$$

Due to the approximations analyzed above, there must be an error between the original model and the fuzzy model. This error will be considered in the simulation. The initial states are set to be $\phi(k) = x(0)$, $k = -d_2, -d_2 + 1, \ldots, 0$, where $x(0)$ are listed in Table 8.3 for different cases.

Table 8.3. Initial state conditions of the model car

	$x_1(0)$ (deg)	$x_2(0)$ (m)	$x_3(0)$ (m)
Case 1	0	6	2
Case 2	0	–6	2
Case 3	40	6	12
Case 4	40	–6	12
Case 5	80	6	22
Case 6	80	–6	22

Let the time-varying delay $d(k)$ randomly change between $d_1 = 1$ and $d_2 = 2$. Moreover, to model the wind disturbance, we suppose that $\omega(k) = \sin(k)$. The trajectories of the uncontrolled car are shown in Fig. 8.5, from which we can see that this car cannot be kept along the straight line $x_2(k) = 0$. Our aim is to design a fuzzy controller in the form of (8.25), such that the model car will be driven to the desired trajectories and be kept moving along it.

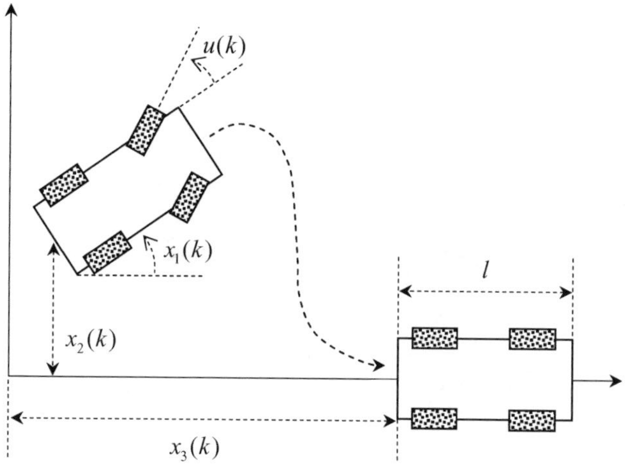

Fig. 8.3. Model of the car and its coordinate system

Applying Theorem 8.18 and solving (8.23)–(8.24) with $\mathcal{X} = -0.5$, $\mathcal{Y} = 1$, $\mathcal{Z} = 1$ and $\alpha = 0.01$, we have

$$G = \begin{bmatrix} 0.1101 & -0.2053 \\ -0.2189 & 0.5185 \end{bmatrix}, \quad \begin{array}{l} H_1 = \begin{bmatrix} -0.1787 & 0.3000 \end{bmatrix}, \\ H_2 = \begin{bmatrix} -0.0581 & 0.0703 \end{bmatrix}. \end{array}$$

Constructing a fuzzy controller by (8.25), we can get the closed-loop system. Fig. 8.6 shows the driven trajectories of the fuzzy model, and Fig. 8.7 depicts the trajectories of the original model. These results manifest that, both models can be steered to the desired trajectories. Then the controller gotten from the approximated fuzzy model can be applied to regulate the original system, although there may exist errors in the approximations.

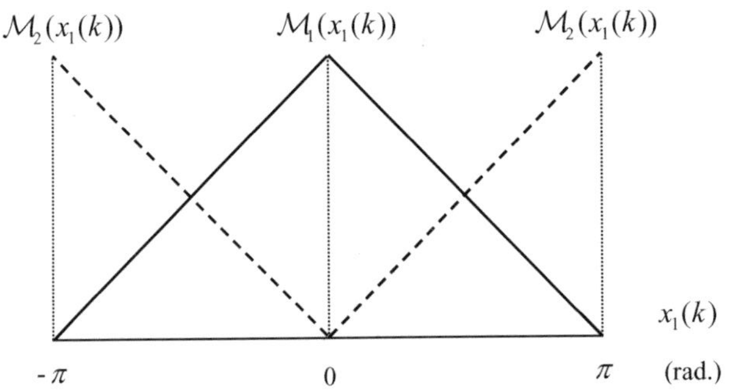

Fig. 8.4. Membership functions

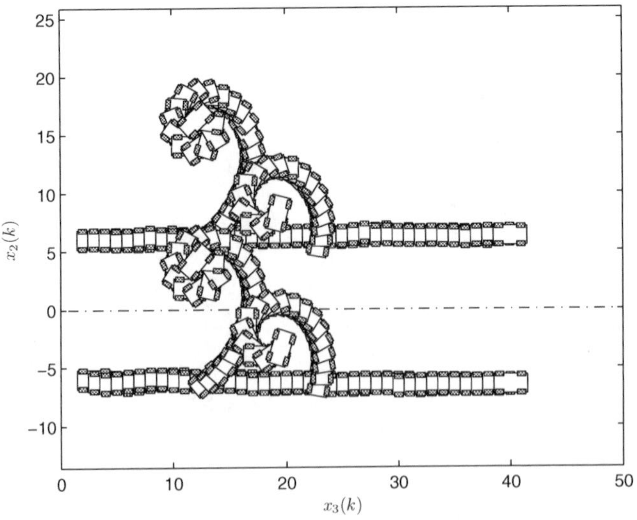

Fig. 8.5. Trajectories of the uncontrolled car in Cases 1–6

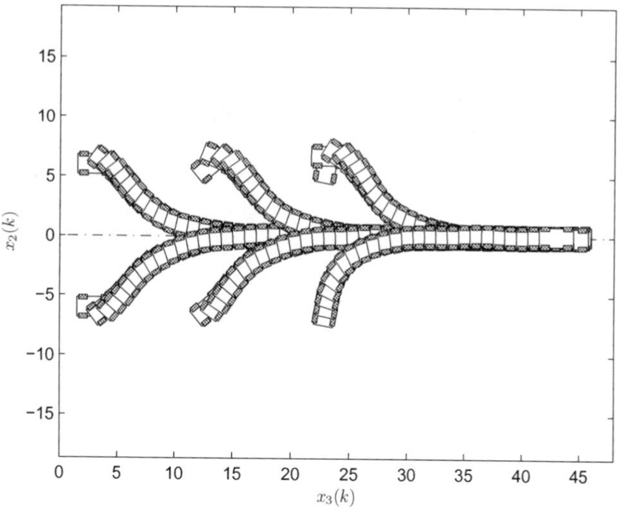

Fig. 8.6. Control results for Cases 1–6 (fuzzy model)

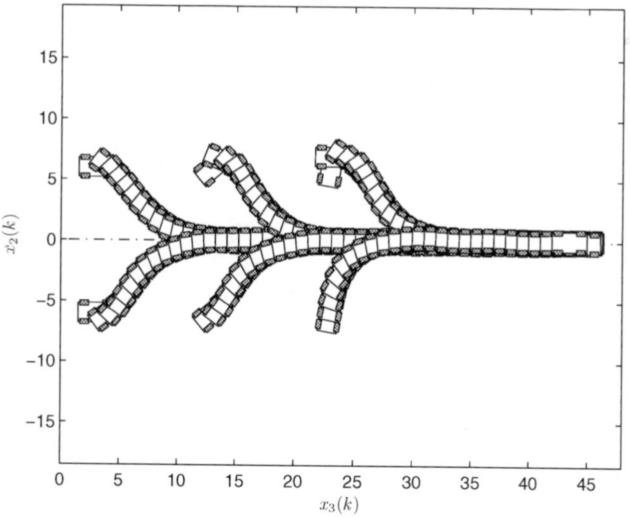

Fig. 8.7. Control results for Cases 1–6 (original model)

8.5 Conclusion

The problems of dissipativity analysis and dissipative controller design have been investigated in this chapter. Based on a novel approximation of the delayed state, a new system with constant time-delays has been obtained via model transformation. The uncertainty of the original time-delay is confined to another subsystem. Dissipativity condition, in the form of LMIs, has been formulated by constructing an LKF which can be regarded as a storage function of the original system. In light of the former conditions, a fuzzy controller has been designed to ensure the required dissipative performance of closed-loop system. Finally, examples have demonstrated the effectiveness of the proposed design scheme.

Chapter 9

Robust $\mathcal{L}_2$-$\mathcal{L}_\infty$ DOF Control of Continuous-Time T-S Fuzzy Stochastic Systems

9.1 Introduction

This chapter aims to investigate the $\mathcal{L}_2$-$\mathcal{L}_\infty$ DOF control for T-S fuzzy stochastic systems with time-varying delay. The slack matrix approach is used to derive a delay-dependent sufficient condition to guarantee the mean-square asymptotic stability with an $\mathcal{L}_2$-$\mathcal{L}_\infty$ performance for the closed-loop system. The corresponding solvability condition for a desired $\mathcal{L}_2$-$\mathcal{L}_\infty$ DOF controller is then established. These obtained conditions, which are not all expressed in terms of LMIs, are cast into sequential minimization problems subject to LMI constraints by applying the CCL method. This enables an easy numerical solution method for the problem under study.

9.2 System Description and Preliminaries

Consider the following T-S fuzzy stochastic time-delay system:

♦**Plant Form:**

Rule i: IF $\theta_1(t)$ is $\mathcal{M}_{i1}$ and $\theta_2(t)$ is $\mathcal{M}_{i2}$ and $\cdots$ and $\theta_p(t)$ is $\mathcal{M}_{ip}$ THEN

$$\begin{aligned}
dx(t) &= [A_i x(t) + A_{di}x(t-d(t)) + B_i u(t) + B_{1i}\omega(t) + F_i f(t)]\,dt \\
&\quad + F_{di}x(t-d(t))d\varpi, && (9.1a)\\
dy(t) &= [C_i x(t) + C_{di}x(t-d(t)) + D_i u(t) + D_{1i}\omega(t) + G_i g(t)]\,dt \\
&\quad + G_{di}x(t-d(t))d\varpi, && (9.1b)\\
z(t) &= E_i x(t) + E_{di}x(t-d(t)) + H_i u(t), && (9.1c)\\
x(t) &= \phi(t), \quad t\in[-d,0], && (9.1d)
\end{aligned}$$

where $i = 1,2,\ldots,r$, and r is the number of IF-THEN rules; $\mathcal{M}_{ij}(i = 1,2,\ldots,r; j = 1,2,\ldots,p)$ are the fuzzy sets; $\theta(t) = \begin{bmatrix}\theta_1(t)\ \theta_2(t)\ \cdots\ \theta_p(t)\end{bmatrix}^T$

L. Wu et al., *Fuzzy Control Systems with Time-Delay and Stochastic Perturbation*, Studies in Systems, Decision and Control 12, DOI: 10.1007/978-3-319-11316-6_9

is the premise variable vector. $x(t) \in \mathbf{R}^n$ is the state vector; $u(t) \in \mathbf{R}^m$ is the control input; $\omega(t) \in \mathbf{R}^l$ is the exogenous disturbance input with $\omega(t) \in \mathcal{L}_2[0,\infty)$; $y(t) \in \mathbf{R}^p$ is the measured output; $z(t) \in \mathbf{R}^q$ is the controlled output; $\varpi(t)$ is a one-dimensional Brownian motion satisfying $\mathbf{E}\{d\varpi(t)\} = 0$ and $\mathbf{E}\{d\varpi^2(t)\} = dt$; $d(t)$ is the time-varying delay which satisfies $0 \le d(t) \le d$ and $\dot{d}(t) \le \tau$; $\phi(t)$ is the initial condition. $A_i, A_{di}, B_i, B_{1i}, F_i, F_{di}, C_i, C_{di}, D_i, D_{1i}, G_i, G_{di}, E_i, E_{di}, H_i$ are matrices of compatible dimensions.

Assumption 9.1 *The nonlinear functions $f(\bullet)$ and $g(\bullet)$ satisfy $f(0,0) = 0$, $g(0,0) = 0$ and Lipschitz conditions, that is, there exist known real matrices M_1, N_1, M_2 and N_2 such that*

$$\begin{aligned}\|f(x,x_d) - f(y,y_d)\| &\le \|M_1(x-y)\| + \|N_1(x_d - y_d)\|,\\ \|g(x,x_d) - g(y,y_d)\| &\le \|M_2(x-y)\| + \|N_2(x_d - y_d)\|.\end{aligned}$$

It is assumed that the premise variables do not depend on the input variable $u(t)$ explicitly. Given a pair of $(x(t), u(t))$, the final output of the nonlinear fuzzy stochastic delay system is inferred as

$$\begin{aligned}dx(t) = \sum_{i=1}^{r} h_i(\theta)\{&[A_i x(t) + A_{di}x(t-d(t)) + B_i u(t) + B_{1i}\omega(t) + F_i f(t)]\,dt\\ &+F_{di}x(t-d(t))d\varpi\},\end{aligned} \tag{9.2a}$$

$$\begin{aligned}dy(t) = \sum_{i=1}^{r} h_i(\theta)\{&[C_i x(t) + C_{di}x(t-d(t)) + D_i u(t) + D_{1i}\omega(t) + G_i g(t)]\,dt\\ &+G_{di}x(t-d(t))d\varpi\},\end{aligned} \tag{9.2b}$$

$$z(t) = \sum_{i=1}^{r} h_i(\theta)\,[E_i x(t) + E_{di}x(t-d(t)) + H_i u(t)], \tag{9.2c}$$

where $h_i(\theta)$, $i = 1, 2, \ldots, r$ are the normalized membership functions, which are defined as that of (1.1) in Chapter 1.

Assume that $h_i(\theta)$ is available for feedback, and suppose the controller's premise variable be the same as the plant's premise variable. Based on the PDC method, the fuzzy controller is designed to share the same IF parts with the following structure:

♦ **Controller Form:**

Rule i: IF $\theta_1(t)$ is $\mathcal{M}_{i1}$ and $\theta_2(t)$ is $\mathcal{M}_{i2}$ and $\cdots$ and $\theta_p(t)$ is $\mathcal{M}_{ip}$ THEN

$$dx_c(t) = A_{ci}x_c(t)dt + B_{ci}dy(t), \tag{9.3a}$$
$$u(t) = C_{ci}x_c(t). \tag{9.3b}$$

The DOF control plant (9.3) can also be represented by the following form:

$$dx_c(t) = \sum_{i=1}^{r} h_i(\theta) \left[A_{ci} x_c(t) dt + B_{ci} dy(t) \right], \tag{9.4a}$$

$$u(t) = \sum_{i=1}^{r} h_i(\theta) C_{ci} x_c(t). \tag{9.4b}$$

Augmenting the model of (9.2) to include the states of (9.4), the resulting closed-loop system can be formulated as

$$d\xi(t) = \sum_{i=1}^{r} \sum_{j=1}^{r} h_i(\theta) h_j(\theta) \left\{ \left[\tilde{A}_{ij} \xi(t) + \tilde{A}_{dij} K \xi(t - d(t)) + \tilde{B}_{ij} \omega(t) + \tilde{F}_{ij} \eta(t) \right] dt \right.$$
$$\left. + \tilde{F}_{dij} K \xi(t - d(t)) d\varpi(t) \right\}, \tag{9.5a}$$

$$z(t) = \sum_{i=1}^{r} \sum_{j=1}^{r} h_i(\theta) h_j(\theta) \left[\tilde{C}_{ij} \xi(t) + \tilde{C}_{dij} K \xi(t - d(t)) \right], \tag{9.5b}$$

where $\xi(t) \triangleq \begin{bmatrix} x(t) \\ x_c(t) \end{bmatrix}$, $\eta(t) \triangleq \begin{bmatrix} f(t) \\ g(t) \end{bmatrix}$, $K \triangleq \begin{bmatrix} I & 0 \end{bmatrix}$ and

$$\tilde{A}_{ij} \triangleq \begin{bmatrix} A_i & B_i C_{cj} \\ B_{cj} C_i & A_{cj} + B_{cj} D_i C_{cj} \end{bmatrix}, \; \tilde{A}_{dij} \triangleq \begin{bmatrix} A_{di} \\ B_{cj} C_{di} \end{bmatrix}, \; \tilde{B}_{ij} \triangleq \begin{bmatrix} B_{1i} \\ B_{cj} D_{1i} \end{bmatrix},$$
$$\tilde{F}_{ij} \triangleq \begin{bmatrix} F_i & 0 \\ 0 & B_{cj} G_i \end{bmatrix}, \; \tilde{F}_{dij} \triangleq \begin{bmatrix} F_{di} \\ B_{cj} G_{di} \end{bmatrix}, \; \tilde{C}_{ij} \triangleq \begin{bmatrix} E_i & H_i C_{cj} \end{bmatrix}, \; \tilde{C}_{dij} \triangleq E_{di}.$$

Before proceeding further, we introduce the following definitions.

Definition 9.1. The closed-loop system in (9.5) with $\omega(t) = 0$ is said to be mean-square asymptotically stable if its solution $\xi(t)$ satisfies

$$\lim_{t \to \infty} \mathbf{E} \left\{ \|\xi(t)\|^2 \right\} = 0.$$

Definition 9.2. Given a scalar $\gamma > 0$, the closed-loop system in (9.5) is said to be mean-square asymptotically stable with an $\mathcal{L}_2$-$\mathcal{L}_\infty$ performance level γ if it is mean-square asymptotically stable when $\omega(t) = 0$, and under zero initial condition, for all nonzero $\omega(t) \in \mathcal{L}_2[0, \infty)$, it holds that

$$\sup_{\forall t} \mathbf{E} \left\{ z^T(t) z(t) \right\} < \gamma^2 \int_0^\infty \omega^T(t) \omega(t) dt. \tag{9.6}$$

Therefore, the $\mathcal{L}_2$-$\mathcal{L}_\infty$ DOF control problem addressed in this chapter can be stated as follows: given a scalar $\gamma > 0$, for system (9.1), design a fuzzy DOF controller in the form of (9.4), such that the closed-loop system in (9.5) is mean-square asymptotically stable with an $\mathcal{L}_2$-$\mathcal{L}_\infty$ performance level γ.

Lemma 9.3. [81] *For any real matrices X_{ij}, $i, j = 1, 2, \ldots, r$ and $\Pi > 0$ with appropriate dimensions, we have*

$$\sum_{i=1}^{r}\sum_{j=1}^{r}\sum_{l=1}^{r}\sum_{m=1}^{r} h_i h_j h_l h_m X_{ij}^T \Pi X_{lm} \leq \sum_{i=1}^{r}\sum_{j=1}^{r} h_i h_j X_{ij}^T \Pi X_{ij},$$

where $h_i \geq 0$ *for* $i = 1, 2, \ldots, r$ *and* $\sum_{i=1}^{r} h_i = 1$.

For stochastic systems, the following Itô's formula plays an important role in the stability analysis.

Lemma 9.4. [147] *Let $x(t)$ be an n-dimensional Itô's process on $t \geq 0$ with the stochastic differential*

$$dx(t) = f(t)dt + g(t)d\omega(t),$$

where $f(t) \in \mathbf{R}^n$ and $g(t) \in \mathbf{R}^{n\times m}$. Let $V(x,t) \in \mathcal{C}^{2,1}(\mathbf{R}^n \times \mathbf{R}^+; \mathbf{R}^+)$. Then $V(x,t)$ is a real-valued Itô's process with its stochastic differential given by

$$\begin{aligned} dV(x,t) &= \mathscr{L}V(x,t)dt + V_x(x,t)g(t)d\omega(t),\\ \mathscr{L}V(x,t) &= V_t(x,t) + V_x(x,t)f(t) + \frac{1}{2}\mathrm{trace}\left(g^T(t)V_{xx}(x,t)g(t)\right), \end{aligned}$$

where $\mathcal{C}^{2,1}(\mathbf{R}^n \times \mathbf{R}^+; \mathbf{R}^+)$ ($\mathcal{C}^{2,1}$ for simplicity) denotes the family of all real-valued functions $V(x,t)$ defined on $\mathbf{R}^n \times \mathbf{R}^+$ such that they are continuously twice differentiable in x and t. if $V(x,t) \in \mathcal{C}^{2,1}$, we set

$$\begin{aligned} V_t(x,t) &\triangleq \frac{\partial V(x,t)}{\partial t},\\ V_{xx}(x,t) &\triangleq \left(\frac{\partial^2 V(x,t)}{\partial x_i x_j}\right)_{n\times n},\\ V_x(x,t) &\triangleq \left(\frac{\partial V(x,t)}{\partial x_1}, \cdots, \frac{\partial V(x,t)}{\partial x_n}\right). \end{aligned}$$

9.3 Main Results

We first analyze the $\mathcal{L}_2$-$\mathcal{L}_\infty$ performance for closed-loop system (9.5).

Theorem 9.5. *Given scalars $\gamma > 0$, $d > 0$ and $\tau > 0$, the closed-loop system in (9.5) is mean-square asymptotically stable with an $\mathcal{L}_2$-$\mathcal{L}_\infty$ performance level γ if there exist matrices $P > 0$, $Q > 0$, $R > 0$, X, Y, Z and a scalar $\varepsilon > 0$ such that the following LMIs hold:*

$$\Psi_{ii} < 0, \quad i = 1, 2, \ldots, r, \tag{9.7a}$$

$$\Psi_{ij} + \Psi_{ji} < 0, \quad i < j \leq r, \tag{9.7b}$$

$$\Phi_{ii} < 0, \quad i = 1, 2, \ldots, r, \tag{9.7c}$$

$$\Phi_{ij} + \Phi_{ji} < 0, \quad i < j \leq r, \tag{9.7d}$$

where

$$\Psi_{ij} \triangleq \begin{bmatrix} \breve{\Pi}_{1ij} + \bar{\Pi}_{2ij} + \bar{\Pi}_{2ij}^T & d\bar{\Pi}_{3ij}^T K^T R & d\bar{W} & \bar{\Pi}_{4ij} & \bar{\Pi}_{5ij} \\ \star & -dR & 0 & 0 & 0 \\ \star & \star & -dR & 0 & 0 \\ \star & \star & \star & -\varepsilon I & 0 \\ \star & \star & \star & \star & -P \end{bmatrix}, \ \bar{W} \triangleq \begin{bmatrix} X \\ Y \\ Z \end{bmatrix},$$

$$\Phi_{ij} \triangleq \begin{bmatrix} -P & 0 & \tilde{C}_{ij}^T \\ \star & -P & K^T \tilde{C}_{dij}^T \\ \star & \star & -\frac{1}{2}\gamma^2 I \end{bmatrix}, \quad \bar{\Pi}_{4ij} \triangleq \begin{bmatrix} P\tilde{F}_{ij} \\ 0 \\ 0 \end{bmatrix}, \quad \bar{\Pi}_{5ij} \triangleq \begin{bmatrix} 0 \\ \tilde{F}_{dij}^T P \\ 0 \end{bmatrix},$$

$$\breve{\Pi}_{1ij} \triangleq \begin{bmatrix} \breve{\Pi}_{111ij} & P\tilde{A}_{dij} & P\tilde{B}_{ij} \\ \star & -(1-\tau)Q + 2\varepsilon N & 0 \\ \star & \star & -I \end{bmatrix}, \quad \begin{array}{l} \bar{\Pi}_{2ij} \triangleq \begin{bmatrix} \bar{W}K & -\bar{W} & 0 \end{bmatrix}, \\ \bar{\Pi}_{3ij} \triangleq \begin{bmatrix} \tilde{A}_{ij} & \tilde{A}_{dij} & \tilde{B}_{ij} \end{bmatrix}, \end{array}$$

with $\breve{\Pi}_{111ij} \triangleq P\tilde{A}_{ij} + \tilde{A}_{ij}^T P + K^T (Q + 2\varepsilon M) K$, $M \triangleq M_1^T M_1 + M_2^T M_2$ *and* $N \triangleq N_1^T N_1 + N_2^T N_2$.

Proof. Choose the following LKF:

$$V(\xi,t) \triangleq \xi^T(t)P\xi(t) + \int_{t-d(t)}^{t} \xi^T(s)K^T QK\xi(s)ds + \int_{-d}^{0}\int_{t+\theta}^{t} \xi^T(s)K^T RK\xi(s)dsd\theta, \tag{9.8}$$

where $P > 0$, $Q > 0$, $R > 0$. By Itô's formula in Lemma 9.4, we have

$$\begin{aligned} \mathscr{L}V(\xi,t) = {} & 2\sum_{i=1}^{r}\sum_{j=1}^{r} h_i(\theta)h_j(\theta) \\ & \times \xi^T(t)P\left[\tilde{A}_{ij}\xi(t) + \tilde{A}_{dij}K\xi(t-d(t)) + \tilde{B}_{ij}\omega(t) + \tilde{F}_{ij}\eta(t)\right] \\ & + \sum_{i=1}^{r}\sum_{j=1}^{r}\sum_{l=1}^{r}\sum_{m=1}^{r} h_i(\theta)h_j(\theta)h_l(\theta)h_m(\theta) \\ & \times \xi^T(t-d(t))K^T\tilde{F}_{dij}^T P\tilde{F}_{dlm}K\xi(t-d(t)) \\ & + \sum_{i=1}^{r}\sum_{j=1}^{r} h_i(\theta)h_j(\theta) \\ & \times \left[\xi^T(t)K^T QK\xi(t) - (1-\tau)\xi^T(t-d(t))K^T QK\xi(t-d(t))\right] \\ & + \sum_{i=1}^{r}\sum_{j=1}^{r} h_i(\theta)h_j(\theta) \\ & \times \left[d\dot{\xi}^T(t)K^T RK\dot{\xi}(t) - \int_{t-d(t)}^{t} \dot{\xi}^T(s)K^T RK\dot{\xi}(s)ds\right]. \end{aligned} \tag{9.9}$$

It follows from Lemma 2 in [212] and Lemma 9.3 that, for a scalar $\varepsilon > 0$,

$$\begin{aligned}&\sum_{i=1}^{r}\sum_{j=1}^{r} 2h_i(\theta)h_j(\theta)\xi^T(t)P\tilde{F}_{ij}\eta(t)\\ &\leq \sum_{i=1}^{r}\sum_{j=1}^{r} \varepsilon^{-1}h_i(\theta)h_j(\theta)\xi^T(t)P\tilde{F}_{ij}\tilde{F}_{ij}^T P\xi(t) + \varepsilon\eta^T(t)\eta(t). \end{aligned} \tag{9.10}$$

Notice from Assumption 9.1 that

$$\begin{aligned}\|f(x, x_d)\| &\leq \|M_1 x(t)\| + \|N_1 x(t-d(t))\|,\\ \|g(x, x_d)\| &\leq \|M_2 x(t)\| + \|N_2 x(t-d(t))\|.\end{aligned}$$

Thus, by defining M and N as in Theorem 9.5, it is easy to see that

$$\eta^T(t)\eta(t) \leq 2\xi^T(t)K^T MK\xi(t) + 2\xi^T(t-d(t))K^T NK\xi(t-d(t)). \tag{9.11}$$

On the other hand, Newton-Leibniz formula gives

$$\xi(t) - \xi(t-d(t)) = \int_{t-d(t)}^{t} \dot{\xi}(s)ds.$$

Then, for $\chi(t) \triangleq \begin{bmatrix} \xi(t) \\ K\xi(t-d(t)) \end{bmatrix}$ and any matrix $W \triangleq \begin{bmatrix} X \\ Y \end{bmatrix}$, it holds that

$$\chi^T(t)WK\left[\xi(t) - \xi(t-d(t)) - \int_{t-d(t)}^{t} \dot{\xi}(s)ds\right] = 0. \tag{9.12}$$

Firstly, we show the mean-square asymptotic stability of the closed-loop system in (9.5) with $\omega(t) = 0$. By (9.9)–(9.12) and Lemma 9.3, we have

$$\begin{aligned}\mathscr{L}V(\xi, t) \leq &\sum_{i=1}^{r}\sum_{j=1}^{r} h_i(\theta)h_j(\theta)\\ &\times\xi^T(t)\left[P\tilde{A}_{ij} + \tilde{A}_{ij}^T P + K^T(Q + 2\varepsilon M)K + \varepsilon^{-1}P\tilde{F}_{ij}\tilde{F}_{ij}^T P\right]\xi(t)\\ &+\sum_{i=1}^{r}\sum_{j=1}^{r} h_i(\theta)h_j(\theta)\\ &\times\xi^T(t-d(t))K^T\left[2\varepsilon N - (1-\tau)Q + \tilde{F}_{dij}^T P\tilde{F}_{dij}\right]K\xi(t-d(t))\\ &+2\sum_{i=1}^{r}\sum_{j=1}^{r} h_i(\theta)h_j(\theta)\chi^T(t)WK\left[\xi(t) - \xi(t-d(t)) - \int_{t-d(t)}^{t} \dot{\xi}(s)ds\right]\\ &+2\sum_{i=1}^{r}\sum_{j=1}^{r} h_i(\theta)h_j(\theta)\xi^T(t)P\tilde{A}_{dij}K\xi(t-d(t))\end{aligned}$$

$$-\sum_{i=1}^{r}\sum_{j=1}^{r}h_i(\theta)h_j(\theta)\int_{t-d(t)}^{t}\dot{\xi}^T(s)K^TRK\dot{\xi}(s)ds$$
$$+d\sum_{i=1}^{r}\sum_{j=1}^{r}h_i(\theta)h_j(\theta)\dot{\xi}^T(t)K^TRK\dot{\xi}(t).$$

Then, by some mathematical operations, we have

$$\begin{aligned}\mathscr{L}V(\xi,t) \leq & \sum_{i=1}^{r}\sum_{j=1}^{r}h_i(\theta)h_j(\theta)\chi^T(t)\left(\Pi_{ij}+dWR^{-1}W^T\right)\chi(t)\\ &-\int_{t-d(t)}^{t}\Big[W^T\chi(t)+RK\dot{\xi}(s)\Big]^T R^{-1}\Big[W^T\chi(t)+RK\dot{\xi}(s)\Big]\,ds,\end{aligned} \quad (9.13)$$

where

$$\Pi_{ij} \triangleq \Pi_{1ij}+\Pi_{2ij}+\Pi_{2ij}^T+d\Pi_{3ij}^TK^TRK\Pi_{3ij},$$
$$\Pi_{1ij} \triangleq \begin{bmatrix}\breve{\Pi}_{111ij}+\varepsilon^{-1}P\tilde{F}_{ij}\tilde{F}_{ij}^TP & P\tilde{A}_{dij}\\ \star & -(1-\tau)Q+2\varepsilon N+\tilde{F}_{dij}^TP\tilde{F}_{dij}\end{bmatrix},$$
$$\Pi_{2ij} \triangleq \begin{bmatrix}WK & -W\end{bmatrix},\quad \Pi_{3ij} \triangleq \begin{bmatrix}\tilde{A}_{ij} & \tilde{A}_{dij}\end{bmatrix}.$$

By Schur complement, (9.7a)–(9.7b) imply $\Pi_{ij}+dWR^{-1}W^T<0$. Notice that

$$\Big[W^T\chi(t)+RK\dot{\xi}(s)\Big]^T R^{-1}\Big[W^T\chi(t)+RK\dot{\xi}(s)\Big]\geq 0. \quad (9.14)$$

Thus, taking expectations to (9.13) and considering (9.14), we have

$$\mathbf{E}\left\{\mathscr{L}V(\xi,t)\right\}\leq \mathbf{E}\left\{\sum_{i=1}^{r}\sum_{j=1}^{r}h_i(\theta)h_j(\theta)\chi^T(t)\left(\Pi_{ij}+dWR^{-1}W^T\right)\chi(t)\right\}<0.$$

This implies, by [147], that the closed-loop system in (9.5) with $\omega(t)=0$ is mean-square asymptotically stable.

Now, we establish the $\mathcal{L}_2$-$\mathcal{L}_\infty$ performance for closed-loop system (9.5). For any

$$\bar{\chi}(t) \triangleq \begin{bmatrix}\xi(t)\\ \xi(t-d(t))K^T\\ \omega(t)\end{bmatrix},\quad \bar{W} \triangleq \begin{bmatrix}X\\ Y\\ Z\end{bmatrix},$$

we have

$$2\bar{\chi}^T(t)\bar{W}K\left[\xi(t)-\xi(t-d(t))-\int_{t-d(t)}^{t}\dot{\xi}(s)ds\right]=0. \quad (9.15)$$

Assume the zero initial condition. Considering (9.9) and (9.15), we have

$$\mathscr{L}V(\xi,t) - \omega^T(t)\omega(t) \leq \sum_{i=1}^{r}\sum_{j=1}^{r} h_i(\theta)h_j(\theta)\bar{\chi}^T(t)\left[\bar{\Pi}_{ij} + d\bar{W}R^{-1}\bar{W}^T\right]\bar{\chi}(t)$$
$$- \int_{t-d(t)}^{t}\left[\bar{W}^T\bar{\chi}(t) + RK\dot{\xi}(s)\right]^T R^{-1}\left[\bar{W}^T\bar{\chi}(t) + RK\dot{\xi}(s)\right] ds,$$

where $\bar{\Pi}_{ij} \triangleq \bar{\Pi}_{1ij} + \bar{\Pi}_{2ij} + \bar{\Pi}_{2ij}^T + d\bar{\Pi}_{3ij}^T K^T RK\bar{\Pi}_{3ij}$ with

$$\bar{\Pi}_{1ij} \triangleq \begin{bmatrix} \Pi_{11ij} & P\tilde{A}_{dij} & P\tilde{B}_{ij} \\ \star & \Pi_{22ij} & 0 \\ \star & \star & -I \end{bmatrix}, \quad \begin{aligned} \bar{\Pi}_{2ij} &\triangleq \begin{bmatrix} \bar{W}K & -\bar{W} & 0 \end{bmatrix}, \\ \bar{\Pi}_{3ij} &\triangleq \begin{bmatrix} \tilde{A}_{ij} & \tilde{A}_{dij} & \tilde{B}_{ij} \end{bmatrix}. \end{aligned}$$

By Schur complement, (9.7a)–(9.7b) imply $\bar{\Pi}_{ij} + d\bar{W}R^{-1}\bar{W}^T < 0$, thus

$$\mathscr{L}V(\xi,t) - \omega^T(t)\omega(t) < 0. \tag{9.16}$$

Integrating both sides of (9.16) from 0 to t^* and then taking expectations give

$$\mathbf{E}\left\{\int_0^{t^*} \mathscr{L}V(\xi,t)dt\right\} < \int_0^{t^*} \omega^T(t)\omega(t)dt < \int_0^{\infty} \omega^T(t)\omega(t)dt.$$

Considering the zero initial condition and (9.8), we have

$$\mathbf{E}\left\{\xi^T(t^*)P\xi(t^*)\right\} < \mathbf{E}\left\{V(\xi,t^*)\right\} < \int_0^{\infty} \omega^T(t)\omega(t)dt. \tag{9.17}$$

Since t^* denotes any time, it is also true that

$$\mathbf{E}\left\{\xi^T(t^* - d(t^*))P\xi(t^* - d(t^*))\right\} < \int_0^{\infty} \omega^T(t)\omega(t)dt. \tag{9.18}$$

From (9.17)–(9.18), we have

$$\mathbf{E}\left\{\begin{bmatrix} \xi(t^*) \\ \xi(t^* - d(t^*)) \end{bmatrix}^T \begin{bmatrix} P & 0 \\ 0 & P \end{bmatrix} \begin{bmatrix} \xi(t^*) \\ \xi(t^* - d(t^*)) \end{bmatrix}\right\} < 2\int_0^{\infty} \omega^T(t)\omega(t)dt. \tag{9.19}$$

By Schur complement, LMI (9.7c)–(9.7d) yield

$$2\gamma^{-2}\begin{bmatrix} \tilde{C}_{ij}^T \\ K^T\tilde{C}_{dij}^T \end{bmatrix}\begin{bmatrix} \tilde{C}_{ij} & \tilde{C}_{dij}K \end{bmatrix} < \begin{bmatrix} P & 0 \\ 0 & P \end{bmatrix}. \tag{9.20}$$

Combining (9.19) with (9.20) gives

$$\mathbf{E}\left\{z^T(t^*)z(t^*)\right\} = \mathbf{E}\left\{\left[\tilde{C}_{ij}\xi(t^*) + \tilde{C}_{dij}K\xi(t^* - d(t^*))\right]^T \times \left[\tilde{C}_{ij}\xi(t^*) + \tilde{C}_{dij}K\xi(t^* - d(t^*))\right]\right\} < \gamma^2 \int_0^\infty \omega^T(t)\omega(t)dt.$$

Taking the supremum over $t^* \geq 0$ yields (9.6), thus the $\mathcal{L}_2$-$\mathcal{L}_\infty$ performance has been established. The proof is completed. ■

Now, we present a solution to the $\mathcal{L}_2$-$\mathcal{L}_\infty$ DOF control problem.

Theorem 9.6. *For given constants $\gamma > 0$, $d > 0$ and $\tau > 0$, suppose that there exist matrices $\mathcal{P} > 0$, $\mathcal{G} > 0$, $\mathcal{R} > 0$, $R > 0$, $\mathcal{Q}_1 > 0$, $\mathcal{Q}_3 > 0$, $\mathcal{Q}_2$, $\mathcal{X}_1$, $\mathcal{X}_2$, $\mathcal{Y}$, $\mathcal{Z}$, $\mathcal{A}_{cij}$, $\mathcal{B}_{cj}$, $\mathcal{C}_{cj}$, $i, j = 1, 2, \ldots, r$ and a scalar $\varepsilon > 0$ satisfying*

$$\tilde{\Psi}_{ii} < 0, \quad i = 1, 2, \ldots, r, \tag{9.21a}$$

$$\tilde{\Psi}_{ij} + \tilde{\Psi}_{ji} < 0, \quad i < j \leq r, \tag{9.21b}$$

$$\tilde{\Phi}_{ii} < 0, \quad i = 1, 2, \ldots, r \tag{9.21c}$$

$$\tilde{\Phi}_{ij} + \tilde{\Phi}_{ji} < 0, \quad i < j \leq r, \tag{9.21d}$$

$$R\mathcal{R} = I, \tag{9.21e}$$

where

$$\tilde{\Psi}_{ij} \triangleq \begin{bmatrix} \tilde{\Psi}_{11ij} & \tilde{\Psi}_{12ij} & \tilde{\Psi}_{13ij} & \tilde{\Psi}_{14ij} & dA_i^T & d\mathcal{X}_1 & \mathcal{P}F_i & \mathcal{B}_{cj}G_i & 0 & 0 \\ \star & \tilde{\Psi}_{22ij} & A_{di} & B_{1i} & d\tilde{\Psi}_{25ij} & d\mathcal{X}_2 & F_i & 0 & 0 & 0 \\ \star & \star & \tilde{\Psi}_{33ij} & 0 & dA_{di}^T & d\mathcal{Y} & 0 & 0 & \tilde{\Psi}_{39ij}^T & F_{di}^T \\ \star & \star & \star & -I & dB_{1i}^T & d\mathcal{Z} & 0 & 0 & 0 & 0 \\ \star & \star & \star & \star & -d\mathcal{R} & 0 & 0 & 0 & 0 & 0 \\ \star & \star & \star & \star & \star & -dR & 0 & 0 & 0 & 0 \\ \star & \star & \star & \star & \star & \star & -\varepsilon I & 0 & 0 & 0 \\ \star & \star & \star & \star & \star & \star & \star & -\varepsilon I & 0 & 0 \\ \star & \star & \star & \star & \star & \star & \star & \star & -\mathcal{P} & -I \\ \star & \star & \star & \star & \star & \star & \star & \star & \star & -\mathcal{G} \end{bmatrix},$$

$$\tilde{\Phi}_{ij} \triangleq \begin{bmatrix} -\mathcal{P} & -I & 0 & 0 & E_i^T \\ \star & -\mathcal{G} & 0 & 0 & (E_i\mathcal{G} + H_i\mathcal{C}_{cj})^T \\ \star & \star & -\mathcal{P} & -I & E_{di}^T \\ \star & \star & \star & -\mathcal{G} & \mathcal{G}E_{di}^T \\ \star & \star & \star & \star & -\frac{1}{2}\gamma^2 I \end{bmatrix},$$

with

$$\tilde{\Psi}_{11ij} \triangleq \mathcal{P}A_i + \mathcal{B}_{cj}C_i + A_i^T\mathcal{P} + C_i^T\mathcal{B}_{cj}^T + \mathcal{Q}_1 + 2\varepsilon M,$$

$$\tilde{\Psi}_{22ij} \triangleq A_i\mathcal{G} + B_i\mathcal{C}_{cj} + \mathcal{G}A_i^T + \mathcal{C}_{cj}^T B_i^T + \mathcal{Q}_3,$$

$$\tilde{\Psi}_{12ij} \triangleq \mathcal{A}_{cij} + A_i^T + \mathcal{Q}_2, \quad \tilde{\Psi}_{33ij} \triangleq -(1-\tau)\mathcal{Q}_1 + 2\varepsilon N,$$

$$\tilde{\Psi}_{13ij} \triangleq \mathcal{P}A_{di} + \mathcal{B}_{cj}C_{di}, \quad \tilde{\Psi}_{14ij} \triangleq \mathcal{P}B_{1i} + \mathcal{B}_{cj}D_{1i},$$

$$\tilde{\Psi}_{25ij} \triangleq \mathcal{G}A_i^T + \mathcal{C}_{cj}^T B_i^T, \quad \tilde{\Psi}_{39ij} \triangleq \mathcal{P}F_{di} + \mathcal{B}_{cj}G_{di}.$$

Then there exists a DOF controller in the form of (9.4) such that the closed-loop system in (9.5) is mean-square asymptotically stable with an $\mathcal{L}_2$-$\mathcal{L}_\infty$ performance level γ. Moreover, if the above conditions are feasible, then a desired DOF controller can be computed by

$$\begin{cases} \mathcal{A}_{cij} \triangleq \mathcal{P}A_i\mathcal{G} + P_2B_{cj}C_i\mathcal{G} + \mathcal{P}B_iC_{cj}G_2^T + P_2A_{cj}G_2^T + P_2B_{cj}D_iC_{cj}G_2^T, \\ \mathcal{B}_{cj} \triangleq P_2B_{cj}, \\ \mathcal{C}_{cj} \triangleq C_{cj}G_2^T. \end{cases} \tag{9.22}$$

Proof. By Theorem 9.5, if (9.7a)–(9.7d) hold then matrix P is nonsingular since $P > 0$. Partition P as

$$P \triangleq \begin{bmatrix} P_1 & P_2 \\ \star & P_3 \end{bmatrix}, \quad G = P^{-1} \triangleq \begin{bmatrix} G_1 & G_2 \\ \star & G_3 \end{bmatrix}. \tag{9.23}$$

Without loss of generality, we assume P_2 and G_2 are nonsingular (if not, P_2 and G_2 may be perturbed respectively by matrices ΔP_2 and ΔG_2 with sufficiently small norms such that $P_2 + \Delta P_2$ and $G_2 + \Delta G_2$ are nonsingular and satisfy (9.7a)–(9.7d)). Define the following matrices which are also nonsingular,

$$\mathcal{J}_P \triangleq \begin{bmatrix} P_1 & I \\ P_2^T & 0 \end{bmatrix}, \quad \mathcal{J}_G \triangleq \begin{bmatrix} I & G_1 \\ 0 & G_2^T \end{bmatrix}. \tag{9.24}$$

Notice that $P\mathcal{J}_G = \mathcal{J}_P$, $G\mathcal{J}_P = \mathcal{J}_G$ and $P_1G_1 + P_2G_2^T = I$. Performing a congruence transformation to $\Psi_{ij} < 0$ by matrix $\mathrm{diag}\{\mathcal{J}_1, R^{-1}, I, I, \mathcal{J}_G\}$ with $\mathcal{J}_1 \triangleq \mathrm{diag}\{\mathcal{J}_G, I, I\}$, we have

$$\begin{bmatrix} \tilde{\Pi}_{ij} & d\tilde{\Pi}_{3ij}^T K^T & d\mathcal{W} & \tilde{\Pi}_{4ij} & \tilde{\Pi}_{5ij} \\ \star & -dR^{-1} & 0 & 0 & 0 \\ \star & \star & -dR & 0 & 0 \\ \star & \star & \star & -\varepsilon I & 0 \\ \star & \star & \star & \star & -\mathcal{J}_G^T P \mathcal{J}_G \end{bmatrix} < 0, \tag{9.25}$$

where $\tilde{\Pi}_{ij} \triangleq \tilde{\Pi}_{1ij} + \tilde{\Pi}_{2ij} + \tilde{\Pi}_{2ij}^T$ and

$$\tilde{\Pi}_{1ij} \triangleq \begin{bmatrix} \mathcal{J}_G^T \breve{\Pi}_{111ij} \mathcal{J}_G & \mathcal{J}_G^T P \tilde{A}_{dij} & \mathcal{J}_G^T P \tilde{B}_{ij} \\ \star & -(1-\tau)Q + 2\varepsilon N & 0 \\ \star & \star & -I \end{bmatrix},$$

$$\tilde{\Pi}_{2ij} \triangleq \begin{bmatrix} \mathcal{W}K & -\mathcal{W} & 0 \end{bmatrix}, \quad \tilde{\Pi}_{3ij} \triangleq \begin{bmatrix} \tilde{A}_{ij}\mathcal{J}_G & \tilde{A}_{dij} & \tilde{B}_{ij} \end{bmatrix},$$

$$\tilde{\Pi}_{4ij} \triangleq \begin{bmatrix} \mathcal{J}_G^T P \tilde{F}_{ij} \\ 0 \\ 0 \end{bmatrix}, \quad \tilde{\Pi}_{5ij} \triangleq \begin{bmatrix} 0 \\ \tilde{F}_{dij}^T P \mathcal{J}_G \\ 0 \end{bmatrix}, \quad \mathcal{W} \triangleq \begin{bmatrix} \mathcal{X} \\ \mathcal{Y} \\ \mathcal{Z} \end{bmatrix}, \quad \mathcal{X} \triangleq \begin{bmatrix} \mathcal{X}_1 \\ \mathcal{X}_2 \end{bmatrix}.$$

Moreover, performing a congruence transformation to $\Phi_{ij} < 0$ by $\text{diag}(\mathcal{J}_G, \mathcal{J}_G, I)$, we have

$$\begin{bmatrix} -\mathcal{J}_G^T P \mathcal{J}_G & 0 & \mathcal{J}_G^T \tilde{C}_{ij}^T \\ \star & -\mathcal{J}_G^T P \mathcal{J}_G & \mathcal{J}_G^T K^T \tilde{C}_{dij}^T \\ \star & \star & -0.5\gamma^2 I \end{bmatrix} < 0. \tag{9.26}$$

Define the following matrices:

$$\begin{cases} \mathcal{P} \triangleq P_1, \quad \mathcal{G} \triangleq G_1, \quad \mathcal{R} \triangleq R^{-1}, \\ \mathcal{Q} \triangleq \begin{bmatrix} I \\ \mathcal{G} \end{bmatrix} Q \begin{bmatrix} I & \mathcal{G} \end{bmatrix} = \begin{bmatrix} \mathcal{Q}_1 & \mathcal{Q}_2 \\ \star & \mathcal{Q}_3 \end{bmatrix} > 0. \end{cases} \tag{9.27}$$

And considering (9.22), we have

$$\begin{cases} \mathcal{J}_G^T P \tilde{A}_{ij} \mathcal{J}_G \triangleq \begin{bmatrix} \mathcal{P} A_i + \mathcal{B}_{cj} C_i & \mathcal{A}_{cij} \\ A_i & A_i \mathcal{G} + B_i \mathcal{C}_{cj} \end{bmatrix}, \\ \mathcal{J}_G^T P \tilde{A}_{dij} \triangleq \begin{bmatrix} \mathcal{P} A_{di} + \mathcal{B}_{cj} C_{di} \\ A_{di} \end{bmatrix}, \\ \mathcal{J}_G^T P \tilde{B}_{ij} \triangleq \begin{bmatrix} \mathcal{P} B_{1i} + \mathcal{B}_{cj} D_{1i} \\ B_{1i} \end{bmatrix}, \\ \mathcal{J}_G^T P \tilde{F}_{dij} \triangleq \begin{bmatrix} \mathcal{P} F_{di} + \mathcal{B}_{cj} G_{di} \\ F_{di} \end{bmatrix}, \\ \mathcal{J}_G^T P \tilde{F}_{ij} \triangleq \begin{bmatrix} \mathcal{P} F_i & \mathcal{B}_{cj} G_i \\ F_i & 0 \end{bmatrix}, \\ \mathcal{J}_G^T P \mathcal{J}_G \triangleq \begin{bmatrix} \mathcal{P} & I \\ I & \mathcal{G} \end{bmatrix}, \\ \tilde{C}_{ij} \mathcal{J}_G \triangleq \begin{bmatrix} E_i & E_i \mathcal{G} + H_i \mathcal{C}_{cj} \end{bmatrix}, \\ \tilde{C}_{dij} K \mathcal{J}_G \triangleq \begin{bmatrix} E_{di} & E_{di} \mathcal{G} \end{bmatrix}. \end{cases} \tag{9.28}$$

LMIs (9.21a)–(9.21b) and (9.21e) imply (9.25) and LMIs (9.21c)–(9.21d) imply (9.26) by considering (9.27)–(9.28). This completes the proof. ∎

Remark 9.7. To solve the output feedback controller by (9.22), matrices P_2 and G_2 should be available in advance, which can be obtained by taking any full rank factorization of $P_2 G_2^T = I - \mathcal{P}\mathcal{G}$ (derived from $P_1 G_1 + P_2 G_2^T = I$). ♦

Remark 9.8. Notice that the conditions in Theorem 9.6 are not all of LMI form due to (9.21e), which cannot be solved directly using LMI procedures. However, with the CCL approach in [39], we can solve these nonconvex feasibility problems by formulating them into some sequential optimization problems subject to LMI constraints. ♦

Now using the CCL method [39], we defined the following minimization problem involving LMI conditions instead of the original nonconvex feasibility problem formulated in Theorem 9.6.

Problem DCNS (DOF Control of NFSD Systems):

$$\begin{aligned} &\min\ \operatorname{trace}(R\mathcal{R}), \\ &\text{subject to}\quad (9.21\text{a})\text{–}(9.21\text{d}) \text{ and } \begin{bmatrix} R & I \\ I & \mathcal{R} \end{bmatrix} \geq 0. \end{aligned} \tag{9.29}$$

Remark 9.9. According to [39], if the solution of the above minimization problem is n, that is, $\min\operatorname{trace}(R\mathcal{R}) = n$, then the conditions in Theorem 9.6 are solvable. Although it is still not guaranteed to always find a global optimal solution, the proposed nonlinear minimization problem is easier to solve than the original nonconvex feasibility problem. Actually, we can readily modify Algorithm 1 in [39] to solve the above minimization problem. ♦

Algorithm DCNS

Step 1. Find a feasible set $\left(\mathcal{P}^{(0)}, \mathcal{G}^{(0)}, \mathcal{R}^{(0)}, R^{(0)}, \mathcal{Q}_1^{(0)}, \mathcal{Q}_3^{(0)}, \mathcal{Q}_2^{(0)}, \mathcal{X}_1^{(0)}, \mathcal{X}_2^{(0)}, \mathcal{Y}^{(0)}, \mathcal{Z}^{(0)}, \mathcal{A}_{cij}^{(0)}, \mathcal{B}_{cj}^{(0)}, \mathcal{C}_{cj}^{(0)}, \varepsilon^{(0)}\right)$ satisfying (9.21a)–(9.21d) and (9.29). Set $\kappa = 0$.

Step 2. Solve the following optimization problem:

$$\begin{aligned} &\min \quad \operatorname{trace}\left(R^{(\kappa)}\mathcal{R} + R\mathcal{R}^{(\kappa)}\right) \\ &\text{subject to}\quad (9.21\text{a})\text{–}(9.21\text{d}) \text{ and } (9.29). \end{aligned}$$

and denote the optimized value by f^*.

Step 3. Substitute the obtained matrix variables $(\mathcal{P}, \mathcal{G}, \mathcal{R}, R, \mathcal{Q}_1, \mathcal{Q}_3, \mathcal{Q}_2, \mathcal{X}_1, \mathcal{X}_2, \mathcal{Y}, \mathcal{Z}, \mathcal{A}_{cij}, \mathcal{B}_{cj}, \mathcal{C}_{cj}, \varepsilon)$ into (9.25). If (9.25) is satisfied, with

$$|f^* - 2n| < \delta,$$

for a sufficiently small scalar $\delta > 0$, then output the feasible solutions $(\mathcal{P}, \mathcal{G}, \mathcal{R}, R, \mathcal{Q}_1, \mathcal{Q}_3, \mathcal{Q}_2, \mathcal{X}_1, \mathcal{X}_2, \mathcal{Y}, \mathcal{Z}, \mathcal{A}_{cij}, \mathcal{B}_{cj}, \mathcal{C}_{cj}, \varepsilon)$. EXIT.

Step 4. If $\kappa > \mathbb{N}$, where $\mathbb{N}$ is the maximum number of iterations allowed, EXIT.

Step 5. Set $\kappa = \kappa + 1$, $\left(\mathcal{P}^{(\kappa)}, \mathcal{G}^{(\kappa)}, \mathcal{R}^{(\kappa)}, R^{(\kappa)}, \mathcal{Q}_1^{(\kappa)}, \mathcal{Q}_3^{(\kappa)}, \mathcal{Q}_2^{(\kappa)}, \mathcal{X}_1^{(\kappa)}, \mathcal{X}_2^{(\kappa)}, \mathcal{Y}^{(\kappa)}, \mathcal{Z}^{(\kappa)}, \mathcal{A}_{cij}^{(\kappa)}, \mathcal{B}_{cj}^{(\kappa)}, \mathcal{C}_{cj}^{(\kappa)}, \varepsilon^{(\kappa)}\right) = (\mathcal{P}, \mathcal{G}, \mathcal{R}, R, \mathcal{Q}_1, \mathcal{Q}_3, \mathcal{Q}_2, \mathcal{X}_1, \mathcal{X}_2, \mathcal{Y}, \mathcal{Z}, \mathcal{A}_{cij}, \mathcal{B}_{cj}, \mathcal{C}_{cj}, \varepsilon)$, and go to Step 2.

9.4 Illustrative Example

Example 9.10. Consider system (9.2) with the parameters given as follows:

$$A_1 = \begin{bmatrix} -1.2 & 0.3 & -0.1 \\ 0.6 & -0.4 & 0.3 \\ 0.2 & 0.6 & 0.5 \end{bmatrix}, \ A_{d1} = \begin{bmatrix} -0.2 & -0.1 & 0.3 \\ 0.1 & 0.3 & 0.2 \\ 0.1 & 0.2 & -0.1 \end{bmatrix}, \ B_1 = \begin{bmatrix} 1.3 \\ 0.9 \\ 1.5 \end{bmatrix},$$

$$A_2 = \begin{bmatrix} -1.1 & 0.2 & 0.4 \\ 0.3 & 0.6 & -0.2 \\ 0.2 & -0.1 & -0.2 \end{bmatrix}, \ A_{d2} = \begin{bmatrix} -0.3 & 0.1 & 0.2 \\ 0.2 & -0.1 & 0.1 \\ 0.0 & 0.2 & -0.2 \end{bmatrix}, \ B_2 = \begin{bmatrix} 1.3 \\ 0.7 \\ 0.9 \end{bmatrix},$$

$$F_1 = \begin{bmatrix} 0.2 & 0.0 & 0.1 \\ 0.3 & 0.2 & 0.1 \\ 0.1 & 0.1 & 0.2 \end{bmatrix}, \ F_{d1} = \begin{bmatrix} 0.2 & 0.0 & 0.1 \\ 0.1 & 0.1 & 0.2 \\ 0.0 & 0.1 & 0.1 \end{bmatrix}, \ B_{11} = \begin{bmatrix} 0.3 \\ 0.5 \\ 0.4 \end{bmatrix}, \ \begin{matrix} D_1 = 0.3, \\ D_2 = 0.4, \end{matrix}$$

$$F_2 = \begin{bmatrix} 0.5 & 0.2 & 0.3 \\ 0.1 & 0.4 & 0.2 \\ 0.4 & 0.1 & 0.3 \end{bmatrix}, \ F_{d2} = \begin{bmatrix} 0.1 & 0.1 & 0.1 \\ 0.1 & 0.2 & 0.0 \\ 0.1 & 0.0 & 0.2 \end{bmatrix}, \ B_{12} = \begin{bmatrix} 0.3 \\ 0.6 \\ 0.4 \end{bmatrix}, \ \begin{matrix} D_{11} = 0.3, \\ D_{12} = 0.4, \end{matrix}$$

$$C_1 = \begin{bmatrix} 1.1 & 1.6 & 1.0 \end{bmatrix}, \ C_{d1} = \begin{bmatrix} 0.4 & 0.6 & 0.3 \end{bmatrix}, \ G_1 = H_1 = 0.4,$$
$$C_2 = \begin{bmatrix} 1.3 & 0.6 & 1.1 \end{bmatrix}, \ C_{d2} = \begin{bmatrix} 0.3 & 0.2 & 0.4 \end{bmatrix}, \ G_2 = H_2 = 0.3,$$
$$E_1 = \begin{bmatrix} 1.2 & 0.8 & 1.5 \end{bmatrix}, \ E_{d1} = \begin{bmatrix} 0.2 & 0.5 & 0.3 \end{bmatrix}, \ G_{d1} = \begin{bmatrix} 0.5 & 0.3 & 0.4 \end{bmatrix},$$
$$E_2 = \begin{bmatrix} 0.7 & 0.8 & 1.0 \end{bmatrix}, \ E_{d2} = \begin{bmatrix} 0.4 & 0.7 & 0.5 \end{bmatrix}, \ G_{d2} = \begin{bmatrix} 0.2 & 0.4 & 0.3 \end{bmatrix},$$

and the nonlinearities $f(t)$ and $g(t)$ are as follows:

$$\begin{aligned} f(t) = & \begin{bmatrix} 0.2x_1(t) + 0.1x_2(t) + 0.1x_3(t) \\ 0.1x_1(t) + 0.1x_2(t) \\ 0.2x_2(t) + 0.2x_3(t) \end{bmatrix} \sin(t) \\ & + \begin{bmatrix} 0.2x_1(t-d(t)) + 0.1x_2(t-d(t)) + 0.2x_3(t-d(t)) \\ 0.1x_1(t-d(t)) + 0.2x_2(t-d(t)) + 0.1x_3(t-d(t)) \\ 0.1x_1(t-d(t)) + 0.1x_3(t-d(t)) \end{bmatrix} \sin(t), \\ g(t) = & \begin{bmatrix} 0.2x_1(t) + 0.1x_2(t) + 0.2x_3(t) \end{bmatrix} \sin(t) \\ & + \begin{bmatrix} 0.2x_1(t-d(t)) + 0.1x_2(t-d(t)) + 0.1x_3(t-d(t)) \end{bmatrix} \sin(t), \end{aligned}$$

which satisfy Assumption 9.1 with

$$M_1 = \begin{bmatrix} 0.2 & 0.1 & 0.1 \\ 0.1 & 0.1 & 0.0 \\ 0.0 & 0.2 & 0.2 \end{bmatrix}, \ N_1 = \begin{bmatrix} 0.2 & 0.1 & 0.2 \\ 0.1 & 0.2 & 0.1 \\ 0.1 & 0.0 & 0.1 \end{bmatrix}, \ \begin{matrix} M_2 = \begin{bmatrix} 0.2 & 0.1 & 0.1 \end{bmatrix}, \\ N_2 = \begin{bmatrix} 0.2 & 0.1 & 0.2 \end{bmatrix}. \end{matrix}$$

The time-delay is given by $d(t) = 0.9 + 0.3\sin(t)$, and a straightforward calculation gives $d = 1.2$ and $\tau = 0.3$. The system with the above parameters is not stable. Our aim is to design an $\mathcal{L}_2$-$\mathcal{L}_\infty$ DOFC in the form of (9.4) such that the closed-loop system is stable. By solving Problem DCNS by using Algorithm DCNS, we obtain the minimized feasible γ is $\gamma^* = 0.9250$, and

$$A_{c1} = \begin{bmatrix} -8.9342 & 5.6923 & 5.5320 \\ 4.6455 & -28.9643 & -23.2153 \\ -6.7595 & 30.6952 & 24.5970 \end{bmatrix}, \quad B_{c1} = \begin{bmatrix} 1.8568 \\ -3.8332 \\ 4.9419 \end{bmatrix},$$
$$A_{c2} = \begin{bmatrix} -3.7632 & -8.8221 & -8.9372 \\ 3.2183 & -23.7433 & -19.1626 \\ -4.1525 & 16.9529 & 21.4357 \end{bmatrix}, \quad B_{c2} = \begin{bmatrix} 1.9649 \\ -2.9962 \\ 4.0793 \end{bmatrix},$$
$$C_{c1} = \begin{bmatrix} -1.4133 \ 2.4389 \ 2.4049 \end{bmatrix}, \quad C_{c2} = \begin{bmatrix} -1.5121 \ 2.3631 \ 2.0615 \end{bmatrix}.$$

Let the membership function be

$$h_1(x_1(t)) = \frac{1 - \sin(x_1(t))}{2}, \quad h_2(x_1(t)) = \frac{1 + \sin(x_1(t))}{2}$$

and suppose the disturbance input be $\omega(t) = \exp(-t)\sin(t)$. We use the discretization approach [86] to simulate the standard Brownian motion. Fig. 9.1 shows the states of the closed-loop system with the initial condition given by $x(t) = \begin{bmatrix} 1.0 & 0.5 & -1.0 \end{bmatrix}^T$, $t \in [-1.2, 0]$. The states of the DOF controller in (9.4) are depicted in Fig. 9.2 and the control input is shown in Fig. 9.3.

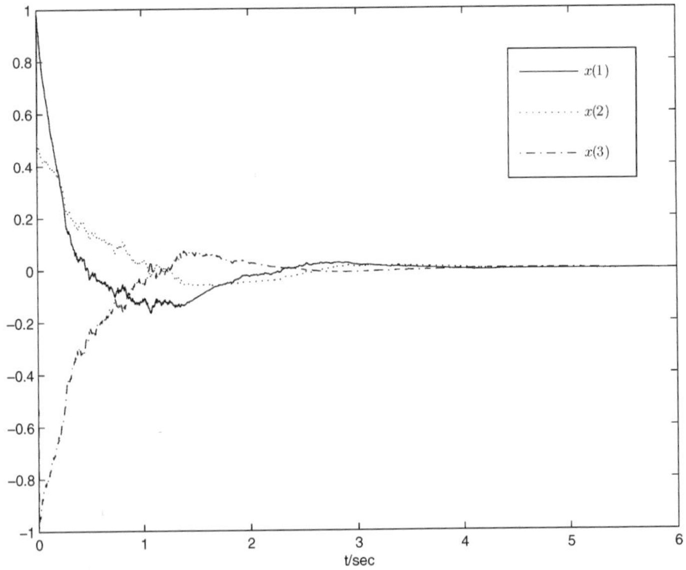

Fig. 9.1. States of the closed-loop system

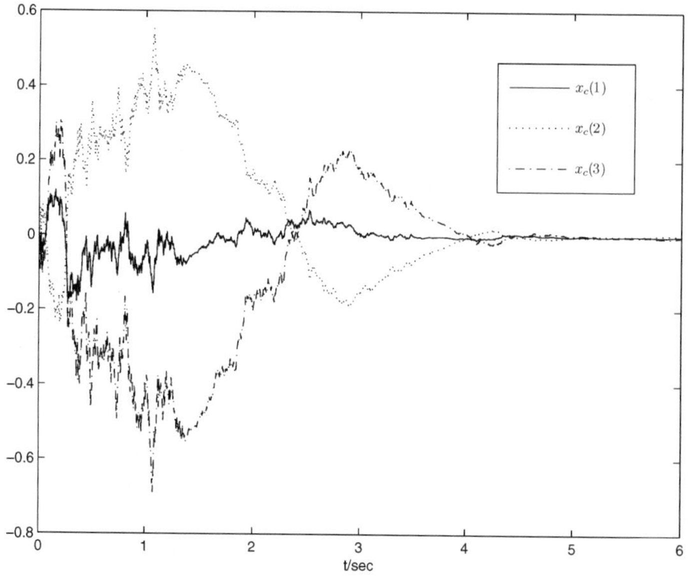

Fig. 9.2. States of the DOF controller

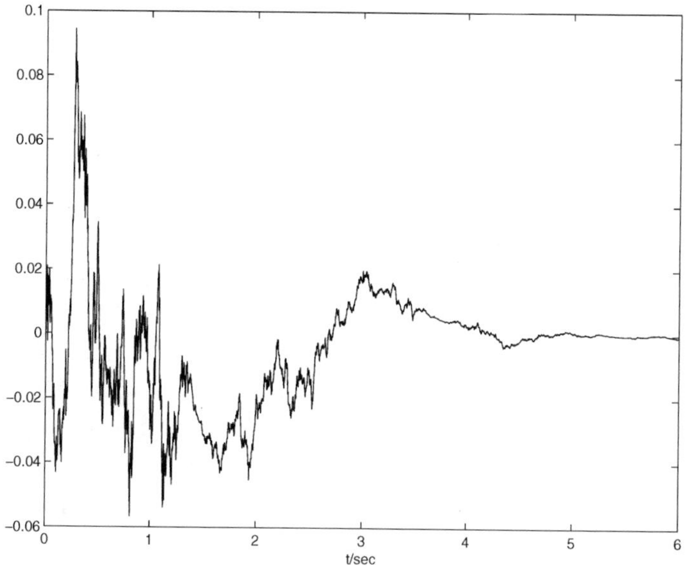

Fig. 9.3. Control input $u(t)$

9.5 Conclusion

In this chapter, the $\mathcal{L}_2$-$\mathcal{L}_\infty$ fuzzy DOF control problem has been investigated for T-S fuzzy stochastic systems with time-varying delay. By using the slack matrix approach, the delay-dependent sufficient condition has been derived to guarantee the mean-square asymptotic stability with an $\mathcal{L}_2$-$\mathcal{L}_\infty$ performance for the closed-loop system. The solvability condition for a desired $\mathcal{L}_2$-$\mathcal{L}_\infty$ fuzzy DOF controller has been established and the corresponding solution algorithm has also been given. A numerical example has been provided to illustrate the effectiveness of the proposed theory.

Chapter 10
Robust $\mathcal{H}_\infty$ Filtering of Discrete-Time T-S Fuzzy Stochastic Systems

10.1 Introduction

In this work, a novel model transformation will be analyzed and applied for the $\mathcal{H}_\infty$ desired filter design of discrete-time T-S fuzzy Itô stochastic systems with time-varying delays in the state. First, model transformation of the original discrete-time T-S fuzzy Itô stochastic time-varying delay system is studied by way of that of a comparison system consisting of two subsystems, which are a constant time-delay forward subsystem and a delay "uncertainty" feed-back subsystem. The forward subsystem needs to be under consideration to ensure the stability of the original systems by applying the scaled small gain theorem. A sufficient condition for the mean-square asymptotically stability of the filter error system is obtained, while an $\mathcal{H}_\infty$ performance is guaranteed. The explicit expression of the desired filter parameters is also derived by applying convex linearization approach, which casts the $\mathcal{H}_\infty$ desired filter design into a convex optimization problem. Simulation examples are used to demonstrate the effectiveness of the proposed approaches in this chapter. The main contributions of this work can be listed as follows. 1) A new model transformation will be analyzed and applied for the $\mathcal{H}_\infty$ desired filter design of discrete-time T-S fuzzy Itô stochastic systems with time-varying delays in the state. Based on this new model, the scaled small gain theorem can be employed to tackle discrete-time T-S fuzzy Itô stochastic systems with time-varying delays in the state; 2) A novel two-term approximation idea [187], $[x(k-d_1)+x(k-d_2)]/2$, will be extended to apply for the $\mathcal{H}_\infty$ desired filter design of T-S fuzzy Itô stochastic time-varying delay systems to reduce the resulting approximation error in new model transformation. The two-term approximation method is employed to pull out the "delay uncertainty" or to estimate the uncertain delay $x(k-d(k))$ precisely to reduce the conservatism in the $\mathcal{H}_\infty$ desired filter design; and 3) The $\mathcal{H}_\infty$ filtering problems will be solved by applying convex linearization approach, which casts the filter design into a convex optimization problem.

L. Wu et al., *Fuzzy Control Systems with Time-Delay and Stochastic Perturbation*, Studies in Systems, Decision and Control 12, DOI: 10.1007/978-3-319-11316-6_10

10.2 System Description and Preliminaries

Consider a class of nonlinear stochastic systems with time-varying delay, which can be represented by the following discrete-time T-S fuzzy model:

♦ **Plant Form:**

Rule i: IF $\theta_1(k)$ is $\mathcal{M}_{i1}$ and $\theta_2(k)$ is $\mathcal{M}_{i2}$ and ... and $\theta_p(k)$ is $\mathcal{M}_{ip}$, THEN

$$\begin{aligned} x(k+1) &= A_i x(k) + A_{di} x(k-d(k)) + B_i \omega(k) + E_i x(k)\varpi(k) + C_i x(k), \\ &\quad + C_{di} x(k-d(k)) + D_i \omega(k) + G_i x(k)\varpi(k), \\ z(k) &= L_i x(k) + L_{di} x(k-d(k)) + F_i \omega(k), \\ x(k) &= \phi(k), \quad k = -d_2, -d_2+1, \ldots, 0, \end{aligned}$$

where $i = 1, 2, \ldots, r$, and r is the number of IF-THEN rules; $\mathcal{M}_{ij} (i = 1, 2, \ldots, r; j = 1, 2, \ldots, p)$ are the fuzzy sets; $\theta(k) = \begin{bmatrix} \theta_1(k) \; \theta_2(k) \; \cdots \; \theta_p(k) \end{bmatrix}^T$ is the premise variable vector. $x(k) \in \mathbf{R}^n$ is the state vector; $y(k) \in \mathbf{R}^p$ is the measured output; $\omega(k) \in \mathbf{R}^l$ is the disturbance input that belongs to $\ell_2[0, \infty)$; $z(k) \in \mathbf{R}^q$ is the signal to be estimated, and $d(k)$ is the time-varying delay which is time-varying in the whole dynamic process and satisfies $1 \leqslant d_1 \leqslant d(k) \leqslant d_2$, where d_1 and d_2 are constant positive scalars representing the minimum and maximum delays, respectively. $\varpi(k)$ is a scalar Brownian motion, which is independent and satisfies $\mathbf{E}\{\varpi(k)\} = 0$ and $\mathbf{E}\{\varpi^2(k)\} = \delta$. A_i, A_{di}, B_i, C_i, C_{di}, D_i, L_i, L_{di}, F_i, E_i and G_i are known real constant matrices; $\phi(k)$ denotes the initial condition.

A more compact presentation of the discrete-time T-S fuzzy stochastic model is given by

$$x(k+1) = \sum_{i=1}^{r} h_i(\theta) \left[A_i x(k) + A_{di} x(k-d(k)) + B_i \omega(k) + E_i x(k)\varpi(k) \right], \tag{10.1a}$$

$$y(k) = \sum_{i=1}^{r} h_i(\theta) \left[C_i x(k) + C_{di} x(k-d(k)) + D_i \omega(k) + G_i x(k)\varpi(k) \right], \tag{10.1b}$$

$$z(k) = \sum_{i=1}^{r} h_i(\theta) \left[L_i x(k) + L_{di} x(k-d(k)) + F_i \omega(k) \right]. \tag{10.1c}$$

where $h_i(\theta(k))$, $i = 1, 2, \ldots, r$ are the normalized membership functions, which are defined as that of (1.2) in Chapter 1.

Here, we design a filter of the following form:

$$x_c(k+1) = A_c x_c(k) + B_c y(k), \tag{10.2a}$$

$$z_c(k) = C_c x_c(k) + D_c y(k), \tag{10.2b}$$

where $x_c(k) \in \mathbf{R}^n$ is the state vector; $z_c(k) \in \mathbf{R}^q$ is an estimation of $z(k)$; and A_c, B_c, C_c and D_c are filter parameter matrices to be determined.

Augmenting the model of (10.1) to include the filter dynamical model (10.2), we obtain the filtering error system as

$$\xi(k+1) = \sum_{i=1}^{r} h_i(\theta)\left[\bar{A}_i\xi(k)+\bar{A}_{di}\xi(k-d(k))+\bar{B}_i\omega(k)+\bar{E}_i\xi(k)\varpi(k)\right], \quad (10.3a)$$

$$e(k) = \sum_{i=1}^{r} h_i(\theta)\left[\bar{L}_i\xi(k)+\bar{L}_{di}\xi(k-d(k))+\bar{F}_i\omega(k)+\bar{G}_i\xi(k)\varpi(k)\right], \quad (10.3b)$$

where $\xi(k) \triangleq \begin{bmatrix} x(k) \\ x_c(k) \end{bmatrix}$, $e(k) \triangleq z(k) - z_c(k)$ and

$$\bar{A}_i \triangleq \begin{bmatrix} A_i & 0 \\ B_cC_i & A_c \end{bmatrix}, \quad \bar{A}_{di} \triangleq \begin{bmatrix} A_{di} & 0 \\ B_cC_{di} & 0 \end{bmatrix}, \quad \bar{B}_i \triangleq \begin{bmatrix} B_i \\ B_cD_i \end{bmatrix},$$
$$\bar{E}_i \triangleq \begin{bmatrix} E_i & 0 \\ B_cG_i & 0 \end{bmatrix}, \quad \bar{F}_i \triangleq F_i - D_cD_i, \quad \bar{G}_i \triangleq \begin{bmatrix} -D_cG_i & 0 \end{bmatrix},$$
$$\bar{L}_i \triangleq \begin{bmatrix} L_i - D_cC_i & -C_c \end{bmatrix}, \quad \bar{L}_{di} \triangleq \begin{bmatrix} L_{di} - D_cC_{di} & 0 \end{bmatrix}.$$

Moreover, we define

$$\bar{A}(k) \triangleq \sum_{i=1}^{r} h_i(\theta)\bar{A}_i, \quad \bar{A}_d(k) \triangleq \sum_{i=1}^{r} h_i(\theta)\bar{A}_{di}, \quad \bar{B}(k) \triangleq \sum_{i=1}^{r} h_i(\theta)\bar{B}_i,$$
$$\bar{E}(k) \triangleq \sum_{i=1}^{r} h_i(\theta)\bar{E}_i, \quad \bar{F}(k) \triangleq \sum_{i=1}^{r} h_i(\theta)\bar{F}_i, \quad \bar{G}(k) \triangleq \sum_{i=1}^{r} h_i(\theta)\bar{G}_i,$$
$$\bar{L}(k) \triangleq \sum_{i=1}^{r} h_i(\theta)\bar{L}_i, \quad \bar{L}_d(k) \triangleq \sum_{i=1}^{r} h_i(\theta)\bar{L}_{di}.$$

Definition 10.1. The filtering error system in (10.3) is said to be mean-square asymptotically stable if under $\omega(k) = 0$,

$$\lim_{k\to\infty} \mathbf{E}\left\{\|\xi(k)\|\right\} = 0.$$

Definition 10.2. Given a scalar $\gamma > 0$, the filtering error system in (10.3) is said to have an $\mathcal{H}_\infty$ performance level γ, if it is mean-square asymptotically stable under $\omega(k) = 0$, and under zero initial condition, it satisfies

$$\|e(k)\|_{\mathbf{E}_2} < \gamma \|\omega(k)\|_2, \quad \forall\, 0 \neq \omega(k) \in \ell_2[0,\infty), \quad (10.4)$$

where

$$\|e(k)\|_{\mathbf{E}_2} \triangleq \sqrt{\mathbf{E}\left\{\sum_{k=0}^{\infty} e^T(k)e(k)\right\}}, \quad \|\omega(k)\|_2 \triangleq \sqrt{\sum_{k=0}^{\infty} \omega^T(k)\omega(k)}.$$

Define $\hat{\omega}(k) \triangleq \gamma\omega(k)$, then the filtering error system in (10.3) becomes

$$\xi(k+1) = \sum_{i=1}^{r} h_i(\theta)\big[\bar{A}_i\xi(k) + \bar{A}_{di}\xi(k-d(k)) + \gamma^{-1}\bar{B}_i\hat{\omega}(k) + \bar{E}_i\xi(k)\varpi(k)\big], \quad (10.5a)$$

$$e(k) = \sum_{i=1}^{r} h_i(\theta)\big[\bar{L}_i\xi(k) + \bar{L}_{di}\xi(k-d(k)) + \gamma^{-1}\bar{F}_i\hat{\omega}(k) + \bar{G}_i\xi(k)\varpi(k)\big]. \quad (10.5b)$$

Clearly, the $\mathcal{H}_\infty$ performance defined in (10.4) is equivalent to

$$\|e(k)\|_{\mathbf{E}_2} < \|\hat{\omega}(k)\|_2 . \quad (10.6)$$

10.3 Main Results

10.3.1 Filtering Analysis

Considering the filtering error system in (10.5), we now estimate the delayed state term $\xi(k-d(k))$ using its lower bound d_1 and upper bound d_2. The two-term approximation $\frac{\xi(k-d_1)+\xi(k-d_2)}{2}$ results in the estimation error:

$$\begin{aligned} \sigma(k) &= \frac{2}{d}\left\{\xi(k-d(k)) - \frac{1}{2}\left[\xi(k-d_1) + \xi(k-d_2)\right]\right\} \\ &= \frac{1}{d}\left(\sum_{i=k-d_2}^{k-d_1-1} \beta(i)\varsigma(i)\right), \end{aligned} \quad (10.7)$$

where $d \triangleq d_2 - d_1$, $\varsigma(i) \triangleq \xi(i+1) - \xi(i)$ and

$$\beta(i) \triangleq \begin{cases} 1, & \text{when } i \le k-d(k)-1, \\ -1, & \text{when } i > k-d(k)-1. \end{cases}$$

To employ the input-output approach, we replace (10.5) with

$$\begin{aligned} \xi(k+1) = \sum_{i=1}^{r} h_i(\theta)\Big\{ & \bar{A}_i\xi(k) + \frac{1}{2}\bar{A}_{di}\left[\xi(k-d_1) + \xi(k-d_2)\right] + \frac{d}{2}\bar{A}_{di}\sigma(k) \\ & + \gamma^{-1}\bar{B}_i\hat{\omega}(k) + \bar{E}_i\xi(k)\varpi(k)\Big\}, \end{aligned} \quad (10.8a)$$

$$\begin{aligned} e(k) = \sum_{i=1}^{r} h_i(\theta)\Big\{ & \bar{L}_i\xi(k) + \frac{1}{2}\bar{L}_{di}\left[\xi(k-d_1) + \xi(k-d_2)\right] + \frac{d}{2}\bar{L}_{di}\sigma(k) \\ & + \gamma^{-1}\bar{F}_i\hat{\omega}(k) + \bar{G}_i\xi(k)\varpi(k)\Big\}. \end{aligned} \quad (10.8b)$$

The following model can formulate system (10.8) in the interconnection frame in Fig. 1.1:

$$(\mathcal{S}_1): \quad \begin{bmatrix} \xi(k+1) \\ \varsigma(k) \\ e(k) \end{bmatrix} = \begin{bmatrix} \Sigma_1(k) & \frac{d}{2}\bar{A}_d(k) & \gamma^{-1}\bar{B}(k) \\ \Sigma_2(k) & \frac{d}{2}\bar{A}_d(k) & \gamma^{-1}\bar{B}(k) \\ \Sigma_3(k) & \frac{d}{2}\bar{L}_d(k) & \gamma^{-1}\bar{F}(k) \end{bmatrix} \begin{bmatrix} \bar{\xi}(k) \\ \sigma(k) \\ \hat{\omega}(k) \end{bmatrix} + \begin{bmatrix} \bar{E}(k) \\ \bar{E}(k) \\ \bar{G}(k) \end{bmatrix} \xi(k)\varpi(k), \tag{10.9a}$$

$$(\mathcal{S}_2): \quad \sigma(k) = \mathcal{K}\varsigma(k), \tag{10.9b}$$

where $\bar{\xi}(k) \triangleq \begin{bmatrix} \xi(k) \\ \xi(k-d_1) \\ \xi(k-d_2) \end{bmatrix}$ and

$$\Sigma_1(k) \triangleq \begin{bmatrix} \bar{A}(k) & \frac{1}{2}\bar{A}_d(k) & \frac{1}{2}\bar{A}_d(k) \end{bmatrix},$$

$$\Sigma_2(k) \triangleq \begin{bmatrix} \bar{A}(k) - I & \frac{1}{2}\bar{A}_d(k) & \frac{1}{2}\bar{A}_d(k) \end{bmatrix},$$

$$\Sigma_3(k) \triangleq \begin{bmatrix} \bar{L}(k) & \frac{1}{2}\bar{L}_d(k) & \frac{1}{2}\bar{L}_d(k) \end{bmatrix}.$$

For brevity, let us use the following operator:

$$(\mathcal{K}): \quad \varsigma(k) \to \sigma(k) = \frac{1}{d}\left(\sum_{i=k-d_2}^{k-d_1-1} \beta(i)\varsigma(i)\right), \tag{10.10}$$

to denote the relation $(\mathcal{S}_2)$ from $\varsigma(k)$ to $\sigma(k)$ in Fig. 1.1. The following result gives an upper bound of the ℓ_2 norm of $(\mathcal{K})$.

Lemma 10.3. [187] *Operator $(\mathcal{K})$ in (10.10) bears the property that $\|\mathcal{K}\|_\infty \leq 1$.*

Proof. In view of the formulation $\sigma(k)$ in (10.7) and using Jensen inequality [77], we can obtain the following inequality under the zero initial condition.

$$\begin{aligned} \|\sigma(k)\|_{\mathbf{E}_2}^2 &= \mathbf{E}\left\{\sum_{i=0}^{\infty} \sigma^T(i)\sigma(i)\right\} \\ &= \frac{1}{d^2}\mathbf{E}\left\{\sum_{i=0}^{\infty}\left(\sum_{i=k-d_2}^{k-d_1-1} \beta(i)\varsigma^T(i)\right)\left(\sum_{i=k-d_2}^{k-d_1-1} \beta(i)\varsigma(i)\right)\right\} \end{aligned}$$

$$
\begin{aligned}
&\leq \frac{1}{d^2}\mathbf{E}\left\{\sum_{i=0}^{\infty}\left((d_2-d_1)\sum_{i=k-d_2}^{k-d_1-1}\beta^2(i)\varsigma^T(i)\varsigma(i)\right)\right\} \\
&= \frac{1}{d}\mathbf{E}\left\{\sum_{j=-d_2}^{-d_1-1}\sum_{i=0}^{\infty}\varsigma^T(i+j)\varsigma(i+j)\right\} \\
&\leq \frac{1}{d}\mathbf{E}\left\{\sum_{j=-d_2}^{-d_1-1}\sum_{i=0}^{\infty}\varsigma^T(i)\varsigma(i)\right\} \\
&= \mathbf{E}\left\{\sum_{i=0}^{\infty}\varsigma^T(i)\varsigma(i)\right\} = \|\varsigma(k)\|_{\mathbf{E}_2}^2,
\end{aligned}
\tag{10.11}
$$

which implies

$$
\|\mathcal{K}\|_\infty = \sup_{\|\varsigma(k)\|_{\mathbf{E}_2}\neq 0}\frac{\|\sigma(k)\|_{\mathbf{E}_2}}{\|\varsigma(k)\|_{\mathbf{E}_2}} \leq 1.
$$

This completes the proof. ■

In view of Lemma 10.3, we can see that the ℓ_2 norm of $(\mathcal{S}_2)$ in (10.9b) from input to output is bounded by one. Then, based on Lemma 1.33, we focus on researching the scaled small gain of $(\mathcal{S}_1)$ for the interconnection frame (10.9a).

Lemma 10.4. *Assume $(\mathcal{S}_1)$ is internally stable in (10.9), the closed-loop system of interconnection system described by (10.9) is mean-square asymptotically stable and has an $\mathcal{H}_\infty$ performance level γ for $(\mathcal{K})$ if there exists a matrix $\hat{\mathcal{X}} \triangleq \mathrm{diag}\{\bar{\mathcal{X}}, I\} > 0$ such that*

$$
\|\hat{\mathcal{X}} \circ \mathcal{G} \circ \hat{\mathcal{X}}^{-1}\|_\infty < 1,
\tag{10.12}
$$

where

$$
\mathcal{G} \triangleq \left[\begin{array}{ccc|c}
\Sigma_1(k) & \frac{d}{2}\bar{A}_d(k) & \gamma^{-1}\bar{B}(k) & \bar{E}(k) \\
\Sigma_2(k) & \frac{d}{2}\bar{A}_d(k) & \gamma^{-1}\bar{B}(k) & \bar{E}(k) \\
\Sigma_3(k) & \frac{d}{2}\bar{L}_d(k) & \gamma^{-1}\bar{F}(k) & \bar{G}(k)
\end{array}\right].
$$

Remark 10.5. Along the interconnection frame (10.9), the sufficient condition in Lemma 10.4 can be converted to the following condition: assumed that $(\mathcal{S}_1)$ is internally stable in (10.9a), the closed-loop system of interconnection system described by (10.9) is mean-square asymptotically stable with an $\mathcal{H}_\infty$ performance level γ for $\mathcal{K}$ if there exist exists a matrix $\mathcal{X} \triangleq \bar{\mathcal{X}}^T\bar{\mathcal{X}}$ such that $\mathcal{J} \triangleq \mathbf{E}\left\{\sum_{k=0}^{\infty}\left[\varsigma^T(k)\mathcal{X}\varsigma(k) - \sigma^T(k)\mathcal{X}\sigma(k) + e^T(k)e(k) - \hat{\omega}^T(k)\hat{\omega}(k)\right]\right\} < 0.$ ♦

In the view of Lemma 10.3, it follows that the ℓ_2 norm of $(\mathcal{S}_2)$ in (10.9b) from input to output is bounded by one. Then based on Lemma 10.4, we focus on studying the $\mathcal{H}_\infty$ performance of $(\mathcal{S}_1)$ by the input-output approach.

In the following, by applying the input-output approach, we will derive an LMI formulation of the $\mathcal{H}_\infty$ performance for system (10.9). Firstly, we let $d = d_2 - d_1$ and

$$\mathcal{T}_1(k) \triangleq \sum_{i=1}^{r} h_i(\theta)\mathcal{T}_{1i}, \quad \mathcal{T}_2(k) \triangleq \sum_{i=1}^{r} h_i(\theta)\mathcal{T}_{2i}, \quad \mathcal{T}_3(k) \triangleq \sum_{i=1}^{r} h_i(\theta)\mathcal{T}_{3i},$$
$$\mathcal{S}_1(k) \triangleq \sum_{i=1}^{r} h_i(\theta)\mathcal{S}_{1i}, \quad \mathcal{S}_2(k) \triangleq \sum_{i=1}^{r} h_i(\theta)\mathcal{S}_{2i}, \quad \mathcal{S}_3(k) \triangleq \sum_{i=1}^{r} h_i(\theta)\mathcal{S}_{3i},$$
$$\mathcal{N}_1(k) \triangleq \sum_{i=1}^{r} h_i(\theta)\mathcal{N}_{1i}, \quad \mathcal{N}_2(k) \triangleq \sum_{i=1}^{r} h_i(\theta)\mathcal{N}_{2i}, \quad \mathcal{N}_3(k) \triangleq \sum_{i=1}^{r} h_i(\theta)\mathcal{N}_{3i},$$
$$\mathcal{M}_1(k) \triangleq \sum_{i=1}^{r} h_i(\theta)\mathcal{M}_{1i}, \mathcal{M}_2(k) \triangleq \sum_{i=1}^{r} h_i(\theta)\mathcal{M}_{2i}, \mathcal{M}_3(k) \triangleq \sum_{i=1}^{r} h_i(\theta)\mathcal{M}_{3i},$$
$$\mathcal{Q}_1(k) \triangleq \sum_{i=1}^{r} h_i(\theta)\mathcal{Q}_{1i}, \quad \mathcal{Q}_2(k) \triangleq \sum_{i=1}^{r} h_i(\theta)\mathcal{Q}_{2i},$$

where $\mathcal{Q}_{1i} > 0$, $\mathcal{Q}_{2i} > 0$, $\mathcal{S}_{1i} > 0$, $\mathcal{S}_{2i} > 0$, $\mathcal{S}_{3i} > 0$, $\mathcal{T}_{1i} > 0$, $\mathcal{T}_{2i} > 0$, $\mathcal{T}_{3i} > 0$, $i = 1, 2, \ldots, r$, are all $(n+k) \times (n+k)$ matrices.

Thus, we construct the following LKF:

$$V(k) \triangleq \sum_{i=1}^{3} V_i(k), \tag{10.13}$$

with

$$\begin{cases} V_1(k) \triangleq \xi^T(k)\mathcal{P}\xi(k), \\ V_2(k) \triangleq \sum\limits_{i=k-d_1}^{k-1} \xi^T(i)\mathcal{Q}_1(i)\xi(i) + \sum\limits_{i=k-d_2}^{k-1} \xi^T(i)\mathcal{Q}_2(i)\xi(i), \\ V_3(k) \triangleq \sum\limits_{i=-d_1}^{-1} \sum\limits_{j=k+i}^{k-1} \varsigma^T(j)\mathcal{Z}_1\varsigma(j) + \sum\limits_{i=-d_2}^{-1} \sum\limits_{j=k+i}^{k-1} \varsigma^T(j)\mathcal{Z}_2\varsigma(j). \end{cases}$$

Then, based on (10.13), we can obtain the following result.

Theorem 10.6. *The filtering error system in (10.9) is mean-square asymptotically stable with an $\mathcal{H}_\infty$ performance level γ if there exist matrices $\mathcal{P} > 0$, $\mathcal{X} > 0$, $\mathcal{Z}_1 > 0$, $\mathcal{Z}_2 > 0$, $\mathcal{Q}_{1i} > 0$, $\mathcal{Q}_{2i} > 0$, $\mathcal{S}_{1i} > 0$, $\mathcal{S}_{2i} > 0$, $\mathcal{S}_{3i} > 0$, $\mathcal{T}_{1i} > 0$, $\mathcal{T}_{2i} > 0$, $\mathcal{T}_{3i} > 0$, $\mathcal{M}_{1i}$, $\mathcal{M}_{2i}$, $\mathcal{M}_{3i}$, $\mathcal{N}_{1i}$, $\mathcal{N}_{2i}$, and $\mathcal{N}_{3i}$, such that for $i, j, s = 1, \ldots, r$,*

$$\Psi_{ijs} \triangleq \begin{bmatrix} \Psi_{11} & 0 & (\Psi_{13i} - \mathcal{I}) & \frac{1}{2}\Psi_{14i} & \frac{1}{2}\Psi_{14i} & \frac{d}{2}\Psi_{14i} & \Psi_{17i} \\ \star & \delta^{-1}\Psi_{11} & \Psi_{23i} & 0 & 0 & 0 & 0 \\ \star & \star & \Psi_{33i} & \Psi_{34i} & \Psi_{35i} & 0 & 0 \\ \star & \star & \star & \Psi_{44ij} & \Psi_{45i} & 0 & 0 \\ \star & \star & \star & \star & \Psi_{55is} & 0 & 0 \\ \star & \star & \star & \star & \star & -\mathcal{X} & 0 \\ \star & \star & \star & \star & \star & \star & -\gamma^2 I \end{bmatrix} < 0, (10.14a)$$

$$\Omega_{1i} \triangleq \begin{bmatrix} \mathcal{S}_{1i} & 0 & 0 & \mathcal{M}_{1i} \\ \star & \mathcal{S}_{2i} & 0 & \mathcal{M}_{2i} \\ \star & \star & \mathcal{S}_{3i} & \mathcal{M}_{3i} \\ \star & \star & \star & \mathcal{Z}_1 \end{bmatrix} \geq 0, (10.14b)$$

$$\Omega_{2i} \triangleq \begin{bmatrix} \mathcal{T}_{1i} & 0 & 0 & \mathcal{N}_{1i} \\ \star & \mathcal{T}_{2i} & 0 & \mathcal{N}_{2i} \\ \star & \star & \mathcal{T}_{3i} & \mathcal{N}_{3i} \\ \star & \star & \star & \mathcal{Z}_2 \end{bmatrix} \geq 0, (10.14c)$$

where

$$\Psi_{13i} \triangleq \begin{bmatrix} \bar{A}_i \\ \bar{A}_i \\ \bar{A}_i \\ \bar{A}_i \\ \bar{L}_i \end{bmatrix}, \Psi_{14i} \triangleq \begin{bmatrix} \bar{A}_{di} \\ \bar{A}_{di} \\ \bar{A}_{di} \\ \bar{A}_{di} \\ \bar{L}_{di} \end{bmatrix}, \Psi_{17i} \triangleq \begin{bmatrix} \bar{B}_i \\ \bar{B}_i \\ \bar{B}_i \\ \bar{B}_i \\ \bar{F}_i \end{bmatrix}, \Psi_{23i} \triangleq \begin{bmatrix} \bar{E}_i \\ \bar{E}_i \\ \bar{E}_i \\ \bar{E}_i \\ \bar{G}_i \end{bmatrix}, \mathcal{I} \triangleq \begin{bmatrix} 0 \\ I \\ I \\ I \\ 0 \end{bmatrix},$$

$$\Psi_{11} \triangleq \mathrm{diag}\{-\mathcal{P}^{-1}, -d_1^{-1}\mathcal{Z}_1^{-1}, -d_2^{-1}\mathcal{Z}_2^{-1}, -\mathcal{X}^{-1}, -I\},\ \Psi_{45i} \triangleq -\mathcal{M}_{3i} - \mathcal{N}_{2i}^T,$$
$$\Psi_{34i} \triangleq -\mathcal{M}_{1i}^T + \mathcal{M}_{2i} + \mathcal{N}_{2i}, \quad \Psi_{35i} \triangleq -\mathcal{N}_{1i}^T + \mathcal{M}_{3i} + \mathcal{N}_{3i},$$
$$\Psi_{33i} \triangleq -\mathcal{P} + \mathcal{Q}_{1i} + \mathcal{Q}_{2i} + \mathcal{M}_{1i} + \mathcal{M}_{1i}^T + \mathcal{N}_{1i} + \mathcal{N}_{1i}^T + d_1\mathcal{S}_{1i} + d_2\mathcal{T}_{1i},$$
$$\Psi_{44ij} \triangleq -\mathcal{Q}_{1j} - \mathcal{M}_{2i} - \mathcal{M}_{2i}^T + d_1\mathcal{S}_{2i} + d_2\mathcal{T}_{2i},$$
$$\Psi_{55is} \triangleq -\mathcal{Q}_{2s} - \mathcal{N}_{3i} - \mathcal{N}_{3i}^T + d_1\mathcal{S}_{3i} + d_2\mathcal{T}_{3i}.$$

Proof. Based on the fuzzy basis functions, from (10.14a)–(10.14c) we obtain

$$\sum_{i=1}^{r}\sum_{j=1}^{r}\sum_{s=1}^{r} h_i(\theta)h_j(\theta(k-d_1))h_s(\theta(k-d_2))\Psi_{ijs} < 0,$$

$$\sum_{i=1}^{r} h_i(\theta)\Omega_{1i} \geq 0,$$

$$\sum_{i=1}^{r} h_i(\theta)\Omega_{2i} \geq 0.$$

A more compact presentation of the above equalities is given by

$$\begin{bmatrix} \Psi_{11} & 0 & (\bar{\Psi}_{13}(k)-\mathcal{I}) & \frac{1}{2}\bar{\Psi}_{14}(k) & \frac{1}{2}\bar{\Psi}_{14}(k) & \frac{d}{2}\bar{\Psi}_{14}(k) & \bar{\Psi}_{17}(k) \\ \star & \delta^{-1}\Psi_{11} & \bar{\Psi}_{23}(k) & 0 & 0 & 0 & 0 \\ \star & \star & \bar{\Psi}_{33}(k) & \bar{\Psi}_{34}(k) & \bar{\Psi}_{35}(k) & 0 & 0 \\ \star & \star & \star & \bar{\Psi}_{44}(k) & \bar{\Psi}_{45}(k) & 0 & 0 \\ \star & \star & \star & \star & \bar{\Psi}_{55}(k) & 0 & 0 \\ \star & \star & \star & \star & \star & -\mathcal{X} & 0 \\ \star & \star & \star & \star & \star & \star & -\gamma^2 I \end{bmatrix} < 0, \quad (10.15a)$$

$$\bar{\Omega}_1(k) \triangleq \begin{bmatrix} \mathcal{S}_1(k) & 0 & 0 & \mathcal{M}_1(k) \\ \star & \mathcal{S}_2(k) & 0 & \mathcal{M}_2(k) \\ \star & \star & \mathcal{S}_3(k) & \mathcal{M}_3(k) \\ \star & \star & \star & \mathcal{Z}_1 \end{bmatrix} \geq 0, \quad (10.15b)$$

$$\bar{\Omega}_2(k) \triangleq \begin{bmatrix} \mathcal{T}_1(k) & 0 & 0 & \mathcal{N}_1(k) \\ \star & \mathcal{T}_2(k) & 0 & \mathcal{N}_2(k) \\ \star & \star & \mathcal{T}_3(k) & \mathcal{N}_3(k) \\ \star & \star & \star & \mathcal{Z}_2 \end{bmatrix} \geq 0, \quad (10.15c)$$

where

$$\bar{\Psi}_{13}(k) \triangleq \begin{bmatrix} \bar{A}(k) \\ \bar{A}(k) \\ \bar{A}(k) \\ \bar{A}(k) \\ \bar{L}(k) \end{bmatrix}, \bar{\Psi}_{14}(k) \triangleq \begin{bmatrix} \bar{A}_d(k) \\ \bar{A}_d(k) \\ \bar{A}_d(k) \\ \bar{A}_d(k) \\ \bar{L}_d(k) \end{bmatrix}, \bar{\Psi}_{17}(k) \triangleq \begin{bmatrix} \bar{B}(k) \\ \bar{B}(k) \\ \bar{B}(k) \\ \bar{B}(k) \\ \bar{F}(k) \end{bmatrix}, \bar{\Psi}_{23}(k) \triangleq \begin{bmatrix} \bar{E}(k) \\ \bar{E}(k) \\ \bar{E}(k) \\ \bar{E}(k) \\ \bar{G}(k) \end{bmatrix},$$

$$\begin{aligned}
\bar{\Psi}_{33}(k) &\triangleq -\mathcal{P} + \mathcal{Q}_1(k) + \mathcal{Q}_2(k) + \mathcal{M}_1(k) + \mathcal{M}_1^T(k) + \mathcal{N}_1(k) + \mathcal{N}_1^T(k) \\
&\quad + d_1\mathcal{S}_1(k) + d_2\mathcal{T}_1(k), \\
\bar{\Psi}_{34}(k) &\triangleq -\mathcal{M}_1^T(k) + \mathcal{M}_2(k) + \mathcal{N}_2(k), \\
\bar{\Psi}_{35}(k) &\triangleq -\mathcal{N}_1^T(k) + \mathcal{M}_3(k) + \mathcal{N}_3(k), \\
\bar{\Psi}_{44}(k) &\triangleq -\mathcal{Q}_1(k) - \mathcal{M}_2(k) - \mathcal{M}_2^T(k) + d_1\mathcal{S}_2(k) + d_2\mathcal{T}_2(k), \\
\bar{\Psi}_{45}(k) &\triangleq -\mathcal{M}_3(k) - \mathcal{N}_2^T(k), \\
\bar{\Psi}_{55}(k) &\triangleq -\mathcal{Q}_2(k) - \mathcal{N}_3(k) - \mathcal{N}_3^T(k) + d_1\mathcal{S}_3(k) + d_2\mathcal{T}_3(k).
\end{aligned}$$

By Schur complement, inequality (10.15a) implies

$$\hat{\Psi}(k) \triangleq \begin{bmatrix} \hat{\Psi}_{33}(k) & \frac{1}{2}\hat{\Psi}_{34}(k) & \frac{1}{2}\hat{\Psi}_{35}(k) & \frac{d}{2}\hat{\Psi}_{36}(k) & \gamma^{-1}\hat{\Psi}_{37}(k) \\ \star & \frac{1}{4}\hat{\Psi}_{44}(k) & \frac{1}{4}\hat{\Psi}_{45}(k) & \frac{d}{4}\hat{\Psi}_{66}(k) & \frac{\gamma^{-1}}{2}\hat{\Psi}_{47}(k) \\ \star & \star & \frac{1}{4}\hat{\Psi}_{55}(k) & \frac{d}{4}\hat{\Psi}_{66}(k) & \frac{\gamma^{-1}}{2}\hat{\Psi}_{47}(k) \\ \star & \star & \star & \frac{d^2}{4}\hat{\Psi}_{66}(k) - \mathcal{X} & \frac{\gamma^{-1}d}{2}\hat{\Psi}_{47}(k) \\ \star & \star & \star & \star & \gamma^{-2}\hat{\Psi}_{77}(k) - I \end{bmatrix} < 0 \quad (10.16)$$

where $\hat{\mathcal{Z}} \triangleq d_1\mathcal{Z}_1 + d_2\mathcal{Z}_2 + \mathcal{X}$, $\tilde{\mathcal{Z}} \triangleq \hat{\mathcal{Z}} + \mathcal{P}$ and

$$\begin{aligned}
\hat{\Psi}_{33}(k) &\triangleq \bar{A}^T(k)\hat{\mathcal{Z}}\bar{A}(k) - \hat{\mathcal{Z}}\bar{A}(k) - \bar{A}^T(k)\hat{\mathcal{Z}} + \hat{\mathcal{Z}} + \bar{A}^T(k)\mathcal{P}\bar{A}(k) + \bar{L}^T(k)\bar{L}(k) \\
&\quad +\delta\bar{G}^T(k)\bar{G}(k) + \delta\bar{E}^T(k)\tilde{\mathcal{Z}}\bar{E}(k) + \bar{\Psi}_{33}(k), \\
\hat{\Psi}_{36}(k) &\triangleq \bar{A}^T(k)\hat{\mathcal{Z}}\bar{A}_d(k) - \hat{\mathcal{Z}}\bar{A}_d(k) + \bar{A}^T(k)\mathcal{P}\bar{A}_d(k) + \bar{L}^T(k)\bar{L}_d(k), \\
\hat{\Psi}_{37}(k) &\triangleq \bar{A}^T(k)\hat{\mathcal{Z}}\bar{B}(k) - \hat{\mathcal{Z}}\bar{B}(k) + \bar{A}^T(k)\mathcal{P}\bar{B}(k) + \bar{L}^T(k)\bar{F}(k), \\
\hat{\Psi}_{47}(k) &\triangleq \bar{A}_d^T(k)\tilde{\mathcal{Z}}\bar{B}(k) + \bar{L}_d^T(k)\bar{F}(k), \quad \hat{\Psi}_{34}(k) \triangleq \hat{\Psi}_{36}(k) + 2\bar{\Psi}_{34}(k), \\
\hat{\Psi}_{66}(k) &\triangleq \bar{A}_d^T(k)\tilde{\mathcal{Z}}\bar{A}_d(k) + \bar{L}_d^T(k)\bar{L}_d(k), \quad \hat{\Psi}_{35}(k) \triangleq \hat{\Psi}_{36}(k) + 2\bar{\Psi}_{35}(k), \\
\hat{\Psi}_{77}(k) &\triangleq \bar{B}^T(k)\tilde{\mathcal{Z}}\bar{B}(k) + \bar{F}^T(k)\bar{F}(k), \quad \hat{\Psi}_{44}(k) \triangleq \hat{\Psi}_{66}(k) + 4\bar{\Psi}_{44}(k), \\
\hat{\Psi}_{45}(k) &\triangleq \hat{\Psi}_{66}(k) + 4\bar{\Psi}_{45}(k), \quad \hat{\Psi}_{55}(k) \triangleq \hat{\Psi}_{66}(k) + 4\bar{\Psi}_{55}(k).
\end{aligned}$$

Along the trajectories of the filter error system (10.9), and considering the mathematical expectation and the difference of the LKF in (10.13), we have

$$\begin{aligned}
\mathbf{E}\{\Delta V(k)\} &= \mathbf{E}\left\{V(k+1)\right\} - V(k) \\
&= \sum_{i=1}^{3} \mathbf{E}\{\Delta V_i(k)\},
\end{aligned} \tag{10.17}$$

where

$$\begin{aligned}
\mathbf{E}\{\Delta V_1(k)\} &= \mathbf{E}\left\{\xi^T(k+1)\mathcal{P}\xi(k+1)\right\} - \mathbf{E}\left\{\xi^T(k)\mathcal{P}\xi(k)\right\}, \\
\mathbf{E}\{\Delta V_2(k)\} &= \mathbf{E}\left\{\xi^T(k)\left[\mathcal{Q}_1(k) + \mathcal{Q}_2(k)\right]\xi(k)\right\} \\
&\quad -\mathbf{E}\left\{\xi^T(k-d_1)\mathcal{Q}_1(k-d_1)\xi(k-d_1)\right\} \\
&\quad -\mathbf{E}\left\{\xi^T(k-d_2)\mathcal{Q}_2(k-d_2)\xi(k-d_2)\right\}, \\
\mathbf{E}\{\Delta V_3(k)\} &= \mathbf{E}\left\{\varsigma^T(k)\left[d_1\mathcal{Z}_1 + d_2\mathcal{Z}_2\right]\varsigma(k)\right\} \\
&\quad -\mathbf{E}\left\{\sum_{i=k-d_1}^{k-1} \varsigma^T(i)\mathcal{Z}_1\varsigma(i)\right\} - \mathbf{E}\left\{\sum_{i=k-d_2}^{k-1} \varsigma^T(i)\mathcal{Z}_2\varsigma(i)\right\}.
\end{aligned}$$

According to the definitions of $\varsigma(k)$ and $\bar{\xi}(k)$, for matrices

$$\begin{aligned}
\mathscr{M}(k) &\triangleq \begin{bmatrix} \mathcal{M}_1(k) & \mathcal{M}_2(k) & \mathcal{M}_3(k) \end{bmatrix}, \\
\mathscr{N}(k) &\triangleq \begin{bmatrix} \mathcal{N}_1(k) & \mathcal{N}_2(k) & \mathcal{N}_3(k) \end{bmatrix},
\end{aligned}$$

the following equations always hold:

$$2\bar{\xi}^T(k)\mathscr{M}^T(k)\left[\xi(k) - \xi(k-d_1) - \sum_{s=k-d_1}^{k-1} \varsigma(s)\right] = 0, \tag{10.18a}$$

$$2\bar{\xi}^T(k)\mathscr{N}^T(k)\left[\xi(k) - \xi(k-d_2) - \sum_{s=k-d_2}^{k-1} \varsigma(s)\right] = 0. \tag{10.18b}$$

Moreover, for any appropriately dimensioned matrices $\mathcal{S}(k) \triangleq \text{diag}\{\mathcal{S}_1(k), \mathcal{S}_2(k), \mathcal{S}_3(k)\} > 0$ and $\mathcal{T}(k) \triangleq \text{diag}\{\mathcal{T}_1(k), \mathcal{T}_2(k), \mathcal{T}_3(k)\} > 0$, we have

$$d_1\bar{\xi}^T(k)\mathcal{S}(k)\bar{\xi}(k) - \sum_{s=k-d_1}^{k-1} \bar{\xi}^T(k)\mathcal{S}(k)\bar{\xi}(k) = 0, \tag{10.19a}$$

$$d_2\bar{\xi}^T(k)\mathcal{T}(k)\bar{\xi}(k) - \sum_{s=k-d_2}^{k-1} \bar{\xi}^T(k)\mathcal{T}(k)\bar{\xi}(k) = 0, \tag{10.19b}$$

Therefore, from (10.17)–(10.19b) we obtain

$$\begin{aligned}
\mathbf{E}\{\Delta V(k)\} &= \sum_{i=1}^{3} \mathbf{E}\{\Delta V_i(k)\} \\
&= \begin{bmatrix} \bar{\xi}(k) \\ \sigma(k) \\ \hat{\omega}(k) \end{bmatrix}^T \tilde{\Psi}(k) \begin{bmatrix} \bar{\xi}(k) \\ \sigma(k) \\ \hat{\omega}(k) \end{bmatrix} - \sum_{s=k-d_2}^{k-1} \begin{bmatrix} \bar{\xi}(k) \\ \varsigma(s) \end{bmatrix}^T \bar{\Omega}_2(k) \begin{bmatrix} \bar{\xi}(k) \\ \varsigma(s) \end{bmatrix} \\
&\quad - \sum_{s=k-d_1}^{k-1} \begin{bmatrix} \bar{\xi}(k) \\ \varsigma(s) \end{bmatrix}^T \bar{\Omega}_1(k) \begin{bmatrix} \bar{\xi}(k) \\ \varsigma(s) \end{bmatrix},
\end{aligned} \tag{10.20}$$

where

$$\tilde{\Psi}(k) \triangleq \begin{bmatrix}
\tilde{\Psi}_{33}(k) & \frac{1}{2}\tilde{\Psi}_{34}(k) & \frac{1}{2}\tilde{\Psi}_{35}(k) & \frac{d}{2}\tilde{\Psi}_{36}(k) & \gamma^{-1}\tilde{\Psi}_{37}(k) \\
\star & \frac{1}{4}\tilde{\Psi}_{44}(k) & \frac{1}{4}\tilde{\Psi}_{45}(k) & \frac{d}{4}\tilde{\Psi}_{66}(k) & \frac{\gamma^{-1}}{2}\tilde{\Psi}_{47}(k) \\
\star & \star & \frac{1}{4}\tilde{\Psi}_{55}(k) & \frac{d}{4}\tilde{\Psi}_{66}(k) & \frac{\gamma^{-1}}{2}\tilde{\Psi}_{47}(k) \\
\star & \star & \star & \frac{d^2}{4}\tilde{\Psi}_{66}(k) & \frac{\gamma^{-1}d}{2}\tilde{\Psi}_{47}(k) \\
\star & \star & \star & \star & \gamma^{-2}\tilde{\Psi}_{77}(k)
\end{bmatrix},$$

with $\boldsymbol{\mathcal{Z}} \triangleq d_1\mathcal{Z}_1 + d_2\mathcal{Z}_2$, $\breve{\mathcal{Z}} \triangleq \boldsymbol{\mathcal{Z}} + \mathcal{P}$ and

$$\begin{aligned}
\tilde{\Psi}_{33}(k) &\triangleq \bar{A}^T(k)\boldsymbol{\mathcal{Z}}\bar{A}(k) - \boldsymbol{\mathcal{Z}}\bar{A}(k) - \bar{A}^T(k)\boldsymbol{\mathcal{Z}} + \boldsymbol{\mathcal{Z}} + \delta\bar{E}^T(k)\breve{\mathcal{Z}}\bar{E}(k) \\
&\quad + \bar{A}^T(k)\mathcal{P}\bar{A}(k) + \bar{\Psi}_{33}(k), \\
\tilde{\Psi}_{36}(k) &\triangleq \bar{A}^T(k)\boldsymbol{\mathcal{Z}}\bar{A}_d(k) - \boldsymbol{\mathcal{Z}}\bar{A}_d(k) + \bar{A}^T(k)\mathcal{P}\bar{A}_d(k), \\
\tilde{\Psi}_{37}(k) &\triangleq \bar{A}^T(k)\boldsymbol{\mathcal{Z}}\bar{B}(k) - \boldsymbol{\mathcal{Z}}\bar{B}(k) + \bar{A}^T(k)\mathcal{P}\bar{B}(k),
\end{aligned}$$

$$\begin{aligned}
\tilde{\Psi}_{47}(k) &\triangleq \bar{A}_d^T(k)\breve{\mathcal{Z}}\bar{B}(k), & \tilde{\Psi}_{34}(k) &\triangleq \tilde{\Psi}_{36}(k) + 2\bar{\Psi}_{34}(k), \\
\tilde{\Psi}_{66}(k) &\triangleq \bar{A}_d^T(k)\breve{\mathcal{Z}}\bar{A}_d(k), & \tilde{\Psi}_{35}(k) &\triangleq \tilde{\Psi}_{36}(k) + 2\bar{\Psi}_{35}(k), \\
\tilde{\Psi}_{77}(k) &\triangleq \bar{B}^T(k)\breve{\mathcal{Z}}\bar{B}(k), & \tilde{\Psi}_{44}(k) &\triangleq \tilde{\Psi}_{66}(k) + 4\bar{\Psi}_{44}(k), \\
\tilde{\Psi}_{45}(k) &\triangleq \tilde{\Psi}_{66}(k) + 4\bar{\Psi}_{45}(k), & \tilde{\Psi}_{55}(k) &\triangleq \tilde{\Psi}_{66}(k) + 4\bar{\Psi}_{55}(k).
\end{aligned}$$

By (10.15b)–(10.16) and the zero input, that is, $\begin{bmatrix} \sigma(k) \\ \hat{\omega}(k) \end{bmatrix} = 0$, we have $\mathbf{E}\{\Delta V(k)\} < 0$, thus the system $(\mathcal{S}_1)$ is mean-square asymptotically stable.

Let $\mathcal{X} > 0$ and consider the following index:

$$\mathcal{J} \triangleq \mathbf{E}\left\{\sum_{k=0}^{\infty} \left[\varsigma^T(k)\mathcal{X}\varsigma(k) - \sigma^T(k)\mathcal{X}\sigma(k) + e^T(k)e(k) - \hat{\omega}^T(k)\hat{\omega}(k)\right]\right\}.$$

Under zero initial condition, that is, $V(k, \mathscr{X}_k)|_{k=0} = 0$, we have

$$\begin{aligned}
\mathcal{J} \leq\; & \mathbf{E}\left\{\sum_{k=0}^{\infty} \left[\varsigma^T(k)\mathcal{X}\varsigma(k) - \sigma^T(k)\mathcal{X}\sigma(k) + e^T(k)e(k) - \hat{\omega}^T(k)\hat{\omega}(k)\right]\right\} \\
& + \mathbf{E}\left\{V(k+1)\,|_{k\to\infty}\right\} - \mathbf{E}\left\{V(k)|_{k=0}\right\} \\
=\; & \mathbf{E}\left\{\sum_{k=0}^{\infty} \left[\varsigma^T(k)\mathcal{X}\varsigma(k) - \sigma^T(k)\mathcal{X}\sigma(k) + e^T(k)e(k) - \hat{\omega}^T(k)\hat{\omega}(k) + \Delta V(k)\right]\right\} \\
=\; & \sum_{k=0}^{\infty} \begin{bmatrix} \bar{\xi}(k) \\ \sigma(k) \\ \hat{\omega}(k) \end{bmatrix}^T \hat{\Psi}(k) \begin{bmatrix} \bar{\xi}(k) \\ \sigma(k) \\ \hat{\omega}(k) \end{bmatrix} \\
& - \sum_{k=0}^{\infty} \sum_{s=k-d_1}^{k-1} \begin{bmatrix} \bar{\xi}(k) \\ \varsigma(s) \end{bmatrix}^T \bar{\Omega}_1(k) \begin{bmatrix} \bar{\xi}(k) \\ \varsigma(s) \end{bmatrix} - \sum_{k=0}^{\infty} \sum_{s=k-d_2}^{k-1} \begin{bmatrix} \bar{\xi}(k) \\ \varsigma(s) \end{bmatrix}^T \bar{\Omega}_2(k) \begin{bmatrix} \bar{\xi}(k) \\ \varsigma(s) \end{bmatrix}.
\end{aligned}$$

Therefore, for all nonzero $\hat{\omega}(k) \in \ell_2[0, \infty)$, we have $\mathcal{J} < 0$ and $e(k) \in \ell_2[0, \infty)$, which means $\|e(k)\|_{\mathbf{E}_2} < \|\hat{\omega}(k)\|_{\mathbf{E}_2}$, thus the proof is completed. ■

Remark 10.7. Theorem 10.6 presents a sufficient condition of the stability and the $\mathcal{H}_\infty$ performance for the filtering error system. Here, to reduce the conservativeness of the result, we employed the input-output method combining with a novel LKF. ♦

10.3.2 $\mathcal{H}_\infty$ Filter Design

In this section, we present a solution to the desired $\mathcal{H}_\infty$ filter design problem.

Theorem 10.8. *The filtering error system in (10.3) is mean-square asymptotically stable with an $\mathcal{H}_\infty$ performance if there exist matrices $\mathscr{P} > 0$, $\mathscr{X} > 0$, $\bar{\mathcal{X}} \triangleq \begin{bmatrix} \mathcal{X}_1 & \mathcal{X}_2 \\ \star & \mathcal{X}_3 \end{bmatrix} > 0$, $\bar{\mathcal{Z}}_1 \triangleq \begin{bmatrix} \mathcal{Z}_{11} & \mathcal{Z}_{12} \\ \star & \mathcal{Z}_{13} \end{bmatrix} > 0$, $\bar{\mathcal{Z}}_2 \triangleq \begin{bmatrix} \mathcal{Z}_{21} & \mathcal{Z}_{22} \\ \star & \mathcal{Z}_{23} \end{bmatrix} > 0$, $\bar{\mathcal{Q}}_{1i} \triangleq \begin{bmatrix} \mathcal{Q}_{1i1} & \mathcal{Q}_{1i2} \\ \star & \mathcal{Q}_{1i3} \end{bmatrix} > 0$, $\bar{\mathcal{Q}}_{2i} \triangleq \begin{bmatrix} \mathcal{Q}_{2i1} & \mathcal{Q}_{2i2} \\ \star & \mathcal{Q}_{2i3} \end{bmatrix} > 0$, $\bar{\mathcal{S}}_{mi} \triangleq \begin{bmatrix} \mathcal{S}_{mi1} & \mathcal{S}_{mi2} \\ \star & \mathcal{S}_{mi3} \end{bmatrix} > 0$, $\bar{\mathcal{T}}_{mi} \triangleq \begin{bmatrix} \mathcal{T}_{mi1} & \mathcal{T}_{mi2} \\ \star & \mathcal{T}_{mi3} \end{bmatrix} > 0$, $\bar{\mathcal{M}}_{mi} \triangleq \begin{bmatrix} \mathcal{M}_{mi1} & \mathcal{M}_{mi2} \\ \mathcal{M}_{mi3} & \mathcal{M}_{mi4} \end{bmatrix}$, $\bar{\mathcal{N}}_{mi} \triangleq \begin{bmatrix} \mathcal{N}_{mi1} & \mathcal{N}_{mi2} \\ \mathcal{N}_{mi3} & \mathcal{N}_{mi4} \end{bmatrix}$, $\mathscr{A}$, $\mathscr{B}$, $\mathscr{C}$, $\mathscr{D}$, $\mathcal{W}_1$ and $\mathcal{W}_2$ such that for $i, j, s = 1, 2, \ldots, r$, $m = 1, 2, 3$,*

$$\begin{bmatrix} \Pi_{11} & 0 & \Pi_{13i} & \frac{1}{2}\Pi_{14i} & \frac{1}{2}\Pi_{14i} & \frac{d}{2}\Pi_{14i} & \Pi_{17i} \\ \star & \delta^{-1}\Pi_{11} & \Pi_{23i} & 0 & 0 & 0 & 0 \\ \star & \star & \Pi_{33i} & \Pi_{34i} & \Pi_{35i} & 0 & 0 \\ \star & \star & \star & \Pi_{44ij} & \Pi_{45i} & 0 & 0 \\ \star & \star & \star & \star & \Pi_{55is} & 0 & 0 \\ \star & \star & \star & \star & \star & \Pi_{66} & 0 \\ \star & \star & \star & \star & \star & \star & -\gamma^2 I \end{bmatrix} < 0, \quad (10.21a)$$

$$\begin{bmatrix} \mathcal{S}_{1i1} & \mathcal{S}_{1i2} & 0 & 0 & 0 & 0 & \mathcal{M}_{1i1} & \mathcal{M}_{1i2} \\ \star & \mathcal{S}_{1i3} & 0 & 0 & 0 & 0 & \mathcal{M}_{1i3} & \mathcal{M}_{1i4} \\ \star & \star & \mathcal{S}_{2i1} & \mathcal{S}_{2i2} & 0 & 0 & \mathcal{M}_{2i1} & \mathcal{M}_{2i2} \\ \star & \star & \star & \mathcal{S}_{2i3} & 0 & 0 & \mathcal{M}_{2i3} & \mathcal{M}_{2i4} \\ \star & \star & \star & \star & \mathcal{S}_{3i1} & \mathcal{S}_{3i2} & \mathcal{M}_{3i1} & \mathcal{M}_{3i2} \\ \star & \star & \star & \star & \star & \mathcal{S}_{3i3} & \mathcal{M}_{3i3} & \mathcal{M}_{3i4} \\ \star & \star & \star & \star & \star & \star & \mathcal{Z}_{11} & \mathcal{Z}_{12} \\ \star & \star & \star & \star & \star & \star & \star & \mathcal{Z}_{13} \end{bmatrix} \geq 0, \quad (10.21b)$$

$$\begin{bmatrix} \mathcal{T}_{1i1} & \mathcal{T}_{1i2} & 0 & 0 & 0 & 0 & \mathcal{N}_{1i1} & \mathcal{N}_{1i2} \\ \star & \mathcal{T}_{1i3} & 0 & 0 & 0 & 0 & \mathcal{N}_{1i3} & \mathcal{N}_{1i4} \\ \star & \star & \mathcal{T}_{2i1} & \mathcal{T}_{2i2} & 0 & 0 & \mathcal{N}_{2i1} & \mathcal{N}_{2i2} \\ \star & \star & \star & \mathcal{T}_{2i3} & 0 & 0 & \mathcal{N}_{2i3} & \mathcal{N}_{2i4} \\ \star & \star & \star & \star & \mathcal{T}_{3i1} & \mathcal{T}_{3i2} & \mathcal{N}_{3i1} & \mathcal{N}_{3i2} \\ \star & \star & \star & \star & \star & \mathcal{T}_{3i3} & \mathcal{N}_{3i3} & \mathcal{N}_{3i4} \\ \star & \star & \star & \star & \star & \star & \mathcal{Z}_{21} & \mathcal{Z}_{22} \\ \star & \star & \star & \star & \star & \star & \star & \mathcal{Z}_{23} \end{bmatrix} \geq 0, \quad (10.21c)$$

where

$$\Pi_{11} \triangleq \mathrm{diag}\{-\Pi_{111}, d_1^{-1}\Pi_{112}, d_2^{-1}\Pi_{113}, \Pi_{114}, -I\},$$

$$\Pi_{13i} \triangleq \begin{bmatrix} \mathscr{P}A_i + \mathscr{B}C_i & \mathscr{A} \\ \mathscr{X}^T A_i + \mathscr{B}C_i & \mathscr{A} \\ \mathcal{W}_1^T A_i + \mathscr{B}C_i - \mathcal{W}_1^T & \mathscr{A} - \mathscr{X} \\ \mathcal{W}_2^T A_i + \mathscr{B}C_i - \mathcal{W}_2^T & \mathscr{A} - \mathscr{X} \\ \mathcal{W}_1^T A_i + \mathscr{B}C_i - \mathcal{W}_1^T & \mathscr{A} - \mathscr{X} \\ \mathcal{W}_2^T A_i + \mathscr{B}C_i - \mathcal{W}_2^T & \mathscr{A} - \mathscr{X} \\ \mathcal{W}_1^T A_i + \mathscr{B}C_i - \mathcal{W}_1^T & \mathscr{A} - \mathscr{X} \\ \mathcal{W}_2^T A_i + \mathscr{B}C_i - \mathcal{W}_2^T & \mathscr{A} - \mathscr{X} \\ L_i - \mathscr{D}C_i & -\mathscr{C} \end{bmatrix}, \quad \Pi_{14i} \triangleq \begin{bmatrix} \mathscr{P}A_{di} + \mathscr{B}C_{di} & 0 \\ \mathscr{X}^T A_{di} + \mathscr{B}C_{di} & 0 \\ \mathcal{W}_1^T A_{di} + \mathscr{B}C_{di} & 0 \\ \mathcal{W}_2^T A_{di} + \mathscr{B}C_{di} & 0 \\ \mathcal{W}_1^T A_{di} + \mathscr{B}C_{di} & 0 \\ \mathcal{W}_2^T A_{di} + \mathscr{B}C_{di} & 0 \\ \mathcal{W}_1^T A_{di} + \mathscr{B}C_{di} & 0 \\ \mathcal{W}_2^T A_{di} + \mathscr{B}C_{di} & 0 \\ L_{di} - \mathscr{D}C_{di} & 0 \end{bmatrix},$$

$$\Pi_{23i} \triangleq \begin{bmatrix} \mathscr{P}E_i + \mathscr{B}G_i & 0 \\ \mathscr{X}^T E_i + \mathscr{B}G_i & 0 \\ \mathcal{W}_1^T E_i + \mathscr{B}G_i & 0 \\ \mathcal{W}_2^T E_i + \mathscr{B}G_i & 0 \\ \mathcal{W}_1^T E_i + \mathscr{B}G_i & 0 \\ \mathcal{W}_2^T E_i + \mathscr{B}G_i & 0 \\ \mathcal{W}_1^T E_i + \mathscr{B}G_i & 0 \\ \mathcal{W}_2^T E_i + \mathscr{B}G_i & 0 \\ -\mathscr{D}G_i & 0 \end{bmatrix}, \quad \Pi_{17i} \triangleq \begin{bmatrix} \mathscr{P}B_i + \mathscr{B}D_i \\ \mathscr{X}^T B_i + \mathscr{B}D_i \\ \mathcal{W}_1^T B_i + \mathscr{B}D_i \\ \mathcal{W}_2^T B_i + \mathscr{B}D_i \\ \mathcal{W}_1^T B_i + \mathscr{B}D_i \\ \mathcal{W}_2^T B_i + \mathscr{B}D_i \\ \mathcal{W}_1^T B_i + \mathscr{B}D_i \\ \mathcal{W}_2^T B_i + \mathscr{B}D_i \\ F_i - \mathscr{D}D_i \end{bmatrix},$$

$$\Pi_{33i} \triangleq \begin{bmatrix} \Pi_{33i1} & \Pi_{33i2} \\ \star & \Pi_{33i3} \end{bmatrix}, \quad \Pi_{44ij} \triangleq \begin{bmatrix} \Pi_{44ij1} & \Pi_{44ij2} \\ \star & \Pi_{44ij3} \end{bmatrix}, \quad \Pi_{66} \triangleq \begin{bmatrix} -\mathcal{X}_1 & -\mathcal{X}_2 \\ \star & -\mathcal{X}_3 \end{bmatrix},$$

$$\Pi_{34i} \triangleq \begin{bmatrix} -\mathcal{M}_{1i1}^T + \mathcal{M}_{2i1} + \mathcal{N}_{2i1} & -\mathcal{M}_{1i3}^T + \mathcal{M}_{2i2} + \mathcal{N}_{2i2} \\ -\mathcal{M}_{1i2}^T + \mathcal{M}_{2i3} + \mathcal{N}_{2i3} & -\mathcal{M}_{1i4}^T + \mathcal{M}_{2i4} + \mathcal{N}_{2i4} \end{bmatrix},$$

$$\Pi_{35i} \triangleq \begin{bmatrix} \mathcal{M}_{3i1} + \mathcal{N}_{3i1} - \mathcal{N}_{1i1}^T & \mathcal{M}_{3i2} + \mathcal{N}_{3i2} - \mathcal{N}_{1i3}^T \\ \mathcal{M}_{3i3} + \mathcal{N}_{3i3} - \mathcal{N}_{1i2}^T & \mathcal{M}_{3i4} + \mathcal{N}_{3i4} - \mathcal{N}_{1i4}^T \end{bmatrix},$$

$$\Pi_{45i} \triangleq \begin{bmatrix} -\mathcal{M}_{3i1} - \mathcal{N}_{2i1}^T & -\mathcal{M}_{3i2} - \mathcal{N}_{2i3}^T \\ -\mathcal{M}_{3i3} - \mathcal{N}_{2i2}^T & -\mathcal{M}_{3i4} - \mathcal{N}_{2i4}^T \end{bmatrix}, \quad \Pi_{55is} \triangleq \begin{bmatrix} \Pi_{55is1} & \Pi_{55is2} \\ \star & \Pi_{55is3} \end{bmatrix},$$

with

$$\Pi_{112} \triangleq \begin{bmatrix} \mathcal{Z}_{11} - \mathcal{W}_1 - \mathcal{W}_1^T & \mathcal{Z}_{12} - \mathcal{W}_2 - \mathscr{X} \\ \star & \mathcal{Z}_{13} - \mathscr{X} - \mathscr{X}^T \end{bmatrix},$$

$$\Pi_{113} \triangleq \begin{bmatrix} \mathcal{Z}_{21} - \mathcal{W}_1 - \mathcal{W}_1^T & \mathcal{Z}_{22} - \mathcal{W}_2 - \mathscr{X} \\ \star & \mathcal{Z}_{23} - \mathscr{X} - \mathscr{X}^T \end{bmatrix},$$

$$\Pi_{114} \triangleq \begin{bmatrix} \mathcal{X}_1 - \mathcal{W}_1 - \mathcal{W}_1^T & \mathcal{X}_2 - \mathcal{W}_2 - \mathscr{X} \\ \star & \mathcal{X}_3 - \mathscr{X} - \mathscr{X}^T \end{bmatrix}, \quad \Pi_{111} \triangleq \begin{bmatrix} \mathscr{P} & \mathscr{X} \\ \star & \mathscr{X}^T \end{bmatrix},$$

$$\Pi_{33i1} \triangleq -\mathscr{P} + \mathcal{Q}_{1i1} + \mathcal{Q}_{2i1} + \mathcal{M}_{1i1} + \mathcal{M}_{1i1}^T + \mathcal{N}_{1i1} + \mathcal{N}_{1i1}^T + d_1\mathcal{S}_{1i1} + d_2\mathcal{T}_{1i1},$$
$$\Pi_{33i2} \triangleq -\mathscr{X} + \mathcal{Q}_{1i2} + \mathcal{Q}_{2i2} + \mathcal{M}_{1i2} + \mathcal{M}_{1i3}^T + \mathcal{N}_{1i2} + \mathcal{N}_{1i3}^T + d_1\mathcal{S}_{1i2} + d_2\mathcal{T}_{1i2},$$
$$\Pi_{33i3} \triangleq -\mathscr{X}^T + \mathcal{Q}_{1i3} + \mathcal{Q}_{2i3} + \mathcal{M}_{1i4} + \mathcal{M}_{1i4}^T + \mathcal{N}_{1i4} + \mathcal{N}_{1i4}^T + d_1\mathcal{S}_{1i3} + d_2\mathcal{T}_{1i3},$$
$$\Pi_{44ij1} \triangleq -\mathcal{Q}_{1j1} - \mathcal{M}_{2i1} - \mathcal{M}_{2i1}^T + d_1\mathcal{S}_{2i1} + d_2\mathcal{T}_{2i1},$$
$$\Pi_{44ij2} \triangleq -\mathcal{Q}_{1j2} - \mathcal{M}_{2i2} - \mathcal{M}_{2i3}^T + d_1\mathcal{S}_{2i2} + d_2\mathcal{T}_{2i2},$$
$$\Pi_{44ij3} \triangleq -\mathcal{Q}_{1j3} - \mathcal{M}_{2i4} - \mathcal{M}_{2i4}^T + d_1\mathcal{S}_{2i3} + d_2\mathcal{T}_{2i3},$$
$$\Pi_{55is1} \triangleq -\mathcal{Q}_{2s1} - \mathcal{N}_{3i1} - \mathcal{N}_{3i1}^T + d_1\mathcal{S}_{3i1} + d_2\mathcal{T}_{3i1},$$
$$\Pi_{55is2} \triangleq -\mathcal{Q}_{2s2} - \mathcal{N}_{3i2} - \mathcal{N}_{3i3}^T + d_1\mathcal{S}_{3i2} + d_2\mathcal{T}_{3i2},$$
$$\Pi_{55is3} \triangleq -\mathcal{Q}_{2s3} - \mathcal{N}_{3i4} - \mathcal{N}_{3i4}^T + d_1\mathcal{S}_{3i3} + d_2\mathcal{T}_{3i3}.$$

Moreover, if the above conditions have a set of feasible solution, then the matrices for an $\mathcal{H}_\infty$ desired filter in the form of (10.2) are given by

$$\begin{bmatrix} A_c & B_c \\ C_c & D_c \end{bmatrix} = \begin{bmatrix} \mathscr{X}^{-1} & 0 \\ 0 & I \end{bmatrix} \begin{bmatrix} \mathscr{A} & \mathscr{B} \\ \mathscr{C} & \mathscr{D} \end{bmatrix}. \tag{10.22}$$

Proof. According to Theorem 10.6, it is easy to prove that the filtering error system in (10.3) is mean-square asymptotically stable with an $\mathcal{H}_\infty$ performance level γ if there exist matrices $\mathcal{P} > 0$, $\mathcal{X} > 0$, $\mathcal{Z}_1 > 0$, $\mathcal{Z}_2 > 0$, $\mathcal{Q}_{1i} > 0$, $\mathcal{Q}_{2i} > 0$, $\mathcal{S}_{1i} > 0$, $\mathcal{S}_{2i} > 0$, $\mathcal{S}_{3i} > 0$, $\mathcal{T}_{1i} > 0$, $\mathcal{T}_{2i} > 0$, $\mathcal{T}_{3i} > 0$, $\mathcal{M}_{1i}$, $\mathcal{M}_{2i}$, $\mathcal{M}_{3i}$, $\mathcal{N}_{1i}$, $\mathcal{N}_{2i}$, and $\mathcal{N}_{3i}$, $(i = 1, 2, \ldots, r)$, and $\mathcal{W}$ satisfying (10.14b)–(10.14c) and

$$\begin{bmatrix} \bar{\Psi}_{11} & 0 & \bar{\Psi}_{13i} - \mathcal{W}^T\mathcal{I} & \frac{1}{2}\bar{\Psi}_{14i} & \frac{1}{2}\bar{\Psi}_{14i} & \frac{d}{2}\bar{\Psi}_{14i} & \bar{\Psi}_{17i} \\ \star & \delta^{-1}\bar{\Psi}_{11} & \bar{\Psi}_{23i} & 0 & 0 & 0 & 0 \\ \star & \star & \Psi_{33i} & \Psi_{34i} & \Psi_{35i} & 0 & 0 \\ \star & \star & \star & \Psi_{44ij} & \Psi_{45i} & 0 & 0 \\ \star & \star & \star & \star & \Psi_{55is} & 0 & 0 \\ \star & \star & \star & \star & \star & -\mathcal{X} & 0 \\ \star & \star & \star & \star & \star & \star & -\gamma^2 I \end{bmatrix} < 0, \quad (10.23)$$

where

$$\bar{\Psi}_{13i} \triangleq \begin{bmatrix} \mathcal{P}\bar{A}_i \\ \mathcal{W}^T\bar{A}_i \\ \mathcal{W}^T\bar{A}_i \\ \mathcal{W}^T\bar{A}_i \\ \bar{L}_i \end{bmatrix}, \quad \bar{\Psi}_{14i} \triangleq \begin{bmatrix} \mathcal{P}\bar{A}_{di} \\ \mathcal{W}^T\bar{A}_{di} \\ \mathcal{W}^T\bar{A}_{di} \\ \mathcal{W}^T\bar{A}_{di} \\ \bar{L}_{di} \end{bmatrix}, \quad \bar{\Psi}_{23i} \triangleq \begin{bmatrix} \mathcal{P}\bar{E}_i \\ \mathcal{W}^T\bar{E}_i \\ \mathcal{W}^T\bar{E}_i \\ \mathcal{W}^T\bar{E}_i \\ \bar{G}_i \end{bmatrix}, \quad \bar{\Psi}_{17i} \triangleq \begin{bmatrix} \mathcal{P}\bar{B}_i \\ \mathcal{W}^T\bar{B}_i \\ \mathcal{W}^T\bar{B}_i \\ \mathcal{W}^T\bar{B}_i \\ \bar{F}_i \end{bmatrix},$$

$$\bar{\Psi}_{11} \triangleq \mathrm{diag}\left\{-\mathcal{P}, d_1^{-1}\left(\mathcal{Z}_1 - \mathcal{W} - \mathcal{W}^T\right), d_2^{-1}\left(\mathcal{Z}_2 - \mathcal{W} - \mathcal{W}^T\right), \left(\mathcal{X} - \mathcal{W} - \mathcal{W}^T\right), -I\right\}.$$

Let the matrix $\mathcal{P}$ be partitioned as

$$\mathcal{P} \triangleq \begin{bmatrix} \mathcal{P}_1 & \mathcal{P}_2 \\ \star & \mathcal{P}_3 \end{bmatrix} > 0, \quad (10.24)$$

where $\mathcal{P}_1 \in \mathbf{R}^{n\times n}$ and $\mathcal{P}_3 \in \mathbf{R}^{n\times n}$ are symmetric positive definite matrices, and $\mathcal{P}_2 \in \mathbf{R}^{n\times n}$.

Define the following matrices which are also nonsingular:

$$\begin{cases} \mathscr{L} \triangleq \begin{bmatrix} I & 0 \\ 0 & \mathcal{P}_3^{-1}\mathcal{P}_2^T \end{bmatrix}, \quad W \triangleq \begin{bmatrix} \mathcal{W}_1 & \mathcal{W}_2\mathcal{P}_2^{-T}\mathcal{P}_3 \\ \mathcal{P}_2^T & \mathcal{P}_3 \end{bmatrix}, \\ \mathscr{X} \triangleq \mathcal{P}_2\mathcal{P}_3^{-1}\mathcal{P}_2^T, \quad \mathscr{P} \triangleq \mathcal{P}_1, \quad \mathcal{X} \triangleq \mathscr{L}^{-T}\bar{\mathcal{X}}\mathscr{L}^{-1}, \\ \mathcal{Z}_1 \triangleq \mathscr{L}^{-T}\bar{\mathcal{Z}}_1\mathscr{L}^{-1}, \quad \mathcal{Z}_2 \triangleq \mathscr{L}^{-T}\bar{\mathcal{Z}}_2\mathscr{L}^{-1}, \\ \mathcal{M}_{mi} \triangleq \mathscr{L}^{-T}\bar{\mathcal{M}}_{mi}\mathscr{L}^{-1}, \quad \mathcal{N}_{mi} \triangleq \mathscr{L}^{-T}\bar{\mathcal{N}}_{mi}\mathscr{L}^{-1}, \\ \mathcal{Q}_{1i} \triangleq \mathscr{L}^{-T}\bar{\mathcal{Q}}_{1i}\mathscr{L}^{-1}, \quad \mathcal{Q}_{2i} \triangleq \mathscr{L}^{-T}\bar{\mathcal{Q}}_{2i}\mathscr{L}^{-1}, \\ \mathcal{S}_{mi} \triangleq \mathscr{L}^{-T}\bar{\mathcal{S}}_{mi}\mathscr{L}^{-1}, \quad \mathcal{T}_{mi} \triangleq \mathscr{L}^{-T}\bar{\mathcal{T}}_{mi}\mathscr{L}^{-1}, \end{cases} \quad (10.25)$$

and

$$\begin{bmatrix} \mathscr{A} & \mathscr{B} \\ \mathscr{C} & \mathscr{D} \end{bmatrix} \triangleq \begin{bmatrix} \mathcal{P}_2 & 0 \\ 0 & I \end{bmatrix} \begin{bmatrix} A_c & B_c \\ C_c & D_c \end{bmatrix} \begin{bmatrix} \mathcal{P}_3^{-1}\mathcal{P}_2^T & 0 \\ 0 & I \end{bmatrix}. \tag{10.26}$$

Performing congruence transformations to (10.14b)–(10.14c) and (10.23) by $\mathrm{diag}\{\mathscr{L}, \mathscr{L}, \mathscr{L}, \mathscr{L}\}$, $\mathrm{diag}\{\mathscr{L}, \mathscr{L}, \mathscr{L}, \mathscr{L}\}$ and $\mathrm{diag}\{\mathscr{L}, \mathscr{L}, \mathscr{L}, \mathscr{L}, I, \mathscr{L}, \mathscr{L}, \mathscr{L}, \mathscr{L}, I, \mathscr{L}, \mathscr{L}, \mathscr{L}, \mathscr{L}, I\}$, respectively, and considering (10.25)–(10.26), we can obtain inequalities (10.21a)–(10.21c). Moreover, notice that (10.26) is equivalent to

$$\begin{aligned} \begin{bmatrix} A_c & B_c \\ C_c & D_c \end{bmatrix} &\triangleq \begin{bmatrix} \mathcal{P}_2^{-1} & 0 \\ 0 & I \end{bmatrix} \begin{bmatrix} \mathscr{A} & \mathscr{B} \\ \mathscr{C} & \mathscr{D} \end{bmatrix} \begin{bmatrix} \mathcal{P}_2^{-T}\mathcal{P}_3 & 0 \\ 0 & I \end{bmatrix} \\ &= \begin{bmatrix} (\mathcal{P}_2^{-T}\mathcal{P}_3)^{-1}\mathscr{L}^{-1} & 0 \\ 0 & I \end{bmatrix} \begin{bmatrix} \mathscr{A} & \mathscr{B} \\ \mathscr{C} & \mathscr{D} \end{bmatrix} \begin{bmatrix} \mathcal{P}_2^{-T}\mathcal{P}_3 & 0 \\ 0 & I \end{bmatrix}. \end{aligned} \tag{10.27}$$

Notice also that the matrices A_c, B_c, C_c and D_c in (10.2) can be written as (10.27), which implies that $\mathcal{P}_2^{-T}\mathcal{P}_3$ can be viewed as a similarity transformation on the state-space realization of the filter and, as such, has no effect on the filter mapping from y to z_c. Without loss of generality, we may set $\mathcal{P}_2^{-T}\mathcal{P}_3 = I$, thus obtain (10.22). Therefore, the filter mode in (10.2) can be constructed by (10.22). This completes the proof. ■

Remark 10.9. Note that Theorem 10.8 provides a sufficient condition for the solvability of $\mathcal{H}_\infty$ desired filter in terms of strict LMIs, thus a desired filter can be designed by solving the following convex optimization problem:

$\min\ \sigma$ subject to (10.21a)–(10.21c) with $\sigma = \gamma^2$. ♦

10.4 Illustrative Example

Some real-world systems such as chemical processes, robotics systems, automotive systems and communication industries can be modeled as the T-S fuzzy Itô stochastic time-varying delay system. In engineering applications, the filter design needs to be employed to estimate system state for the concerned system by measurement noise inputs. For example, the modified Henon mapping model [194] can be approximated by a T-S fuzzy Itô stochastic system with time-varying delay, and system state needs to be estimated by the filter design. Now, consider the modified Henon mapping system with time-varying state delay:

$$\begin{cases} x_1(k+1) = -\left[\mu x_1(k) + (1-\mu)x_1(k-d(k))\right]^2 + \omega(k) + 0.3x_2(k) + 0.01x_1(k)\varpi(k), \\ x_2(k+1) = \mu x_1(k) + (1-\mu)x_1(k-d(k)) + 0.01x_2(k)\varpi(k), \\ y(k) = \mu x_1(k) + (1-\mu)x_1(k-d(k)) + \omega(k) + 0.1x_1(k)\varpi(k), \\ z(k) = x_1(k), \end{cases}$$

where $\omega(k)$ is the disturbance input; $\delta \triangleq \mathbf{E}\{\varpi^2(k)\}$; and the constant $\mu \in [0,1]$ is the retarded coefficient.

Let $\theta = \mu x_1(k) + (1-\mu)x_1(k-d)$ and assume $\theta \in [-\zeta, \zeta]$, $\zeta > 0$. By using the same procedure as in [194], the nonlinear term of θ^2 can be exactly represented as

$$\theta^2 = -h_1(\theta)\zeta\theta + h_2(\theta)\zeta\theta,$$

where $h_1(\theta), h_2(\theta) \in [0,1]$, and $h_1(\theta) + h_2(\theta) = 1$. Thus, the membership functions $h_1(\theta)$ and $h_2(\theta)$ can be chosen as

$$h_1(\theta) = \frac{1}{2}\left(1 - \frac{\theta}{\zeta}\right), \quad h_2(\theta) = \frac{1}{2}\left(1 + \frac{\theta}{\zeta}\right).$$

It can be seen from the aforementioned expressions that $h_1(\theta) = 1$ and $h_2(\theta) = 0$ when $\theta = -\zeta$, and $h_1(\theta) = 0$ and $h_2(\theta) = 1$ when $\theta = \zeta$. Then, the above nonlinear system can be approximately represented by the following T-S fuzzy model:

♦ **Plant Form:**

Rule 1: IF θ is $-\zeta$, THEN

$$\begin{cases} x(k+1) = A_1x(k) + A_{d1}x(k-d(k)) + B_1\omega(k) + E_1x(k)\varpi(k), \\ y(k) = C_1x(k) + C_{d1}x(k-d(k)) + D_1\omega(k) + G_1x(k)\varpi(k), \\ z(k) = L_1x(k), \end{cases}$$

Rule 2: IF θ is ζ, THEN

$$\begin{cases} x(k+1) = A_2x(k) + A_{d2}x(k-d(k)) + B_2\omega(k) + E_2x(k)\varpi(k), \\ y(k) = C_2x(k) + C_{d2}x(k-d(k)) + D_2\omega(k) + G_2x(k)\varpi(k), \\ z(k) = L_2x(k), \end{cases}$$

where

$$A_1 = \begin{bmatrix} \mu\zeta & 0.3 \\ \mu & 0 \end{bmatrix}, \quad A_{d1} = \begin{bmatrix} (1-\mu)\zeta & 0 \\ 1-\mu & 0 \end{bmatrix}, \quad E_1 = E_2 = \begin{bmatrix} 0.01 & 0.1 \\ 0 & 0.01 \end{bmatrix},$$
$$A_2 = \begin{bmatrix} -\mu\zeta & 0.3 \\ \mu & 0 \end{bmatrix}, \quad A_{d2} = \begin{bmatrix} -(1-\mu)\zeta & 0 \\ 1-\mu & 0 \end{bmatrix}, \quad B_1 = B_2 = \begin{bmatrix} 1 \\ 0 \end{bmatrix},$$
$$C_1 = C_2 = \begin{bmatrix} \mu & 0 \end{bmatrix}, \quad C_{d1} = C_{d2} = \begin{bmatrix} 1-\mu & 0 \end{bmatrix}, \quad D_1 = 1,$$
$$G_1 = G_2 = \begin{bmatrix} 0.1 & 0 \end{bmatrix}, \quad L_1 = L_2 = \begin{bmatrix} 1 & 0 \end{bmatrix}, \quad D_2 = 0.5,.$$

In the example, $\mu = 0.8$, $\zeta = 0.2$, $\delta = 0.1$, and $1 \leq d(k) \leq 3$ represents time-varying delay. Our aim is to design a filter in the form of (10.2) such that the filter error system is mean-square asymptotically stable with an $\mathcal{H}_\infty$

performance level γ. Solving the conditions in Theorem 10.8, we obtain that the minimized feasible γ is $\gamma = 2.0694$ and

$$\mathscr{X} = \begin{bmatrix} 0.0078 & 0.0051 \\ 0.0051 & 0.0038 \end{bmatrix}, \quad \mathscr{A} = \begin{bmatrix} 0.0058 & 0.0037 \\ 0.0036 & 0.0027 \end{bmatrix}, \quad \mathscr{B} = \begin{bmatrix} 0.0013 \\ 0.0007 \end{bmatrix},$$
$$\mathscr{C} = \begin{bmatrix} -0.0080 & -0.0041 \end{bmatrix}, \quad \mathscr{D} = 0.5870. \tag{10.28}$$

Thus, by (10.22) and (10.28), the parameters of the desired filter are as follows:

$$A_c = \begin{bmatrix} 1.0138 & 0.0799 \\ -0.4132 & 0.6033 \end{bmatrix}, \quad B_c = \begin{bmatrix} 0.3774 \\ -0.3223 \end{bmatrix},$$
$$C_c = \begin{bmatrix} -0.0080 & -0.0041 \end{bmatrix}, \quad D_c = 0.5870. \tag{10.29}$$

Let the initial condition be zero, that is, $x(0) = 0$and $\hat{x}(0) = 0$, and suppose the disturbance input $\omega(k)$ be

$$\omega(k) = 0.89e^{(-0.18k)} \sin(0.75k).$$

The simulation results are shown in Figs. 10.1–10.3. Among them, Fig. 10.1 shows the time-varying delay $d(k)$ which changes randomly between $d_1 = 1$ and $d_2 = 3$. Fig. 10.2 plots the signal $z(k)$ (solid line), and its estimations $z_c(k)$ with the designed filter of (10.29) (dash-dot line). The corresponding estimation error $e(k)$ is shown in Fig. 10.3.

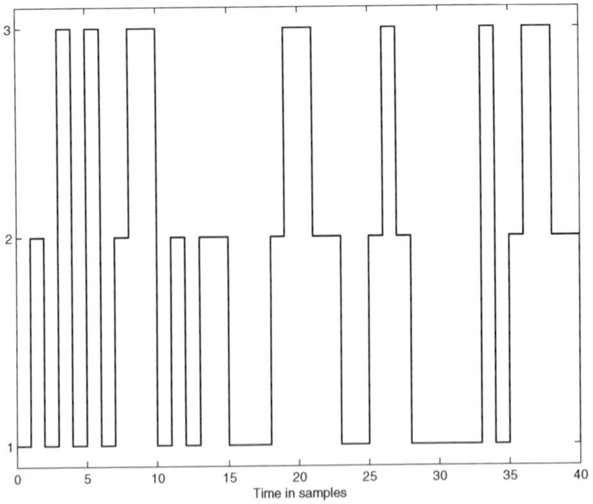

Fig. 10.1. Time-varying delays $d(k)$

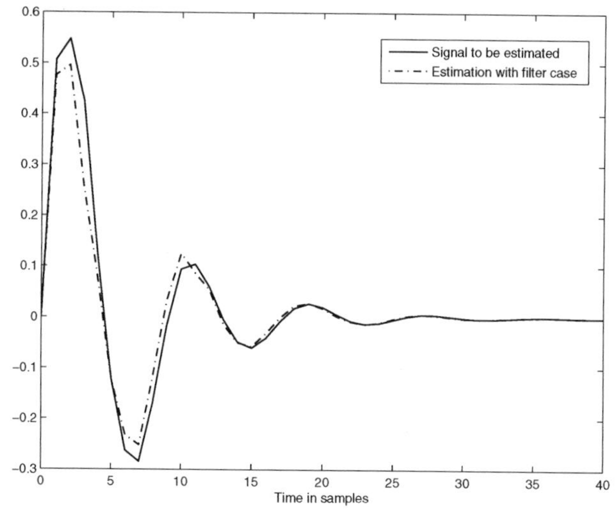

Fig. 10.2. Signal $z(k)$ and its estimation $z_c(k)$ of the desired filter

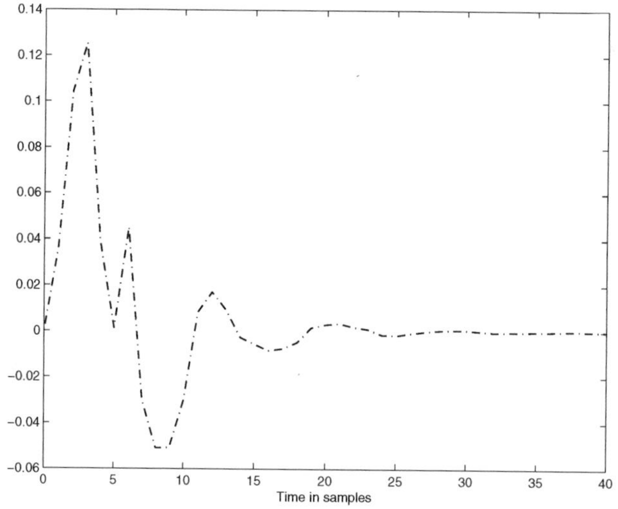

Fig. 10.3. Estimation error $e(k)$

10.5 Conclusion

In this chapter, the $\mathcal{H}_\infty$ filtering problem has been investigated for a class of discrete-time T-S fuzzy Itô stochastic time-varying delay systems. A new comparison model has been presented by employing a novel approximation for delayed state with smaller approximation error than the existing ones. Based on the scaled small gain theorem, a sufficient condition has been proposed to guarantee the mean-square asymptotically stability with an $\mathcal{H}_\infty$ performance for the filtering error system. Then, the desired filter design has been cast into a convex optimization problem. Finally, a numerical example has been provided to illustrate the effectiveness of the proposed theory.

Chapter 11
Fault Detection of Continuous-Time T-S Fuzzy Stochastic Systems

11.1 Introduction

The fault detection problem is an important topic in systems science and control engineering from the viewpoint of improving system reliability. The basic idea of fault detection is to construct a residual signal and, based on this, to determine a residual evaluation function to compare with a predefined threshold. When the residual evaluation function has a value larger than the threshold, an alarm of faults is generated. Since accurate mathematical models are not always available, unavoidable modelling errors and external disturbances may seriously affect the performance of model-based fault-detection systems. To overcome this, fault detection systems have to be robust to such modelling errors or disturbances. A system designed to provide both sensitivity to faults and robustness to modelling errors or disturbances is called a robust fault detection scheme.

In this chapter, we investigate the robust $\mathcal{H}_\infty$ fault detection for T-S fuzzy stochastic systems. Research in this area is interesting yet challenging as it involves the combination of stochastic systems and fuzzy systems, both of which are highly important. We consider a stochastic system with a Brownian motion in this work, which differs from a general stochastic system. Hence, for robust fault detection problem, there exists totally an external disturbance, a fault signal and a Brownian motion in one system. Therefore, the effect Brownian motion has on fault detection processes, and the detection of a fault signal from such a complicated system are unexploited research problems which need to be investigated. We shall aim to solve the fault detection by designing a robust filter which generates a residual signal to estimate the fault signal. The main aim is to make the error between residual and fault as small as possible. Both the fuzzy-rule-independent and the fuzzy-rule-dependent fault detection filters are designed, and the corresponding solvability conditions for desired fault detection filters are also been established. The detection threshold of the filter is also discussed.

L. Wu et al., *Fuzzy Control Systems with Time-Delay and Stochastic Perturbation*, Studies in Systems, Decision and Control 12, DOI: 10.1007/978-3-319-11316-6_11

11.2 System Description and Preliminaries

We consider the following T-S fuzzy stochastic system:

♦ **Plant Form:**

Rule i: IF $\theta_1(t)$ is $\mathcal{M}_{i1}$ and $\theta_2(t)$ is $\mathcal{M}_{i2}$ and $\cdots$ and $\theta_p(t)$ is $\mathcal{M}_{ip}$ THEN

$$dx(t) = [A_i x(t) + B_{0i}u(t) + B_i\omega(t) + B_{1i}f(t)]\,dt + E_i x(t)d\varpi, \quad (11.1a)$$
$$dy(t) = [C_i x(t) + D_{0i}u(t) + D_i\omega(t) + D_{1i}f(t)]\,dt + F_i x(t)d\varpi, \quad (11.1b)$$

where $i = 1, 2, \ldots, r$, and r is the number of IF-THEN rules; $\mathcal{M}_{ij}(i = 1, 2, \ldots, r; j = 1, 2, \ldots, p)$ are the fuzzy sets; $\theta(t) = \left[\theta_1(t)\ \theta_2(t)\ \cdots\ \theta_p(t)\right]^T$ is the premise variable vector. $x(t) \in \mathbf{R}^n$ is the state vector; $u(t) \in \mathbf{R}^m$ is the known input; $\omega(t) \in \mathbf{R}^q$ is the unknown disturbance input; $f(t) \in \mathbf{R}^l$ is the fault signal to be detected; $u(t)$, $\omega(t)$ and $f(t)$ are all assumed to be energy-bounded, that is, they all belong to $\mathcal{L}_2[0,\infty)$. $y(t) \in \mathbf{R}^p$ is the measured output; $\varpi(t)$ is a one-dimensional Brownian motion which satisfies $\mathbf{E}\{d\varpi(t)\} = 0$ and $\mathbf{E}\{d\varpi^2(t)\} = dt$. A_i, B_{0i}, B_i, B_{1i}, E_i, C_i, D_{0i}, D_i, D_{1i} and F_i are real constant matrices.

It is assumed that the premise variables do not depend on the input variables $u(t)$. Given a pair of $(x(t), u(t))$, the final output of the fuzzy stochastic systems is inferred as follows:

$$dx(t) = \sum_{i=1}^{r} h_i(\theta)\left\{[A_i x(t) + B_{0i}u(t) + B_i\omega(t) + B_{1i}f(t)]\,dt + E_i x(t)d\varpi\right\}, \quad (11.2a)$$
$$dy(t) = \sum_{i=1}^{r} h_i(\theta)\left\{[C_i x(t) + D_{0i}u(t) + D_i\omega(t) + D_{1i}f(t)]\,dt + F_i x(t)d\varpi\right\}, \quad (11.2b)$$

where $h_i(\theta)$, $i = 1, 2, \ldots, r$ are the normalized membership functions, which are defined as that of (1.1) in Chapter 1.

Throughout this chapter, the nominal system of (11.2) is assumed to be stable. Typically fault detection schemes are concern with construction of a dynamical system called a residual generator. This auxiliary system takes the known input and output of a system and generates a signal called the residual. This signal is then processed to decide whether or not a fault has occurred in the system [264]. Therefore, a typical fault detection system consists of a residual generator and a residual evaluation stage including an evaluation function and a prescribed threshold.

In this chapter, for the plant represented by (11.1) or (11.2), we consider the following two kinds of fault detection filters.

◊ **Fuzzy-Rule-Independent Filter:**

In the case that the premise variable of the original fuzzy model $\theta(t)$ is *unavailable* in filter implementation, the filter structure will have to be independent of the fuzzy rules. In other words, a fixed filter is to be designed for the fuzzy time-delay stochastic model $\theta(t)$. In this case, we like to design a fault detection filter of the form:

$$dx_c(t) = A_c x_c(t)dt + B_c dy(t), \quad (11.3a)$$
$$\chi_c(t) = C_c x_c(t), \quad (11.3b)$$

where $x_c(t) \in \mathbf{R}^n$ is the state vector of the fault detection filter; $\chi_c(t) \in \mathbf{R}^l$ is the so-called residual signal; A_c, B_c and C_c are the filter parameters to be designed. Fig. 11.1 shows the block diagram of the fuzzy-rule-independent filter design.

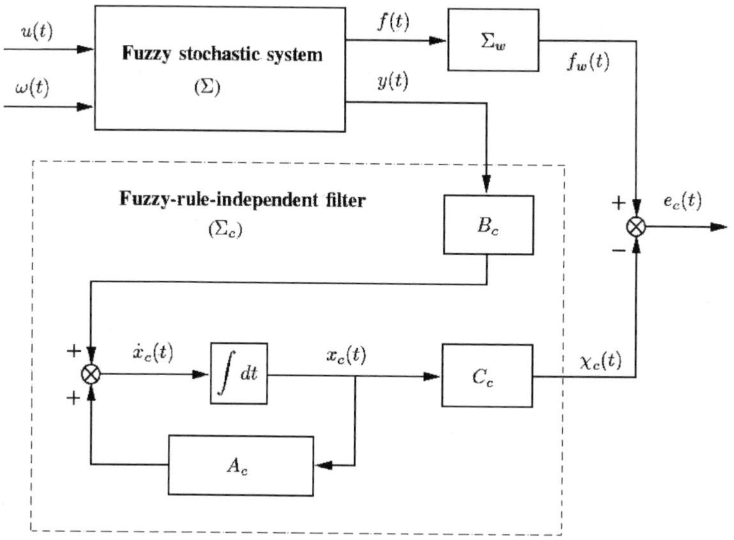

Fig. 11.1. Block diagram of the fuzzy-rule-independent filter design

To improve the performance of fault detection system, we add a weighting matrix function into the fault $f(s)$, that is, $f_w(s) = W(s)f(s)$, where $f(s)$ and $f_w(s)$ denote respectively the Laplace transforms of $f(t)$ and $f_w(t)$. One state space realization of $f_w(s) = W(s)f(s)$ can be

$$\dot{x}_w(t) = A_w x_w(t) + B_w f(t), \quad (11.4a)$$
$$f_w(t) = C_w x_w(t), \quad (11.4b)$$

where $x_w(t) \in \mathbf{R}^k$ is the state vector with $x_w(0) = 0$, and A_w, B_w, C_w are constant matrices.

Denoting $e_c(t) \triangleq \chi_c(t) - f_w(t)$, and augmenting the model of (11.2) to include the states of (11.3) and (11.4), then the overall dynamics of fault detection system is governed by

$$d\xi(t) = \sum_{i=1}^{r} h_i(\theta) \left\{ \left[\tilde{A}_i \xi(t) + \tilde{B}_i \upsilon(t) \right] dt + \tilde{E}_i K \xi(t) d\varpi(t) \right\}, \quad (11.5a)$$

$$e_c(t) = \sum_{i=1}^{r} h_i(\theta) \tilde{C}_i \xi(t), \quad (11.5b)$$

where $\xi(t) \triangleq \begin{bmatrix} x(t) \\ x_c(t) \\ x_w(t) \end{bmatrix}$ and

$$\begin{cases} \tilde{A}_i \triangleq \left[\begin{array}{cc|c} A_i & 0 & 0 \\ B_c C_i & A_c & 0 \\ \hline 0 & 0 & A_w \end{array}\right], \quad \tilde{E}_i \triangleq \left[\begin{array}{c} E_i \\ B_c F_i \\ \hline 0 \end{array}\right], \\ \tilde{B}_i \triangleq \left[\begin{array}{ccc} B_{0i} & B_i & B_{1i} \\ B_c D_{0i} & B_c D_i & B_c D_{1i} \\ \hline 0 & 0 & B_w \end{array}\right], \quad \upsilon(t) \triangleq \begin{bmatrix} u(t) \\ \omega(t) \\ f(t) \end{bmatrix}, \\ \tilde{C}_i \triangleq \left[0 \;\; C_c | \; -C_w \right], \quad K \triangleq \left[I \;\; 0 \;\; 0 \right]. \end{cases} \quad (11.6)$$

◊ **Fuzzy-Rule-Dependent Filter:**

Now, assume that the premise variable of the fuzzy model $\theta(t)$ is available for feedback which implies that $h_i(\theta)$ is available for feedback. Supposed that the filter's premise variable is the same as the plant's premise variable. Based on the parallel distributed compensation, the fuzzy-rule-dependent filter is designed to share the same IF parts with the following structure:

Rule i: IF $\theta_1(t)$ is $\mathcal{M}_{i1}$ and $\theta_2(t)$ is $\mathcal{M}_{i2}$ and $\cdots$ and $\theta_p(t)$ is $\mathcal{M}_{ip}$ THEN

$$dx_f(t) = A_{fi} x_f(t) dt + B_{fi} dy(t), \quad (11.7a)$$

$$\chi_f(t) = C_{fi} x_f(t), \quad i \in \mathbb{R}. \quad (11.7b)$$

The filter plant (11.7) can also be represented by

$$dx_f(t) = \sum_{i=1}^{r} h_i(\theta) \left[A_{fi} x_f(t) dt + B_{fi} dy(t) \right], \quad (11.8a)$$

$$\chi_f(t) = \sum_{i=1}^{r} h_i(\theta) C_{fi} x_f(t). \quad (11.8b)$$

Fig. 11.2 shows the block diagram of the fuzzy-rule-dependent filter design. Denote $e_f(t) \triangleq \chi_f(t) - f_w(t)$ and consider (11.2), (11.4) and (11.8), the fault detection system is governed by

$$d\zeta(t) = \sum_{i=1}^{r} \sum_{j=1}^{r} h_i(\theta) h_j(\theta) \left\{ \left[\tilde{A}_{ij} \zeta(t) + \tilde{B}_{ij} \upsilon(t) \right] dt + \tilde{E}_{ij} K \zeta(t) d\varpi(t) \right\}, \quad (11.9a)$$

$$e_f(t) = \sum_{i=1}^{r}\sum_{j=1}^{r} h_i(\theta)h_j(\theta)\tilde{C}_{ij}\zeta(t), \tag{11.9b}$$

where $\zeta(t) \triangleq \begin{bmatrix} x(t) \\ x_f(t) \\ x_w(t) \end{bmatrix}$ and

$$\begin{cases} \tilde{A}_{ij} \triangleq \left[\begin{array}{cc|c} A_i & 0 & 0 \\ B_{fj}C_i & A_{fj} & 0 \\ \hline 0 & 0 & A_w \end{array}\right], \quad \tilde{E}_{ij} \triangleq \left[\begin{array}{c} E_i \\ B_{fj}F_i \\ \hline 0 \end{array}\right], \\ \tilde{B}_{ij} \triangleq \left[\begin{array}{ccc} B_{0i} & B_i & B_{1i} \\ B_{fj}D_{0i} & B_{fj}D_i & B_{fj}D_{1i} \\ \hline 0 & 0 & B_w \end{array}\right], \quad \tilde{C}_{ij} \triangleq \left[\begin{array}{cc|c} 0 & C_{fj} & -C_w \end{array}\right]. \end{cases} \tag{11.10}$$

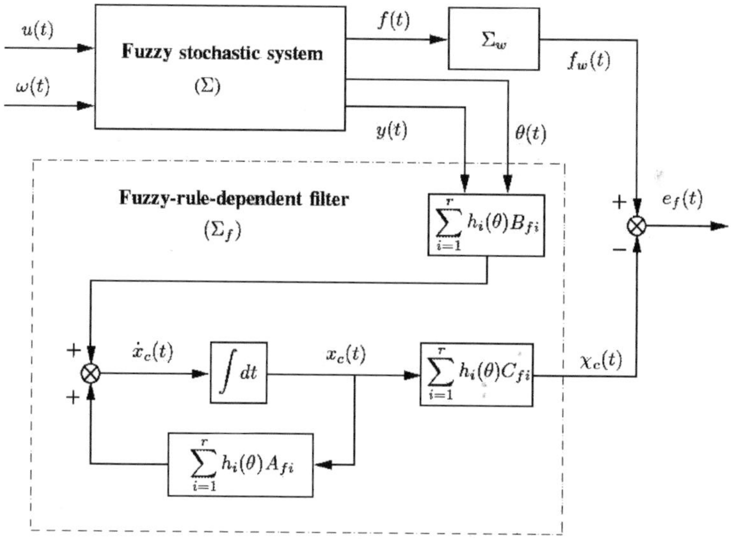

Fig. 11.2. Block diagram of the fuzzy-rule-dependent filter design

Remark 11.1. In this chapter, two approaches are presented to the fault detection filter design: one is fuzzy-rule-independent approach, which is suitable for the case that the premise variable of the original fuzzy model $\theta(t)$ is *unavailable*. The other is fuzzy-rule-dependent approach, which is applicable to the case that $\theta(t)$ is *available*. Since the information of the premise variable $\theta(t)$ is fully taken into account in filter design by using the fuzzy-rule-dependent approach, the result obtained is less conservative. However, by the fuzzy-rule-dependent approach the filter will become more complicated in filter implementation. There should be a trade-off between the fuzzy-rule-independent and fuzzy-rule-dependent approaches through the above analysis. ♦

Definition 11.2. The fault detection system in (11.5a) with $\upsilon(t) = 0$ is said to be mean-square asymptotically stable if its solution $\xi(t)$ satisfies

$$\lim_{t\to\infty} \mathbf{E}\left\{\|\xi(t,\xi(0))\|^2\right\} = 0.$$

Definition 11.3. Given a scalar $\gamma > 0$, the fault detection system in (11.5) is said to be mean-square asymptotically stable with an $\mathcal{H}_\infty$ performance level γ if it is mean-square asymptotically stable when $\upsilon(t) \equiv 0$ and, under zero initial condition and for all nonzero $\upsilon(t) \in \mathcal{L}_2[0,\infty)$, the following holds:

$$\mathbf{E}\left\{\int_0^\infty e_c^T(t)e_c(t)dt\right\} < \gamma^2 \int_0^\infty \upsilon^T(t)\upsilon(t)dt. \tag{11.11}$$

Therefore, the fault detection problem to be addressed in this chapter can be stated as the following two steps:

Step 1. Generate a residual signal: for fuzzy stochastic system (11.2), develop a robust $\mathcal{H}_\infty$ filter in the form of (11.3) (and (11.8) for fuzzy-rule-dependent case) to generate a residual signal. Meanwhile, the filter is designed to assure that the resulting overall fault detection system (11.5) (and (11.9)) to be mean-square asymptotically stable with an $\mathcal{H}_\infty$ performance level $\gamma > 0$.

Step 2. Set up a fault detection measure: select an evaluation function and a threshold. In this work, a residual evaluation function $\mathcal{J}(\chi)$ (where χ denotes $\chi_c(t)$ or $\chi_f(t)$) and a threshold $\mathcal{J}_{th}$ are selected as

$$\mathcal{J}(\chi) \triangleq \sqrt{\int_{t_0}^{t_0+t^\star} \chi^T(t)\chi(t)dt} \tag{11.12}$$

$$\mathcal{J}_{th} \triangleq \sup_{0\neq\omega\in\mathcal{L}_2, 0\neq u\in\mathcal{L}_2, f=0} \mathcal{J}(\chi) \tag{11.13}$$

where t_0 denotes the initial evaluation time instant, $t^\star$ stands for the evaluation time. Based on this, the occurrence of faults can be detected by comparing $\mathcal{J}(\chi)$ and $\mathcal{J}_{th}$ according to the following test:

$$\mathcal{J}(\chi) > \mathcal{J}_{th} \quad \Rightarrow \quad \text{with faults} \quad \Rightarrow \quad \text{alarm},$$
$$\mathcal{J}(\chi) \leq \mathcal{J}_{th} \quad \Rightarrow \quad \text{no faults}.$$

11.3 Main Results

11.3.1 Fuzzy-Rule-Independent Case

Firstly, we analyze the $\mathcal{H}_\infty$ performance for system (11.5).

Theorem 11.4. *Given a scalar $\gamma > 0$, the fuzzy stochastic fault detection system (11.5) is mean-square asymptotically stable with an $\mathcal{H}_\infty$ performance level γ if there exists matrix $P > 0$ such that the following LMIs hold:*

$$\begin{bmatrix} P\tilde{A}_i + \tilde{A}_i^T P & P\tilde{B}_i & K^T\tilde{E}_i^T P & \tilde{C}_i^T \\ \star & -\gamma^2 I & 0 & 0 \\ \star & \star & -P & 0 \\ \star & \star & \star & -I \end{bmatrix} < 0. \tag{11.14}$$

Proof. Choose a Lyapunov function as

$$V(\xi, t) \triangleq \xi^T(t)P\xi(t), \quad P > 0.$$

By Itô's formula in Lemma 9.4 of Chapter 9, we have

$$\begin{aligned} \mathscr{L}V(\xi, t) &= 2\sum_{i=1}^{r} h_i(\theta)\xi^T(t)P\left(\tilde{A}_i\xi(t) + \tilde{B}_i\upsilon(t)\right) \\ &\quad + \sum_{i=1}^{r}\sum_{j=1}^{r} h_i(\theta)h_j(\theta)\xi^T(t)K^T\tilde{E}_i^T P\tilde{E}_j K\xi(t) \\ &\le \sum_{i=1}^{r} h_i(\theta)\xi^T(t)\left(P\tilde{A}_i + \tilde{A}_i^T P + K^T\tilde{E}_i^T P\tilde{E}_i K\right)\xi(t) \\ &\quad + 2\sum_{i=1}^{r} h_i(\theta)\xi^T(t)P\tilde{B}_i\upsilon(t). \end{aligned} \tag{11.15}$$

Here, we use θ to represent $\theta(t)$ for simplicity. Therefore, when assuming the zero input, that is, $\upsilon(t) = 0$, we have from (11.15) that

$$\mathscr{L}V(\xi, t) \le \sum_{i=1}^{r} h_i(\theta)\xi^T(t)\left(P\tilde{A}_i + \tilde{A}_i^T P + K^T\tilde{E}_i^T P\tilde{E}_i K\right)\xi(t).$$

LMI (11.14) implies

$$P\tilde{A}_i + \tilde{A}_i^T P + K^T\tilde{E}_i^T P\tilde{E}_i K < 0,$$

thus by Schur complement, we have $\mathbf{E}\{\mathscr{L}V(\xi, t)\} < 0$. This implies, by [212], that the fuzzy stochastic fault detection system (11.5) with $\upsilon(t) = 0$ is mean-square asymptotically stable.

Now, we will establish the $\mathcal{H}_\infty$ performance for the fuzzy stochastic fault detection system (11.5). Assume zero initial condition, we have

$$\begin{aligned}\mathbf{J} &= \mathbf{E}\left\{\int_0^\infty \left[e_c^T(t)e_c(t) - \gamma^2 \upsilon^T(t)\upsilon(t)\right] dt\right\} \\ &\leq \mathbf{E}\left\{\int_0^\infty \left[e_c^T(t)e_c(t) - \gamma^2 \upsilon^T(t)\upsilon(t)\right] dt\right\} + \mathbf{E}\left\{V(\xi, t)\right\} - \mathbf{E}\left\{V(0,0)\right\} \\ &= \mathbf{E}\left\{\int_0^\infty \left[e_c^T(t)e_c(t) - \gamma^2 \upsilon^T(t)\upsilon(t) + \mathscr{L}V(\xi,t)\right] dt\right\}\end{aligned}$$

It follows from (11.15) that

$$e_c^T(t)e_c(t) - \gamma^2 \upsilon^T(t)\upsilon(t) + \mathscr{L}V(\xi,t) \leq \sum_{i=1}^r h_i(\theta) F^T(t)\Omega_i F(t)$$

where $F(t) \triangleq \begin{bmatrix} \xi(t) \\ \upsilon(t) \end{bmatrix}$ and

$$\Omega_i \triangleq \begin{bmatrix} P\tilde{A}_i + \tilde{A}_i^T P + K^T \tilde{E}_i^T P \tilde{E}_i K + \tilde{C}_i^T \tilde{C}_i & P\tilde{B}_i \\ \star & -\gamma^2 I \end{bmatrix}.$$

By Schur complement, (11.14) implies $\Omega_i < 0$, thus

$$\mathbf{J} \triangleq \mathbf{E}\left\{\int_0^\infty \left[e_c^T(t)e_c(t) - \gamma^2 \upsilon^T(t)\upsilon(t)\right] dt\right\} < 0,$$

which implies (11.11). The $\mathcal{H}_\infty$ performance has been established and the proof is completed. ∎

Now, we present a solution to the $\mathcal{H}_\infty$ fault detection filter design for system (11.2).

Theorem 11.5. *Consider the system (11.2). For a given scalar $\gamma > 0$, suppose there exist matrices $\mathcal{U} > 0$, $\mathcal{V} > 0$, $V > 0$, $\mathcal{A}_c$, $\mathcal{B}_c$ and $\mathcal{C}_c$ such that for $i = 1, 2, \ldots, r$, the following LMIs hold:*

$$\begin{bmatrix} \Pi_{11i} & \Pi_{12i} & 0 & \Pi_{14i} & \Pi_{15i} & \Pi_{16i} & \Pi_{17i}^T & \Pi_{18i}^T & 0 \\ \star & \Pi_{22} & 0 & \Pi_{24i} & \Pi_{25i} & \Pi_{26i} & 0 & 0 & \mathcal{C}_c^T \\ \star & \star & \Pi_{33} & 0 & 0 & VB_w & 0 & 0 & -C_w^T \\ \star & \star & \star & -\gamma^2 I & 0 & 0 & 0 & 0 & 0 \\ \star & \star & \star & \star & -\gamma^2 I & 0 & 0 & 0 & 0 \\ \star & \star & \star & \star & \star & -\gamma^2 I & 0 & 0 & 0 \\ \star & \star & \star & \star & \star & \star & -\mathcal{U} & -\mathcal{V} & 0 \\ \star & \star & \star & \star & \star & \star & \star & -\mathcal{V} & 0 \\ \star & \star & \star & \star & \star & \star & \star & \star & -I \end{bmatrix} < 0, \quad (11.16a)$$

$$\begin{bmatrix} \mathcal{U} & \mathcal{V} \\ \star & \mathcal{V} \end{bmatrix} > 0, \quad (11.16b)$$

where

$$\begin{aligned}
&\Pi_{11i} \triangleq \mathcal{U}A_i + \mathcal{B}_c C_i + A_i^T \mathcal{U} + C_i^T \mathcal{B}_c^T, \quad \Pi_{22} \triangleq \mathcal{A}_c + \mathcal{A}_c^T, \\
&\Pi_{12i} \triangleq \mathcal{A}_c + A_i^T \mathcal{V} + C_i^T \mathcal{B}_c^T, \quad \Pi_{33} \triangleq V A_w + A_w^T V, \\
&\Pi_{14i} \triangleq \mathcal{U}B_{0i} + \mathcal{B}_c D_{0i}, \quad \Pi_{15i} \triangleq \mathcal{U}B_i + \mathcal{B}_c D_i, \\
&\Pi_{24i} \triangleq \mathcal{V}B_{0i} + \mathcal{B}_c D_{0i}, \quad \Pi_{25i} \triangleq \mathcal{V}B_i + \mathcal{B}_c D_i, \\
&\Pi_{16i} \triangleq \mathcal{U}B_{1i} + \mathcal{B}_c D_{1i}, \quad \Pi_{17i} \triangleq \mathcal{U}E_i + \mathcal{B}_c F_i, \\
&\Pi_{26i} \triangleq \mathcal{V}B_{1i} + \mathcal{B}_c D_{1i}, \quad \Pi_{18i} \triangleq \mathcal{V}E_i + \mathcal{B}_c F_i.
\end{aligned}$$

Then, there exists a fuzzy-rule-independent fault detection filter (11.3) such that the fuzzy stochastic fault detection system in (11.5) is mean-square asymptotically stable with an $\mathcal{H}_\infty$ performance level γ. Moreover, if the above LMI conditions are feasible, then a desired $\mathcal{H}_\infty$ filter realization is given by

$$\begin{bmatrix} A_c & B_c \\ C_c & 0 \end{bmatrix} = \begin{bmatrix} \mathcal{V}^{-1} & 0 \\ 0 & I \end{bmatrix} \begin{bmatrix} \mathcal{A}_c & \mathcal{B}_c \\ \mathcal{C}_c & 0 \end{bmatrix}. \tag{11.17}$$

Proof. By Theorem 11.4, let $P \triangleq \text{diag}(U, V) > 0$ in (11.14), where $U \in \mathbf{R}^{2n\times 2n}$ and $V \in \mathbf{R}^{k\times k}$, we get a new result. Specifically, given a scalar $\gamma > 0$, the fuzzy stochastic filtering error system (11.5) is mean-square asymptotically stable with an $\mathcal{H}_\infty$ performance level γ if there exist matrices $U > 0$ and $V > 0$ such that the following LMI holds:

$$\begin{bmatrix} U\hat{A}_i + \hat{A}_i^T U & 0 & U\hat{B}_i & \hat{E}_i^T U & \hat{C}_i^T \\ \star & VA_w + A_w^T V & V\hat{B}_w & 0 & -C_w^T \\ \star & \star & -\gamma^2 I & 0 & 0 \\ \star & \star & \star & -U & 0 \\ \star & \star & \star & \star & -I \end{bmatrix} < 0, \tag{11.18}$$

where

$$\hat{A}_i \triangleq \begin{bmatrix} A_i & 0 \\ B_c C_i & A_c \end{bmatrix}, \quad \hat{B}_i \triangleq \begin{bmatrix} B_{0i} & B_i & B_{1i} \\ B_c D_{0i} & B_c D_i & B_c D_{1i} \end{bmatrix}, \quad \hat{E}_i \triangleq \begin{bmatrix} E_i & 0 \\ B_c F_i & 0 \end{bmatrix},$$
$$\hat{B}_w \triangleq \begin{bmatrix} 0 & 0 & B_w \end{bmatrix}, \quad \hat{C}_i \triangleq \begin{bmatrix} 0 & C_c \end{bmatrix}.$$

Now, partition U as

$$U \triangleq \begin{bmatrix} U_1 & U_2 \\ \star & U_3 \end{bmatrix} > 0, \tag{11.19}$$

where $U_j \in \mathbf{R}^{n\times n}, j = 1, 2, 3$.

As we are considering a full-order filter, U_2 is square. Without loss of generality, we assume U_2 is nonsingular (if not, U_2 may be perturbed by ΔU_2 with sufficiently small norm such that $U_2 + \Delta U_2$ is nonsingular and satisfying (11.18)). Define the following matrices which are also nonsingular:

$$\mathcal{J} \triangleq \begin{bmatrix} I & 0 \\ 0 & U_3^{-1}U_2^T \end{bmatrix}, \quad \mathcal{U} \triangleq U_1, \quad \mathcal{V} \triangleq U_2U_3^{-1}U_2^T, \tag{11.20}$$

and

$$\begin{bmatrix} \mathcal{A}_c & \mathcal{B}_c \\ \mathcal{C}_c & 0 \end{bmatrix} \triangleq \begin{bmatrix} U_2 & 0 \\ 0 & I \end{bmatrix} \begin{bmatrix} A_c & B_c \\ C_c & 0 \end{bmatrix} \begin{bmatrix} U_3^{-1}U_2^T & 0 \\ 0 & I \end{bmatrix}. \tag{11.21}$$

Performing a congruence transformation to (11.18) by diag $\{\mathcal{J}, I, I, \mathcal{J}, I\}$, it follows that

$$\begin{bmatrix} \tilde{\Pi}_{11i} & 0 & \mathcal{J}^T U\hat{B}_i & \mathcal{J}^T\hat{E}_i^T U\mathcal{J} & \mathcal{J}^T\hat{C}_i^T \\ \star & \Pi_{33} & V\hat{B}_w & 0 & -C_w^T \\ \star & \star & -\gamma^2 I & 0 & 0 \\ \star & \star & \star & -\mathcal{J}^T U\mathcal{J} & 0 \\ \star & \star & \star & \star & -I \end{bmatrix} < 0, \tag{11.22}$$

where $\tilde{\Pi}_{11i} \triangleq \mathcal{J}^T(U\hat{A}_i + \hat{A}_i^T U)\mathcal{J}$ and

$$\begin{cases} \mathcal{J}^T U\hat{A}_i\mathcal{J} \triangleq \begin{bmatrix} \mathcal{U}A_i + \mathcal{B}_cC_i & \mathcal{A}_c \\ \mathcal{V}A_i + \mathcal{B}_cC_i & \mathcal{A}_c \end{bmatrix}, & \hat{C}_i\mathcal{J} \triangleq \begin{bmatrix} 0 & \mathcal{C}_c \end{bmatrix}, \\ \mathcal{J}^T U\hat{E}_i\mathcal{J} \triangleq \begin{bmatrix} \mathcal{U}E_i + \mathcal{B}_cF_i & 0 \\ \mathcal{V}E_i + \mathcal{B}_cF_i & 0 \end{bmatrix}, & \mathcal{J}^T U\mathcal{J} \triangleq \begin{bmatrix} \mathcal{U} & \mathcal{V} \\ \mathcal{V} & \mathcal{V} \end{bmatrix}, \\ \mathcal{J}^T U\hat{B}_i \triangleq \begin{bmatrix} \mathcal{U}B_{0i} + \mathcal{B}_cD_{0i} & \mathcal{U}B_i + \mathcal{B}_cD_i & \mathcal{U}B_{1i} + \mathcal{B}_cD_{1i} \\ \mathcal{V}B_{0i} + \mathcal{B}_cD_{0i} & \mathcal{V}B_i + \mathcal{B}_cD_i & \mathcal{V}B_{1i} + \mathcal{B}_cD_{1i} \end{bmatrix}. \end{cases}$$

Considering (11.23), we can obtain LMI (11.16a) from (11.22).

Moreover, notice that (11.21) is equivalent to

$$\begin{aligned} \begin{bmatrix} A_c & B_c \\ C_c & 0 \end{bmatrix} &= \begin{bmatrix} U_2^{-1} & 0 \\ 0 & I \end{bmatrix} \begin{bmatrix} \mathcal{A}_c & \mathcal{B}_c \\ \mathcal{C}_c & 0 \end{bmatrix} \begin{bmatrix} U_2^{-T}U_3 & 0 \\ 0 & I \end{bmatrix} \\ &= \begin{bmatrix} \left(U_2^{-T}U_3\right)^{-1}\mathcal{V}^{-1} & 0 \\ 0 & I \end{bmatrix} \begin{bmatrix} \mathcal{A}_c & \mathcal{B}_c \\ \mathcal{C}_c & 0 \end{bmatrix} \begin{bmatrix} U_2^{-T}U_3 & 0 \\ 0 & I \end{bmatrix}. \end{aligned} \tag{11.23}$$

Notice also that the filter matrices A_c, B_c and C_c in (11.3) can be written as (11.23), which implies that $U_2^{-T}U_3$ can be viewed as a similarity transformation on the state-space realization of the filter and, as such, has no effect on the filter mapping from y to χ_c. Without loss of generality, we may set $U_2^{-T}U_3 = I$, thus obtain (11.17). Therefore, the filter (11.3) can be constructed by (11.17). This completes the proof. ■

Remark 11.6. Notice that the obtained conditions in Theorem 11.5 are all in LMI form, a desired fuzzy-rule-independent $\mathcal{H}_\infty$ fault detection filter can be determined by solving the following convex optimization problem:

$$\min\ \delta \quad \text{subject to (11.16a)–(11.16b)} \quad (\text{where } \delta = \gamma^2).$$ ♦

Remark 11.7. By solving the above convex optimization problem, we can obtain the parameters of the filter in (11.3) by (11.17), and then the residual signal $\chi_c(t)$ is generated. The next work is to set up a fault detection measure, a residual evaluation function $\mathcal{J}(\chi)$ and a threshold $\mathcal{J}_{th}$ are selected respectively as (11.12) and (11.13), by which the fault can be detected. ♦

11.3.2 Fuzzy-Rule-Dependent Case

In this section, we consider the fuzzy-rule-dependent case.

Theorem 11.8. *Given a scalar $\gamma > 0$, the fuzzy stochastic fault detection system (11.9) is mean-square asymptotically stable with an $\mathcal{H}_\infty$ performance level γ if there exists matrix $P > 0$ such that the following LMIs hold:*

$$\Phi_{ii} < 0, \quad i = 1, 2, \ldots, r, \tag{11.24a}$$
$$\Phi_{ij} + \Phi_{ji} < 0, \quad i < j \leq r, \tag{11.24b}$$

where

$$\Phi_{ij} \triangleq \begin{bmatrix} P\tilde{A}_{ij} + \tilde{A}_{ij}^T P & P\tilde{B}_{ij} & K^T \tilde{E}_{ij}^T P & \tilde{C}_{ij}^T \\ \star & -\gamma^2 I & 0 & 0 \\ \star & \star & -P & 0 \\ \star & \star & \star & -I \end{bmatrix}.$$

Proof. This theorem can be proved by employing the same techniques as in the proof of Theorem 11.4, hence the detailed procedure is omitted here. ■

Theorem 11.9. *Consider system (11.2). For a given scalar $\gamma > 0$, suppose there exist matrices $\mathcal{U} > 0$, $\mathcal{V} > 0$, $V > 0$, $\mathcal{A}_{fi}$, $\mathcal{B}_{fi}$ and $\mathcal{C}_{fi}$ such that (11.16b) and the following LMIs hold:*

$$\Psi_{ii} < 0, \quad i = 1, 2, \ldots, r, \tag{11.25a}$$
$$\Psi_{ij} + \Psi_{ji} < 0, \quad i < j \leq r, \tag{11.25b}$$

where

$$\Psi_{ij} \triangleq \begin{bmatrix} \Psi_{11ij} & \Psi_{12ij} & 0 & \Psi_{14ij} & \Psi_{15ij} & \Psi_{16ij} & \Psi_{17ij}^T & \Psi_{18ij}^T & 0 \\ \star & \Psi_{22ij} & 0 & \Psi_{24ij} & \Psi_{25ij} & \Psi_{26ij} & 0 & 0 & C_{cj}^T \\ \star & \star & \Pi_{33} & 0 & 0 & VB_w & 0 & 0 & -C_w^T \\ \star & \star & \star & -\gamma^2 I & 0 & 0 & 0 & 0 & 0 \\ \star & \star & \star & \star & -\gamma^2 I & 0 & 0 & 0 & 0 \\ \star & \star & \star & \star & \star & -\gamma^2 I & 0 & 0 & 0 \\ \star & \star & \star & \star & \star & \star & -\mathcal{U} & -\mathcal{V} & 0 \\ \star & \star & \star & \star & \star & \star & \star & -\mathcal{V} & 0 \\ \star & \star & \star & \star & \star & \star & \star & \star & -I \end{bmatrix},$$

with the notations in Ψ_{ij} are given as follows:

$$\begin{aligned}
&\Psi_{11ij} \triangleq \mathcal{U}A_i + \mathcal{B}_{cj}C_i + A_i^T\mathcal{U} + C_i^T\mathcal{B}_{cj}^T,\\
&\Psi_{12ij} \triangleq \mathcal{A}_{cj} + A_i^T\mathcal{V} + C_i^T\mathcal{B}_{cj}^T, \quad \Psi_{22ij} \triangleq \mathcal{A}_{cj} + \mathcal{A}_{cj}^T,\\
&\Psi_{14ij} \triangleq \mathcal{U}B_{0i} + \mathcal{B}_{cj}D_{0i}, \quad \Psi_{15ij} \triangleq \mathcal{U}B_i + \mathcal{B}_{cj}D_i,\\
&\Psi_{24ij} \triangleq \mathcal{V}B_{0i} + \mathcal{B}_{cj}D_{0i}, \quad \Psi_{25ij} \triangleq \mathcal{V}B_i + \mathcal{B}_{cj}D_i,\\
&\Psi_{16ij} \triangleq \mathcal{U}B_{1i} + \mathcal{B}_{cj}D_{1i}, \quad \Psi_{17ij} \triangleq \mathcal{U}E_i + \mathcal{B}_{cj}F_i,\\
&\Psi_{26ij} \triangleq \mathcal{V}B_{1i} + \mathcal{B}_{cj}D_{1i}, \quad \Psi_{18ij} \triangleq \mathcal{V}E_i + \mathcal{B}_{cj}F_i.
\end{aligned}$$

Then, there exists a fuzzy-rule-dependent fault detection filter (11.8), such that the fuzzy stochastic fault detection system (11.9) is mean-square asymptotically stable with an $\mathcal{H}_\infty$ performance level γ. Moreover, a desired $\mathcal{H}_\infty$ filter realization is given by

$$\begin{bmatrix} A_{fi} & B_{fi} \\ C_{fi} & 0 \end{bmatrix} = \begin{bmatrix} \mathcal{V}^{-1} & 0 \\ 0 & I \end{bmatrix} \begin{bmatrix} \mathcal{A}_{fi} & \mathcal{B}_{fi} \\ \mathcal{C}_{fi} & 0 \end{bmatrix}, \quad i = 1, 2, \ldots, r. \tag{11.26}$$

Proof. The theorem can be proved by following the same line of the proof of Theorem 11.5. ∎

11.4 Illustrative Example

Consider the T-S fuzzy stochastic system in (11.2) with the model parameters given as follows:

$$\begin{aligned}
A_1 &= \begin{bmatrix} -3.0 & 0.2 & 0.4 \\ 0.3 & -1.7 & 0.5 \\ 0.2 & 0.5 & -2.5 \end{bmatrix}, \; B_1 = \begin{bmatrix} 0.5 \\ 0.8 \\ 0.6 \end{bmatrix}, \; B_{01} = \begin{bmatrix} 0.3 \\ 0.6 \\ 0.5 \end{bmatrix}, \; B_{11} = \begin{bmatrix} 0.4 \\ 0.5 \\ 0.4 \end{bmatrix},\\
A_2 &= \begin{bmatrix} -2.7 & 0.3 & 0.6 \\ 0.2 & -1.5 & 0.8 \\ 0.3 & 0.4 & -2.4 \end{bmatrix}, \; B_2 = \begin{bmatrix} 0.5 \\ 0.6 \\ 0.3 \end{bmatrix}, \; B_{02} = \begin{bmatrix} 0.4 \\ 0.6 \\ 0.7 \end{bmatrix}, \; B_{12} = \begin{bmatrix} 0.6 \\ 0.4 \\ 0.3 \end{bmatrix},\\
E_1 &= \begin{bmatrix} 0.2 & 0 & 0.1 \\ 0.3 & 0.1 & 0.2 \\ 0.0 & 0.1 & 0.2 \end{bmatrix}, \; E_2 = \begin{bmatrix} 0.3 & 0.1 & 0.2 \\ 0.1 & 0.2 & 0.2 \\ 0.0 & 0.3 & 0.2 \end{bmatrix}, \; \begin{array}{l} F_1 = \begin{bmatrix} 0.1 & 0.2 & 0.4 \end{bmatrix}, \\ F_2 = \begin{bmatrix} 0.1 & 0.2 & 0.2 \end{bmatrix}, \end{array}\\
C_1 &= \begin{bmatrix} 1.5 & 0.6 & 1.3 \end{bmatrix}, \; D_{01} = 0.6, \; D_1 = 0.2, \; D_{11} = 0.5,\\
C_2 &= \begin{bmatrix} 1.0 & 1.3 & 0.7 \end{bmatrix}, \; D_{02} = 0.4, \; D_2 = 0.3, \; D_{12} = 0.4.
\end{aligned}$$

The weighting matrix $W(s)$ in $f_w(s) = W(s)f(s)$ is supposed to be $W(s) = 5/(s+5)$. Its state space realization is given as (11.4) with $A_w = -5$, $B_w = 5$ and $C_w = 1$.

Firstly, we consider the fuzzy-rule-independent case. By solving LMIs (11.16a)–(11.16b) and (11.17) in Theorem 11.5, we have that the minimized feasible γ is $\gamma^* = 1.0027$, and

$$A_c = \begin{bmatrix} -6.3205 & -1.7435 & -0.3712 \\ -1.3305 & -1.7739 & -2.3382 \\ -1.0170 & 0.7815 & -5.6619 \end{bmatrix}, \quad B_c = \begin{bmatrix} -1.5255 \\ -1.3681 \\ -1.1309 \end{bmatrix},$$
$$C_c = \begin{bmatrix} -0.1604 & 0.0152 & -0.3919 \end{bmatrix}.$$

We further consider the fuzzy-rule-dependent case. Solving LMIs (11.16b), (11.25a)–(11.25b) and (11.26) in Theorem 11.9, we obtain that the minimized feasible γ is $\gamma^* = 1.0007$, and

$$A_{f1} = \begin{bmatrix} -6.2608 & 0.1738 & -0.8381 \\ -1.7425 & -1.4227 & -2.9189 \\ -1.0463 & 1.8460 & -7.6312 \end{bmatrix}, \quad B_{f1} = \begin{bmatrix} -1.1069 \\ -1.3844 \\ -1.1725 \end{bmatrix},$$
$$A_{f2} = \begin{bmatrix} -6.4503 & -1.0140 & -1.2947 \\ -1.0104 & -1.8160 & -1.8330 \\ -0.2993 & 0.8286 & -5.9038 \end{bmatrix}, \quad B_{f2} = \begin{bmatrix} -1.8552 \\ -1.1598 \\ -0.8975 \end{bmatrix},$$
$$C_{f1} = \begin{bmatrix} -0.0602 & 0.0627 & -0.6547 \end{bmatrix}, \quad C_{f2} = \begin{bmatrix} -0.5500 & -0.0592 & 0.1034 \end{bmatrix}.$$

Notice that the minimized feasible γ for the fuzzy-rule-independent case is $\gamma^* = 1.0027$, for the fuzzy-rule-dependent case is $\gamma^* = 1.0007$. The maximum singular values of the filtering error systems for the two cases are depicted in Fig. 11.3, which has illustrated that the fuzzy-rule-dependent filter is less conservative than the fuzzy-rule-independent filter in the sense of the disturbance attenuation performance level.

In the following, we shall further show the effectiveness of the designed robust $\mathcal{H}_\infty$ fault detection filters (the fuzzy-rule-independent and the fuzzy-rule-dependent cases) through simulation. Let the initial condition be $x(0) = \begin{bmatrix} -1.0 & 0.5 & 1.0 \end{bmatrix}^T$ and choose the membership function as

$$h_1(x_1(t)) = \frac{1 - \sin(x_1(t))}{2}, \quad h_2(x_1(t)) = \frac{1 + \sin(x_1(t))}{2}.$$

Suppose the unknown disturbance input $\omega(t)$ to be random noise, as shown in Fig. 11.4. The known input is given as $u(t) = \sin(t)$, $0 \le t \le 10$; and the fault signal is set up as:

$$f(t) = \begin{cases} 1, & 2.5 \le t \le 5, \\ 0, & \text{otherwise}. \end{cases}$$

Thus, the weighting fault signal $f_w(t)$ is shown in Fig. 11.5.

We select the evaluation function and the threshold as (11.12)–(11.13). By using the discretization approach in [86], we simulate the standard Brownian motion. Some initial parameters are given as follows: the simulation time $t \in [0, T^*]$ with $T^* = 10$, the normally distributed variance $\delta t = \frac{T^*}{N^*}$ with $N^* = 2^{11}$, step size $\Delta t = \rho \delta t$ with $\rho = 2$. For the designed $\mathcal{H}_\infty$ fault detection filter with the fuzzy-rule-independent case, the simulation results along an individual discretized Brownian path are given in Figs. 11.6–11.8. Among them, Fig. 11.6 shows the states of the designed $\mathcal{H}_\infty$ fault detection filter under zero disturbance. Fig. 11.7 depicts the generated residual signal $\chi_c(t)$; Fig. 11.8 presents the evaluation function of $\mathcal{J}(\chi)$ for both the fault case (solid line) and fault-free case (dash-dot line). The corresponding simulation results for fuzzy-rule-dependent case are depicted in Figs. 11.9–11.11, respectively.

When the residual signal is generated, the next step is to set up the fault detection measure. For the delay-rule-independent case, with a selected threshold of

$$\mathcal{J}_{th} = \sup_{\omega \neq 0, u \neq 0, f=0} \sqrt{\int_0^{10} \chi^T(t)\chi(t)dt} = 0.1895,$$

the simulation results show that $\sqrt{\int_0^{2.8} \chi^T(t)\chi(t)dt} = 0.2375 > \mathcal{J}_{th}$. Thus, the appeared fault can be detected after 0.3 sec. By the same way, we can detect the fault with the delay-rule-dependent method.

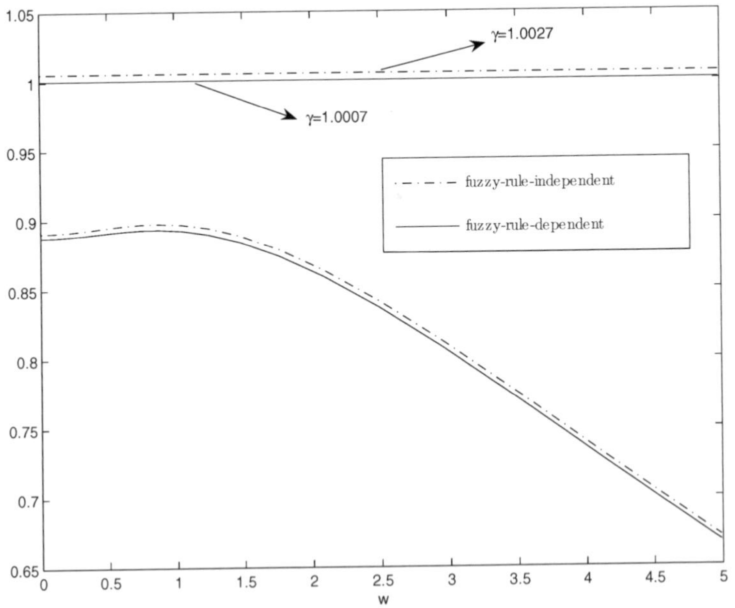

Fig. 11.3. Maximum singular values of the filtering error systems

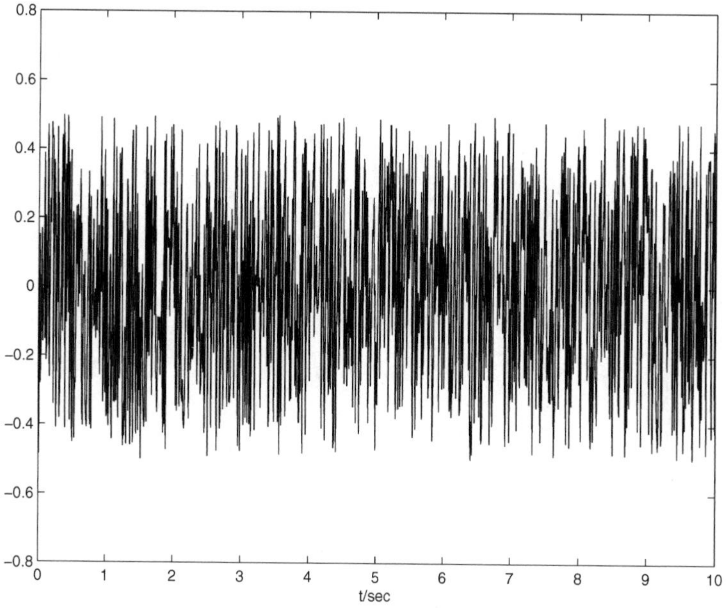

Fig. 11.4. Unknown disturbance input $\omega(t)$

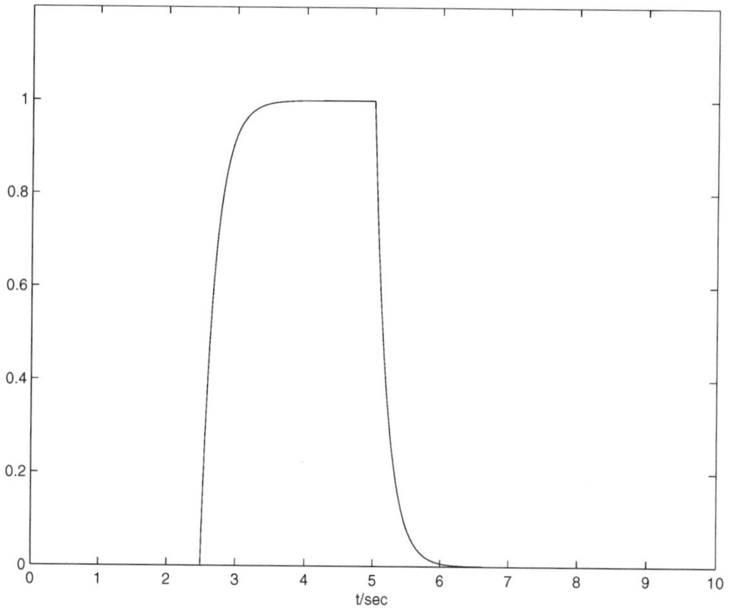

Fig. 11.5. Weighting fault signal $f_w(t)$

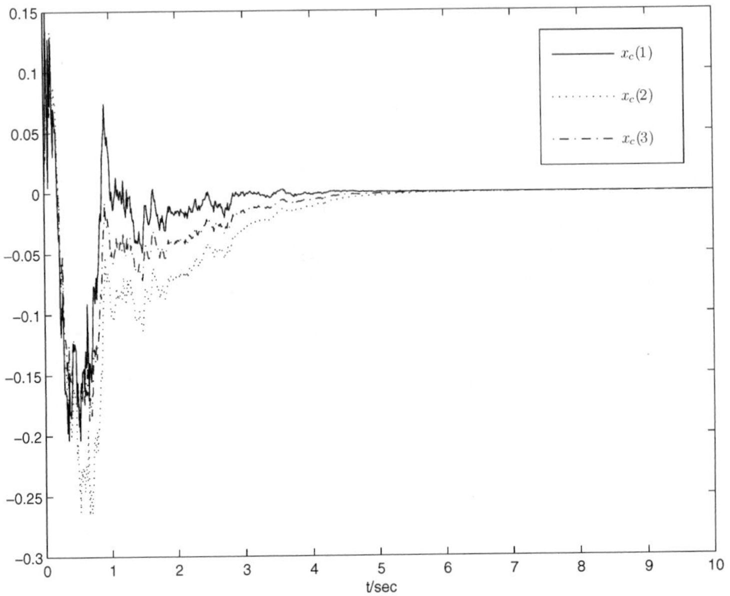

Fig. 11.6. States of the fault detection filter of the fuzzy-rule-independent case

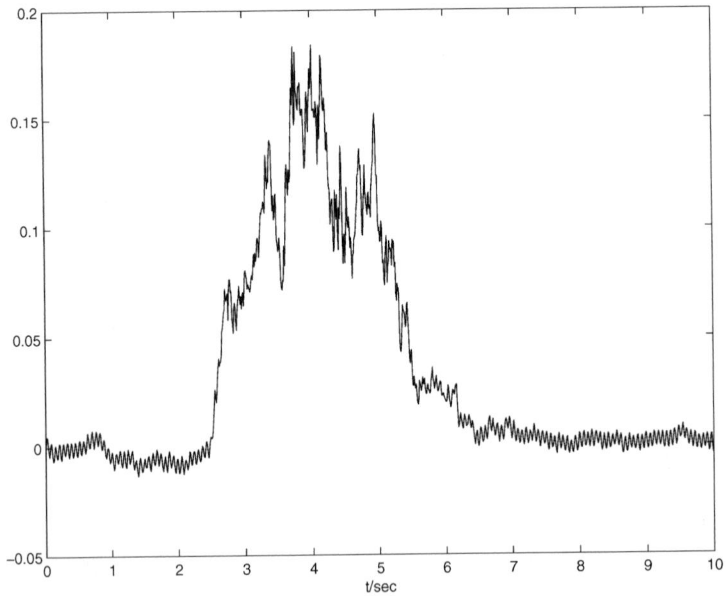

Fig. 11.7. Residual signal $\chi_c(t)$ of the fuzzy-rule-independent case

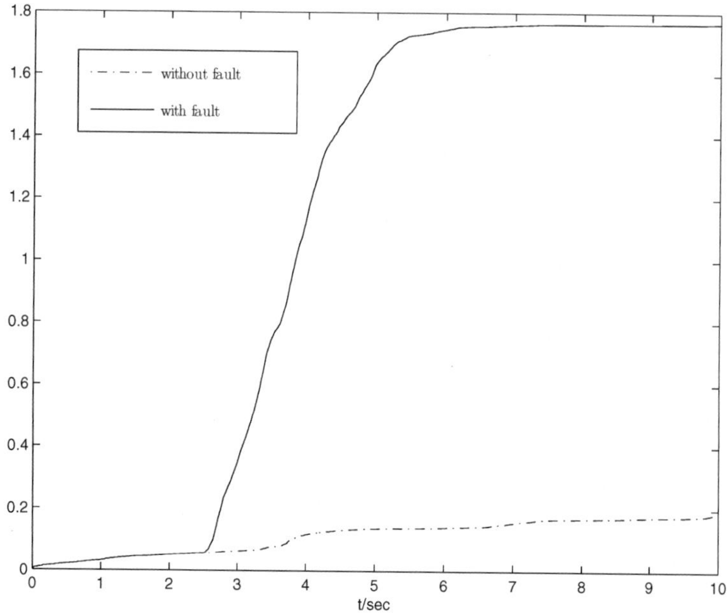

Fig. 11.8. Evaluation function of $\mathcal{J}(\chi)$ of the fuzzy-rule-independent case

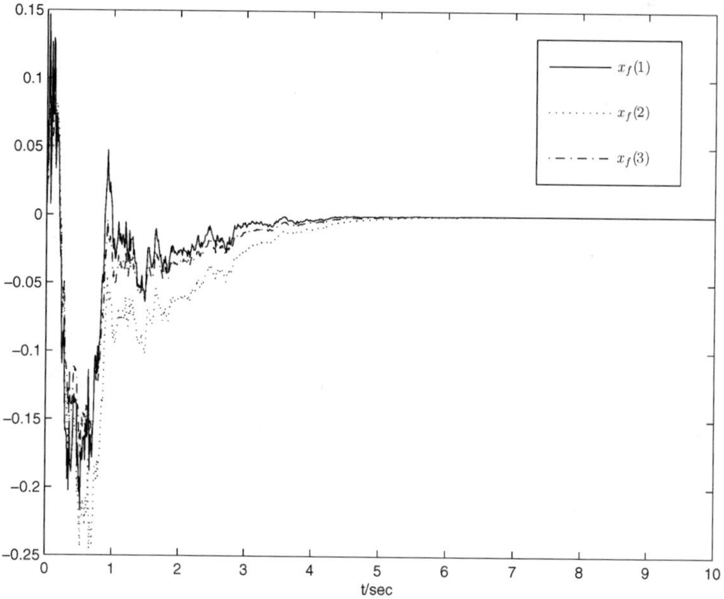

Fig. 11.9. States of the fault detection filter of the fuzzy-rule-dependent case

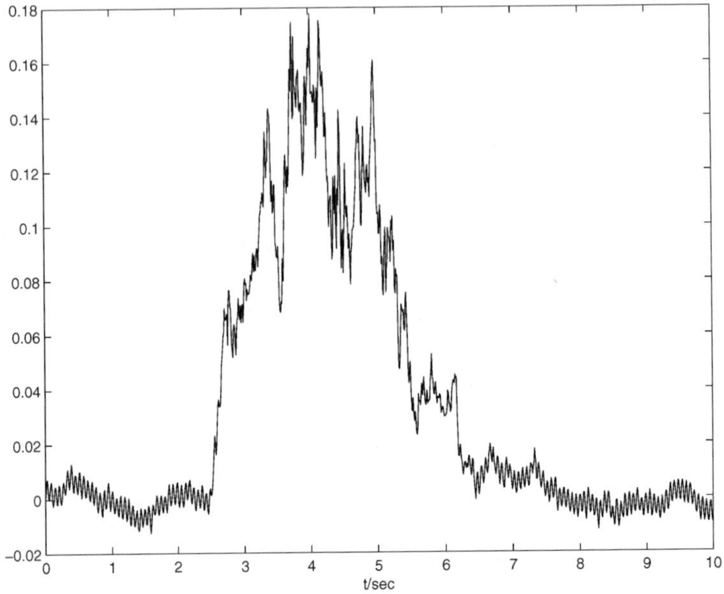

Fig. 11.10. Residual signal $\chi_f(t)$ of the fuzzy-rule-dependent case

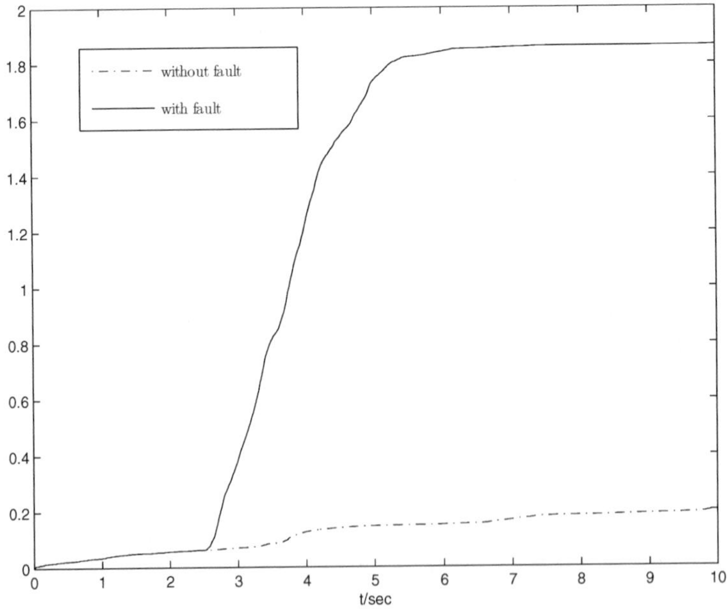

Fig. 11.11. Evaluation function of $\mathcal{J}(\chi)$ of the fuzzy-rule-dependent case

11.5 Conclusion

In this chapter, the robust $\mathcal{H}_\infty$ fuzzy fault detection has been studied for a class of nonlinear stochastic systems. Both the fuzzy-rule-independent and the fuzzy-rule-dependent fault detection filters have been designed. Sufficient conditions have been proposed to guarantee the mean-square asymptotic stability with an $\mathcal{H}_\infty$ performance for the fuzzy stochastic fault detection system. Then, the corresponding solvability conditions for desired fault detection filters have also been established. A numerical example has been provided to illustrate the effectiveness of the proposed theory.

Chapter 12
Model Approximation of Continuous-Time T-S Fuzzy Stochastic Systems

12.1 Introduction

In this chapter, we consider the $\mathcal{H}_\infty$ model approximation problem for continuous-time T-S fuzzy stochastic systems. For a given high-order T-S fuzzy stochastic system, our attention is focused on the construction of a reduced-order model, which guarantees the corresponding approximation error system to be mean-square asymptotically stable and has a specified $\mathcal{H}_\infty$ norm error performance. A sufficient condition is firstly proposed for the existence of the desired reduced-order model in terms of LMIs. Then, two different approaches are proposed to solve the model approximation problem. One casts the model approximation into a convex optimization problem by using a linearization procedure (namely convex linearization approach), and the other is the projection approach based on CCL idea, which casts the model approximation into a sequential minimization problem subject to LMI constraints.

12.2 System Description and Preliminaries

In this chapter, we consider a class of nonlinear stochastic systems which can be described by the following T-S fuzzy stochastic model:

♦ **Plant Form:**

Rule i: IF $\theta_1(t)$ is $\mathcal{M}_{i1}$ and $\theta_2(t)$ is $\mathcal{M}_{i2}$ and $\cdots$ and $\theta_p(t)$ is $\mathcal{M}_{ip}$ THEN

$$dx(t) = [A_i x(t) + B_i u(t)]\,dt + E_i x(t) d\varpi, \tag{12.1a}$$
$$y(t) = C_i x(t) + D_i u(t), \tag{12.1b}$$

where $i = 1, 2, \ldots, r$, and r is the number of IF-THEN rules; $\mathcal{M}_{ij}(i = 1, 2, \ldots, r; j = 1, 2, \ldots, p)$ are the fuzzy sets; $\theta(t) = \begin{bmatrix} \theta_1(t)\ \theta_2(t) \cdots \theta_p(t) \end{bmatrix}^T$ is the premise variable vector. $x(t) \in \mathbf{R}^n$ is the state vector; $u(t) \in \mathbf{R}^m$ is the

L. Wu et al., *Fuzzy Control Systems with Time-Delay and Stochastic Perturbation*, Studies in Systems, Decision and Control 12, DOI: 10.1007/978-3-319-11316-6_12

input which belongs to $\mathcal{L}_2[0,\infty)$; $y(t) \in \mathbf{R}^p$ is the output; $\varpi(t)$ is a scalar Brownian motion defined on the probability space $(\Omega, \mathcal{F}, \{\mathcal{F}_t\}_{t\geq 0}, \mathcal{P})$, and it satisfies $\mathbf{E}\{d\varpi(t)\} = 0$ and $\mathbf{E}\{d\varpi^2(t)\} = dt$. A_i, B_i, C_i, D_i and E_i are real constant matrices.

It is assumed that the premise variables do not depend on the input variables $u(t)$. Given a pair of $(x(t), u(t))$, the final output of the fuzzy stochastic systems is inferred as follows:

$$dx(t) = \sum_{i=1}^{r} h_i(\theta)\{[A_i x(t) + B_i u(t)]\, dt + E_i x(t) d\varpi\}, \tag{12.2a}$$

$$y(t) = \sum_{i=1}^{r} h_i(\theta)\{C_i x(t) + D_i u(t)\}, \tag{12.2b}$$

where $h_i(\theta)$, $i = 1, 2, \ldots, r$ are the normalized membership functions, which are defined as that of (1.1) in Chapter 1.

Here, we approximate system (12.2) by a reduced-order model described by

$$d\hat{x}(t) = \left[\hat{A}\hat{x}(t) + \hat{B}u(t)\right] dt + \hat{E}\hat{x}(t)d\varpi, \tag{12.3a}$$

$$\hat{y}(t) = \hat{C}\hat{x}(t) + \hat{D}u(t), \tag{12.3b}$$

where $\hat{x}(t) \in \mathbf{R}^k$ is the state vector of the reduced-order model with $k < n$; $\hat{A}$, $\hat{B}$, $\hat{C}$, $\hat{D}$ and $\hat{E}$ are matrices to be determined.

Augmenting the model of (12.2) to include the states of (12.3), we obtain the approximation error system as

$$d\xi(t) = \sum_{i=1}^{r} h_i(\theta)\left\{\left[\tilde{A}_i\xi(t) + \tilde{B}_i u(t)\right] dt + \tilde{E}_i\xi(t)d\varpi(t)\right\}, \tag{12.4a}$$

$$e(t) = \sum_{i=1}^{r} h_i(\theta)\left\{\tilde{C}_i\xi(t) + \tilde{D}_i u(t)\right\}, \tag{12.4b}$$

where $\xi(t) \triangleq \begin{bmatrix} x(t) \\ \hat{x}(t) \end{bmatrix}$, $e(t) \triangleq y(t) - \hat{y}(t)$ and

$$\begin{cases} \tilde{A}_i \triangleq \begin{bmatrix} A_i & 0 \\ 0 & \hat{A} \end{bmatrix}, \quad \tilde{E}_i \triangleq \begin{bmatrix} E_i & 0 \\ 0 & \hat{E} \end{bmatrix}, \quad \tilde{B}_i \triangleq \begin{bmatrix} B_i \\ \hat{B} \end{bmatrix}, \\ \tilde{C}_i \triangleq \begin{bmatrix} C_i & -\hat{C} \end{bmatrix}, \quad \tilde{D}_i \triangleq D_i - \hat{D}. \end{cases} \tag{12.5}$$

Definition 12.1. The approximation error system in (12.4) with $u(t) = 0$ is said to be mean-square asymptotically stable if its solution $\xi(t)$ satisfies

$$\lim_{t\to\infty} \mathbf{E}\left\{\|\xi(t, \xi(0))\|^2\right\} = 0.$$

Definition 12.2. Given a scalar $\gamma > 0$, the approximation error system in (12.4) is said to be mean-square asymptotically stable with an $\mathcal{H}_\infty$ performance level γ if it is mean-square asymptotically stable when $u(t) \equiv 0$ and, under zero initial condition and for all nonzero $u(t) \in \mathcal{L}_2[0, \infty)$, the following holds:

$$\mathbf{E}\left\{\int_0^\infty e^T(t)e(t)dt\right\} < \gamma^2 \int_0^\infty u^T(t)u(t)dt. \tag{12.6}$$

Therefore, the $\mathcal{H}_\infty$ model approximation problem addressed in this chapter can be formulated as follows: given the T-S fuzzy stochastic system in (12.2) and a scalar $\gamma > 0$, determine a reduced-order model of (12.3) such that the resulting approximation error system in (12.4) is mean-square asymptotically stable with an $\mathcal{H}_\infty$ performance level γ.

12.3 Main Results

Firstly, we give the following result without proof, the detailed proof can be found in [219].

Lemma 12.3. *Given a scalar $\gamma > 0$, the approximation error system in (12.4) is mean-square asymptotically stable with an $\mathcal{H}_\infty$ performance level γ if there exists a matrix $P > 0$ such that for $i = 1, 2, \ldots, r$,*

$$\begin{bmatrix} P\tilde{A}_i + \tilde{A}_i^T P & P\tilde{B}_i & \tilde{E}_i^T P & \tilde{C}_i^T \\ \star & -\gamma^2 I & 0 & \tilde{D}_i^T \\ \star & \star & -P & 0 \\ \star & \star & \star & -I \end{bmatrix} < 0. \tag{12.7}$$

Based on the result in Lemma 12.3, in what follows, we will provide two different approaches to solve the $\mathcal{H}_\infty$ model approximation for system (12.2): one makes use of the convex linearization procedure, and the other is based on the projection lemma.

12.3.1 Convex Linearization Approach

We present a solution to the $\mathcal{H}_\infty$ model approximation by the convex linearization approach.

Theorem 12.4. *Consider the T-S fuzzy stochastic system in (12.2). There exists a reduced-order model in the form of (12.3) that solves the $\mathcal{H}_\infty$ model approximation problem with (12.6) satisfied, if there exist matrices $\mathcal{P} > 0$, $\mathcal{Q} > 0$, $\mathcal{A}$, $\mathcal{B}$, $\mathcal{C}$, $\mathcal{D}$ and $\mathcal{E}$ such that for $i = 1, 2, \ldots, r$,*

$$\begin{bmatrix} \Pi_{11i} & \Pi_{12i} & \Pi_{13i} & E_i^T \mathcal{P} & E_i^T \mathscr{H} \mathcal{Q} & C_i^T \\ \star & \mathcal{A}+\mathcal{A}^T & \Pi_{23i} & \mathcal{E}^T \mathscr{H}^T & \mathcal{E}^T & -\mathcal{C}^T \\ \star & \star & -\gamma^2 I & 0 & 0 & D_i^T - \mathcal{D}^T \\ \star & \star & \star & -\mathcal{P} & -\mathscr{H}\mathcal{Q} & 0 \\ \star & \star & \star & \star & -\mathcal{Q} & 0 \\ \star & \star & \star & \star & \star & -I \end{bmatrix} < 0, \tag{12.8}$$

where $\mathscr{H} \triangleq \begin{bmatrix} I_{k\times k} \\ 0_{(n-k)\times k} \end{bmatrix}$ *and*

$$\Pi_{11i} \triangleq \mathcal{P}A_i + A_i^T \mathcal{P}, \ \Pi_{12i} \triangleq \mathscr{H}\mathcal{A} + A_i^T \mathscr{H}\mathcal{Q},$$
$$\Pi_{13i} \triangleq \mathcal{P}B_i + \mathscr{H}\mathcal{B}, \ \Pi_{23i} \triangleq \mathcal{Q}\mathscr{H}^T B_i + \mathcal{B}.$$

Moreover, if the above conditions are feasible, then the parameters of an admissible reduced-order model in the form of (12.3) can be calculated from

$$\begin{bmatrix} \hat{A} & \hat{B} \\ \hat{C} & \hat{D} \\ \hat{E} & 0 \end{bmatrix} = \begin{bmatrix} \mathcal{Q}^{-1} & 0 & 0 \\ 0 & I & 0 \\ 0 & 0 & \mathcal{Q}^{-1} \end{bmatrix} \begin{bmatrix} \mathcal{A} & \mathcal{B} \\ \mathcal{C} & \mathcal{D} \\ \mathcal{E} & 0 \end{bmatrix}. \tag{12.9}$$

Proof. According to Lemma 12.3, P is nonsingular since $P > 0$. Now, partition P as

$$P \triangleq \begin{bmatrix} P_1 & P_2 \\ \star & P_3 \end{bmatrix}, \ P_2 \triangleq \begin{bmatrix} P_4 \\ 0_{(n-k)\times k} \end{bmatrix}, \tag{12.10}$$

where $P_1 \in \mathbf{R}^{n\times n}$ and $P_3 \in \mathbf{R}^{k\times k}$ are symmetric positive definite matrices; $P_2 \in \mathbf{R}^{n\times k}$ and $P_4 \in \mathbf{R}^{k\times k}$. Without loss of generality, we assume P_4 is nonsingular. To see this, let the matrix $M \triangleq P + \alpha N$, where α is a positive scalar and

$$N \triangleq \left[\begin{array}{c|c} 0_{n\times n} & \mathscr{H} \\ \hline \star & 0_{k\times k} \end{array}\right], \ M \triangleq \begin{bmatrix} M_1 & M_2 \\ \star & M_3 \end{bmatrix}, \ M_2 \triangleq \begin{bmatrix} M_4 \\ 0_{(n-k)\times k} \end{bmatrix}.$$

Observe that since $P > 0$, we have that $M > 0$ for $\alpha > 0$ in the neighborhood of the origin. Thus, it can be easily verified that there exists an arbitrarily small $\alpha > 0$ such that M_4 is nonsingular and (12.7) is feasible with P replaced by M. Since M_4 is nonsingular, we thus conclude that there is no loss of generality to assume the matrix P_4 to be nonsingular.

Define the following matrices which are also nonsingular:

$$\mathcal{J} \triangleq \begin{bmatrix} I & 0 \\ 0 & P_3^{-1}P_4^T \end{bmatrix}, \quad \mathcal{P} \triangleq P_1, \quad \mathcal{Q} \triangleq P_4 P_3^{-1} P_4^T, \tag{12.11}$$

and

$$\begin{bmatrix} \mathcal{A} & \mathcal{B} \\ \mathcal{C} & \mathcal{D} \\ \mathcal{E} & 0 \end{bmatrix} \triangleq \begin{bmatrix} P_4 & 0 & 0 \\ 0 & I & 0 \\ 0 & 0 & P_4 \end{bmatrix} \begin{bmatrix} \hat{A} & \hat{B} \\ \hat{C} & \hat{D} \\ \hat{E} & 0 \end{bmatrix} \begin{bmatrix} P_3^{-1}P_4^T & 0 \\ 0 & I \end{bmatrix}. \tag{12.12}$$

Performing a congruence transformation to (12.7) by diag $\{\mathcal{J}, I, \mathcal{J}, I\}$, we obtain

$$\begin{bmatrix} \mathcal{J}^T \left(P\tilde{A}_i + \tilde{A}_i^T P\right) \mathcal{J} & \mathcal{J}^T P\tilde{B}_i & \mathcal{J}^T \tilde{E}_i^T P\mathcal{J} & \mathcal{J}^T \tilde{C}_i^T \\ \star & -\gamma^2 I & 0 & \tilde{D}_i^T \\ \star & \star & -\mathcal{J}^T P\mathcal{J} & 0 \\ \star & \star & \star & -I \end{bmatrix} < 0, \tag{12.13}$$

where

$$\begin{cases} \mathcal{J}^T P\tilde{A}_i\mathcal{J} \triangleq \begin{bmatrix} \mathcal{P}A_i & \mathscr{H}\mathcal{A} \\ \mathcal{Q}\mathscr{H}^T A_i & \mathcal{A} \end{bmatrix}, \\ \mathcal{J}^T P\tilde{E}_i\mathcal{J} \triangleq \begin{bmatrix} \mathcal{P}E_i & \mathscr{H}\mathcal{E} \\ \mathcal{Q}\mathscr{H}^T E_i & \mathcal{E} \end{bmatrix}, \\ \mathcal{J}^T P\tilde{B}_i \triangleq \begin{bmatrix} \mathcal{P}B_i + \mathscr{H}\mathcal{B} \\ \mathcal{Q}\mathscr{H}^T B_i + \mathcal{B} \end{bmatrix}, \\ \mathcal{J}^T P\mathcal{J} \triangleq \begin{bmatrix} \mathcal{P} & \mathscr{H}\mathcal{Q} \\ \star & \mathcal{Q} \end{bmatrix}, \\ \tilde{C}_i\mathcal{J} \triangleq \begin{bmatrix} C_i & -\mathcal{C} \end{bmatrix}, \\ \tilde{D}_i \triangleq D_i - \mathcal{D}. \end{cases} \tag{12.14}$$

Considering (12.14), we can obtain LMI (12.8) from (12.13). Moreover, notice that (12.12) is equivalent to

$$\begin{aligned} \begin{bmatrix} \hat{A} & \hat{B} \\ \hat{C} & \hat{D} \\ \hat{E} & 0 \end{bmatrix} &\triangleq \begin{bmatrix} P_4^{-1} & 0 & 0 \\ 0 & I & 0 \\ 0 & 0 & P_4^{-1} \end{bmatrix} \begin{bmatrix} \mathcal{A} & \mathcal{B} \\ \mathcal{C} & \mathcal{D} \\ \mathcal{E} & 0 \end{bmatrix} \begin{bmatrix} P_4^{-T}P_3 & 0 \\ 0 & I \end{bmatrix} \\ &= \begin{bmatrix} (P_4^{-T}P_3)^{-1}\mathcal{Q}^{-1} & 0 & 0 \\ 0 & I & 0 \\ 0 & 0 & (P_4^{-T}P_3)^{-1}\mathcal{Q}^{-1} \end{bmatrix} \begin{bmatrix} \mathcal{A} & \mathcal{B} \\ \mathcal{C} & \mathcal{D} \\ \mathcal{E} & 0 \end{bmatrix} \begin{bmatrix} P_4^{-T}P_3 & 0 \\ 0 & I \end{bmatrix}. \end{aligned}$$

Notice also that the matrices $\hat{A}$, $\hat{B}$, $\hat{C}$, $\hat{D}$ and $\hat{E}$ in (12.3) can be written as the above equation, which implies that $P_4^{-T}P_3$ can be viewed as a similarity transformation on the state-space realization of the reduced-order model and, as such, has no effect on the reduced-order model mapping from u to $\hat{y}$. Without loss of generality, we may set $P_4^{-T}P_3 = I$, thus obtain (12.9). Therefore, the reduced-order model (12.3) can be constructed by (12.9). This completes the proof. ■

Remark 12.5. Notice that the obtained conditions in Theorem 12.4 are all in LMI form, the fuzzy-rule-independent $\mathcal{H}_\infty$ model approximation can be determined by solving the following convex optimization problem:

$$\min\ \delta \quad \text{subject to (12.8)} \quad (\text{where } \delta = \gamma^2). \qquad \blacklozenge$$

When $E_i = 0$ in (12.2), that is, there is no Brownian motion, and system (12.2) becomes a common T-S fuzzy system which has the following form:

$$\dot{x}(t) = \sum_{i=1}^{r} h_i(\theta)\left[A_i x(t) + B_i u(t)\right], \tag{12.15a}$$

$$y(t) = \sum_{i=1}^{r} h_i(\theta)\left[C_i x(t) + D_i u(t)\right]. \tag{12.15b}$$

Then we have the following corollary for system (12.15).

Corollary 12.6. *Consider the T-S fuzzy system in (12.15). There exists a reduced-order model in the form of (12.3) with $\hat{E} = 0$ that solves the $\mathcal{H}_\infty$ model approximation problem with (12.6) satisfied, if there exist matrices $\mathcal{P} > 0$, $\mathcal{Q} > 0$, $\mathcal{A}$, $\mathcal{B}$, $\mathcal{C}$ and $\mathcal{D}$ such that for $i = 1, 2, \ldots, r$,*

$$\begin{bmatrix} \Pi_{11i} & \Pi_{12i} & \Pi_{13i} & C_i^T \\ \star & \Pi_{22} & \Pi_{23i} & -\mathcal{C}^T \\ \star & \star & -\gamma^2 I & \Pi_{36i} \\ \star & \star & \star & -I \end{bmatrix} < 0, \tag{12.16}$$

where the notations are defined in Theorem 12.4. Moreover, if the above conditions are feasible, then the parameters of an admissible $\mathcal{H}_\infty$ reduced-order model can be calculated from

$$\begin{bmatrix} \hat{A} & \hat{B} \\ \hat{C} & \hat{D} \end{bmatrix} = \begin{bmatrix} \mathcal{Q}^{-1} & 0 \\ 0 & I \end{bmatrix} \begin{bmatrix} \mathcal{A} & \mathcal{B} \\ \mathcal{C} & \mathcal{D} \end{bmatrix}.$$

12.3.2 Projection Approach

In what follows, based on projection lemma in Lemma 6.3, we will solve the $\mathcal{H}_\infty$ model approximation problem.

Theorem 12.7. *Consider the T-S fuzzy stochastic system in (12.2). There exists a reduced-order model in the form of (12.3) that solves the $\mathcal{H}_\infty$ model approximation problem with (12.6) satisfied, if there exist matrices $P > 0$ and $\mathscr{P} > 0$ such that for $i = 1, 2, \ldots, r$,*

$$\begin{bmatrix} H\left(\bar{A}_i \mathscr{P} + \mathscr{P} \bar{A}_i^T\right) H^T & H\bar{B}_i & H\mathscr{P}\bar{E}_i^T H^T \\ \star & -\gamma^2 I & 0 \\ \star & \star & -H\mathscr{P}H^T \end{bmatrix} < 0, \tag{12.17a}$$

$$\begin{bmatrix} H\left(P\bar{A}_i + \bar{A}_i^T P\right) H^T & H\bar{E}_i^T P & H\bar{C}_i^T \\ \star & -P & 0 \\ \star & \star & -I \end{bmatrix} < 0, \tag{12.17b}$$

$$P\mathscr{P} = I. \tag{12.17c}$$

Moreover, if the above conditions are feasible, then the system matrices of an admissible reduced-order model in the form of (12.3) can be calculated from

$$\mathcal{G} \triangleq \begin{bmatrix} \hat{D} & \hat{C} \\ \hat{B} & \hat{A} \\ 0 & \hat{E} \end{bmatrix}, \tag{12.18}$$

where

$$\begin{cases} \mathcal{G} = -\Pi^{-1}U^T \Lambda V^T \left(V\Lambda V^T\right)^{-1} + \Pi^{-1}\Xi^{1/2} L \left(V\Lambda V^T\right)^{-1/2}, \\ \Lambda = \left(U\Pi^{-1}U^T - W\right)^{-1} > 0, \\ \Xi = \Pi - U^T \left[\Lambda - \Lambda V^T \left(V\Lambda V^T\right)^{-1} V\Lambda\right] U > 0, \end{cases} \tag{12.19}$$

with Π and L are any matrices satisfying $\Pi > 0$, $\|L\| < 1$ and

$$\begin{cases} W \triangleq \begin{bmatrix} P\bar{A}_i + \bar{A}_i^T P & P\bar{B}_i & \bar{E}_i^T P & \bar{C}_i^T \\ \star & -\gamma^2 I & 0 & \bar{D}_i^T \\ \star & \star & -P & 0 \\ \star & \star & \star & -I \end{bmatrix}, \quad U \triangleq \begin{bmatrix} PX_1 \\ 0_{m\times(p+2k)} \\ PX_3 \\ X_2 \end{bmatrix}, \\ V \triangleq \begin{bmatrix} Y_1 & Y_2 & 0_{(m+k)\times(n+k)} & 0_{(m+k)\times p} \end{bmatrix}, \quad \bar{C}_i \triangleq \begin{bmatrix} C_i & 0_{p\times k} \end{bmatrix}, \\ \bar{A}_i \triangleq \begin{bmatrix} A_i & 0_{n\times k} \\ 0_{k\times n} & 0_{k\times k} \end{bmatrix}, \quad \bar{B}_i \triangleq \begin{bmatrix} B_i \\ 0_{k\times m} \end{bmatrix}, \quad \bar{E}_i \triangleq \begin{bmatrix} E_i & 0_{n\times k} \\ 0_{k\times n} & 0_{k\times k} \end{bmatrix}, \\ X_1 \triangleq \begin{bmatrix} 0_{n\times p} & 0_{n\times k} & 0_{n\times k} \\ 0_{k\times p} & I_{k\times k} & 0_{k\times k} \end{bmatrix}, \quad X_3 \triangleq \begin{bmatrix} 0_{n\times p} & 0_{n\times k} & 0_{n\times k} \\ 0_{k\times p} & 0_{k\times k} & I_{k\times k} \end{bmatrix}, \\ Y_1 \triangleq \begin{bmatrix} 0_{m\times n} & 0_{m\times k} \\ 0_{k\times n} & I_{k\times k} \end{bmatrix}, \quad Y_2 \triangleq \begin{bmatrix} I_{m\times m} \\ 0_{k\times m} \end{bmatrix}, \quad \bar{D}_i \triangleq D_i, \\ X_2 \triangleq \begin{bmatrix} -I_{p\times p} & 0_{p\times k} & 0_{p\times k} \end{bmatrix}, \quad H \triangleq \begin{bmatrix} I_{n\times n} & 0_{n\times k} \end{bmatrix}. \end{cases} \tag{12.20}$$

Proof. Rewrite $\tilde{A}_i$, $\tilde{B}_i$, $\tilde{C}_i$, $\tilde{D}_i$ and $\tilde{E}_i$ in (12.5) in the following form:

$$\begin{cases} \tilde{A}_i \triangleq \bar{A}_i + X_1\mathcal{G}Y_1, \quad \tilde{B}_i \triangleq \bar{B}_i + X_1\mathcal{G}Y_2, \\ \tilde{C}_i \triangleq \bar{C}_i + X_2\mathcal{G}Y_1, \quad \tilde{D}_i \triangleq \bar{D}_i + X_2\mathcal{G}Y_2, \quad \tilde{E}_i \triangleq \bar{E}_i + X_3\mathcal{G}Y_1, \end{cases} \tag{12.21}$$

where $\mathcal{G}$, $\bar{A}_i$, $\bar{B}_i$, $\bar{C}_i$, $\bar{D}_i$, $\bar{E}_i$ and X_j, Y_k, $j = 1, 2, 3$, $k = 1, 2$ are defined in (12.18) and (12.20). With (12.21), LMI (12.7) in Lemma 12.3 can be rewritten as

$$W + U\mathcal{G}V + (U\mathcal{G}V)^T < 0, \tag{12.22}$$

where W, U and V are defined in (12.20). We choose

$$U^{\perp} \triangleq \begin{bmatrix} HP^{-1} & 0 & 0 & 0 \\ 0 & I & 0 & 0 \\ 0 & 0 & HP^{-1} & 0 \end{bmatrix}, \quad V^{T\perp} \triangleq \begin{bmatrix} H & 0 & 0 & 0 \\ 0 & 0 & I & 0 \\ 0 & 0 & 0 & I \end{bmatrix},$$

where H is defined in (12.20). Then, by projection lemma in Lemma 6.3, inequality (12.22) is solvable for $\mathcal{G}$ if and only if

$$U^{\perp}WU^{T\perp} < 0, \quad V^{T\perp}WV^{\perp} < 0,$$

which can be formulated specifically as

$$\begin{bmatrix} H\left(\bar{A}_iP^{-1} + P^{-1}\bar{A}_i^T\right)H^T & H\bar{B}_i & HP^{-1}\bar{E}_i^TH^T \\ \star & -\gamma^2 I & 0 \\ \star & \star & -HP^{-1}H^T \end{bmatrix} < 0, \tag{12.23}$$

$$\begin{bmatrix} H\left(P\bar{A}_i + \bar{A}_i^TP\right)H^T & H\bar{E}_i^TP & H\bar{C}_i^T \\ \star & -P & 0 \\ \star & \star & -I \end{bmatrix} < 0. \tag{12.24}$$

By noting $\mathscr{P} \triangleq P^{-1}$, it follows that (12.23)–(12.24) imply respectively (12.17a)–(12.17b). In addition, when (12.17a)–(12.17c) are satisfied, the parametrization (12.18) of all reduced-order models corresponding to a feasible solution can be obtained by using the results in [62] and [92]. This completes the proof. ■

By using the projection approach, in the following, we will give the corresponding result for system (12.15). The result can be proved by following the same line as the proof of Theorem 12.7.

Corollary 12.8. *Consider the T-S fuzzy system in (12.15). There exists a reduced-order model in the form of (12.3) with $\hat{E} = 0$ that solves the $\mathcal{H}_\infty$ model approximation problem with (12.6) satisfied, if there exist matrices $P > 0$ and $\mathscr{P} > 0$ such that for $i = 1, 2, \ldots, r$,*

$$\begin{bmatrix} H\left(\bar{A}_i\mathscr{P} + \mathscr{P}\bar{A}_i^T\right)H^T & H\bar{B}_i \\ \star & -\gamma^2 I \end{bmatrix} < 0,$$

$$\begin{bmatrix} H\left(P\bar{A}_i + \bar{A}_i^TP\right)H^T & H\bar{C}_i^T \\ \star & -I \end{bmatrix} < 0,$$

$$P\mathscr{P} = I.$$

Moreover, if the above conditions are feasible, then the parameters of an admissible $\mathcal{H}_\infty$ reduced-order model can be calculated from

$$\mathcal{G} \triangleq \begin{bmatrix} \hat{D} & \hat{C} \\ \hat{B} & \hat{A} \end{bmatrix},$$

where

$$\begin{cases} \mathcal{G} = -\Pi^{-1}U^T\Lambda V^T\left(V\Lambda V^T\right)^{-1} + \Pi^{-1}\Xi^{1/2}L\left(V\Lambda V^T\right)^{-1/2}, \\ \Lambda = \left(U\Pi^{-1}U^T - W\right)^{-1} > 0, \\ \Xi = \Pi - U^T\left[\Lambda - \Lambda V^T\left(V\Lambda V^T\right)^{-1}V\Lambda\right]U > 0, \end{cases}$$

with Π and L are any matrices satisfying $\Pi > 0$, $\|L\| < 1$, and the notations are defined as in Theorem 12.7.

Notice that the solvability conditions in Theorem 12.7 are not all in LMI form due to the matrix equality (12.17c). We suggest the following minimization problem involving LMI conditions instead of the original nonconvex feasibility problem formulated in Theorem 12.7.

Problem $\mathcal{H}_\infty$-MRTSFSS ($\mathcal{H}_\infty$ Model Reduction for T-S Fuzzy Stochastic Systems):

$$\begin{aligned} \min \quad & \text{trace}\,(P\mathscr{P}) \\ \text{subject to} \quad & \text{(12.17a)–(12.17b) and} \\ & \begin{bmatrix} P & I \\ I & \mathscr{P} \end{bmatrix} \geq 0. \end{aligned} \tag{12.26}$$

We suggest the following algorithm to solve the above minimization problem.

Algorithm $\mathcal{H}_\infty$-MRTSFSS

Step 1. Find a feasible set $\left(P^{(0)}, \mathscr{P}^{(0)}\right)$ satisfying (12.17a)–(12.17b) and (12.26). Set $\kappa = 0$.

Step 2. Solve the following optimization problem:

$$\begin{aligned} \min \ & \text{trace}\left(P^{(\kappa)}\mathscr{P} + P\mathscr{P}^{(\kappa)}\right) \\ \text{subject to} \ & \text{(12.17a)–(12.17b) and (12.26)}, \end{aligned}$$

and denote f^* to be the optimized value.

Step 3. Substitute the obtained matrix variables $(P, \mathscr{P})$ into (12.23). If (12.23) is satisfied, with $|f^* - 2(n+k)| < \delta$, for a sufficiently small scalar $\delta > 0$, then output the feasible solutions $(P, \mathscr{P})$. EXIT.

Step 4. If $\kappa > \mathbb{N}$ where $\mathbb{N}$ is the maximum number of iterations allowed, EXIT.

Step 5. Set $\kappa = \kappa + 1$, $\left(P^{(\kappa)}, \mathscr{P}^{(\kappa)}\right) = (P, \mathscr{P})$, and go to Step 2.

Remark 12.9. Notice from Theorem 12.4 that to cast the considered $\mathcal{H}_\infty$ reduced-order model design into a convex optimization problem in Remark 12.5, the matrix $\mathscr{H}$, called here as an order reduction factor, plays a key role in the reduced-order model design. However, some conservativeness have been introduced due to the fact that the matrix $\mathscr{H}$ defined in Theorem 12.4 has a fixed structure, which can be seen as a main disadvantage of using the convex linearization approach to the model approximation problem. To conquer the disadvantage of the convex linearization approach, the projection approach is then used in Theorem 12.7 to solve the $\mathcal{H}_\infty$ model approximation problem. The projection approach does not need to introduce such an order reduction factor of $\mathscr{H}$, but the compromise is that the solvability conditions in Theorem 12.7 are not all in the LMI form. ♦

12.4 Illustrative Example

Example 12.10. Consider the T-S fuzzy stochastic system in (12.2) with the parameters given as follows:

$$A_1 = \begin{bmatrix} -3.2 & 0.2 & 0.4 & 0.2 \\ 0.1 & -2.2 & 0.1 & 0.3 \\ 0.4 & 0.0 & -3.4 & 0.3 \\ 0.2 & 0.3 & 0.2 & -1.8 \end{bmatrix}, \quad B_1 = \begin{bmatrix} 2.2 \\ 1.0 \\ 1.2 \\ 1.0 \end{bmatrix}, \quad E_1 = \begin{bmatrix} 0.2 & 0.1 & 0.2 & 0.0 \\ 0.0 & 0.03 & 0.1 & 0.2 \\ 0.02 & 0.1 & 0.2 & 0.0 \\ 0.1 & 0.0 & 0.1 & 0.3 \end{bmatrix},$$

$$A_2 = \begin{bmatrix} -2.2 & 0.2 & 0.0 & 0.2 \\ 0.4 & -3.5 & 0.1 & 0.6 \\ 0.1 & 0.3 & -2.0 & 0.4 \\ 0.0 & 0.2 & 0.0 & -1.5 \end{bmatrix}, \quad B_2 = \begin{bmatrix} 1.2 \\ 1.0 \\ 1.0 \\ 1.5 \end{bmatrix}, \quad E_2 = \begin{bmatrix} 0.2 & 0.1 & 0.2 & 0.0 \\ 0.0 & 0.2 & 0.0 & 0.02 \\ 0.1 & 0.0 & 0.04 & 0.1 \\ 0.0 & 0.2 & 0.0 & 0.2 \end{bmatrix},$$

$$C_1 = \begin{bmatrix} 1.0 & 1.2 & 0.8 & 0.6 \end{bmatrix}, \; C_2 = \begin{bmatrix} 0.6 & 1.0 & 0.6 & 0.8 \end{bmatrix}, \; D_1 = 0.1, \; D_2 = 0.2.$$

Here, we are interested in finding reduced-order systems (Case1: $k = 3$; Case 2: $k = 2$; Case 3: $k = 1$) in the form of (12.3) to approximate the above system in an $\mathcal{H}_\infty$ sense by using the convex linearization and the projection approaches presented in this chapter, respectively.

Firstly, we consider the convex linearization approach. Solving LMIs (12.8)–(12.9) in Theorem 12.4, we have the results for different cases are as follows:

Case 1. with $k = 3$, the minimum γ is $\gamma^* = 0.277$ and

$$\begin{bmatrix} \hat{A} & \hat{B} \\ \hline \hat{C} & \hat{D} \\ \hline \hat{E} & \end{bmatrix} = \left[\begin{array}{ccc|c} -8.1122 & -10.5155 & 12.5026 & -4.6490 \\ -5.0081 & -13.6797 & 12.9744 & -3.8166 \\ 9.5566 & 21.4011 & -25.3080 & 4.1098 \\ \hline -0.5723 & -2.2353 & -0.9852 & 0.0952 \\ \hline 0.1735 & 0.1828 & 0.1565 & \\ -0.0162 & 0.2089 & 0.1175 & \\ 0.0435 & 0.0380 & 0.2094 & \end{array}\right]. \tag{12.27}$$

Case 2. with $k = 2$, the minimum γ is $\gamma^* = 0.509$ and

$$\left[\begin{array}{c|c} \hat{A} & \hat{B} \\ \hline \hat{C} & \hat{D} \\ \hline \hat{E} & \end{array}\right] = \left[\begin{array}{cc|c} -6.7105 & 6.5227 & -1.2957 \\ 3.7870 & -7.3100 & -1.3653 \\ \hline -1.6037 & -1.9139 & 0.1260 \\ \hline 0.3824 & 0.0726 & \\ -0.0538 & 0.3938 & \end{array}\right]. \tag{12.28}$$

Case 3. with $k = 1$, the minimum γ is $\gamma^* = 0.949$ and

$$\left[\begin{array}{c|c} \hat{A} & \hat{B} \\ \hline \hat{C} & \hat{D} \\ \hline \hat{E} & \end{array}\right] = \left[\begin{array}{c|c} -2.4736 & -1.7787 \\ \hline -3.1264 & 0.0464 \\ \hline 0.4407 & \end{array}\right]. \tag{12.29}$$

Now, we will consider the projection approach. Solve the nonconvex feasibility problem of Problem $\mathcal{H}_\infty$-MRTSFSS with Algorithm $\mathcal{H}_\infty$-MRTSFSS, the obtained results for different cases are as follows:

Case 1. with $k = 3$, the minimum γ is $\gamma^* = 0.0065$ and

$$\left[\begin{array}{c|c} \hat{A} & \hat{B} \\ \hline \hat{C} & \hat{D} \\ \hline \hat{E} & \end{array}\right] = \left[\begin{array}{ccc|c} -3.4032 & -0.7695 & -0.0620 & 1.2498 \\ 0.8221 & -1.5918 & -0.3875 & 0.1512 \\ -0.6165 & -0.7075 & -2.0727 & 0.0771 \\ \hline 3.4541 & 4.6134 & -0.7422 & 0.1004 \\ \hline 0.2834 & 0.3115 & 0.3499 & \\ 0.0786 & 0.0755 & -0.4198 & \\ -0.0936 & -0.0973 & 0.2285 & \end{array}\right]. \tag{12.30}$$

Case 2. with $k = 2$, the minimum γ is $\gamma^* = 0.025$ and

$$\left[\begin{array}{c|c} \hat{A} & \hat{B} \\ \hline \hat{C} & \hat{D} \\ \hline \hat{E} & \end{array}\right] = \left[\begin{array}{cc|c} -1.4719 & 0.1319 & 1.4005 \\ -1.3926 & -2.8291 & -2.3615 \\ \hline -0.7456 & -2.4906 & 0.1091 \\ \hline 0.3507 & -0.0521 & \\ -0.4041 & 0.1298 & \end{array}\right]. \tag{12.31}$$

Case 3. with $k = 1$, the minimum γ is $\gamma^* = 0.359$ and

$$\left[\begin{array}{c|c} \hat{A} & \hat{B} \\ \hline \hat{C} & \hat{D} \\ \hline \hat{E} & \end{array}\right] = \left[\begin{array}{c|c} -1.7418 & 0.5947 \\ \hline 6.3782 & 0.2060 \\ \hline 0.3234 & \end{array}\right]. \tag{12.32}$$

The achieved γ^* for these two approaches are presented in Table 12.1, from which we can see that the projection approach is less conservative than the convex linearization approach, as stated in Remark 12.9.

Table 12.1. Achieved γ^* for two approaches

Methods	$k=3$	$k=2$	$k=1$
Convex linearization approach	$\gamma^*=0.277$	$\gamma^*=0.509$	$\gamma^*=0.949$
Projection approach	$\gamma^*=0.0065$	$\gamma^*=0.025$	$\gamma^*=0.359$

In addition, to show the model approximation performances of the obtained reduced-order models, let the initial condition be zero, that is, $\tilde{x}(0)=0$ ($x(0)=0$, $\hat{x}(0)=0$), and choose the membership functions to be

$$h_1(x_1(t)) = \exp\left(-\frac{x_1(t)-\vartheta}{2\sigma^2}\right), \quad h_2(x_1(t)) = 1-\exp\left(-\frac{x_1(t)-\vartheta}{2\sigma^2}\right),$$

which are shown in Fig. 12.1, and the exogenous input $u(t)$ is supposed be

$$u(t)=\exp(-t)\sin(t), \quad t\geq 0.$$

Fig. 12.2 shows the outputs of the original system (12.2) (solid line), the third-order reduced model (12.27) (dotted line), the second-order reduced model (12.28) (dashed line) and the first-order reduced model (12.29) (dash-dot line) due to the above input signal. The output errors between the original system and the reduced models are shown in Fig. 12.3. Fig. 12.4 shows the outputs of the original system (12.2) (solid line), the third-order reduced model (12.30) (dotted line), the second-order reduced model (12.31) (dashed line) and the first-order reduced model (12.32) (dash-dot line) due to the above input signal. The output errors between the original system and the reduced models are shown in Fig. 12.5.

Example 12.11. Consider a tunnel diode circuit shown in Fig. 12.6, and its fuzzy modeling was presented in [3], where $x_1(t)=v_C(t)$, $x_2(t)=i_L(t)$; $u(t)$ is the disturbance input, and $y(t)$ is the measurement output. Thus the system can be approximated by the following T-S fuzzy model with two fuzzy rules:

♦ **Plant Form:**

Rule 1: IF $x_1(t)$ is $\mathcal{M}_1(x_1(t))$ THEN

$$\begin{cases} \dot{x}(t) = A_1x(t)+B_1u(t), \\ y(t) = C_1x(t)+D_1u(t). \end{cases}$$

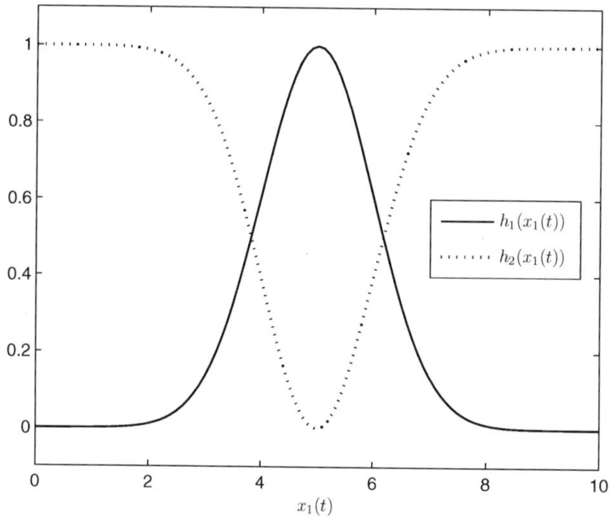

Fig. 12.1. Membership functions

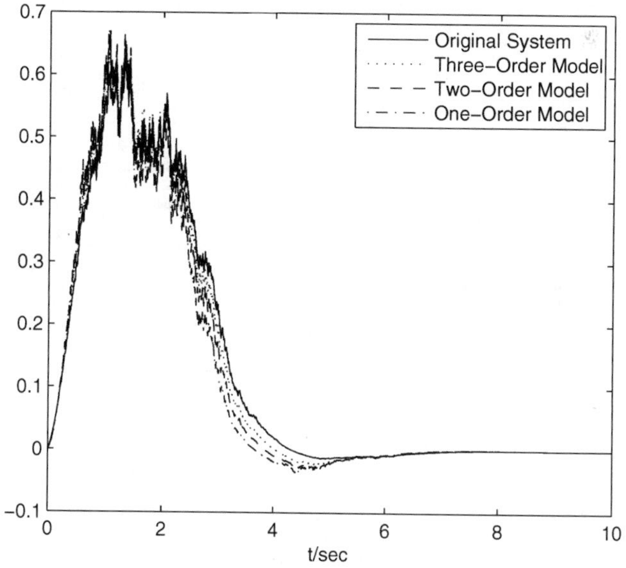

Fig. 12.2. Outputs of the original system and the reduced-order models (convex linearization approach)

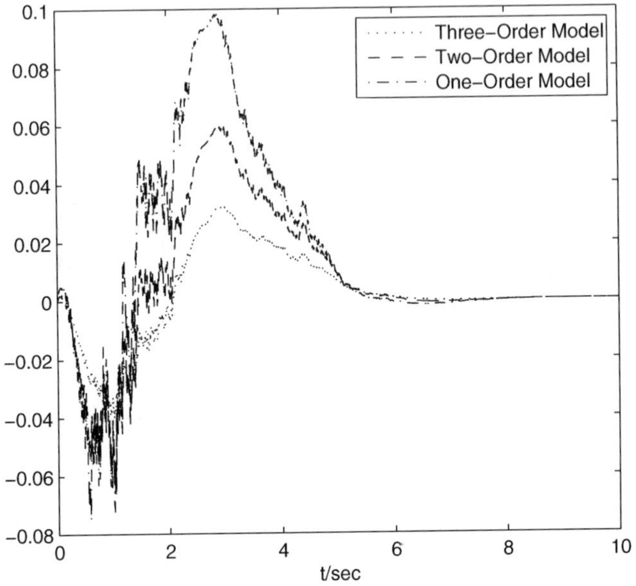

Fig. 12.3. Output errors between the original system and the reduced-order models (convex linearization approach)

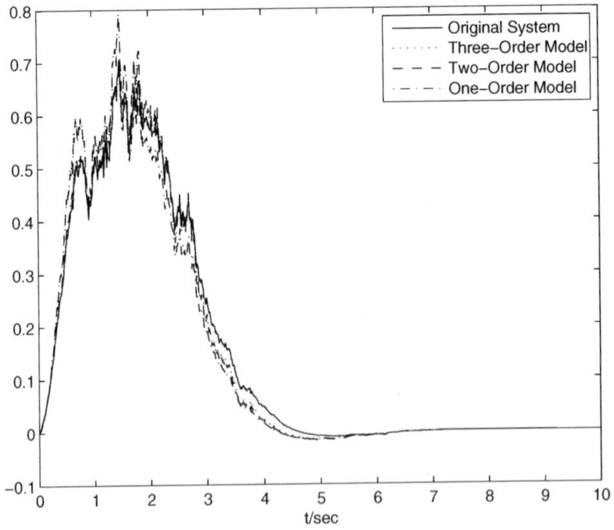

Fig. 12.4. Outputs of the original system and the reduced-order models (projection approach)

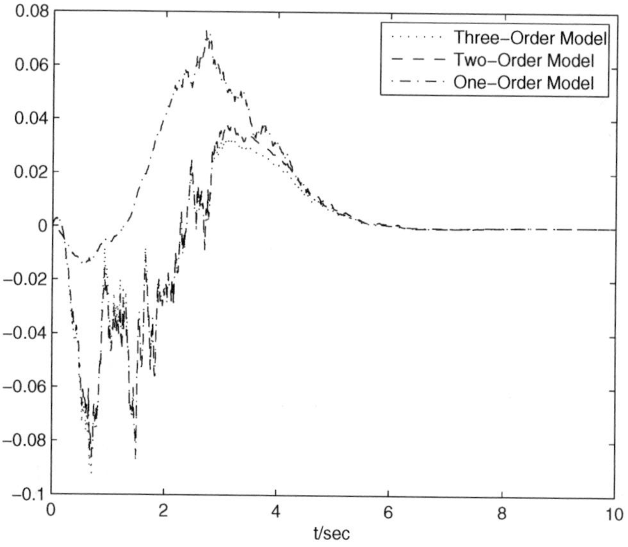

Fig. 12.5. Output errors between the original system and the reduced-order models (projection approach)

Rule 2: IF $x_1(t)$ is $\mathcal{M}_2(x_1(t))$ THEN

$$\begin{cases} \dot{x}(t) = A_2 x(t) + B_2 u(t), \\ y(t) = C_2 x(t) + D_2 u(t), \end{cases}$$

where

$$A_1 = \begin{bmatrix} -0.1 & 50 \\ -1 & -10 \end{bmatrix}, \quad B_1 = \begin{bmatrix} 0 \\ 1 \end{bmatrix}, \quad C_1 = \begin{bmatrix} 1 & 0 \end{bmatrix}, \quad D_1 = 1,$$
$$A_2 = \begin{bmatrix} -4.6 & 50 \\ -1 & -10 \end{bmatrix}, \quad B_2 = \begin{bmatrix} 0 \\ 1 \end{bmatrix}, \quad C_2 = \begin{bmatrix} 1 & 0 \end{bmatrix}, \quad D_2 = 1.$$

Fig. 12.7 shows the membership functions for Rules 1 and 2.

Here, we are interested in finding reduced-order models in the form of (12.3) with $\hat{E} = 0$, to approximate the above system in an $\mathcal{H}_\infty$ sense by using the convex linearization approach. Solve the (12.16) in Corollary 12.6, we obtain that the minimum performance level γ is $\gamma^* = 0.711$ and the parameters of the reduced-order model is given as follows:

$$\left[\begin{array}{c|c} \hat{A} & \hat{B} \\ \hline \hat{C} & \hat{D} \end{array}\right] = \left[\begin{array}{c|c} -10.7123 & -5.3724 \\ \hline -1.0008 & 1.0001 \end{array}\right].$$

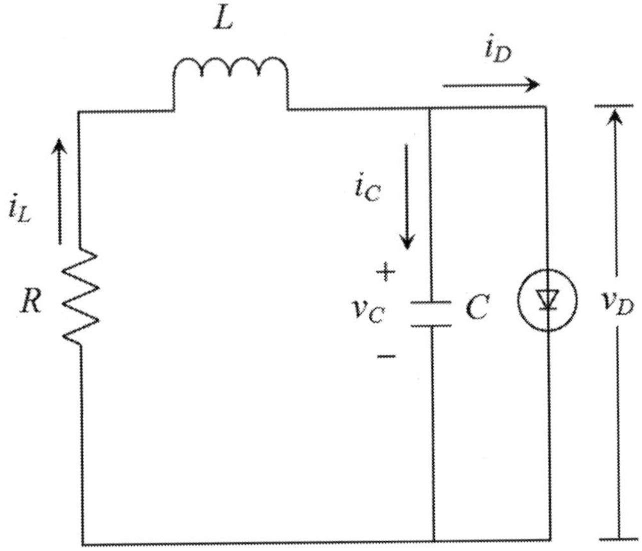

Fig. 12.6. Tunnel diode circuit

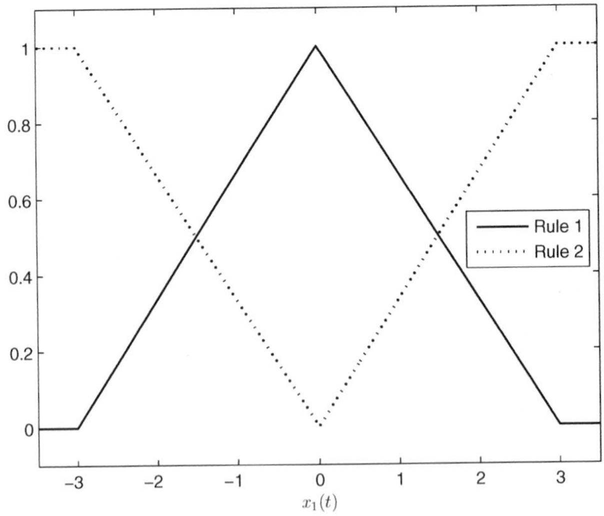

Fig. 12.7. Membership functions

In addition, to show the model approximation performances of the obtained reduced-order model, let the initial condition be zero, that is, $\tilde{x}(0) = 0$ ($x(0) = 0$ and $\hat{x}(0) = 0$), and the disturbance noise input $u(t)$ be $u(t) = \exp(-2t)\sin(3t)$, $t \geq 0$. Fig. 12.8 shows the outputs of the original system (solid line) and the above reduced-order model (dotted line), and the output error between the original system and the reduced model is shown in Fig. 12.9.

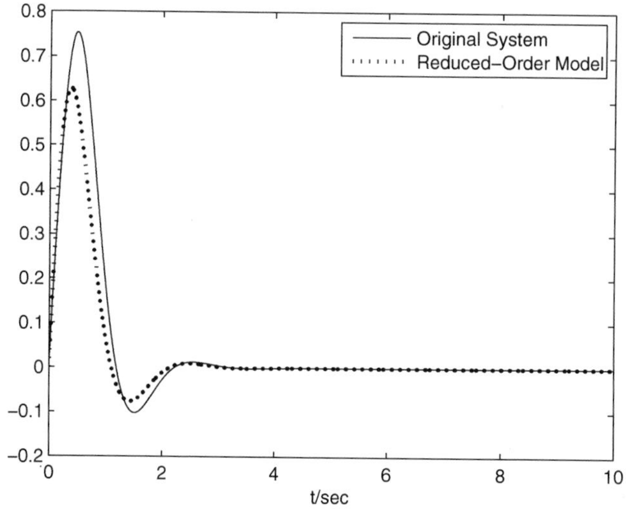

Fig. 12.8. Outputs of the original system and the reduced-order model

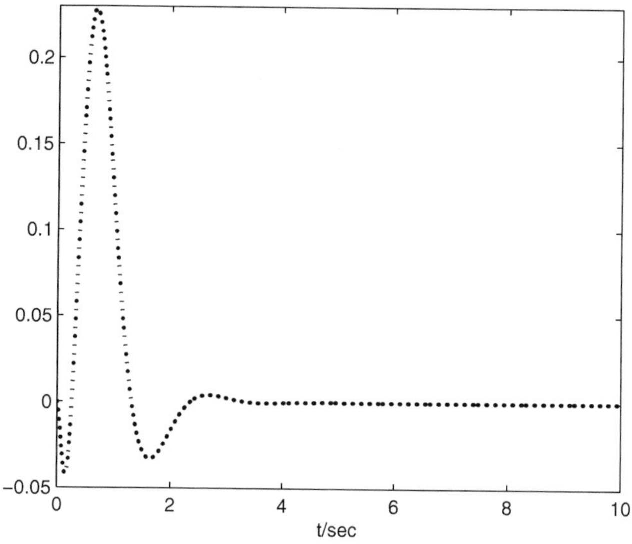

Fig. 12.9. Output error between the original system and the reduced-order model

12.5 Conclusion

In this chapter, the $\mathcal{H}_\infty$ model approximation problem has been studied for T-S fuzzy stochastic systems by using two different approaches, that is, the convex linearization approach and the projection approach. Sufficient conditions have been established to solve the reduced-order models by using these two approaches, respectively. The presented numerical examples have shown the utility of the proposed methods.

Part III
Fuzzy Control Applications

Chapter 13
Fuzzy Control of Nonlinear Electromagnetic Suspension Systems

13.1 Introduction

Owing to its environmental, commercial and technological attractions, the electromagnetic suspension system [91] has been widely adopted in many real applications. Systems, such as that in high-speed maglev passenger trains [231, 105], levitation of wind tunnel models, levitation of molten metal in induction furnaces, vibration isolation and frictionless bearings, are mostly based on electromagnetic suspension systems. Therefore, in this view, electromagnetic suspension systems can be regarded as repulsive system or attractive system which is based on the source of electromagnetic levitation forces. Due to the involvement of magnetic force, these kind of systems are mostly modeled by highly nonlinear differential equations and usually unstable, thus making it difficult when considering controller design. Over the past few years, various controller design schemes have been considered to manipulate electromagnetic suspension systems, see for example, [38, 84, 103, 182, 183, 191, 204].

In this chapter, motivated by the fact that fuzzy logic based controller design is an effective approach for the manipulation of complex nonlinear systems, we design an ℓ_2-ℓ_∞ fuzzy controller for nonlinear electromagnetic suspension systems. This controller is designed such that, in presence of energy bounded disturbance, the position offset of maglev train is within an allowable scale. Firstly, the nonlinear dynamic equations of the electromagnetic suspension system are established, and then to facilitate the controller design, a discrete-time T-S fuzzy model is constructed to describe the original nonlinear dynamic equations. Further, by using the fuzzy Lyapunov technique, sufficient ℓ_2-ℓ_∞ performance conditions are proposed in terms of LMIs, based on which the desired controller is then designed.

L. Wu et al., *Fuzzy Control Systems with Time-Delay and Stochastic Perturbation*, Studies in Systems, Decision and Control 12, DOI: 10.1007/978-3-319-11316-6_13

13.2 Modeling of Electromagnetic Suspension Systems

13.2.1 Nonlinear Dynamics

The maglev train is a kind of high-tech transportation means. In the normal working state, this kind of train is levitated by the electromagnetic suspension system. A single module of the electromagnetic suspension system is presented in Fig. 13.1.

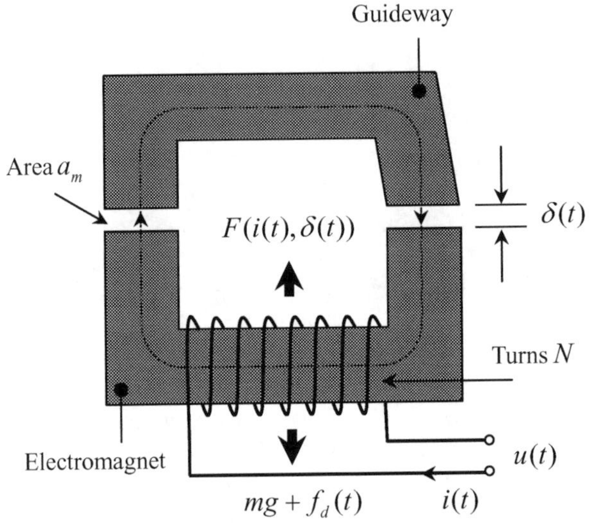

Fig. 13.1. Configuration of the electromagnetic suspension system module

Some variables of the electromagnetic suspension system in Fig. 13.1 are presented as follows:

m	gross mass of carriage and electromagnet
R_m	coil resistance
a_m	valid pole area of the coil
g	gravitational acceleration
N	number of turns in the coil
$f_d(t)$	vertical disturbance force
$\delta(t)$	suspension airgap
$i(t)$	current of the electromagnet coil
$u(t)$	voltage of the electromagnet coil

By Newton's law and Kirchhoff's law, the dynamic motion of the dynamics for the electromagnetic suspension system can be described by

$$\begin{cases} m\dfrac{d^2\delta(t)}{dt^2} = -F(i(t),\delta(t)) + f_d(t) + mg, \\ F(i(t),\delta(t)) = \dfrac{\mu_0 N^2 a_m}{4}\left(\dfrac{i(t)}{\delta(t)}\right)^2, \\ \dfrac{d\Psi(i(t),\delta(t))}{dt} = u(t) - R_m i(t), \\ \Psi(i(t),\delta(t)) = \dfrac{\mu_0 N^2 a_m i(t)}{2\delta(t)}, \end{cases} \tag{13.1}$$

where $F(i(t),\delta(t))$ is the suspension force produced by electromagnet; $\Psi(i(t),\delta(t))$ is the magnetic potential of the electromagnetic system; μ_0 is the permeability of air. By some simple manipulations, dynamics in (13.1) can be changed into the following equivalent equations

$$\begin{cases} m\dfrac{d^2\delta(t)}{dt^2} = -\dfrac{\mu_0 N^2 a_m}{4}\left(\dfrac{i(t)}{\delta(t)}\right)^2 + f_d(t) + mg, \\ \dfrac{di(t)}{dt} = \dfrac{i(t)}{\delta(t)}\dfrac{d\delta(t)}{dt} + \dfrac{2\delta(t)}{\mu_0 N^2 \alpha_m}(u(t) - R_m i(t)). \end{cases} \tag{13.2}$$

Assuming that the system states are $z_1(t) = \delta(t)$, $z_2(t) = \dot{\delta}(t)$ and $z_3(t) = i(t)$, the state-space equations in (13.2) are expressed as

$$\begin{cases} \dot{z}_1(t) = z_2(t), \\ \dot{z}_2(t) = -\dfrac{\mu_0 N^2 a_m}{4m}\left(\dfrac{z_3(t)}{z_1(t)}\right)^2 + g + \dfrac{1}{m}f_d(t), \\ \dot{z}_3(t) = \dfrac{-2R_m}{\mu_0 N^2 \alpha_m} z_1(t)z_3(t) + \dfrac{z_2(t)z_3(t)}{z_1(t)} + \dfrac{2z_1(t)}{\mu_0 N^2 \alpha_m}u(t). \end{cases} \tag{13.3}$$

Since the electromagnet system is required to be kept at the stable position $z_{1e} = \delta_{ref}$, the corresponding equilibrium of the system will be $z_e \triangleq (z_{1e}, z_{2e}, z_{3e})^T = (\delta_{ref}, 0, \sqrt{\kappa}\delta_{ref})^T$ and $u_e = \sqrt{\kappa}R_m\delta_{ref}$, where $\kappa \triangleq \frac{4mg}{\mu_0 N^2 a_m}$. For convenience of the following analysis, we consider the following change of coordinates

$$x(t) \triangleq \begin{bmatrix} x_1(t) \\ x_2(t) \\ x_3(t) \end{bmatrix} = \begin{bmatrix} z_1(t) - z_{1e} \\ z_2(t) \\ z_3(t) - z_{3e} \end{bmatrix}. \tag{13.4}$$

With $v(t) = u(t) - u_e$ and the above transformation of coordinates, it is easy to obtain the following equivalent state-space equations for the system in (13.3),

$$\begin{cases} \dot{x}_1(t) = x_2(t), \\ \dot{x}_2(t) = \dfrac{(2\sqrt{\kappa}\delta_{ref} + x_3(t))gx_3(t)}{\kappa(x_1(t) + \delta_{ref})^2} - \dfrac{(2\delta_{ref} + x_1(t))gx_1(t)}{(x_1(t) + \delta_{ref})^2} + \dfrac{1}{m}f_d(t), \\ \dot{x}_3(t) = \dfrac{x_3(t) + \sqrt{\kappa}\delta_{ref}}{x_1(t) + \delta_{ref}}x_2(t) - \dfrac{\kappa(x_1(t) + \delta_{ref})}{2mg}(R_m x_3(t) - v(t)). \end{cases} \tag{13.5}$$

Obviously, the above state-space equations are highly complex and nonlinear, thus the conventional linear control design scheme is not applicable to regulate such a nonlinear system. To facilitate the controller design, in this chapter, we will adopt the model-based fuzzy control scheme.

13.2.2 T-S Fuzzy Modeling

Before the design of fuzzy controller, firstly we should get the T-S fuzzy model of original nonlinear system. Let's define $\theta_1(t) \triangleq x_1(t)$ and $\theta_2(t) \triangleq x_3(t)$. With the definition

$$\begin{cases} \theta_{1\max} \triangleq \max\limits_{x(t)} \theta_1(t), & \theta_{1\min} \triangleq \min\limits_{x(t)} \theta_1(t), \\ \theta_{2\max} \triangleq \max\limits_{x(t)} \theta_2(t), & \theta_{2\min} \triangleq \min\limits_{x(t)} \theta_2(t), \end{cases} \tag{13.6}$$

where $\theta_1(t)$ and $\theta_2(t)$ can be expressed by

$$\begin{cases} \theta_1(t) = \xi_{11}(\theta_1(t))\theta_{1\max} + \xi_{21}(\theta_1(t))\theta_{1\min}, \\ \theta_2(t) = \xi_{12}(\theta_2(t))\theta_{2\max} + \xi_{22}(\theta_2(t))\theta_{2\min}. \end{cases} \tag{13.7}$$

where $\xi_{ij}(\theta_j(t)) \in [0, 1]$ for all $(i = 1, 2; j = 1, 2)$, and

$$\begin{cases} \xi_{11}(\theta_1(t)) + \xi_{21}(\theta_1(t)) = 1, \\ \xi_{12}(\theta_2(t)) + \xi_{22}(\theta_2(t)) = 1. \end{cases} \tag{13.8}$$

Then, based on (13.6)–(13.8), the membership functions can be calculated as

$$\begin{cases} \xi_{11}(\theta_1(t)) = \dfrac{\theta_1(t) - \theta_{1\min}}{\theta_{1\max} - \theta_{1\min}}, \\ \xi_{12}(\theta_2(t)) = \dfrac{\theta_2(t) - \theta_{2\min}}{\theta_{2\max} - \theta_{2\min}}, \\ \xi_{21}(\theta_1(t)) = \dfrac{\theta_{1\max} - \theta_1(t)}{\theta_{1\max} - \theta_{1\min}}, \\ \xi_{22}(\theta_2(t)) = \dfrac{\theta_{2\max} - \theta_2(t)}{\theta_{2\max} - \theta_{2\min}}, \end{cases} \tag{13.9}$$

which have been shown in Figs. 13.2 and 13.3.

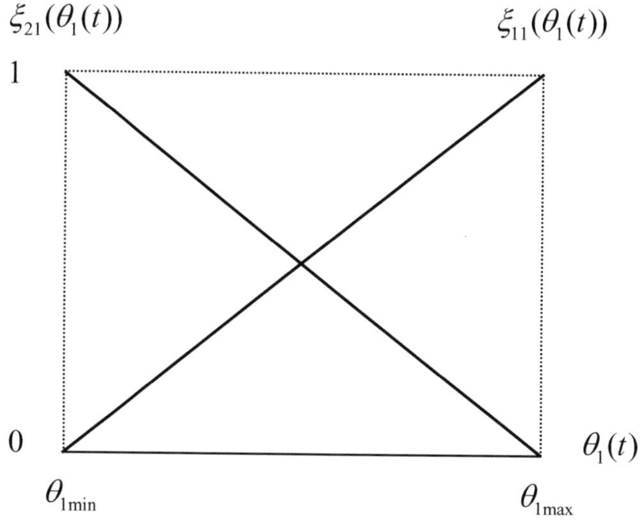

Fig. 13.2. $\xi_{11}(\theta_1(t))$ and $\xi_{21}(\theta_1(t))$

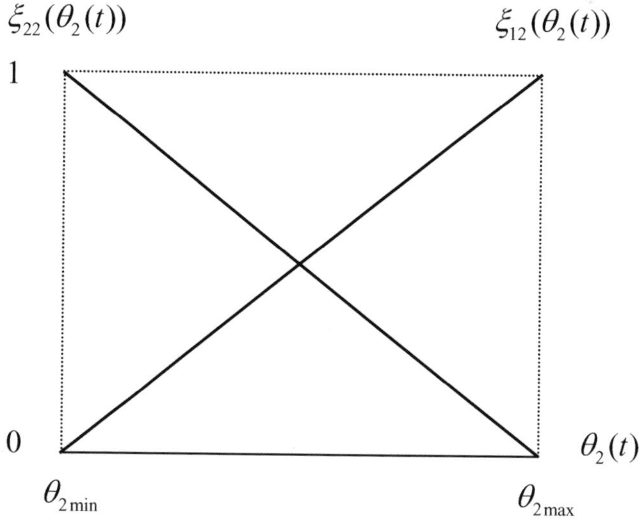

Fig. 13.3. $\xi_{12}(\theta_2(t))$ and $\xi_{22}(\theta_2(t))$

In this work, $f_d(t)$ is regarded as the exogenous noise input, and we are interested in the influence of $f_d(t)$ on $\delta(t)$. Thus, the output of this system is chosen as $y(t) = x_1(t)$. By using the local approximation in fuzzy partition spaces [194], the corresponding four-rules fuzzy model can be obtained for the electromagnetic suspension system, and they can be formulated as follows:

1) When $\theta_1(t)$ is near $\theta_{1\max}$ and $\theta_2(t)$ is near $\theta_{2\max}$, the nonlinear equations can be simplified as

$$\begin{cases} \dot{x}_1(t) = x_2(t), \\ \dot{x}_2(t) = \dfrac{(2\sqrt{\kappa}\delta_{ref} + \theta_{2\max})gx_3(t)}{\kappa(\theta_{1\max} + \delta_{ref})^2} - \dfrac{(2\delta_{ref} + \theta_{1\max})gx_1(t)}{(\theta_{1\max} + \delta_{ref})^2} + \dfrac{1}{m}f_d(t), \\ \dot{x}_3(t) = \dfrac{\theta_{2\max} + \sqrt{\kappa}\delta_{ref}}{\theta_{1\max} + \delta_{ref}}x_2(t) - \dfrac{\kappa(\theta_{1\max} + \delta_{ref})}{2mg}(R_m x_3(t) - v(t)), \\ y(t) = x_1(t). \end{cases}$$

2) When $\theta_1(t)$ is near $\theta_{1\max}$ and $\theta_2(t)$ is near $\theta_{2\min}$, the nonlinear equations can be simplified as

$$\begin{cases} \dot{x}_1(t) = x_2(t), \\ \dot{x}_2(t) = \dfrac{(2\sqrt{\kappa}\delta_{ref} + \theta_{2\min})gx_3(t)}{\kappa(\theta_{1\max} + \delta_{ref})^2} - \dfrac{(2\delta_{ref} + \theta_{1\max})gx_1(t)}{(\theta_{1\max} + \delta_{ref})^2} + \dfrac{1}{m}f_d(t), \\ \dot{x}_3(t) = \dfrac{\theta_{2\min} + \sqrt{\kappa}\delta_{ref}}{\theta_{1\max} + \delta_{ref}}x_2(t) - \dfrac{\kappa(\theta_{1\max} + \delta_{ref})}{2mg}(R_m x_3(t) - v(t)), \\ y(t) = x_1(t). \end{cases}$$

3) When $\theta_1(t)$ is near $\theta_{1\min}$ and $\theta_2(t)$ is near $\theta_{2\max}$, the nonlinear equations can be simplified as

$$\begin{cases} \dot{x}_1(t) = x_2(t), \\ \dot{x}_2(t) = \dfrac{(2\sqrt{\kappa}\delta_{ref} + \theta_{2\max})gx_3(t)}{\kappa(\theta_{1\min} + \delta_{ref})^2} - \dfrac{(2\delta_{ref} + \theta_{1\min})gx_1(t)}{(\theta_{1\min} + \delta_{ref})^2} + \dfrac{1}{m}f_d(t), \\ \dot{x}_3(t) = \dfrac{\theta_{2\max} + \sqrt{\kappa}\delta_{ref}}{\theta_{1\min} + \delta_{ref}}x_2(t) - \dfrac{\kappa(\theta_{1\min} + \delta_{ref})}{2mg}(R_m x_3(t) - v(t)), \\ y(t) = x_1(t). \end{cases}$$

4) When $\theta_1(t)$ is near $\theta_{1\min}$ and $\theta_2(t)$ is near $\theta_{2\min}$, the nonlinear equations can be simplified as

$$\begin{cases} \dot{x}_1(t) = x_2(t), \\ \dot{x}_2(t) = \dfrac{(2\sqrt{\kappa}\delta_{ref} + \theta_{2\min})gx_3(t)}{\kappa(\theta_{1\min} + \delta_{ref})^2} - \dfrac{(2\delta_{ref} + \theta_{1\min})gx_1(t)}{(\theta_{1\min} + \delta_{ref})^2} + \dfrac{1}{m}f_d(t), \\ \dot{x}_3(t) = \dfrac{\theta_{2\min} + \sqrt{\kappa}\delta_{ref}}{\theta_{1\min} + \delta_{ref}}x_2(t) - \dfrac{\kappa(\theta_{1\min} + \delta_{ref})}{2mg}(R_m x_3(t) - v(t)), \\ y(t) = x_1(t). \end{cases}$$

Then, employing the Euler first-order approximation, we obtain the following discrete-time T-S fuzzy model.

♦ **Plant Form:**

Rule 1: IF $\theta_1(k)$ is $\mathcal{M}_{11}$ and $\theta_2(k)$ is $\mathcal{M}_{12}$, THEN

$$\begin{cases} x(k+1) = A_1 x(k) + B_{11}\omega(k) + B_{21}v(k), \\ y(k) = C_1 x(k). \end{cases}$$

Rule 2: IF $\theta_1(k)$ is $\mathcal{M}_{21}$ and $\theta_2(k)$ is $\mathcal{M}_{22}$, THEN

$$\begin{cases} x(k+1) = A_2 x(k) + B_{12}\omega(k) + B_{22}v(k), \\ y(k) = C_2 x(k). \end{cases}$$

Rule 3: IF $\theta_1(k)$ is $\mathcal{M}_{31}$ and $\theta_2(k)$ is $\mathcal{M}_{32}$, THEN

$$\begin{cases} x(k+1) = A_3 x(k) + B_{13}\omega(k) + B_{23}v(k), \\ y(k) = C_3 x(k). \end{cases}$$

Rule 4: IF $\theta_1(k)$ is $\mathcal{M}_{41}$ and $\theta_2(k)$ is $\mathcal{M}_{42}$, THEN

$$\begin{cases} x(k+1) = A_4 x(k) + B_{14}\omega(k) + B_{24}v(k), \\ y(k) = C_4 x(k). \end{cases}$$

where $w(k)$ is an alternative description of $f_d(t)$ in discrete-time domain; $\mathcal{M}_{11}$ and $\mathcal{M}_{21}$ represent "about $\theta_{1\max}$"; $\mathcal{M}_{31}$ and $\mathcal{M}_{41}$ represent "about $\theta_{1\min}$"; $\mathcal{M}_{12}$ and $\mathcal{M}_{32}$ represent "about $\theta_{2\max}$"; $\mathcal{M}_{22}$ and $\mathcal{M}_{42}$ represent "about $\theta_{2\min}$". Correspondingly, their membership functions are as follows

$$\begin{aligned} \mathcal{M}_{11}(\theta_1(k)) &= \mathcal{M}_{21}(\theta_1(k)) = \xi_{11}(\theta_1(k)), \\ \mathcal{M}_{31}(\theta_1(k)) &= \mathcal{M}_{41}(\theta_1(k)) = \xi_{21}(\theta_1(k)), \\ \mathcal{M}_{12}(\theta_2(k)) &= \mathcal{M}_{32}(\theta_2(k)) = \xi_{12}(\theta_2(k)), \\ \mathcal{M}_{22}(\theta_2(k)) &= \mathcal{M}_{42}(\theta_2(k)) = \xi_{22}(\theta_2(k)). \end{aligned}$$

The system matrices are expressed as

$$A_1 = \begin{bmatrix} 1 & T & 0 \\ -\frac{T(2\delta_{ref}+\theta_{1\max})g}{(\theta_{1\max}+\delta_{ref})^2} & 1 & \frac{T(2\sqrt{\kappa}\delta_{ref}+\theta_{2\max})g}{\kappa(\theta_{1\max}+\delta_{ref})^2} \\ 0 & \frac{T(\theta_{2\max}+\sqrt{\kappa}\delta_{ref})}{\theta_{1\max}+\delta_{ref}} & 1-\frac{\kappa T R_m(\theta_{1\max}+\delta_{ref})}{2mg} \end{bmatrix},$$

$$A_2 = \begin{bmatrix} 1 & T & 0 \\ -\frac{T(2\delta_{ref}+\theta_{1\max})g}{(\theta_{1\max}+\delta_{ref})^2} & 1 & \frac{T(2\sqrt{\kappa}\delta_{ref}+\theta_{2\min})g}{\kappa(\theta_{1\max}+\delta_{ref})^2} \\ 0 & \frac{T(\theta_{2\min}+\sqrt{\kappa}\delta_{ref})}{\theta_{1\max}+\delta_{ref}} & 1-\frac{\kappa T R_m(\theta_{1\max}+\delta_{ref})}{2mg} \end{bmatrix},$$

$$A_3 = \begin{bmatrix} 1 & T & 0 \\ -\frac{T(2\delta_{ref}+\theta_{1\min})g}{(\theta_{1\min}+\delta_{ref})^2} & 1 & \frac{T(2\sqrt{\kappa}\delta_{ref}+\theta_{2\max})g}{\kappa(\theta_{1\min}+\delta_{ref})^2} \\ 0 & \frac{T(\theta_{2\max}+\sqrt{\kappa}\delta_{ref})}{\theta_{1\min}+\delta_{ref}} & 1-\frac{\kappa T R_m(\theta_{1\min}+\delta_{ref})}{2mg} \end{bmatrix},$$

$$A_4 = \begin{bmatrix} 1 & T & 0 \\ -\frac{T(2\delta_{ref}+\theta_{1\min})g}{(\theta_{1\min}+\delta_{ref})^2} & 1 & \frac{T(2\sqrt{\kappa}\delta_{ref}+\theta_{2\min})g}{\kappa(\theta_{1\min}+\delta_{ref})^2} \\ 0 & \frac{T(\theta_{2\min}+\sqrt{\kappa}\delta_{ref})}{\theta_{1\min}+\delta_{ref}} & 1-\frac{\kappa T R_m(\theta_{1\min}+\delta_{ref})}{2mg} \end{bmatrix},$$

$$B_{11} = B_{12} = B_{13} = B_{14} = \begin{bmatrix} 0 \\ \frac{T}{m} \\ 0 \end{bmatrix}, \quad C_1 = C_2 = C_3 = C_4 = \begin{bmatrix} 1 & 0 & 0 \end{bmatrix},$$

$$B_{21} = B_{22} = \begin{bmatrix} 0 \\ 0 \\ \frac{\kappa T(\theta_{1\max}+\delta_{ref})}{2mg} \end{bmatrix}, \quad B_{23} = B_{24} = \begin{bmatrix} 0 \\ 0 \\ \frac{\kappa T(\theta_{1\min}+\delta_{ref})}{2mg} \end{bmatrix},$$

with T the sampling time. Noting that,

$$h_i(\theta(k)) \triangleq \frac{\prod_{j=1}^{2} \mathcal{M}_{ij}(\theta_j(k))}{\sum_{i=1}^{4}\prod_{j=1}^{2} \mathcal{M}_{ij}(\theta_j(k))}, \quad i = 1, 2, 3, 4,$$

we will further get the following fuzzy basis functions,

$$\begin{aligned} h_1(\theta(k)) &= \xi_{11}(\theta_1(k))\xi_{12}(\theta_2(k)), \\ h_2(\theta(k)) &= \xi_{11}(\theta_1(k))\xi_{22}(\theta_2(k)), \\ h_3(\theta(k)) &= \xi_{21}(\theta_1(k))\xi_{12}(\theta_2(k)), \\ h_4(\theta(k)) &= \xi_{21}(\theta_1(k))\xi_{22}(\theta_2(k)). \end{aligned}$$

It is assumed that the premise variables do not depend on the input variable $v(k)$ explicitly. Then, the defuzzified T-S fuzzy model can be given by

$$x(k+1) = \sum_{i=1}^{4} h_i(\theta(k)) \left[A_i x(k) + B_{1i}\omega(k) + B_{2i}v(k) \right],$$
$$y(k) = \sum_{i=1}^{4} h_i(\theta(k)) C_i x(k).$$

A more compact presentation of the T-S fuzzy model is given by

$$x(k+1) = \bar{A}(k)x(k) + \bar{B}_1(k)\omega(k) + \bar{B}_2(k)v(k), \tag{13.10a}$$
$$y(k) = \bar{C}(k)x(k), \tag{13.10b}$$

where

$$\bar{A}(k) \triangleq \sum_{i=1}^{4} h_i(\theta(k))A_i, \quad \bar{B}_1(k) \triangleq \sum_{i=1}^{4} h_i(\theta(k))B_{1i},$$
$$\bar{C}(k) \triangleq \sum_{i=1}^{4} h_i(\theta(k))C_i, \quad \bar{B}_2(k) \triangleq \sum_{i=1}^{4} h_i(\theta(k))B_{2i}.$$

13.3 Fuzzy Control

Assume that the premise variable of the fuzzy model $\theta(k)$ is available for feedback, which implies that $h_i(\theta(k))$ is available for feedback. Suppose that the controller's premise variables are the same as those in the plant. The PDC strategy is utilized and the fuzzy state feedback controller obeys the following rules:

♦ **Controller Form:**

Rule i: IF $\theta_1(k)$ is $\mathcal{M}_{i1}$ and $\theta_2(k)$ is $\mathcal{M}_{i2}$, THEN

$$v(k) = K_i x(k), \quad i = 1, 2, 3, 4, \tag{13.11}$$

where K_i is the gain matrix of the state-feedback controller. Thus, the controller in (13.11) can also be represented by

$$v(k) = \sum_{i=1}^{r} h_i(\theta(k))K_i x(k),$$

with its compact form as

$$v(k) = \bar{K}(k)x(k), \tag{13.12}$$

where

$$\bar{K}(k) = \sum_{i=1}^{4} h_i(\theta(k))K_i. \tag{13.13}$$

Therefore, the closed-loop system is obtained as

$$x(k+1) = \sum_{i=1}^{4}\sum_{j=1}^{4} h_i(\theta(k))\left[(A_i + h_j(\theta(k))B_{2i}K_j)\,x(k) + B_{1i}\omega(k)\right],$$
$$y(k) = \sum_{i=1}^{4} h_i(\theta(k))C_i x(k),$$

and its compact form can be expressed as

$$x(k+1) = \hat{A}(k)x(k) + \bar{B}_1(k)\omega(k), \tag{13.14a}$$
$$y(k) = \bar{C}(k)x(k), \tag{13.14b}$$

where

$$\hat{A}(k) \triangleq \bar{A}(k) + \bar{B}_2(k)\bar{K}(k).$$

Definition 13.1. The closed-loop system in (13.14) is said to be asymptotically stable if under $\omega(k) = 0$,

$$\lim_{k\to\infty} |x(k)| = 0,$$

where $|x(k)| \triangleq \sqrt{x^T(k)x(k)}$.

Definition 13.2. Given a scalar $\gamma > 0$, the closed-loop system in (13.14) is said to be asymptotically stable with an ℓ_2-ℓ_∞ performance γ, if it is asymptotically stable under $\omega(k) = 0$, and satisfies

$$\|y(k)\|_\infty < \gamma\|\omega(k)\|_2, \quad \forall\, 0 \neq \omega(k) \in \ell_2[0,\infty),$$

where

$$\|y(k)\|_\infty \triangleq \sup_k \sqrt{y^T(k)y(k)}, \quad \|\omega(k)\|_2 \triangleq \sqrt{\sum_{k=0}^{\infty} \omega^T(k)\omega(k)}.$$

The main objective in this work is to design fuzzy controller (13.12) such that the closed-loop system in (13.14) is asymptotically stable with a guaranteed ℓ_2-ℓ_∞ performance level γ.

13.3.1 Performance Analysis

In this section, a new Lyapunov function for the electromagnetic suspension systems is introduced. Based on this, a new stability criterion with the ℓ_2-ℓ_∞ performance level γ is derived. To this end, let

$$\begin{cases} \hat{P}(k) \triangleq \bar{G}^{-T}(k)\bar{P}(k)\bar{G}^{-1}(k), \\ \bar{P}(k) \triangleq \sum_{i=1}^{4} h_i(\theta(k))P_i, \\ \bar{G}(k) \triangleq \sum_{i=1}^{4} h_i(\theta(k))G_i, \end{cases} \tag{13.15}$$

where $P_i > 0$ and G_i, $i = 1, 2, 3, 4$ are $n \times n$ matrices. We construct the following fuzzy Lyapunov function:

$$V(k) \triangleq x^T(k)\hat{P}(k)x(k). \tag{13.16}$$

Then, based on the above fuzzy Lyapunov, we can obtain the following result.

Theorem 13.3. *The closed-loop system in (13.14) is asymptotically stable with a guaranteed ℓ_2-ℓ_∞ performance level γ if there exist matrices $P_i > 0$ and G_i, $i = 1, 2, 3, 4$, which are defined in (13.15), such that for any integer k, the following matrix inequalities hold:*

$$\begin{bmatrix} -\bar{G}(k+1) - \bar{G}^T(k+1) + \bar{P}(k+1) & \hat{A}(k)\bar{G}(k) & \bar{B}_1(k) \\ \star & -\bar{P}(k) & 0 \\ \star & \star & -I \end{bmatrix} < 0, \tag{13.17a}$$

$$\begin{bmatrix} \gamma^2 I & \bar{C}(k)\bar{G}(k) \\ \star & \bar{P}(k) \end{bmatrix} > 0. \tag{13.17b}$$

Proof. From the fact that

$$\left[\bar{P}(k+1) - \bar{G}(k+1)\right] \bar{P}^{-1}(k+1) \left[\bar{P}(k+1) - \bar{G}(k+1)\right]^T \geq 0,$$

we have

$$-\hat{P}^{-1}(k+1) \leq -\bar{G}(k+1) - \bar{G}^T(k+1) + \bar{P}(k+1). \tag{13.18}$$

Thus, it follows from (13.17a) and (13.18) that

$$\begin{bmatrix} -\hat{P}^{-1}(k+1) & \hat{A}(k)\bar{G}(k) & \bar{B}_1(k) \\ * & -\bar{P}(k) & 0 \\ * & * & -I \end{bmatrix} < 0. \tag{13.19}$$

Define the following matrices:

$$\mathcal{T}(k) \triangleq \text{diag}\left\{I, \bar{G}^{-1}(k), I\right\},$$
$$\mathcal{S}(k) \triangleq \text{diag}\left\{I, \bar{G}^{-1}(k)\right\}.$$

Performing congruence transformations to (13.17b) and (13.19) by matrices $\mathcal{S}(k)$ and $\mathcal{T}(k)$, respectively, and considering (13.15), we have

$$\begin{bmatrix} -\hat{P}^{-1}(k+1) & \hat{A}(k) & \bar{B}_1(k) \\ \star & -\hat{P}(k) & 0 \\ \star & \star & -I \end{bmatrix} < 0, \tag{13.20}$$

$$\begin{bmatrix} \gamma^2 I & \bar{C}(k) \\ \star & \hat{P}(k) \end{bmatrix} > 0. \tag{13.21}$$

By Schur complement, it follows that (13.20) and (13.21) are equivalent respectively to

$$\begin{bmatrix} \hat{A}^T(k)\hat{P}(k+1)\hat{A}(k) - \hat{P}(k) & \hat{A}^T(k)\hat{P}(k+1)\bar{B}_1(k) \\ \star & \bar{B}_1^T(k)\hat{P}(k+1)\bar{B}_1(k) - I \end{bmatrix} < 0, \tag{13.22}$$

$$\bar{C}^T(k)\bar{C}(k) - \gamma^2\hat{P}(k) < 0. \tag{13.23}$$

Considering the fuzzy Lyapunov function in (13.16), and along the trajectories of the closed-loop system in (13.14), we have

$$\begin{aligned} \Delta V(k) &\triangleq V(k+1) - V(k) \\ &= x^T(k+1)\hat{P}(k+1)x(k+1) - x^T(k)\hat{P}(k)x(k) \\ &= \begin{bmatrix} x(k) \\ \omega(k) \end{bmatrix}^T \Upsilon(k) \begin{bmatrix} x(k) \\ \omega(k) \end{bmatrix}, \end{aligned} \tag{13.24}$$

where

$$\Upsilon(k) \triangleq \begin{bmatrix} \hat{A}^T(k)\hat{P}(k+1)\hat{A}(k) - \hat{P}(k) & \hat{A}^T(k)\hat{P}(k+1)\bar{B}_1(k) \\ \star & \bar{B}_1^T(k)\hat{P}(k+1)\bar{B}_1(k) \end{bmatrix}.$$

By (13.22) and (13.24), it follows that $\Delta V(k) < 0$, thus we can conclude that the closed-loop system in (13.14) is asymptotically stable.

In the following, we investigate the ℓ_2-ℓ_∞ performance. Supposing the initial condition be zero, we have $V(0) = 0$. Considering the following index:

$$\mathcal{J}(k) \triangleq V(k) - \sum_{i=0}^{k-1} \omega^T(i)\omega(i).$$

For any nonzero $\omega(k) \in \ell_2[0, \infty)$ and $k > 0$, we have

$$\mathcal{J}(k) = V(k) - V(0) - \sum_{i=0}^{k-1} \omega^T(i)\omega(i)$$

$$= \sum_{i=0}^{k-1} \left[\Delta V(i) - \omega^T(i)\omega(i)\right] = \sum_{i=0}^{k-1} \begin{bmatrix} x(i) \\ \omega(i) \end{bmatrix}^T \Omega(i) \begin{bmatrix} x(i) \\ \omega(i) \end{bmatrix}.$$

where

$$\Omega(i) \triangleq \begin{bmatrix} \hat{A}^T(i)\hat{P}(i+1)\hat{A}(i) - \hat{P}(i) & \hat{A}^T(i)\hat{P}(i+1)\bar{B}_1(i) \\ \star & \bar{B}_1^T(i)\hat{P}(i+1)\bar{B}_1(i) - I \end{bmatrix}.$$

By (13.22), we have $\mathcal{J}(k) < 0$. It further gives rise to

$$x^T(k)\hat{P}(k)x(k) < \sum_{i=0}^{k-1} \omega^T(i)\omega(i). \tag{13.25}$$

Then, based on (13.23) we have

$$y^T(k)y(k) < \gamma^2 x^T(k)\hat{P}(k)x(k). \tag{13.26}$$

Combining (13.25) and (13.26) together, we have

$$y^T(k)y(k) < \gamma^2 \sum_{i=0}^{k-1} \omega^T(i)\omega(i).$$

Taking the supremun of $y^T(k)y(k)$ over k and the limit of $\sum_{i=0}^{k-1} \omega^T(i)\omega(i)$ with $k \to \infty$, we obtain

$$\sup_k\{y^T(k)y(k)\} < \gamma^2 \sum_{i=0}^{\infty} \omega^T(i)\omega(i).$$

Obviously, $\|y(k)\|_\infty < \gamma\|\omega(k)\|_2$ holds, thus the proof is completed. ■

13.3.2 Fuzzy Controller Design

In this section, we present a solution to the controller design based on Theorem 13.3. As [82], the following non-PDC controller is considered:

$$v(k) = \bar{K}(k)\bar{G}^{-1}(k)x(k), \tag{13.27}$$

where $\bar{K}(k)$ and $\bar{G}(k)$ are defined in (13.13) and (13.15), respectively. If we take $G_i = G$ then (13.27) becomes a PDC controller. Substituting (13.27) into (13.10a), the resulted closed-loop system can be formulated by

$$x(k+1) = \left(\bar{A}(k) + \bar{B}_2(k)\bar{K}(k)\bar{G}^{-1}(k)\right) x(k) + \bar{B}_1(k)\omega(k). \tag{13.28}$$

Theorem 13.4. *The closed-loop system in (13.28) is asymptotically stable with an ℓ_2-ℓ_∞ performance if there exist matrices $P_i > 0$ and $G_i (i = 1, 2, 3, 4)$ such that (13.17b) and for any integer k,*

$$\begin{bmatrix} \bar{P}(k+1)-\bar{G}(k+1)-\bar{G}^T(k+1) & \bar{A}(k)\bar{G}(k)+\bar{B}_2(k)\bar{K}(k) & \bar{B}_1(k) \\ \star & -\bar{P}(k) & 0 \\ \star & \star & -I \end{bmatrix} < 0. \quad (13.29)$$

where $\bar{P}(k)$ and $\bar{G}(k)$ are defined in (13.15).

Proof. In the view of the fuzzy closed-loop system (13.28), we replace $\hat{A}(k)$ in (13.14) with $\left(\bar{A}(k) + \bar{B}_2(k)\bar{K}(k)\bar{G}^{-1}(k)\right)$. Following the same line as the proof of Theorem 13.3, the result can be easily derived. ■

Note that the condition in Theorem 13.4 is expressed in the form of fuzzy-basis-dependent matrix inequalities, which cannot be directly implemented for the fuzzy controller design. Our next objective is to convert the above matrix inequalities into a set of LMIs.

Theorem 13.5. *The closed-loop system in (13.14) is asymptotically stable with an ℓ_2-ℓ_∞ performance if there exist matrices $P_i > 0$, G_i and K_i, $i = 1, 2, 3, 4$, which are defined in (13.15), such that for any $s, i, j = 1, 2, 3, 4$,*

$$\frac{1}{3}\Pi_{sii} + \frac{1}{2}\left(\Pi_{sij} + \Pi_{sji}\right) < 0, \quad i \neq j, \quad (13.30a)$$

$$\Pi_{sii} < 0, \quad (13.30b)$$

$$\frac{1}{3}\Phi_{ii} + \frac{1}{2}\left(\Phi_{ij} + \Phi_{ji}\right) > 0, \quad i \neq j, \quad (13.30c)$$

$$\Phi_{ii} > 0, \quad (13.30d)$$

where

$$\Pi_{sij} \triangleq \begin{bmatrix} -G_s - G_s^T + P_s & A_iG_j + B_{2i}K_j & B_{1i} \\ \star & -P_i & 0 \\ \star & \star & -I \end{bmatrix}, \quad \Phi_{ij} \triangleq \begin{bmatrix} \gamma^2 I & C_iG_j \\ \star & P_i \end{bmatrix}.$$

Meanwhile, there exists a fuzzy controller in the form of (13.27).

Proof. Inequalities (13.17b) and (13.29) can be respectively written as

$$\sum_{s=1}^{4}\sum_{j=1}^{4}\sum_{i=1}^{4} h_s(\theta(k+1))h_j(\theta(k))h_i(\theta(k))\Pi_{sij} < 0, \quad (13.31)$$

$$\sum_{j=1}^{4}\sum_{i=1}^{4} h_j(\theta(k))h_i(\theta(k))\Phi_{ij} < 0. \quad (13.32)$$

By [203], if (13.30a)–(13.30d) hold then (13.31) and (13.32) are fulfilled. Therefore, from Theorem 13.4 we can see that system (13.14) is asymptotically stable with an ℓ_2-ℓ_∞ performance. ■

13.4 Simulation Results

In this section, we apply the proposed methods to design a fuzzy feedback controller for the electromagnetic suspension system. Some parameters of the electromagnetic suspension system are presented in Table 13.1 (other sets of parameters can be found in [103, 183, 231]), and the allowable bounds of $\theta(k)$ are listed in Table 13.2.

Table 13.1. Parameter values of the electromagnetic suspension system

Parameter	m	R_m	g	a_m	N	μ_0	δ_{ref}	T
Unit	Kg	Ω	m/s^2	m^2	kilo	H/m	m	ms
Value	150	1.1	9.8	1.024×10^{-2}	1	$4\pi\times10^{-7}$	0.004	0.5

Table 13.2. Allowable bounds of $\theta(k)$

Parameter	$\theta_{1\max}$	$\theta_{1\min}$	$\theta_{2\max}$	$\theta_{2\min}$
Unit	m	m	A	A
Value	0.001	-0.001	1	-1

With the parameters given in Tables 13.1 and 13.2, we will get the exact expressions of original nonlinear electromagnetic suspension system and its approximated T-S fuzzy model. Firstly, we analyze the stability of the original system with the initial states given by $x(0) = \begin{bmatrix} 0.0001 & 0 & 0 \end{bmatrix}^T$. The states of the open-loop system are plotted in Fig. 13.4, and it is shown that the open-loop electromagnetic suspension system is unstable. Setting $\gamma = 0.005$, it follows by solving conditions (13.30a)–(13.30d) in Theorem 13.5 that

$$G_1 = \begin{bmatrix} 0.0000 & -0.0011 & 0.0001 \\ -0.0011 & 0.2835 & -0.0820 \\ -0.0052 & -3.0222 & 57.9373 \end{bmatrix}, \quad G_2 = \begin{bmatrix} 0.0000 & -0.0011 & 0.0001 \\ -0.0011 & 0.2780 & -0.0325 \\ -0.0053 & -3.0567 & 58.3998 \end{bmatrix},$$

$$G_3 = \begin{bmatrix} 0.0000 & -0.0011 & 0.0002 \\ -0.0010 & 0.2969 & -0.5271 \\ -0.0051 & -2.8855 & 55.9006 \end{bmatrix}, \quad G_4 = \begin{bmatrix} 0.0000 & -0.0011 & 0.0001 \\ -0.0010 & 0.2892 & -0.3933 \\ -0.0051 & -2.9619 & 55.9910 \end{bmatrix},$$

$$K_1 = 10^5 \begin{bmatrix} 0.0001 & -0.0058 & -1.4860 \end{bmatrix}, \quad K_2 = 10^5 \begin{bmatrix} 0.0001 & -0.0031 & -1.4994 \end{bmatrix},$$

$$K_3 = 10^5 \begin{bmatrix} 0.0001 & -0.0159 & -2.1868 \end{bmatrix}, \quad K_4 = 10^5 \begin{bmatrix} 0.0001 & -0.0100 & -2.1925 \end{bmatrix}.$$

Constructing a non-PDC fuzzy controller by (13.27), we can get the system states of the closed-loop system. The states of the controlled T-S fuzzy system are shown in Fig. 13.5, from which we can see that the designed fuzzy controller can stabilize the T-S fuzzy model of the electromagnetic suspension system well. However, note that in the fuzzy modeling of the electromagnetic suspension system, some approximation errors are inevitably introduced in the process of fuzzification, thus the fuzzy model can not accurately represent the original nonlinear model of the electromagnetic suspension system (that is, there is an approximation error between the fuzzy model and the original nonlinear model). To illustrate the effectiveness of the designed fuzzy controller to the original nonlinear model, we also provide the corresponding simulation here. The states of the closed-loop nonlinear system under the proposed fuzzy control is plotted in Fig. 13.6, from which we can see that, compared with the fuzzy model under the fuzzy control, the controlled nonlinear model is also stablizable. Therefore, we can say that the designed fuzzy control strategy can stabilize not only the fuzz model but also the original nonlinear model of the electromagnetic suspension system.

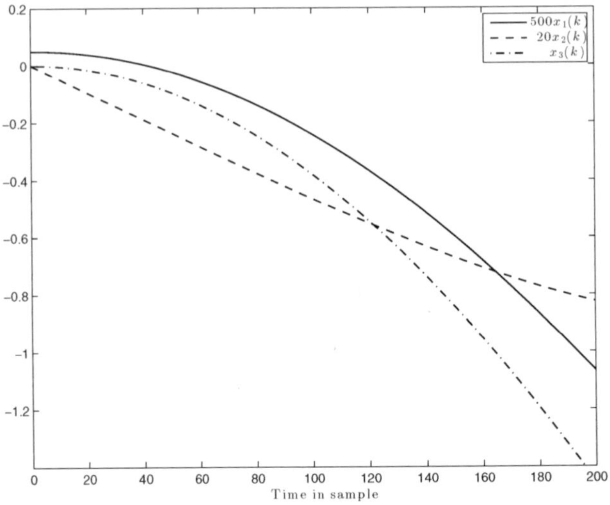

Fig. 13.4. States of the original open-loop system

Furthermore, suppose the exogenous disturbance $\omega(k)$ be

$$\omega(k) = \begin{cases} 90, & \text{when } 0 < k < 20 \\ 0, & \text{when } k \leq 0 \text{ or } k \geq 20. \end{cases}$$

By calculation, we can obtain $\|\omega(k)\|_2 = \sqrt{\sum_{k=0}^{\infty} \omega^T(k)\omega(k)} = 392.3$. The corresponding output of the controlled electromagnetic suspension system is shown in Fig. 13.7, where the peak value of $y(k)$ is about 3×10^{-5}. So the actual value of $\frac{\|y(k)\|_\infty}{\|\omega(k)\|_2}$ is smaller than the bound of $\gamma = 0.005$, which exactly verifies that our designed fuzzy controller ensures the prescript ℓ_2-ℓ_∞ performance level. The fuzzy control input $u(k)$ is presented in Fig. 13.8.

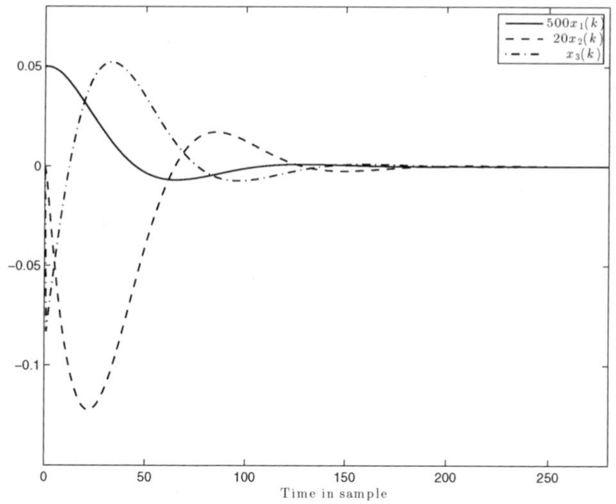

Fig. 13.5. States of the controlled fuzzy model

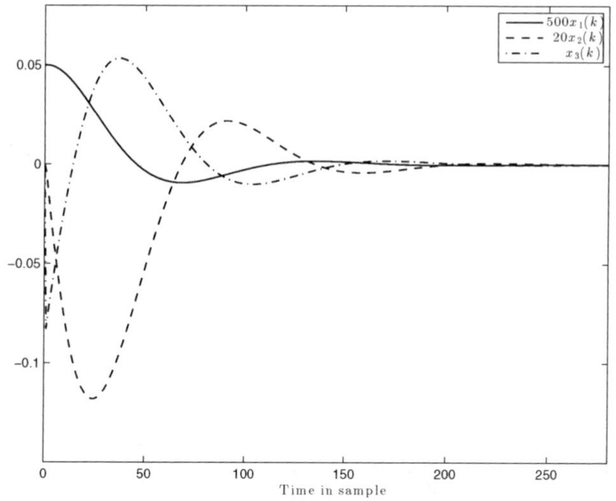

Fig. 13.6. States of the controlled nonlinear system

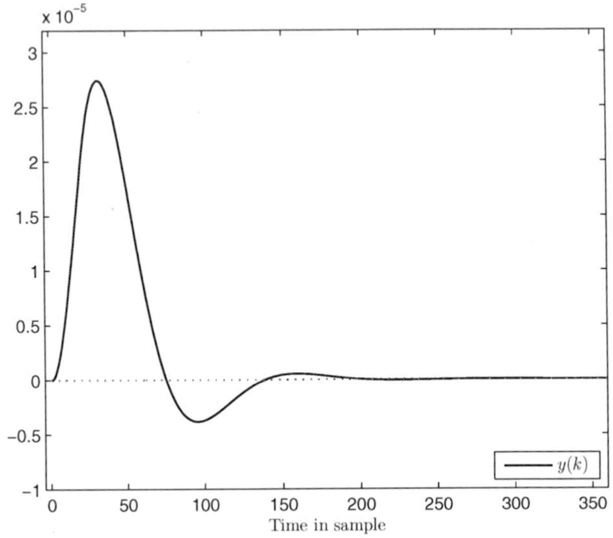

Fig. 13.7. Zero-state response of the controlled nonlinear system

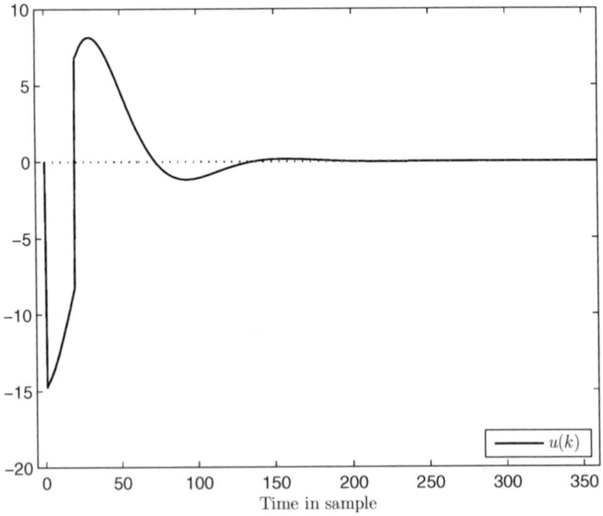

Fig. 13.8. The fuzzy control input

13.5 Conclusion

In this chapter, the T-S fuzzy model based ℓ_2-ℓ_∞ fuzzy control method has been proposed for the nonlinear electromagnetic suspension system. Firstly, the nonlinear dynamic equations of the electromagnetic suspension system have been established from some physical laws. Then, the nonlinear system has been presented by a T-S fuzzy model. Further, ℓ_2-ℓ_∞ performance analysis of the electromagnetic suspension system has been performed based on its approximated T-S fuzzy model, and corresponding fuzzy controller design method has been derived under a non-PDC scheme. Finally, simulations have been presented to demonstrate the effectiveness of the proposed controller design method.

Chapter 14
Fuzzy Control of Nonlinear Air-Breathing Hypersonic Vehicles

14.1 Introduction

Air-breathing hypersonic vehicles (AHVs) are a promising and cost-efficient technology for launching low-earth-orbit satellites and providing rapid global-response capabilities [55]. As the application of air-breathing scramjet engines [30, 85], AHVs can carry more payload than traditional expendable rockets, because the scramjet engines can obtain oxygen from atmosphere rather than carrying it. The design of guidance and control systems for AHVs is challenging due to significant aero-thermo-elastic-propulsion interactions, strong couplings between propulsive and aerodynamic forces, extreme range of operating conditions and rapid change of mass distribution. In addition, the requirements of flight stability and high speed response, the existence of various random interference factors and large uncertainties make it more difficult [162, 216].

Hypersonic flight technologies have been studied for more than half a century [8], the challenges associated with the dynamics and control of AHVs also have drawn the researchers' attention. In order to design reliable and effective controllers for AHVs, it is essential to consider the unique dynamic characteristics and the coupling of these vehicles. AHVs have the uniqueness configuration of a tightly integrated airframe and scramjet propulsion system. Flight control design for AHVs is highly challenging, so the problem of flight control design is one of the key technique for the application of AHVs. Because of the dynamics' enormous complexity, only the longitudinal dynamics models of AHVs have been used for control design. Generally, there are two kinds of longitudinal models for AHVs. One is a rigid model developed in [22, 177], the other is a more complex model developed in [10, 11] which includes the flexible dynamics of the vehicles. Due to the slender geometries and light structures of this generic vehicles, significant flexible effects can not be neglected in control design [166], so the second model approaches the real situation better. A wide range of control laws have been developed for the

L. Wu et al., *Fuzzy Control Systems with Time-Delay and Stochastic Perturbation*,
Studies in Systems, Decision and Control 12, DOI: 10.1007/978-3-319-11316-6_14

rigid AHVs model [15, 150, 229, 230], but when the flexible dynamic is considered in modeling, most of the proposed methods, especially the nonlinear control laws are not suitable for the flexible AHVs [166]. The equations of this models become exceedingly complex when flexibility effects are considered, so these models can be used only for simulations or validation purposes [162, 216]. In [166], a control-oriented model was derived for the flexible AHVs models by using curve fits calculated directly from the forces and moments included in the truth model, then an approximate feedback liberalization example of control design was given to derive a nonlinear controller. In [93], the authors presented two output feedback control design methods for the flexible AHVs models, and adaptive control techniques were also considered in [119]. In [181], dynamic output feedback control techniques was used to provide robust velocity and altitude tracking control in the presence of model uncertainties and varying flight conditions, and in [76, 180], linear controllers with input constraints using on-line optimization and anti-windup techniques were also proposed. More recently, a nonlinear robust adaptive control design method was presented in [56], and in [214] the authors considered the modeling of aerothermoelastic effects and gave a Lyapunov-based tracking controller. In addition, The problem of robust control and output feedback control design for the flexible AHVs has been investigated in [93, 181]. Though much work has been done, the robust control for the high nonlinear dynamics of flexible AHVs is still an open and challenging problem, especially when uncertainties and disturbances exist simultaneously.

Motivated by the fact that T-S fuzzy technique has been widely accepted as an effective approach to the control design of nonlinear systems. It has been proved that any smooth nonlinear function can be approximated by a fuzzy model to any specified accuracy. Using a T-S fuzzy plant model enable one to describe a nonlinear system as a weighted sum of some simple linear subsystems, thus some conventional linear control theories can be applied to analysis and synthesis of nonlinear systems based on the PDC scheme. In this chapter, we propose a T-S fuzzy robust $\mathcal{H}_\infty$ controller design method for the longitudinal nonlinear model of flexible AHVs via dynamic output feedback. We shall focus on the problem of reference output tracking control for longitudinal model of flexible AHVs. T-S fuzzy modeling technique presented in [198] is firstly used to construct a fuzzy model which can represent the complex nonlinear longitudinal model of flexible AHVs. The developed T-S fuzzy model of flexible AHVs include uncertainties and disturbances, so it can approach the dynamics of flexible AHVs better. Then based on the PDC scheme, a full-order fuzzy dynamic output feedback controller is designed to stabilize the closed-loop systems since part of the flexible AHVs models are difficult to measure. Sufficient conditions for the existence of admissible controllers are proposed in terms of LMIs. Simulation results are provided to show the effectiveness of the proposed control design method.

14.2 Nonlinear Model

Nomenclature

$C_D(\alpha, \delta_e)$	drag coefficient
$C_D^{\alpha_i}$	ith order coefficient of α contribution to $C_D(\alpha, \delta_e)$
$C_D^{\delta_e^i}$	ith order coefficient of δ_e contribution to $C_D(\alpha, \delta_e)$
C_D^0	constant term in $C_D(\alpha, \delta_e)$
$C_L(\alpha, \delta_e)$	lift coefficient
$C_L^{\alpha_i}$	ith order coefficient of α contribution to $C_L(\alpha, \delta_e)$
$C_L^{\delta_e}$	coefficient of δ_e contribution to $C_L(\alpha, \delta_e)$
C_L^0	constant term in $C_L(\alpha, \delta_e)$
$C_{M,Q}(\alpha, Q)$	contribution to moment due to pitch rate
$C_{M,\alpha}(\alpha)$	contribution to moment due to angle of attack
$C_{M,\delta_e}(\delta_e, \delta_c)$	control surface contribution to moment
$C_{M,\alpha}^{\alpha_i}$	ith order coefficient of α contribution to $C_{M,\alpha}(\alpha)$
$C_{M,\alpha}^0$	constant term in $C_{M,\alpha}(\alpha)$
$C_T^{\alpha_i}(\Phi)$	ith order coefficient of α in T
$\bar{c}$	mean aerodynamic chord
c_c	canard coefficient in $C_{M,\delta_e}(\delta_e, \delta_c)$
c_e	elevator coefficient in $C_{M,\delta_e}(\delta_e, \delta_c)$
D	drag
g	acceleration due to gravity
h	altitude
I_{yy}	moment of inertia
L	left
L_v	vehicle length
M	pitching moment
m	vehicle mass
N_i	ith generalized force
$N_i^{\alpha_j}$	jth order contribution of α to N_i
N_i^0	constant term in N_i
$N_2^{\delta_e}$	contribution of δ_e to N_2
Q	pitch rate
$\bar{q}$	dynamic pressure
S	reference area
T	trust
V	velocity
x	state of the control-oriented model

α	angle of attack
$\beta_i\,(h, \bar{q})$	ith thrust fit parameter
γ	flight path angle, $\gamma = \theta - \alpha$
δ_c	canard angular deflection
δ_e	elevator angular deflection
ξ	damping ratio for the Φ dynamics
ξ_i	damping ratio for elastic mode η_i
η_i	ith generalized elastic coordinate
θ	pitch angle
λ_i	inertial coupling term of ith elastic mode
ρ	density of air
Φ	stoichiometrically normalized fuel-to-air ratio
ω	natural frequency for the Φ dynamics
ω_i	natural frequency for elastic mode η_i
$1/h_s$	air density decay rate

The hypersonic vehicle model considered in this chapter was developed by Bolender and Doman [10, 11]. Due to the enormous complexity of the vehicle dynamics, only the longitudinal model is adopted for control design. The equations of the longitudinal dynamics of flexible AHVs are derived using Lagrange's equations and compressible flow theory. Flexibility effects are included by modeling the vehicle as a single flexible structure, whereas the scramjet engine model is adopted from Chavez and Schmidt [22]. The nonlinear equations are described as

$$\begin{cases} \dot{h} = V\sin(\theta-\alpha), \\ \dot{V} = \dfrac{1}{m}(T\cos\alpha - D) - g\sin(\theta-\alpha), \\ \dot{\alpha} = \dfrac{1}{mV}(-T\sin\alpha - L) + Q + \dfrac{g}{V}\cos(\theta-\alpha), \\ \dot{\theta} = Q, \\ \dot{Q} = \dfrac{M}{I_{yy}}, \\ \ddot{\eta}_1 = -2\varsigma_1\omega_1\dot{\eta}_1 - \omega_1^2\eta_1 + N_1, \\ \ddot{\eta}_2 = -2\varsigma_2\omega_2\dot{\eta}_2 - \omega_2^2\eta_2 + N_2. \end{cases} \tag{14.1}$$

The above equations are composed of rigid-body state variables and flexible states, the control input $u = \begin{bmatrix}\Phi & \delta_e\end{bmatrix}^T$ does not appear explicitly in these equations. As mentioned in [166], they enter through the forces and moments T, L, D, M, N_1 and N_2 as follows:

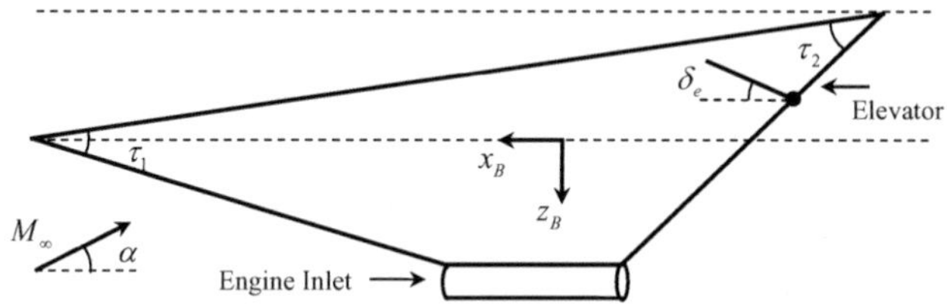

Fig. 14.1. Geometry of the flexible hypersonic vehicle model

$$\begin{cases} L \approx \frac{1}{2}\rho V^2 S C_L(\alpha, \delta_e), \\ D \approx \frac{1}{2}\rho V^2 S C_D(\alpha, \delta_e), \\ M \approx z_T T + \frac{1}{2}\rho V^2 S\bar{c}(C_{M,\alpha}(\alpha) + C_{M,\delta_e}(\delta_e)), \\ T \approx C_T^{\alpha^3}\alpha^3 + C_T^{\alpha^2}\alpha^2 + C_T^{\alpha}\alpha + C_T^0, \\ N_1 \approx N_1^{\alpha^2}\alpha^2 + N_1^{\alpha}\alpha + N_1^0, \\ N_2 \approx N_2^{\alpha^2}\alpha^2 + N_2^{\alpha}\alpha + N_2^{\delta_e}\delta_e + N_2^0, \end{cases} \tag{14.2}$$

with

$$\begin{cases} \rho = \rho_0 e^{\frac{-(h-h_0)}{h_s}}, \\ C_L = C_L^{\alpha}\alpha + C_L^{\delta_e}\delta_e + C_L^0, \\ C_D = C_D^{\alpha^2}\alpha^2 + C_D^{\alpha}\alpha + C_D^{\delta_e^2}\delta_e^2 + C_D^{\delta_e}\delta_e + C_D^0, \\ C_{M,\alpha} = C_{M,\alpha}^{\alpha^2}\alpha^2 + C_{M,\alpha}^{\alpha}\alpha + C_{M,\alpha}^0, \\ C_{M,\delta_e} = c_e\delta_e, \\ \bar{q} = \frac{1}{2}\rho V^2, \\ C_T^{\alpha^3} = \beta_1(h,\bar{q})\Phi + \beta_2(h,\bar{q}), \\ C_T^{\alpha^2} = \beta_3(h,\bar{q})\Phi + \beta_4(h,\bar{q}), \\ C_T^{\alpha} = \beta_5(h,\bar{q})\Phi + \beta_6(h,\bar{q}), \\ C_T^0 = \beta_7(h,\bar{q})\Phi + \beta_8(h,\bar{q}). \end{cases} \tag{14.3}$$

Obviously, the above dynamics of flexible AHVs is high nonlinear and coupling with each other, and the parameters are time-varying due to the flight envelop, which makes the control design difficult. In what follows, we will introduce a T-S fuzzy model to describe flexible AHVs dynamics.

14.3 T-S Fuzzy Modeling

The system in (14.1)–(14.3) is a complex nonlinear one, and the control input Φ and δ_e do not occur explicitly in the equations. To construct a T-S fuzzy mode for the hypersonic vehicle, we employ the same methods as those presented in [198]. First, the nonlinear equations should be transformed into an affine nonlinear form as

$$\dot{x}(t) = f(x,t) + g(x,t)u(t), \tag{14.4}$$

where the input appears directly at the right hand side of the equation. By substituting (14.3) into (14.2), we have

$$\left\{\begin{aligned}
L &\approx \frac{1}{2}\rho V^2 S C_L(\alpha,\delta_e) = \frac{1}{2}\rho V^2 S\left(C_L^{\alpha}\alpha + C_L^{\delta_e}\delta_e + C_L^0\right)\\
&= \frac{1}{2}\rho V^2 S\left(C_L^{\alpha}\alpha + C_L^0\right) + \frac{1}{2}\rho V^2 S C_L^{\delta_e}\delta_e,\\
D &\approx \frac{1}{2}\rho V^2 S C_D(\alpha,\delta_e)\\
&= \frac{1}{2}\rho V^2 S\left(C_D^{\alpha^2}\alpha^2 + C_D^{\alpha}\alpha + C_D^{\delta_e^2}\delta_e^2 + C_D^{\delta_e}\delta_e + C_D^0\right)\\
&= \frac{1}{2}\rho V^2 S\left(C_D^{\alpha^2}\alpha^2 + C_D^{\alpha}\alpha + C_D^0\right) + \frac{1}{2}\rho V^2 S\left(C_D^{\delta_e^2}\delta_e + C_D^{\delta_e}\right)\delta_e,\\
M &\approx z_T T + \frac{1}{2}\rho V^2 S\bar{c}(C_{M,\alpha}(\alpha) + C_{M,\delta_e}(\delta_e))\\
&= z_T\left[\beta_2(h,\bar{q})\alpha^3 + \beta_4(h,\bar{q})\alpha^2 + \beta_6(h,\bar{q})\alpha + \beta_8(h,\bar{q})\right]\\
&\quad + z_T\left[\beta_1(h,\bar{q})\alpha^3 + \beta_3(h,\bar{q})\alpha^2 + \beta_5(h,\bar{q})\alpha + \beta_7(h,\bar{q})\right]\Phi\\
&\quad + \frac{1}{2}\rho V^2 S\bar{c}C_{M,\alpha}(\alpha) + \frac{1}{2}\rho V^2 S\bar{c}c_e\delta_e,\\
T &\approx C_T^{\alpha^3}\alpha^3 + C_T^{\alpha^2}\alpha^2 + C_T^{\alpha}\alpha + C_T^0\\
&= \left[\beta_2(h,\bar{q})\alpha^3 + \beta_4(h,\bar{q})\alpha^2 + \beta_6(h,\bar{q})\alpha + \beta_8(h,\bar{q})\right]\\
&\quad + \left[\beta_1(h,\bar{q})\alpha^3 + \beta_3(h,\bar{q})\alpha^2 + \beta_5(h,\bar{q})\alpha + \beta_7(h,\bar{q})\right]\Phi,\\
N_1 &\approx N_1^{\alpha^2}\alpha^2 + N_1^{\alpha}\alpha + N_1^0,\\
N_2 &\approx N_2^{\alpha^2}\alpha^2 + N_2^{\alpha}\alpha + N_2^{\delta_e}\delta_e + N_2^0\\
&= N_2^{\alpha^2}\alpha^2 + N_2^{\alpha}\alpha + N_2^0 + N_2^{\delta_e}\delta_e.
\end{aligned}\right. \tag{14.5}$$

Then, by devoting (14.5) into (14.1), it follows that

$$\begin{cases}
\dot{h} = V\sin(\theta-\alpha), \\
\dot{V} = \dfrac{1}{m}(T\cos\alpha - D) - g\sin(\theta-\alpha) \\
\quad = \dfrac{1}{m}\left[\beta_2(h,\bar{q})\alpha^3 + \beta_4(h,\bar{q})\alpha^2 + \beta_6(h,\bar{q})\alpha + \beta_8(h,\bar{q})\right]\cos\alpha \\
\qquad -\dfrac{1}{m}\left[\dfrac{1}{2}\rho V^2 S\left(C_D^{\alpha^2}\alpha^2 + C_D^{\alpha}\alpha + C_D^0\right)\right] - g\sin(\theta-\alpha) \\
\qquad +\dfrac{1}{m}\left\{\left[\beta_1(h,\bar{q})\alpha^3 + \beta_3(h,\bar{q})\alpha^2 + \beta_5(h,\bar{q})\alpha + \beta_7(h,\bar{q})\right]\cos\alpha\right\}\Phi \\
\qquad -\dfrac{1}{m}\left[\dfrac{1}{2}\rho V^2 S\left(C_D^{\delta_e^2}\delta_e + C_D^{\delta_e}\right)\right]\delta_e, \\
\dot{\alpha} = \dfrac{-1}{mV}\left[\beta_2(h,\bar{q})\alpha^3 + \beta_4(h,\bar{q})\alpha^2 + \beta_6(h,\bar{q})\alpha + \beta_8(h,\bar{q})\right]\sin\alpha \\
\qquad -\dfrac{1}{mV}\left[\dfrac{1}{2}\rho V^2 S\left(C_L^{\alpha}\alpha + C_L^0\right)\right] + Q + \dfrac{g}{V}\cos(\theta-\alpha) \\
\qquad -\dfrac{1}{mV}\left\{\left[\beta_1(h,\bar{q})\alpha^3 + \beta_3(h,\bar{q})\alpha^2 + \beta_5(h,\bar{q})\alpha + \beta_7(h,\bar{q})\right]\sin\alpha\right\}\Phi \\
\qquad -\dfrac{1}{mV}\left(\dfrac{1}{2}\rho V^2 S C_L^{\delta_e}\right)\delta_e, \\
\dot{\theta} = Q, \\
\dot{Q} = \dfrac{1}{I_{yy}}\left\{z_T\left[\beta_2(h,\bar{q})\alpha^3 + \beta_4(h,\bar{q})\alpha^2 + \beta_6(h,\bar{q})\alpha + \beta_8(h,\bar{q})\right]\right. \\
\qquad \left. + \dfrac{1}{2}\rho V^2 S\bar{c}\left(C_{M,\alpha}^{\alpha^2}\alpha^2 + C_{M,\alpha}^{\alpha}\alpha + C_{M,\alpha}^0\right)\right\} \\
\qquad +\dfrac{1}{I_{yy}}\left\{z_T\left[\beta_1(h,\bar{q})\alpha^3 + \beta_3(h,\bar{q})\alpha^2 + \beta_5(h,\bar{q})\alpha + \beta_7(h,\bar{q})\right]\right\}\Phi \\
\qquad +\dfrac{1}{I_{yy}}\left(\dfrac{1}{2}\rho V^2 S\bar{c}c_e\right)\delta_e, \\
\ddot{\eta}_1 = -2\varsigma_1\omega_1\dot{\eta}_1 - \omega_1^2\eta_1 + N_1^{\alpha^2}\alpha^2 + N_1^{\alpha}\alpha + N_1^0, \\
\ddot{\eta}_2 = -2\varsigma_2\omega_2\dot{\eta}_2 - \omega_2^2\eta_2 + N_2^{\alpha^2}\alpha^2 + N_2^{\alpha}\alpha + N_2^0 + N_2^{\delta_e}\delta_e,
\end{cases}$$

and the affine form of the flexible AHVs can be obtained as

$$f(x,t) = \begin{bmatrix} f_1(x,t) \\ f_2(x,t) \\ f_3(x,t) \\ f_4(x,t) \\ f_5(x,t) \\ f_6(x,t) \\ f_7(x,t) \\ f_8(x,t) \\ f_9(x,t) \end{bmatrix}, \quad g(x,t) = \begin{bmatrix} 0 & 0 \\ b_{21} & b_{22} \\ b_{31} & b_{32} \\ 0 & 0 \\ b_{51} & b_{52} \\ 0 & 0 \\ 0 & 0 \\ 0 & 0 \\ 0 & b_{92} \end{bmatrix}, \quad y = Cx, \ z = Ex,$$

where

$$\begin{cases} f_1(x,t) = V\sin(\theta-\alpha), \\ f_2(x,t) = \dfrac{1}{m}\Big[\beta_2(h,\bar{q})\alpha^3 + \beta_4(h,\bar{q})\alpha^2 + \beta_6(h,\bar{q})\alpha + \beta_8(h,\bar{q})\Big]\cos\alpha \\ \qquad -\dfrac{1}{m}\left[\dfrac{1}{2}\rho V^2 S\left(C_D^{\alpha^2}\alpha^2 + C_D^{\alpha}\alpha + C_D^0\right)\right] - g\sin(\theta-\alpha), \\ f_3(x,t) = \dfrac{-1}{mV}\Big[\beta_2(h,\bar{q})\alpha^3 + \beta_4(h,\bar{q})\alpha^2 + \beta_6(h,\bar{q})\alpha + \beta_8(h,\bar{q})\Big]\sin\alpha \\ \qquad +\dfrac{1}{mV}\left[\dfrac{1}{2}\rho V^2 S\left(C_L^{\alpha}\alpha + C_L^0\right)\right] + Q + \dfrac{g}{V}\cos(\theta-\alpha), \\ f_4(x,t) = Q, \\ f_5(x,t) = \dfrac{1}{I_{yy}}\Big\{z_T\Big[\beta_2(h,\bar{q})\alpha^3 + \beta_4(h,\bar{q})\alpha^2 + \beta_6(h,\bar{q})\alpha + \beta_8(h,\bar{q})\Big] \\ \qquad +\dfrac{1}{2}\rho V^2 S\bar{c}\left(C_{M,\alpha}^{\alpha^2}\alpha^2 + C_{M,\alpha}^{\alpha}\alpha + C_{M,\alpha}^0\right)\Big\}, \\ f_6(x,t) = \dot{\eta}_1, \\ f_7(x,t) = -2\varsigma_1\omega_1\dot{\eta}_1 - \omega_1^2\eta_1 + N_1^{\alpha^2}\alpha^2 + N_1^{\alpha}\alpha + N_1^0, \\ f_8(x,t) = \dot{\eta}_2, \\ f_9(x,t) = -2\varsigma_2\omega_2\dot{\eta}_2 - \omega_2^2\eta_2 + N_2^{\alpha^2}\alpha^2 + N_2^{\alpha}\alpha + N_2^0, \end{cases}$$

and

$$\begin{cases} b_{21} = \dfrac{1}{m}\Big\{\left[\beta_1(h,\bar{q})\alpha^3 + \beta_3(h,\bar{q})\alpha^2 + \beta_5(h,\bar{q})\alpha + \beta_7(h,\bar{q})\right]\cos\alpha\Big\}, \\ b_{22} = -\dfrac{1}{m}\left[\dfrac{1}{2}\rho V^2 S\left(C_D^{\delta_e^2}\delta_e + C_D^{\delta_e}\right)\right], \\ b_{31} = -\dfrac{1}{mV}\Big\{\left[\beta_1(h,\bar{q})\alpha^3 + \beta_3(h,\bar{q})\alpha^2 + \beta_5(h,\bar{q})\alpha + \beta_7(h,\bar{q})\right]\sin\alpha\Big\}, \\ b_{32} = -\dfrac{1}{mV}\left(\dfrac{1}{2}\rho V^2 S C_L^{\delta_e}\right), \\ b_{51} = \dfrac{1}{I_{yy}}\Big\{z_T\left[\beta_1(h,\bar{q})\alpha^3 + \beta_3(h,\bar{q})\alpha^2 + \beta_5(h,\bar{q})\alpha + \beta_7(h,\bar{q})\right]\Big\}, \\ b_{52} = \dfrac{1}{I_{yy}}\left(\dfrac{1}{2}\rho V^2 S\bar{c}c_e\right), \\ b_{92} = N_2^{\delta_e}. \end{cases}$$

where

$$x = \begin{bmatrix} h \\ V \\ \alpha \\ \theta \\ Q \\ \eta_1 \\ \dot{\eta}_1 \\ \eta_2 \\ \dot{\eta}_2 \end{bmatrix}, \quad u = \begin{bmatrix} \Phi \\ \delta_e \end{bmatrix}, \quad z = \begin{bmatrix} V \\ h \end{bmatrix},$$

are the state vector, the control input and the controlled output, respectively; y represents the measured output, and

$$C = \mathrm{diag}\left\{c_1, c_2, c_3, c_4, c_5, c_6, c_7, c_8, c_9\right\},$$
$$E = \begin{bmatrix} 1\,0\,0\,0\,0\,0\,0\,0\,0 \\ 0\,1\,0\,0\,0\,0\,0\,0\,0 \end{bmatrix},$$

where c_i is 1 or 0 ($i = 1, 2, \ldots, 9$). When $c_i = 1$, the ith element of the state x is measured and when $c_i = 0$, the ith element of x is unmeasured.

The output vectors V and h are chosen as the premise variables, and the T-S fuzzy modeling technique expressed in [198] is employed to construct an exact T-S fuzzy model for the nonlinear hypersonic vehicle system (14.1). For the application of the above-mentioned method, three levels are chosen for every premise variable: a lower bound, a upper bound and a equilibrium point, which named as "small (S)", "big (B)" and "middle (M)", respectively. The bigger the envelop between the lower bound and the upper bound is, the better the T-S fuzzy model approaches the nonlinear dynamics of flexible AHVs. The nonlinear model (14.1) can then be represented by a T-S fuzzy model composed of 9 (3^2) fuzzy rules. The T-S fuzzy model can be constructed by the following two ways:

a) When the operation state is the equilibrium point, the local model can be obtained by using Taylor's linearization approach, the model can be written as

$$\dot{x}(t) = A\left(x(t) - x_0\right) + B\left(u(t) - u_0\right), \tag{14.6}$$

where

$$A = \frac{\partial f(x,t)}{\partial x}\bigg|_{\substack{x = x_0 \\ u = u_0}}, \qquad B = \frac{\partial f(x,t)}{\partial u}\bigg|_{\substack{x = x_0 \\ u = u_0}} \tag{14.7}$$

b) When the operation state is not the equilibrium point, that is, $f(x,t) + g(x,t)u(t) \neq 0$. For this case, the local models will be an affine linear model instead of a linear one if the common linearization approach is applied. In [198], an optimum method is expressed for solving this problem, with this method, the linear system at the operating point yields

$$\begin{cases} a_i = \nabla f_i(x_0) + \dfrac{f_i(x_0) - x_0^T \nabla f_i(x_0)}{\|x_0\|^2} x_0, \quad x_0 \neq 0, \\ A = \begin{bmatrix} a_1 \ a_2 \ \cdots \ a_n \end{bmatrix}^T, \\ B = g(x_0). \end{cases} \tag{14.8}$$

Considering the system with parametric uncertainties and disturbances, the T-S fuzzy model for the nonlinear system in (14.1) can be represented by the following nine-rule fuzzy model:

Rule 1) If V is small (V_S) and h is small (h_S), then

$$\begin{cases} \dot{x}(t) = (A_1 + \Delta A_1)\, x(t) + (B_1 + \Delta B_1)\, u(t) + D_1 \varpi(t), \\ y(t) = C_1 x(t), \\ z(t) = E_1 x(t), \end{cases}$$

Rule 2) If V is small (V_S) and h is middle (h_M), then

$$\begin{cases} \dot{x}(t) = (A_2 + \Delta A_2)\, x(t) + (B_2 + \Delta B_2)\, u(t) + D_2 \varpi(t), \\ y(t) = C_2 x(t), \\ z(t) = E_2 x(t), \end{cases}$$

Rule 3) If V is small (V_S) and h is big (h_B), then

$$\begin{cases} \dot{x}(t) = (A_3 + \Delta A_3)\, x(t) + (B_3 + \Delta B_3)\, u(t) + D_3 \varpi(t), \\ y(t) = C_3 x(t), \\ z(t) = E_3 x(t), \end{cases}$$

Rule 4) If V is middle (V_M) and h is small (h_S), then

$$\begin{cases} \dot{x}(t) = (A_4 + \Delta A_4)\, x(t) + (B_4 + \Delta B_4)\, u(t) + D_4 \varpi(t), \\ y(t) = C_4 x(t), \\ z(t) = E_4 x(t), \end{cases}$$

Rule 5) If V is middle (V_M) and h is middle (h_M), then

$$\begin{cases} \dot{x}(t) = (A_5 + \Delta A_5)\, x(t) + (B_5 + \Delta B_5)\, u(t) + D_5 \varpi(t), \\ y(t) = C_5 x(t), \\ z(t) = E_5 x(t), \end{cases}$$

Rule 6) If V is middle (V_M) and h is big (h_B), then

$$\begin{cases} \dot{x}(t) = (A_6 + \Delta A_6)\, x(t) + (B_6 + \Delta B_6)\, u(t) + D_6 \varpi(t), \\ y(t) = C_6 x(t), \\ z(t) = E_6 x(t), \end{cases}$$

Rule 7) If V is big (V_B) and h is small (h_S), then

$$\begin{cases} \dot{x}(t) = (A_7 + \Delta A_7(t))\, x(t) + (B_7 + \Delta B_7)\, u(t) + D_7 \varpi(t), \\ y(t) = C_7 x(t), \\ z(t) = E_7 x(t), \end{cases}$$

Rule 8) If V is big (V_B) and h is middle (h_M), then

$$\begin{cases} \dot{x}(t) = (A_8 + \Delta A_8)\, x(t) + (B_8 + \Delta B_8)\, u(t) + D_8 \varpi(t), \\ y(t) = C_8 x(t), \\ z(t) = E_8 x(t), \end{cases}$$

Rule 9) If V is big (V_B) and h is big (h_B), then

$$\begin{cases} \dot{x}(t) = (A_9 + \Delta A_9)\, x(t) + (B_9 + \Delta B_9)\, u(t) + D_9 \varpi(t), \\ y(t) = C_9 x(t), \\ z(t) = E_9 x(t), \end{cases}$$

where ΔA_i and ΔB_i $(i = 1, 2, \ldots, 9)$ are the unknown parameter uncertainties of the matrices A_i and B_i, $\varpi(t)$ is the uncertain extraneous disturbance or the nonlinearity. $C_1 = C_2 = \cdots = C_9 = C$, and $E_1 = E_2 = \cdots = E_9 = E$. ΔA_i and ΔB_i are assumed to be of the form

$$\begin{bmatrix} \Delta A_i & \Delta B_i \end{bmatrix} = M_i F_i(t) \begin{bmatrix} N_{1i} & N_{2i} \end{bmatrix}, \tag{14.9}$$

where M_i, N_{1i} and N_{2i} are known real constant matrices, and $F_i(t)$ is an unknown matrix function satisfying $F_i^T(t) F_i(t) \leq I$.

The fuzzy membership functions of V and h are defined as

$$\text{if } V > V_M, \quad \begin{cases} h_S(V) = 0, \\ h_M(V) = 1 - h_B(V), \\ h_B(V) = \exp\left(-3.5 \times 10^{-12} \left|V(t) - V_B\right|^4\right), \end{cases}$$

$$\text{if } V < V_M, \quad \begin{cases} h_S(V) = \exp\left(-3.5 \times 10^{-12} \left|V(t) - V_S\right|^4\right), \\ h_M(V) = 1 - h_B(V), \\ h_B(V) = 0, \end{cases}$$

$$\text{if } h > h_M, \quad \begin{cases} h_S(h) = 0, \\ h_M(h) = 1 - h_B(h), \\ h_B(h) = \exp\left(-2.44 \times 10^{-16} \left|h(t) - h_B\right|^4\right), \end{cases} \tag{14.10}$$

$$\text{if } h < h_M, \quad \begin{cases} h_S(h) = \exp\left(-2.44 \times 10^{-16} \left|h(t) - h_S\right|^4\right), \\ h_M(h) = 1 - h_b(h), \\ h_B(h) = 0. \end{cases}$$

where V_S, V_M and V_B represent "small (S)", "middle (M)" and "big (B)" of V, respectively. So do h_S, h_M and h_B.

Therefore, the T-S fuzzy model which represents the nonlinear hypersonic vehicle model (14.1) can be formulated by

$$\dot{x}(t) = \sum_{i=1}^{9} h_i(t) \left[(A_i + \Delta A_i)\, x(t) + (B_i + \Delta B_i)\, u(t)\right] + D_i \varpi(t), \tag{14.11a}$$

$$y(t) = Cx(t), \tag{14.11b}$$

$$z(t) = Ex(t), \tag{14.11c}$$

where

$$\begin{cases} h_1(t) = h_S(t) h_S(t), \\ h_2(t) = h_S(t) h_M(t), \\ \quad \vdots \\ h_9(t) = h_B(t) h_B(t). \end{cases}$$

with $h_i(t) \geq 0,\ i = 1, 2, \ldots, 9$ and $\sum_{i=1}^{9} h_i(t) = 1$.

14.4 Reference Output Tracking Control

The main objective of this chapter is to design a controller such that the output $z(t) = \begin{bmatrix} V(t) \\ h(t) \end{bmatrix}$ can track a reference command. The reference command velocity and altitude vector can be defined as a fixed reference output signal $z_{\text{com}}(t) = \begin{bmatrix} V_{\text{com}}(t) \\ h_{\text{com}}(t) \end{bmatrix}$. For the T-S fuzzy model (14.11), our aim is to design a fuzzy controller such that the output tracking error achieve zero, that is,

$$\lim_{t \to \infty} e(t) = \lim_{t \to \infty} \left(z(t) - z_{\text{com}}(t)\right) = 0. \tag{14.12}$$

In order to eliminate the steady-state tracking error, we introduce the error integral action in the controller. Define

$$d(t) = \int_0^t e(\tau)\mathrm{d}\tau = \int_0^t (z(\tau) - z_{\mathrm{com}}(\tau))\mathrm{d}\tau,$$

then

$$\dot{d}(t) = e(t) = z(t) - z_{\mathrm{com}}(t).$$

The augmented state-space description of the fuzzy model (14.11) can be obtained as follows:

$$\dot{\zeta}(t) = \sum_{i=1}^{9} h_i(t)\left[\left(\bar{A}_i + \Delta\bar{A}_i\right)\zeta(t) + \left(\bar{B}_i + \Delta\bar{B}_i\right)u(t) + \bar{D}w(t)\right], \quad (14.13a)$$

$$\bar{y}(t) = \bar{C}\zeta(t), \quad (14.13b)$$

$$z(t) = \bar{E}\zeta(t), \quad (14.13c)$$

where $w(t) \triangleq \begin{bmatrix} \varpi(t) \\ z_{\mathrm{com}}(t) \end{bmatrix}$, $\zeta(t) \triangleq \begin{bmatrix} x(t) \\ d(t) \end{bmatrix}$ and

$$\bar{A}_i = \begin{bmatrix} A_i & 0 \\ C & 0 \end{bmatrix}, \ \Delta\bar{A}_i = \begin{bmatrix} \Delta A_i & 0 \\ 0 & 0 \end{bmatrix}, \ \bar{B}_i = \begin{bmatrix} B_i \\ 0 \end{bmatrix}, \ \Delta\bar{B}_i = \begin{bmatrix} \Delta B_i \\ 0 \end{bmatrix},$$

$$\bar{C} = \begin{bmatrix} C & 0 \\ 0 & I \end{bmatrix}, \ \bar{D}_i = \begin{bmatrix} D_i & 0 \\ 0 & -I \end{bmatrix}, \ \bar{E} = \begin{bmatrix} E & 0 \end{bmatrix},$$

and $w(t)$ consists of nonlinearities, disturbances $\varpi(t)$, and reference input $z_{\mathrm{com}}(t)$. According to (14.9), we have

$$\begin{bmatrix} \Delta\bar{A}_i & \Delta\bar{B}_i \end{bmatrix} = \bar{M}_i\bar{F}_i(t)\begin{bmatrix} \bar{N}_{1i} & \bar{N}_{2i} \end{bmatrix}, \quad (14.14)$$

where

$$\bar{M}_i = \begin{bmatrix} M_i & 0 \\ 0 & 0 \end{bmatrix}, \ \bar{N}_{1i} = \begin{bmatrix} N_{1i} & 0 \\ 0 & 0 \end{bmatrix}, \ \bar{N}_{2i} = \begin{bmatrix} N_{2i} \\ 0 \end{bmatrix}, \ \bar{F}_i(t) = \begin{bmatrix} F_i(t) & 0 \\ 0 & 0 \end{bmatrix}.$$

It is obviously that $\bar{F}_i^T(t)\bar{F}_i(t) \leq I$.

Therefore, the output tracking controller design problem can be converted into the stability analysis problem for the closed-loop system in (14.11). That is to say, if there exists a fuzzy controller to stabilize the closed-loop system in (14.11), then the output $z(t)$ can track the reference z_{com}.

In the following, a T-S fuzzy dynamic output feedback controller design method for the reference output tracking control of the flexible AHVs is discussed. Compared to state feedback control, the output feedback control is more serviceable because some states may be impossible to obtain in many situations. For flexible AHVs, some states may not be available, especially, the flexible states of the system in (14.1), so an output feedback controller is more suitable in this case. The dynamic output feedback control provides more flexible choice for the controller design than state feedback one, thus it is employed here to constitute a T-S fuzzy controller for the reference output tracking problem (14.13) in the context of unmeasured states.

By the PDC method, a full-order fuzzy dynamic output feedback controller for fuzzy system (14.13) is constructed as:

♦ **Controller Form:**

Rule i: If V is V_i and h is h_i, then

$$\begin{aligned}\dot{\hat{\zeta}}(t) &= A_{ci}\hat{\zeta}(t) + B_{ci}\bar{y}(t),\\ u(t) &= C_{ci}\hat{\zeta}(t),\end{aligned}$$

where $\hat{\zeta}(t)$ is the controller state, $\bar{y}(t)$ is the measurable signal; A_{ci}, B_{ci} and C_{ci} are matrices to be determined later. Then, the overall fuzzy output feedback controller is given by

$$\dot{\hat{\zeta}}(t) = \sum_{i=1}^{9} h_i(t)A_{ci}\hat{\zeta}(t) + B_{ci}\bar{y}(t), \tag{14.15a}$$

$$u(t) = \sum_{i=1}^{9} h_i(t)C_{ci}\hat{\zeta}(t), \tag{14.15b}$$

Substituting the above controller into system (14.13), the closed-loop system can be obtained as

$$\dot{\chi}(t) = \sum_{i=1}^{9}\sum_{j=1}^{9} h_i(t)h_j(t)\left(A_{eij} + \Delta A_{eij}\right)\chi(t) + D_e w(t), \tag{14.16a}$$

$$z(t) = E_e\chi(t), \tag{14.16b}$$

where $\chi(t) \triangleq \begin{bmatrix}\zeta(t)\\ \hat{\zeta}(t)\end{bmatrix}$ and

$$\begin{aligned}
A_{eij} &= \begin{bmatrix}\bar{A}_i & \bar{B}_i C_{cj}\\ B_{cj}\bar{C} & A_{cj}\end{bmatrix}, \quad \Delta A_{eij} = \begin{bmatrix}\Delta\bar{A}_i & \Delta\bar{B}_i C_{cj}\\ 0 & 0\end{bmatrix},\\
D_e &= \begin{bmatrix}\bar{D}_i\\ 0\end{bmatrix}, \quad E_e = \begin{bmatrix}E & 0\end{bmatrix}, \quad \Delta A_{eij} = \tilde{M}_{eij}\tilde{F}_{eij}(t)\tilde{N}_{eij},\\
\tilde{M}_{eij} &= \begin{bmatrix}\bar{M}_i & 0\\ 0 & 0\end{bmatrix}, \quad \tilde{F}_{eij}(t) = \begin{bmatrix}\bar{F}_i(t) & 0\\ 0 & 0\end{bmatrix}, \quad \tilde{N}_{eij} = \begin{bmatrix}\bar{N}_{1i} & \bar{N}_{2i}C_{cj}\\ 0 & 0\end{bmatrix}.
\end{aligned}$$

The aim is to design a robust dynamic output feedback controller, such that the closed-loop system in (14.16) is robustly asymptotically stable and has an $\mathcal{H}_\infty$ performance in presence of parameter uncertainties and an external disturbance. To this end, $\mathcal{H}_\infty$ performance is set as follows

$$\int_0^{t_f} z^T(t)z(t)dt \leq \gamma^2 \int_0^{t_f} w^T(t)w(t)dt, \tag{14.17}$$

where t_f is the terminal time of control.

Lemma 14.1. [168] *Let E, F and H be real matrices of appropriate dimensions, with $F^TF \leq I$, then we have that for any scalar $\delta > 0$*

$$EFH + H^TF^TE^T \leq \delta^{-1}EE^T + \delta H^TH.$$

Lemma 14.2. [203] *The parameterized linear matrix inequalities,*

$$\sum_{i=1}^{9}\sum_{j=1}^{9} h_ih_jM_{ij} < 0,$$

is fulfilled if the following condition holds:

$$M_{ii} < 0,$$
$$\frac{1}{k-1}M_{ii} + \frac{1}{2}(M_{ij} + M_{ji}) < 0, \quad 1 \leq i \neq j \leq k.$$

Theorem 14.3. *For uncertain T-S fuzzy system (14.16), if there exist a matrix $P > 0$ and a scalar $\varepsilon > 0$ satisfying*

$$\sum_{i=1}^{9}\sum_{j=1}^{9} h_ih_j \begin{bmatrix} PA_{eij}+A_{eij}^TP+E_e^TE+\varepsilon\tilde{N}_{eij}^T\tilde{N}_{eij} & PD_e & P\tilde{M}_{eij} \\ \star & -\gamma^2I & 0 \\ \star & \star & -\varepsilon I \end{bmatrix} < 0, \tag{14.18}$$

then system (14.16) is robustly stable and the $\mathcal{H}_\infty$ performance defined in (14.17) is guaranteed.

Proof. For system (14.16), define the following Lyapunov function:

$$V(t) = \chi^T(t)P\chi(t), \tag{14.19}$$

then by taking time derivative of $V(t)$, we have

$$\begin{aligned}
\dot{V}(t) &= 2\chi^T(t)P\dot{\chi}(t) \\
&= 2\chi^T(t)P\left(\sum_{i=1}^{9}\sum_{j=1}^{9} h_i(t)h_j(t)\left(A_{eij} + \Delta A_{eij}\right)\chi(t) + D_ew(t)\right) \\
&= \sum_{i=1}^{9}\sum_{j=1}^{9} h_i(t)h_j(t)\chi^T(t)\left(PA_{eij} + A_{eij}^TP + P\Delta A_{eij} + \Delta A_{eij}^TP\right)\chi(t) \\
&\quad +2\chi^T(t)PD_ew(t).
\end{aligned}$$

Using Lemma 14.1, we have

$$P\Delta A_{eij} + \Delta A_{eij}^T P \le \varepsilon^{-1} P\tilde{M}_{eij}\tilde{M}_{eij}^T P + \varepsilon \tilde{N}_{eij}^T \tilde{N}_{eij},$$

then

$$\begin{aligned}\dot{V}(t) \le & \sum_{i=1}^{9}\sum_{j=1}^{9} h_i(t)h_j(t) \\ & \times\chi^T(t)\left(PA_{eij} + A_{eij}^T P + \varepsilon^{-1}P\tilde{M}_{eij}\tilde{M}_{eij}^T P + \varepsilon\tilde{N}_{eij}^T\tilde{N}_{eij}\right)\chi(t) \\ & +2\chi^T(t)PD_e w(t),\end{aligned}$$

From this and (14.18), we have $\dot{V}(t) < 0$, thus the closed-loop system in (14.16) with $w(t) = 0$ is asymptotically stable.

Next, we establish the $\mathcal{H}_\infty$ performance which is defined in (14.17). It can be shown that for any nonzero $w(t) \in \mathcal{L}_2[0, \infty)$ and $t > 0$,

$$\begin{aligned}\int_0^{t_f} z^T(t)z(t)dt &= V(0) - V(t_f) + \int_0^{t_f}\left(z^T(t)z(t) + \dot{V}(t)\right)dt \\ &\le \chi^T(0)P\chi(0) + \int_0^{t_f}\Bigg\{\sum_{i=1}^{9}\sum_{j=1}^{9}\chi^T(t)\left(PA_{eij} + A_{eij}^T P\right. \\ &\quad \left. +\varepsilon^{-1}P\tilde{M}_{eij}\tilde{M}_{eij}^T P + \varepsilon\tilde{N}_{eij}^T\tilde{N}_{eij}\right)\chi(t) \\ &\quad +\chi^T(t)E_e^T E_e\chi(t) + 2\chi^T(t)PD_e w(t)\Bigg\}dt.\end{aligned}$$

By Lemma 14.1, we have

$$\begin{aligned}\int_0^{t_f}\left\{w^T(t)D_e^T P\chi(t) + \chi^T(t)PD_e w(t)\right\}dt \le{} & \gamma^2\int_0^{t_f} w^T(t)w(t)dt \\ & + \gamma^{-2}\int_0^{t_f}\chi^T(t)PD_e D_e^T P\chi(t)dt.\end{aligned}$$

Applying the Schur complement to LMIs (14.18) results in

$$\begin{aligned}\sum_{i=1}^{9}\sum_{j=1}^{9} h_i(t)h_j(t)\Big(&PA_{eij} + A_{eij}^T P + \varepsilon^{-1}P\tilde{M}_{eij}\tilde{M}_{eij}^T P \\ &+\varepsilon\tilde{N}_{eij}^T\tilde{N}_{eij} + \gamma^{-2}PD_e D_e^T P + E_e^T E\Big) < 0, \quad (14.20)\end{aligned}$$

thus

$$\int_0^{t_f} z^T(t)z(t)dt \le \chi^T(0)P\chi(0) + \gamma^2\int_0^{t_f} w^T(t)w(t)dt.$$

Under the zero initial condition, that is, $\chi(0) = 0$, we have

$$\int_0^{t_f} z^T(t)z(t)dt \le \gamma^2 \int_0^{t_f} w^T(t)w(t)dt.$$

Therefore, the $\mathcal{H}_\infty$ performance is achieved. The proof is completed. ∎

Now, we present a solution to the robust $\mathcal{H}_\infty$ dynamic output feedback control problem.

Theorem 14.4. *Consider the uncertain T-S fuzzy system in (14.16), and for a prescribed constant scalar $\gamma > 0$, if there exist matrices $P_1 > 0$, $G_1 > 0$, X_i, Y_i and a constant $\varepsilon > 0$ such that the following LMIs hold:*

$$\begin{bmatrix} P_1 & I \\ \star & G_1 \end{bmatrix} > 0, \tag{14.21a}$$

$$\Theta_{ii} < 0, \quad i = 1, 2, \ldots, 9, \tag{14.21b}$$

$$\frac{1}{k-1}\Theta_{ii} + \frac{1}{2}(\Theta_{ij} + \Theta_{ji}) < 0, \quad 1 \le i \ne j \le 9, \tag{14.21c}$$

where

$$\Theta_{ij} = \begin{bmatrix} \Pi_{1ij} + \Pi_{1ij}^T & \Pi_{2ij} & \Pi_{3ij} & \Pi_{4ij} & \Pi_{5ij} \\ \star & -\gamma^2 I & 0 & 0 & 0 \\ \star & \star & -\varepsilon I & 0 & 0 \\ \star & \star & \star & -\varepsilon^{-1} I & 0 \\ \star & \star & \star & \star & -I \end{bmatrix},$$

$$\Pi_{ij} = \begin{bmatrix} \bar{A}_i G_1 + \bar{B}_i X_j & \bar{A}_i \\ Z_i & P_1 \bar{A}_i + Y_j \bar{C} \end{bmatrix}, \quad \Pi_{2ij} = \begin{bmatrix} \bar{D}_i \\ P_1 \bar{D}_i \end{bmatrix},$$

$$\Pi_{3ij} = \begin{bmatrix} \bar{M}_i & 0 \\ P_1 \bar{M}_i & 0 \end{bmatrix}, \quad \Pi_{4ij} = \begin{bmatrix} G_1 \bar{N}_{1i}^T + X_j^T \bar{N}_{2i}^T & 0 \\ \bar{N}_{1i}^T & 0 \end{bmatrix}, \quad \Pi_{5ij} = \begin{bmatrix} G_1 \bar{E}^T \\ \bar{E}^T \end{bmatrix},$$

then a desired robust $\mathcal{H}_\infty$ dynamic output feedback controller in the form of (14.15) exists, and its parameters can be given by

$$\begin{cases} A_{ci} = P_2^{-1}\left(Z_i - P_1 \bar{A}_i G_1 - P_2 B_{ci} \bar{C} G_1 - P_1 \bar{B}_i C_{ci} G_2^T\right) G_2^{-T}, \\ B_{ci} = P_2^{-1} Y_i, \\ C_{ci} = X_i G_2^{-T}, \end{cases} \tag{14.22}$$

where P_2 and G_2 are any nonsingular matrices satisfying

$$P_2 G_2^T = I - P_1 G_1. \tag{14.23}$$

Proof. According to Theorem 14.3, the matrix P is nonsingular. Partition P and its inverse respectively as

$$P = \begin{bmatrix} P_1 & P_2 \\ \star & P_3 \end{bmatrix}, \quad P^{-1} = \begin{bmatrix} G_1 & G_2 \\ \star & G_3 \end{bmatrix},$$

Without loss of generality, we can assume that both P_2 and G_2 are full rank matrices. Let

$$T_1 = \begin{bmatrix} G_1 & I \\ G_2^T & 0 \end{bmatrix}, \quad T_2 = \begin{bmatrix} I & P_1 \\ 0 & P_2^T \end{bmatrix}, \tag{14.24}$$

then T_1 and T_2 are nonsingular. Notice that $PP^{-1} = I$ leads to (14.23). It is easy to see from (14.21a) that $P_1 - G_1^{-1} > 0$, therefore $I - P_1 G_1$ is nonsingular. This ensures that there always exist nonsingular matrices P_1 and G_1 such that (14.21a) is satisfied. Then, we can conclude form (14.24) that the following equations always hold:

$$P_1 G_1 + P_2 G_2^T = I, \quad P = T_2 {T_1}^{-1}.$$

Note that the condition in Theorem 14.3 is equivalent to (14.20), thus applying the Schur complement to (14.20) results in

$$\sum_{i=1}^{9}\sum_{j=1}^{9} h_i(t)h_j(t) \begin{bmatrix} PA_{eij} + A_{eij}^T P & PD_e & P\tilde{M}_{eij} & \tilde{N}_{eij}^T & E_e^T \\ \star & -\gamma^2 I & 0 & 0 & 0 \\ \star & \star & -\varepsilon I & 0 & 0 \\ \star & \star & \star & -\varepsilon^{-1} I & 0 \\ \star & \star & \star & \star & -I \end{bmatrix} < 0. \tag{14.25}$$

Performing a congruence transformation to (14.25) by $\text{diag}\{T_1, I, I, I, I\}$, and defining

$$\begin{cases} Z_i = P_2 A_{ci} G_2^T + P_1 \bar{A}_i G_1 + P_2 B_{ci} \bar{C} G_1 + P_1 \bar{B}_i C_{ci} G_2^T, \\ Y_i = P_2 B_{ci}, \\ X_i = C_{ci} G_2^T, \end{cases}$$

we have

$$\sum_{i=1}^{9}\sum_{j=1}^{9} h_i(t)h_j(t)\Theta_{ij} < 0.$$

Then by Lemma 14.2, (14.21b)–(14.21c) can easily be obtained. The proof is completed. ■

14.5 Simulation Results

In this section, a numerical example is provided to test the effectiveness of the robust $\mathcal{H}_\infty$ dynamic output feedback controller design method proposed in Section 3. The hypersonic vehicle model parameter values are borrowed from [166]. The equilibrium point of the nonlinear vehicle dynamics for the fuzzy model is listed in Table 14.1. The low and up bound of V and h are chosen as: $V_B = 9000ft/s$, $V_S = 6400ft/s$, $h_B = 10000ft$, $h_S = 7000ft$, and the other states are chosen according to the flight envelop. The modeling of parameter uncertainties is similar to [15], and in this work, the parameters of $\left(C_L^{\alpha}, C_L^{\delta_e}, C_L^0,\ C_D^{\alpha^2}, C_D^{\alpha}, C_D^{\delta_e^2}, C_D^{\delta_e}, C_D^0, C_{M,\alpha}^{\alpha^2}, C_{M,\alpha}^{\alpha}, C_{M,\alpha}^0, C_{M,\delta_e}\right)$ are assumed to be uncertain, and these uncertainties are assumed to lie within $\pm 10\%$ of nominal values, respectively. The uncertainty of S lies within $\pm 5\%$ of nominal value, so does the mean aerodynamic chord $\bar{c}$. According to [72], the disturbance $f(x(t), t)$ is assumed to be bounded, which can be regarded as a gust of wind in aerospace. Then, by using the T-S fuzzy modeling method proposed in Section 14.3, a T-S fuzzy model can be obtained with the membership functions shown in Figs. 14.2–14.3.

The control objective is to track a step signal (predefined) with respect to a trim condition. The input reference commands are chosen as step inputs, so each command will pass through a prefilter as

$$H(s) = \frac{\omega_n^2}{s^2 + 2\zeta\omega_n s + \omega_n^2},$$

where ζ denotes damping ratio, ω_n stands for natural frequency, and they are assumed to be 0.9 and $0.01rad/s$, respectively. The output of the prefilter is defined as a reference command which is to be tracked. In simulation, to illustrate the effectiveness of the proposed controller, we will use the original nonlinear model (not the T-S fuzzy linear model) to test the performance of the control system. By Section 3, the fuzzy dynamic output feedback controller can be gotten. Here, the reference commands for velocity and altitude are chosen as $1000ft/s$ and $10000ft$, respectively.

We consider the following three cases:

Case (I): $\dot{\eta}_1$ and $\dot{\eta}_2$ are both unmeasurable, which implies $C = \text{diag}\{1,1,1,1,\ 1,1,0,1,0\}$. In this case, the simulation results are shown in Figs. 14.4–14.5. We can see from Fig. 14.4 that the controller provides a stable tracking for the reference trajectories. The control input is shown in Fig. 14.5.

Case (II): The flexible dynamics are all unmeasurable, that is, $C = \text{diag}\{1,1,1,1,1,0,0,0,0\}$. In this case, Fig. 14.6 presents the tracking performance, and Fig. 14.7 shows the input and states, from which we can conclude that even the flexible dynamics are all unmeasure, the dynamic output feedback controller can track the reference command well.

Case (III): Only the output is measurable, that is, $C = \text{diag}\{1,1,0,0,0,0,0,0,0\}$. This is the worst situation. In this case, the useful information is particularly limited. We can design the desired dynamic output feedback controller by Theorem . Figs. 14.8–14.9 show the simulation results for the reference output tracking.

Table 14.1. Equilibrium point

State	Value
h	$85000ft$
V	$7702.0808ft \cdot s^{-1}$
α	1.5153 deg
θ	1.5153 deg
Q	$0deg \cdot s^{-1}$
η_1	1.5122
$\dot{\eta}_1$	0
η_2	1.2144
$\dot{\eta}_2$	0
Φ	0.2514
δ_e	11.4635

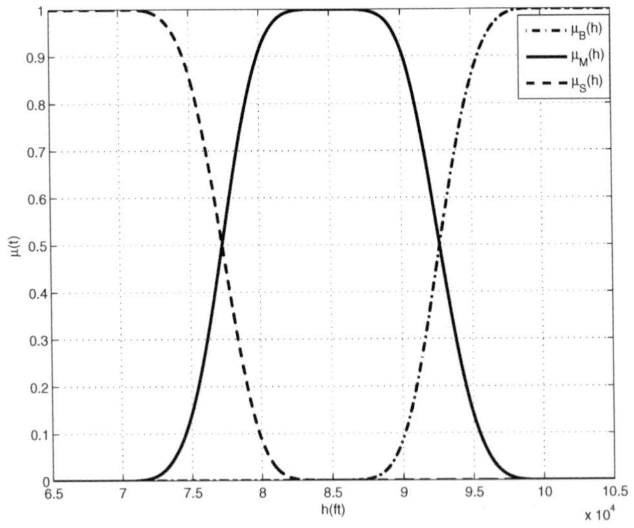

Fig. 14.2. Membership functions of h

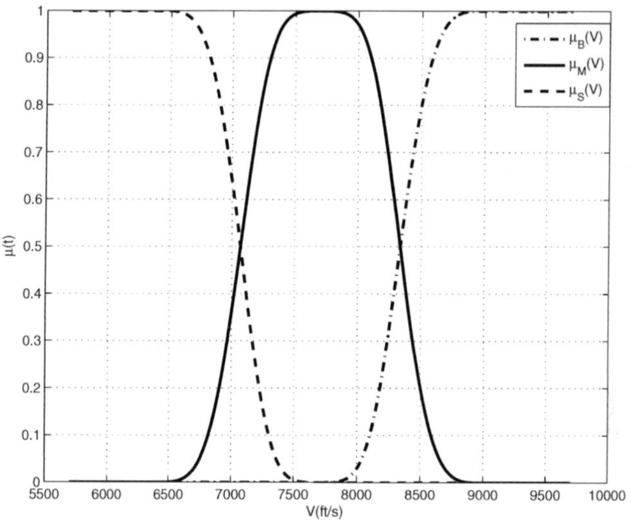

Fig. 14.3. Membership functions of V

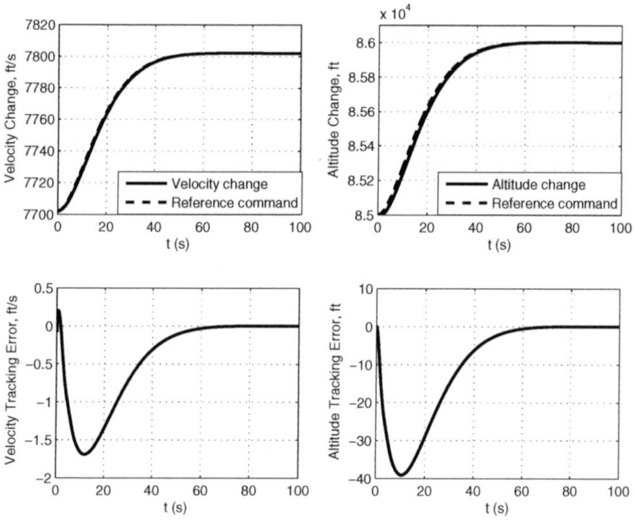

Fig. 14.4. Case I: Tracking performance of closed-loop simulation

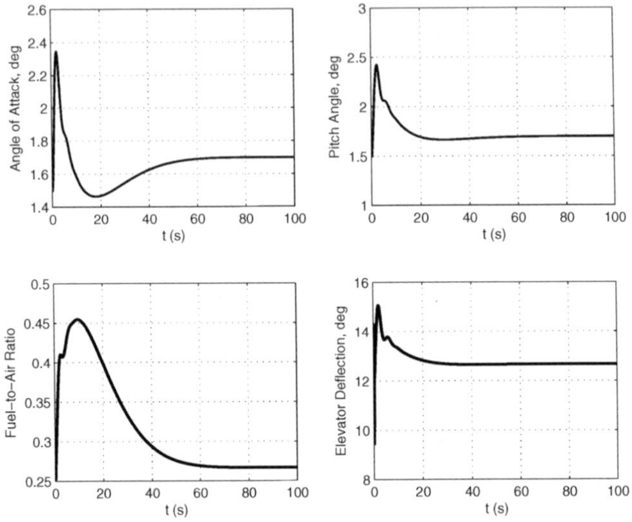

Fig. 14.5. Case I: Angle of attack, flight path angle and the inputs of the plant

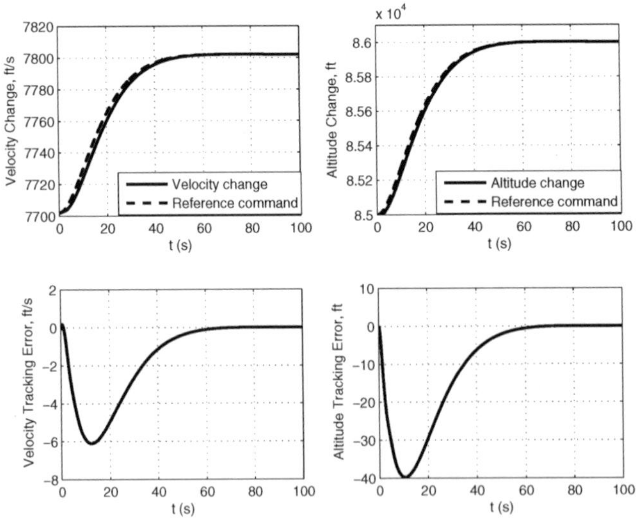

Fig. 14.6. Case II: Tracking performance of closed-loop simulation

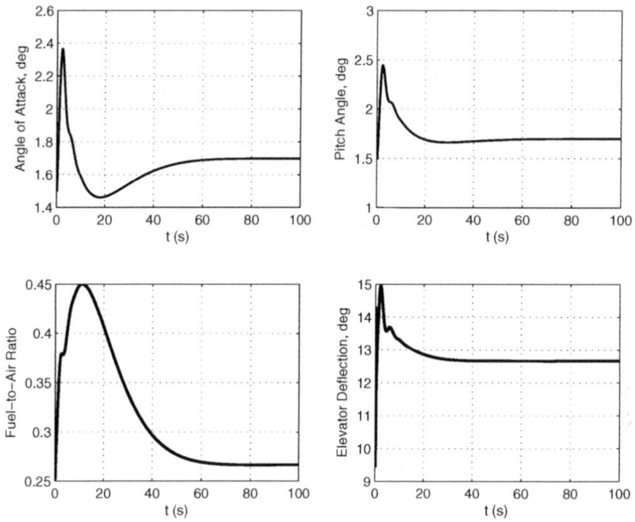

Fig. 14.7. Case II: Angle of attack, flight path angle and the inputs of the plant

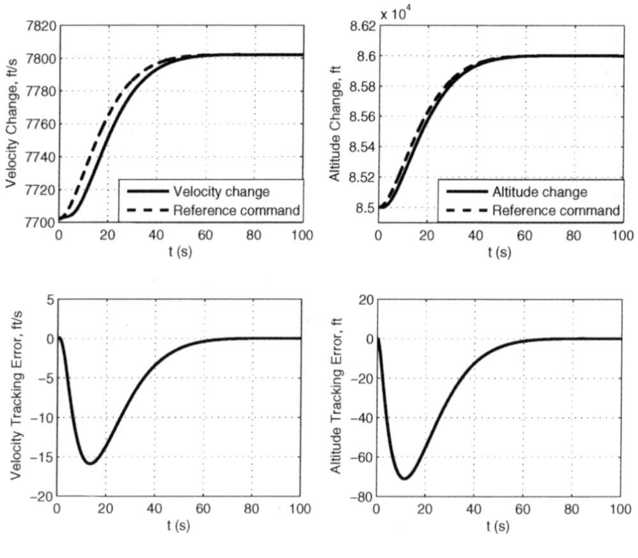

Fig. 14.8. Case III: Tracking performance of closed-loop simulation

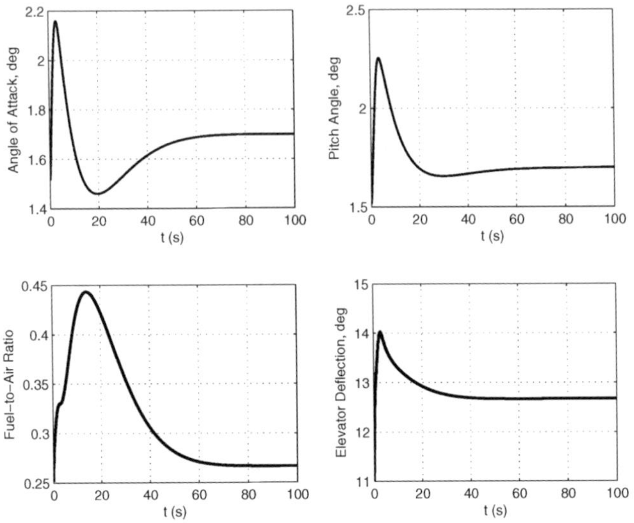

Fig. 14.9. Case III: Angle of attack, flight path angle and the inputs of the plant

14.6 Conclusion

In this chapter, the problem of robust $\mathcal{H}_\infty$ for the nonlinear longitudinal model of flexible AHVs via dynamic output feedback control has been addressed. A T-S fuzzy model has been constructed to represent the nonlinear dynamics of flexible AHVs, which includes parameter uncertainties and extraneous disturbances. Then, a robust $\mathcal{H}_\infty$ dynamic output feedback controller design method has been proposed. Sufficient conditions for designing such a controller have been proposed in terms of LMIs. Moreover, simulations have been carried out to demonstrate the effectiveness of the proposed design scheme.

References

1. An, J., Wen, G., Lin, C.: New results on a delay-derivative-dependent fuzzy $\mathcal{H}_\infty$ filter design for T-S fuzzy systems. IEEE Transactions on Fuzzy Systems 19(4), 770–779 (2011)
2. Antoulas, A.C.: Approximation of Large-Scale Dynamic Systems. SIAM Press (2005)
3. Assawinchaichote, W., Nguang, S.K.: $\mathcal{H}_\infty$ filtering for fuzzy dynamic systems with $\mathcal{D}$ stability constraints. IEEE Transactions on Circuits and Systems, Part I: Fundamental Theory and Applications 50(11), 1503–1508 (2003)
4. Assawinchaichote, W., Nguang, S.K., Shi, P.: $\mathcal{H}_\infty$ output feedback control design for uncertain fuzzy singularly perturbed systems: an LMI approach. Automatica 40(12), 2147–2152 (2004)
5. Bai, Y., Zhuang, H.Q., Roth, Z.S.: Fuzzy logic control to suppress noises and coupling effects in a laser tracking system. IEEE Transactions on Control System Technology 13(1), 113–121 (2005)
6. Baker, G.A.J., Graves-Morris, P.: Padé Approximants. Cambridge University Press, New York (1996)
7. Baturone, I., Moreno-Velo, F.J., Sanchez-Solano, S., Ollero, A.: Automatic design of fuzzy controllers for car-like autonomous robots. IEEE Transactions on Fuzzy Systems 12(4), 447–465 (2004)
8. Bertin, J.J., Cumming, R.M.: Fifty years of hypersonics: where we've been, where we're going. Progress in Aerospace Sciences 39, 511–536 (2003)
9. Bingul, Z., Cook, G.E., Strauss, A.M.: Application of fuzzy logic to spatial thermal control in fusion welding. IEEE Transactions on Industry Applications 36(6), 1523–1530 (2000)
10. Bolender, M., Doman, D.: A non-linear model for the longitudinal dynamics of a hypersonic air-breathing vehicle. Air Force Research Lab Wright-Patterson AFB OH 45433 (2006)
11. Bolender, M., Doman, D.: A nonlinear longitudinal dynamical model of an air-breathing hypersonic vehicle. Journal of Spacecraft and Rockets 44(2), 374–387 (2007)
12. Bonissone, P.P., Badami, V., Chiang, K.H., Khedkar, P.S., Marcelle, K.W., Schutten, M.J.: Industrial applications of fuzzy logic at general electric. Proceedings of the IEEE 38(3), 450–465 (1995)

13. Boroushaki, M., Ghofrani, M.B., Lucas, C., Yazdanpanah, M.J.: Identification and control of a nuclear reactor core (VVER) using recurrent neural networks and fuzzy systems. IEEE Transactions on Nuclear Science 50(1), 159–174 (2003)
14. Boukezzoula, R., Galichet, S., Foulloy, L.: Observer-based fuzzy adaptive control for a class of nonlinear systems: real-time implementation for a robot wrist. IEEE Transactions on Control System Technology 12(3), 340–351 (2004)
15. Buschek, H., Calise, A.J.: Uncertainty modeling and fixed-order controller design for a hypersonic vehicle model. Journal of Guidance, Control, and Dynamics 20(1), 42–48 (1997)
16. Campos, V.C.S., Souza, F.O., Torres, L.A.B., Palhares, R.M.: New stability conditions based on piecewise fuzzy Lyapunov functions and tensor product transformations. IEEE Transactions on Fuzzy Systems 21(4), 748–760 (2013)
17. Cao, S.G., Rees, N.W., Feng, G.: Stability analysis of fuzzy control systems. IEEE Transactions on Systems, Man, and Cybernetics, Part B: Cybernetics 26(1), 201–204 (1996)
18. Cao, Y.-Y., Sun, Y.-X., Lam, J.: Delay-dependent robust $\mathcal{H}_\infty$ control for uncertain systems with time-varying delays. In: IEE Proceedings of Control Theory and Applications, vol. 145(3), pp. 338–344 (1998)
19. Chadli, M., Guerra, T.M.: LMI solution for robust static output feedback control of discrete Takagi-Sugeno fuzzy models. IEEE Transactions on Fuzzy Systems 20(6), 1160–1165 (2012)
20. Chang, W., Sun, C.: Constrained fuzzy controller design of discrete Takagi-Sugeno fuzzy models. Fuzzy Sets and Systems 133(1), 37–55 (2003)
21. Chang, X.-H.: Robust nonfragile $\mathcal{H}_\infty$ filtering of fuzzy systems with linear fractional parametric uncertainties. IEEE Transactions on Fuzzy Systems 20(6), 1001–1011 (2012)
22. Chavez, F.R., Schmidt, D.K.: Analytical aeropropulsive/aeroelastic hypersonic-vehicle model with dynamic analysis. Journal of Guidance, Control, and Dynamics 17(6), 1308–1319 (1994)
23. Chen, B., Liu, X.P.: Delay-dependent robust $\mathcal{H}_\infty$ control for T-S fuzzy systems with time delay. IEEE Transactions on Fuzzy Systems 13(4), 544–556 (2005)
24. Chen, B.S., Tseng, C.S., Uang, H.J.: Mixed $\mathcal{H}_2/\mathcal{H}_\infty$ fuzzy output feedback control design for nonlinear dynamic systems: an LMI approach. IEEE Transactions on Fuzzy Systems 8(3), 249–265 (2000)
25. Chen, C.L., Feng, G., Sun, D., Zhu, Y.: $\mathcal{H}_\infty$ output feedback control of discrete-time fuzzy systems with application to chaos control. IEEE Transactions on Fuzzy Systems 13(4), 531–543 (2005)
26. Chen, S.S., Chang, Y.C., Su, S.F., Chung, S.L., Lee, T.T.: Robust static output-feedback stabilization for nonlinear discrete-time systems with time delay via fuzzy control approach. IEEE Transactions on Fuzzy Systems 13(2), 263–272 (2005)
27. Chen, Y.-J., Ohtake, H., Tanaka, K., Wang, W.-J.: Relaxed stabilization criterion for T-S fuzzy systems by minimum-type piecewise-Lyapunov-function-based switching fuzzy controller. IEEE Transactions on Fuzzy Systems 20(6), 1166–1173 (2012)
28. Choi, D.J., Park, P.G.: $\mathcal{H}_\infty$ state-feedback controller design for discrete-time fuzzy systems using fuzzy weighting-dependent Lyapunov functions. IEEE Transactions on Fuzzy Systems 11(2), 271–278 (2003)

29. Cuesta, F., Gordillo, F., Aracil, J., Ollero, A.: Stability analysis of nonlinear multivariable Takagi-Sugeno fuzzy control systems. IEEE Transactions on Fuzzy Systems 7(5), 508–520 (1999)
30. Curran, E.: Scramjet engines: the first forty years. Journal of Propulsion and Power 17(6), 1138–1148 (2001)
31. Delmotte, F., Guerra, T.M., Kruszewski, A.: Discrete Takagi-Sugeno's fuzzy models: reduction of the number of LMI in fuzzy control techniques. IEEE Transactions on Systems, Man, and Cybernetics, Part B: Cybernetics 38(5), 1423–1427 (2008)
32. Desoer, C.A., Vidyasagar, M.: Feedback Systems: Input-Output Properties. Academic, New York (1975)
33. de Souza, C.E., Li, X.: Delay-dependent robust $\mathcal{H}_\infty$ control of uncertain linear state-delayed systems. Automatica 35(7), 1313–1321 (1999)
34. Ding, B.: Stabilization of Takagi-Sugeno model via nonparallel distributed compensation law. IEEE Transactions on Fuzzy Systems 18(6), 188–194 (2010)
35. Ding, B.: Homogeneous polynomially nonquadratic stabilization of discrete-time Takagi-Sugeno systems via nonparallel distributed compensation law. IEEE Transaction on Fuzzy Systems 18(5), 994–1000 (2010)
36. Dong, H., Wang, Z., Ho, D.W.C., Gao, H.: Robust $\mathcal{H}_\infty$ fuzzy output-feedback control with multiple probabilistic delays and multiple missing measurements. IEEE Transactions on Fuzzy Systems 18(4), 712–725 (2010)
37. Du, B., Lam, J., Shu, Z., Wang, Z.: A delay-partitioning projection approach to stability analysis of continuous systems with multiple delay components. IET Control Theory and Applications 3(4), 383–390 (2009)
38. Du, H., Zhang, N.: Fuzzy control for nonlinear uncertain electrohydraulic active suspensions with input constraint. IEEE Transactions on Fuzzy Systems 17(2), 343–356 (2009)
39. El Ghaoui, L., Oustry, F., Ait Rami, M.: A cone complementarity linearization algorithm for static output-feedback and related problems. IEEE Transactions on Automatic Control 42(8), 1171–1176 (1997)
40. Fang, C.-H., Liu, Y.-S., Kau, S.-W., Lin, H., Lee, C.-H.: A new LMI-based approach to relaxed quadratic stabilization of T-S fuzzy control systems. IEEE Transactions on Fuzzy Systems 14(3), 386–397 (2006)
41. Fei, Z., Gao, H., Shi, P.: New results on stabilization of Markovian jump systems with time delay. Automatica 45(10), 2300–2306 (2009)
42. Feng, G.: Approaches to quadratic stabilization of uncertain fuzzy dynamic systems. IEEE Transactions on Circuits and Systems I: Fundamental Theory and Applications 48(6), 760–769 (2001)
43. Feng, G., Ma, J.: Quadratic stabilization of uncertain discrete-time fuzzy dynamic systems. IEEE Transactions on Circuits and Systems I: Fundamental Theory and Applications 48(11), 1337–1343 (2001)
44. Feng, G., Sun, D.: Generalized $\mathcal{H}_\infty$ controller synthesis of fuzzy dynamic systems based on piecewise Lyapunov functions. IEEE Transactions on Circuits and Systems I: Fundamental Theory and Applications 49(12), 1843–1850 (2002)
45. Feng, G.: Controller synthesis of fuzzy dynamic systems based on piecewise Lyapunov functions. IEEE Transactions on Fuzzy Systems 11(10), 605–612 (2003)
46. Feng, G.: $\mathcal{H}_\infty$ controller design of fuzzy dynamic systems based on piecewise Lyapunov functions. IEEE Transactions on Systems, Man, and Cybernetics, Part B: Cybernetics 34(1), 283–292 (2004)

47. Feng, G.: Stability analysis of discrete time fuzzy dynamic systems based on piecewise Lyapunov functions. IEEE Transactions on Fuzzy Systems 12(1), 22–28 (2004)
48. Feng, G., Chen, C.L., Sun, D., Guan, X.P.: $\mathcal{H}_\infty$ controller synthesis of fuzzy dynamic systems based on piecewise Lyapunov functions and bilinear matrix inequalities. IEEE Transactions on Fuzzy Systems 13(1), 94–103 (2005)
49. Feng, G.: Robust $\mathcal{H}_\infty$ filtering of fuzzy dynamic systems. IEEE Transactions on Aerospace and Electronic Systems 41(2), 658–670 (2005)
50. Feng, G.: A survey on analysis and design of model-based fuzzy control systems. IEEE Transactions on Fuzzy Systems 14(5), 676–697 (2006)
51. Feng, Z., Lam, J., Gao, H.: α-dissipativity analysis of singular time-delay systems. Automatica 47(11), 2548–2552 (2011)
52. Feng, M., Harris, C.J.: Piecewise Lyapunov stability conditions of fuzzy systems. IEEE Transactions on Systems, Man, and Cybernetics, Part B: Cybernetics 31(2), 259–262 (2001)
53. Feng, Z., Lam, J., Gao, H.: Delay-dependent robust $\mathcal{H}_\infty$ controller synthesis for discrete singular delay systems. International Journal of Robust and Nonlinear Control 21(16), 1880–1902 (2011)
54. Feng, Z., Lam, J.: Robust reliable dissipative filtering for discrete delay singular systems. Signal Processing 92(12), 3010–3025 (2012)
55. Fidan, B., Mirmirani, M., Ioannou, P.: Air-breathing hypersonic flght control. In: The 16th IFAC Symposium on Automatic Control in Aerospace, pp. 571–576 (2004)
56. Fiorentini, L., Serrani, A., Bolender, M.A., Doman, D.B.: Nonlinear robust adaptive control of flexible air-breathing hypersonic vehicles. Journal of Guidance Control, and Dynamics 32(2), 401–416 (2009)
57. Fridman, E.: New Lyapunov-Krasovskii functionals for stability of linear retarded and neutral type systems. Systems and Control Letters 43(4), 309–319 (2001)
58. Fridman, E., Shaked, U.: A descriptor system approach to $\mathcal{H}_\infty$ control of linear time-delay systems. IEEE Transactions on Automatic Control 47(2), 253–270 (2002)
59. Fridman, E., Shaked, U.: An improved stabilisation method for linear time-delay systems. IEEE Transactions on Automatic Control 47(11), 1931–1937 (2002)
60. Fridman, E., Shaked, U.: Delay-dependent stability and $\mathcal{H}_\infty$ control: constant and time-varying delays. International Journal of Control 76(1), 48–60 (2003)
61. Fridman, E., Shaked, U.: Input-output approach to stability and $\mathcal{L}_2$-gain analysis of systems with time-varying delays. Systems and Control Letters 55(12), 1041–1053 (2006)
62. Gahinet, P., Apkarian, P.: A linear matrix inequality approach to $\mathcal{H}_\infty$ control. International Journal of Robust and Nonlinear Control 4(4), 421–448 (1994)
63. Gao, H., Wang, C.: Delay-dependent robust $\mathcal{H}_\infty$ and $\mathcal{L}_2$-$\mathcal{L}_\infty$ filtering for a class of uncertain nonlinear time-delay systems. IEEE Transactions on Automatic Control 48(9), 1661–1666 (2003)
64. Gao, H., Lam, J., Wang, C., Xu, S.: $\mathcal{H}_\infty$ model reduction for discrete time-delay systems: delay independent and dependent approaches. International Journal of Control 77(4), 321–335 (2004)
65. Gao, H., Chen, T.: New results on stability of discrete-time systems with time-varying state delay. IEEE Transactions on Automatic Control 52(2), 328–334 (2007)

66. Gao, H., Liu, X., Lam, J.: Stability analysis and stabilization for discrete-time fuzzy systems with time-varying delay. IEEE Transactions on Systems, Man, and Cybernetics, Part B: Cybernetics 39(2), 306–317 (2009)
67. Gao, H., Zhao, Y., Chen, T.: $\mathcal{H}_\infty$ fuzzy control of nonlinear systems under unreliable communication links. IEEE Transactions on Fuzzy Systems 17(2), 265–278 (2009)
68. Gao, H., Zhao, Y., Lam, J., Chen, K.: $\mathcal{H}_\infty$ fuzzy filtering of nonlinear systems with intermittent measurements. IEEE Transactions on Fuzzy Systems 17(2), 291–300 (2009)
69. Gao, H., Fei, Z., Lam, J., Du Further, B.: results on exponential estimates of Markovian jump systems with mode-dependent time-varying delays. IEEE Transactions on Automatic Control 56(1), 223–229 (2011)
70. Gao, Z., Shi, X., Steven, X.D.: Fuzzy state/disturbance observer design for T-S fuzzy systems with application to sensor fault estimation. IEEE Transactions on Systems, Man, and Cybernetics, Part B: Cybernetics 38(3), 875–880 (2008)
71. Gassara, H., El Hajjaji, A., Chaabane, M.: Observer-based robust $\mathcal{H}_\infty$ reliable control for uncertain T-S fuzzy systems with state time delay. IEEE Transactions on Fuzzy Systems 18(6), 1027–1040 (2010)
72. Gibson, T.E., Crespo, L.G., Annaswamy, A.M.: Adaptive control of hypersonic vehicles in the presence of modeling uncertainties. In: Proceedings of American Control Conference, St. Louis, Missouri, USA, June 10-12, pp. 3178–3183 (2009)
73. Glover, K.: All optimal Hankel norm approximations of linear multivariable systems and their $\mathcal{L}_\infty$ error bounds. International of Control AC-39(6), 1115–1193 (1984)
74. Gouaisbaut, F., Peaucelle, D.: Delay-dependent stability analysis of linear time delay systems. In: IFAC Workshop on Time Delay System, L'Aquila, Italy, July 10-12 (2006)
75. Grimme, E.J.: Krylov Projection Methods for Model Reduction. Ph.D. Thesis, ECE Dept., University of Illinois, Urbana-Champaign (1997)
76. Groves, K.P., Serrani, A., Yurkovich, S., Bolender, M.A., Doman, D.B.: Anti-windup control for an air-breathing hypersonic vehicle model. AIAA Paper, pp. 2006–6557 (2006)
77. Gu, K.: An integral inequality in the stability problem of time-delay systems. In: Proceedings of the 39th IEEE Conference on Decision and Control, Sydney, Australia, December 12-15, pp. 2805–2810 (2000)
78. Gu, K., Kharitonov, V., Chen, J.: Stability of Time-Delay Systems. Birkhauser, Cambridge (2003)
79. Gu, K., Niculescu, S.-I.: Survey on recent results in the stability and control of time-delay systems. Journal of Dynamic Systems, Measurement, and Control 125(2), 158–165 (2003)
80. Gu, K., Zhang, Y., Xu, S.: Small gain problem in coupled differential-difference equations, time-varying delays, and direct Lyapunov method. International Journal of Robust and Nonlinear Control 21(4), 429–451 (2010)
81. Guan, X., Chen, C.: Delay-dependent guaranteed cost control for T-S fuzzy systems with time delays. IEEE Transactions on Fuzzy Systems 12(2), 236–249 (2004)
82. Guerra, T.M., Vermeiren, L.: LMI-based relaxed nonquadratic stabilization conditions for nonlinear systems in the Takagi-Sugeno's form. Automatica 40(5), 823–829 (2004)

83. Gugercin, S., Antoulas, A.C.: A survey of model reduction by balanced truncation and some new results. International of Control 77(8), 748–766 (2004)
84. Gysen, B., Paulides, J., Janssen, J., Lomonova, E.: Active eletromagnetic suspension system for improved vehicle dynamics. IEEE Transactions on Vehicular Technology 59(3), 1156–1163 (2010)
85. Heiser, W.H., Pratt, D.T., Daley, D.H., Mehta, U.B.E.: Hypersonic Airbreathing Propulsion. AIAA, Washington, DC (1994)
86. Higham, D.: An algorithmic introduction to numerical simulation of stochastic differential equations. SIAM Review 43(3), 525–546 (2001)
87. Hill, D.J., Moylan, P.J.: Dissipative dynamical systems: basic input-output and state properties. Journal of the Franklin Institute 309(5), 327–357 (1980)
88. Hong, S.K., Langari, R.: An LMI-based $\mathcal{H}_\infty$ fuzzy control system design with T-S framework. Information Sciences 123(3-4), 163–179 (2000)
89. Huang, D., Nguang, S.: Robust $\mathcal{H}_\infty$ static output feedback control of fuzzy systems: an ILMI approach. IEEE Transactions on Systems, Man, and Cybernetics, Part B: Cybernetics 36(1), 216–222 (2006)
90. Huang, S.-J., He, X.-Q., Zhang, N.-N.: New results on $\mathcal{H}_\infty$ filter design for nonlinear systems with time delay via T-S fuzzy models. IEEE Transactions on Fuzzy Systems 19(1), 193–199 (2011)
91. Hurley, W.G., Wolfle, W.H.: Electromagnetic design of magnetic suspension system. IEEE Transactions on Education 40(2), 124–130 (1997)
92. Iwasaki, T., Skelton, R.E.: All controllers for the general $\mathcal{H}_\infty$ control problems: LMI existence conditions and state space formulas. Automatica 30(8), 1307–1317 (1994)
93. Jankovsky, P., Sigthorsson, D., Serrani, A., Yurkovich, S., Bolender, M., Doman, D.: Output feedback control and sensor placement for a hypersonic vehicle model. In: AIAA Guidance, Navigation and Control Conference and Exhibit, pp. 2007–6327 (2007)
94. Jia, X., Zhang, D., Hao, X.: Fuzzy $\mathcal{H}_\infty$ tracking control for nonlinear networked control systems in T-S fuzzy model. IEEE Transactions on Systems, Man, and Cybernetics, Part B: Cybernetics 39(4), 1073–1079 (2009)
95. Jiang, B., Mao, Z., Shi, P.: $\mathcal{H}_\infty$ filter design for a class of networked control systems via T-S fuzzy-model approach. IEEE Transactions on Fuzzy Systems 18(1), 201–208 (2010)
96. Jiang, B., Gao, Z., Shi, P., Xu, Y.: Adaptive fault-tolerant tracking control of near-space vehicle using Takagi-Sugeno fuzzy models. IEEE Transactions on Fuzzy Systems 18(5), 1000–1007 (2010)
97. Jiang, B., Zhang, K., Shi, P.: Integrated fault estimation and accommodation design for discrete-time Takagi-Sugeno fuzzy systems with actuator faults. IEEE Transactions on Fuzzy Systems 19(2), 291–304 (2011)
98. Jiang, X., Han, Q.-L., Yu, X.: Stability criteria for linear discrete- time systems with interval-like time-varying delay. In: Proceedings of the American Control Conference, Portland, OR, USA, June 8-10, pp. 2817–2822 (2005)
99. Jiang, X., Han, Q.-L.: Robust $\mathcal{H}_\infty$ control for uncertain Takagi-Sugeno fuzzy systems with interval time-varying delay. IEEE Transactions on Fuzzy Systems 15(2), 321–331 (2007)
100. Joh, J., Chen, Y.H., Langari, R.: On the stability issues of linear Takagi-Sugeno fuzzy models. IEEE Transactions on Fuzzy Systems 6(3), 402–410 (1998)

101. Johansen, T.A.: Fuzzy model based control: stability robustness and performance issues. IEEE Transactions on Fuzzy Systems 2(1), 221–233 (1994)
102. Johansson, M., Rantzer, A., Arzen, K.E.: Piecewise quadratic stability of fuzzy systems. IEEE Transactions on Fuzzy Systems 7(6), 713–722 (1999)
103. Joo, S., Seo, J.H.: Design and analysis of the nonlinear feedback linearizing control for an electromagnetic suspension system. IEEE Transactions on Control Systems Technology 5(1), 135–144 (1997)
104. Kalman, R.E.: A new approach to linear filtering and prediction problems. Transactions of the ASME – Journal of Basic Engineering 82(Series D), 35–45 (1960)
105. Kang, B.J., Hung, L.S., Kuo, S.K., Lin, S.C., Liaw, C.M.: $\mathcal{H}_\infty$ 2DOF control for the motion of a magnetic suspension positioning stage driven by inverter-fed linear motor. Mechatronics 13(7), 677–696 (2003)
106. Kao, C., Lincoln, B.: Simple stability criteria for systems with time-varying delays. Automatica 40(8), 1429–1434 (2004)
107. Kavranoglu, D.: Zeroth order $\mathcal{H}_\infty$ norm approximation of multivariable systems. Numerical Functional Analysis and Optimization 14(1-2), 89–101 (1993)
108. Kim, E., Lee, H.: New approaches to relaxed quadratic stability condition of fuzzy control systems. IEEE Transactions on Fuzzy Systems 8(5), 523–534 (2000)
109. Kim, E., Kim, D.: Stability analysis and synthesis for an affine fuzzy system via LMI and ILMI: discrete case. IEEE Transactions on Systems, Man, and Cybernetics, Part B: Cybernetics 31(1), 132–140 (2001)
110. Kim, E., Kim, S.: Stability analysis and synthesis for an affine fuzzy control system via LMI and ILMI: continuous case. IEEE Transactions on Fuzzy Systems 10(3), 391–400 (2002)
111. Kim, S.H., Park, P.: $\mathcal{H}_\infty$ state-feedback control design for fuzzy systems using Lyapunov functions with quadratic dependence on fuzzy weighting functions. IEEE Transactions on Fuzzy Systems 16(6), 1655–1663 (2008)
112. Kim, S.H., Park, P.: Observer-based relaxed $\mathcal{H}_\infty$ control for fuzzy systems using a multiple Lyapunov function. IEEE Transactions on Fuzzy Systems 17(2), 476–484 (2009)
113. Kim, S.H., Park, P.: Relaxed $\mathcal{H}_\infty$ stabilization conditions for discrete-time fuzzy systems with interval time-varying delays. IEEE Transactions on Fuzzy Systems 17(6), 1441–1449 (2009)
114. Kim, S.H., Park, P.: $\mathcal{H}_\infty$ state-feedback-control design for discrete-time fuzzy systems using relaxation technique for parameterized LMI. IEEE Transactions on Fuzzy Systems 18(5), 985–993 (2010)
115. Kim, S.H.: Improved approach to robust $\mathcal{H}_\infty$ stabilization of discrete-time T-S fuzzy systems with time-varying delays. IEEE Transactions on Fuzzy Systems 18(5), 1008–1015 (2010)
116. Kiriakidis, K.: Robust stabilization of the Takagi-Sugeno fuzzy model via bilinear matrix inequalities. IEEE Transactions on Fuzzy Systems 9(2), 269–277 (2001)
117. Kolmanovskii, V.B., Niculescu, S.-I., Richard, J.-P.: On the Lyapunov-Krasovskii functionals for stability analysis of linear delay systems. International Journal of Control 72(4), 374–384 (1999)
118. Kolmanovskii, V.B., Richard, J.-P.: Stability of some linear systems with delays. IEEE Transactions on Automatic Control 44(5), 984–989 (1999)

119. Kuipers, M., Mirmirani, M., Ioannou, P., Huo, Y.: Adaptive control of an aeroelastic airbreathing hypersonic cruise vehicle. AIAA Paper, pp. 2007–6326 (2007)
120. Kung, C.C., Chen, T.H., Chen, C.H.: $\mathcal{H}_\infty$ state feedback controller design for T-S fuzzy systems based on piecewise Lyapunov function. In: Proceedings of the 14th IEEE International Conference on Fuzzy Systems, Reno, NV, pp. 708–713 (May 2005)
121. Lam, H.K., Leung, F.H.F., Tam, P.K.S.: Nonlinear state feedback controller for nonlinear systems: stability analysis and design based on fuzzy plant model. IEEE Transactions on Fuzzy Systems 9(4), 657–661 (2001)
122. Lam, H.K., Leung, F.H.E.: LMI-based stability and performance conditions for continuous-time nonlinear systems in Takagi-Sugeno's form. IEEE Transactions on Systems, Man, and Cybernetics, Part B: Cybernetics 37(5), 1396–1406 (2007)
123. Lam, H.K., Seneviratne, L.D.: Stability analysis of interval type-2 fuzzy-model-based control systems. IEEE Transactions on Systems, Man, and Cybernetics, Part B: Cybernetics 38(3), 617–628 (2008)
124. Lam, H.K., Narimani, M.: Stability analysis and performance design for fuzzy-model-based control system under imperfect premise matching. IEEE Transactions on Fuzzy Systems 17(4), 949–961 (2009)
125. Lam, H.K., Narimani, M.: Quadratic-stability analysis of fuzzy-model-based control systems using staircase membership functions. IEEE Transactions on Fuzzy Systems 18(1), 125–137 (2010)
126. Lam, H.K.: LMI-based stability analysis for fuzzy-model-based control systems using artificial T-S fuzzy model. IEEE Transactions on Fuzzy Systems 19(3), 505–513 (2011)
127. Lam, H.K.: Stabilization of nonlinear systems using sampled-data output-feedback fuzzy controller based on polynomial-fuzzy-model-based control approach. IEEE Transactions on Systems, Man, and Cybernetics, Part B: Cybernetics 42(1), 258–267 (2012)
128. Landry, M., Campbell, S.A., Morris, K., Aguilar, C.O.: Dynamics of an inverted pendulum with delayed feedback control. SIAM Journal on Applied Mathematics 4(2), 333–351 (2005)
129. Lee, D., Park, J., Joo, Y.: Improvement on nonquadratic stabilization of discrete-time Takagi-Sugeno fuzzy systems: multiple-parameterization approach. IEEE Transactions on Fuzzy Systems 18(2), 425–429 (2010)
130. Lee, D., Park, J., Joo, Y.: A new fuzzy Lyapunov function for relaxed stability condition of continuous-time Takagi-Sugeno fuzzy systems. IEEE Transactions on Fuzzy Systems 19(4), 785–791 (2011)
131. Lee, H., Park, J., Chen, G.: Robust fuzzy control of nonlinear systems with parametric uncertainties. IEEE Transactions on Fuzzy Systems 9(2), 369–379 (2001)
132. Lee, K., Kim, J., Jeung, E.: Output feedback robust $\mathcal{H}_\infty$ control of uncertain fuzzy dynamic systems with time-varying delay. IEEE Transactions on Fuzzy Systems 8(6), 657–664 (2000)
133. Leephakpreeda, T.: $\mathcal{H}_\infty$ stability robustness of fuzzy control systems. Automatica 35(8), 1467–1470 (1999)
134. Li, T.-H.S., Tsai, S.-H., Lee, J.-Z., Chao, C.-H.: Robust $\mathcal{H}_\infty$ fuzzy control for a class of uncertain discrete fuzzy bilinear systems. IEEE Transactions on Systems, Man, and Cybernetics, Part B: Cybernetics 38(2), 510–527 (2008)

135. Li, X., de Souza, C.E.: Delay-dependent robust stability and stabilisation of uncertain linear delay systems: a linear matrix inequality approach. IEEE Transactions on Automatic Control 42(8), 1144–1148 (1997)
136. Li, X., de Souza, C.E.: Criteria for robust stability and stabilization of uncertain linear systems with state delays. Automatica 33(9), 1657–1662 (1997)
137. Li, X., Gao, H.: A new model transformation of discrete-time systems with time-varying delay and its application to stability analysis. IEEE Transactions on Automatic Control 56(9), 2172–2178 (2011)
138. Li, X., Yang, G.: Switching-type $\mathcal{H}_\infty$ filter design for T-S fuzzy systems with unknown or partially unknown membership functions. IEEE Transactions on Fuzzy Systems 21(2), 385–392 (2013)
139. Lian, K.-Y., Tu, H.-W., Liou, J.-J.: Stability conditions for LMI-based fuzzy control from viewpoint of membership functions. IEEE Transactions on Fuzzy Systems 14(6), 874–884 (2006)
140. Lin, C., Wang, Q.-G., Lee, T., He, Y., Chen, B.: Observer-based $\mathcal{H}_\infty$ control for T-S fuzzy systems with time delay: delay-dependent design method. IEEE Transactions on Systems, Man, and Cybernetics, Part B: Cybernetics 37(4), 1030–1038 (2007)
141. Lin, C., Wang, Q.-G., Lee, T.H.: B Chen, $\mathcal{H}_\infty$ filter design for nonlinear systems with time-delay through T-S fuzzy model approach. IEEE Transactions on Fuzzy Systems 16(3), 739–746 (2008)
142. Liu, J., Zhang, J.: Note on stability of discrete-time time-varying delay systems. IET Control Theory and Applications 6(2), 335–339 (2012)
143. Liu, M., Cao, X., Shi, P.: Fuzzy-model-based fault-tolerant design for nonlinear stochastic systems against simultaneous sensor and actuator faults. IEEE Transactions on Fuzzy Systems 21(5), 789–799 (2013)
144. Liu, M., Cao, X., Shi, P.: Fault estimation and tolerant control for fuzzy stochastic systems. IEEE Transactions on Fuzzy Systems 21(2), 221–229 (2013)
145. Lo, J.C., Chen, Y.M.: Stability issues on Takagi-Sugeno fuzzy model-parametric approach. IEEE Transactions on Fuzzy Systems 7(5), 597–607 (1999)
146. Lo, J.C., Lin, M.L.: Robust $\mathcal{H}_\infty$ control for fuzzy systems with Frobenius norm-bounded uncertainties. IEEE Transactions on Fuzzy Systems 14(1), 1–15 (2006)
147. Mao, X., Yuan, C.: Stochastic Differential Equations with Markovian Switching. Imperial College Press, London (2006)
148. Meng, X., Lam, J., Du, B., Gao, H.: A partial delay-partitioning approach to stability analysis of discrete-time systems. Automatica 46(3), 610–614 (2009)
149. Montagner, V.F., Oliveira, R.C.L.F., Peres, P.L.D.: Convergent LMI relaxations for quadratic stabilizability and $\mathcal{H}_\infty$ control of Takagi-Sugeno fuzzy systems. IEEE Transactions on Fuzzy Systems 17(4), 863–873 (2009)
150. Mooij, E.: Numerical investigation of model reference adaptive control for hypersonic aircraft. Journal of Guidance, Control, and Dynamics 24(2), 315–323 (2001)
151. Moon, Y.S., Park, P., Kwon, W.H., Lee, Y.S.: Delay-dependent robust stabilisation of uncertain state-delayed systems. International Journal of Control 74(14), 1447–1455 (2001)
152. Mou, S., Gao, H., Lam, J., Qiang, W.: A new criterion of delay-dependent asymptotic stability for Hopfield neural networks with time delay. IEEE Transactions on Neural Networks 19(3), 532–535 (2008)

153. Mozelli, L.A., Palhares, R.M., Souza, F.O., Mendes, E.M.A.M.: A systematic approach to improve multiple Lyapunov function stability and stabilization conditions for fuzzy systems. Information Sciences 179(8), 1149–1162 (2009)
154. Nachidi, M., Benzaouia, A., Tadeo, F., Ait Rami, M.: LMI-based approach for output-feedback stabilization for discrete-time Takagi-Sugeno systems. IEEE Transactions on Fuzzy Systems 16(5), 1188–1196 (2008)
155. Narimani, M., Lam, H.K.: Relaxed LMI-based stability conditions for Takagi-Sugeno fuzzy control systems using regional membership function shape dependent analysis approach. IEEE Transactions on Fuzzy Systems 17(5), 1221–1228 (2009)
156. Narimani, M., Lam, H.K.: SOS-based stability analysis of polynomial fuzzy-model-based control systems via polynomial membership functions. IEEE Transactions on Fuzzy Systems 18(5), 862–871 (2010)
157. Narimani, M., Lam, H.K., Dilmaghani, R., Wolfe, C.: LMI-based stability analysis of fuzzy-model-based control systems using approximated polynomial membership functions. IEEE Transactions on Systems, Man, and Cybernetics, Part B: Cybernetics 41(3), 713–724 (2011)
158. Niculescu, S.-I.: Delay Effects on Stability: A Robust Control Approach. LNCIS, vol. 269. Springer, Heidelberg (2001)
159. Niculescu, S.-I.: On delay-dependent stability under model transformations of some neutral linear systems. International Journal of Control 74(6), 609–617 (2001)
160. Niculescu, S.-I., Dion, J.-M., Dugard, L.: Robust stabilisation for uncertain time-delay systems containing saturating actuators. IEEE Transactions on Automatic Control 41(5), 742–747 (1996)
161. Ohtake, H., Tanaka, K., Wang, H.O.: Switching fuzzy controller design based on switching Lyapunov function for a class of nonlinear systems. IEEE Transactions on Systems, Man, and Cybernetics, Part B: Cybernetics 36(1), 13–23 (2006)
162. Oppenheimer, M., Bolender, M., Doman, D.: Effects of unsteady and viscous aerodynamics on the dynamics of a flexible air-breathing hypersonic vehicle. AIAA Paper, pp. 2007–6397 (2007)
163. Pan, J.-T., Guerra, T.M., Fei, S.-M., Jaadari, A.: Nonquadratic stabilization of continuous T-S fuzzy models: LMI solution for a local approach. IEEE Transactions on Fuzzy Systems 20(3), 594–602 (2012)
164. Park, P.: A delay-dependent stability criterion for systems with uncertain time-invariant delays. IEEE Transactions on Automatic Control 44(4), 876–877 (1999)
165. Park, P., Ko, J.W., Jeong, C.: Reciprocally convex approach to stability of systems with time-varying delays. Automatica 47(1), 235–238 (2011)
166. Parker, J.T., Serrani, A., Yurkovich, S., Bolender, M.A., Doman, D.B.: Control-oriented modeling of an air-breathing hypersonic vehicle. Journal of Guidance, Control, and Dynamics 30(3), 856–869 (2007)
167. Peng, C., Yue, D., Tian, Y.-C.: New approach on robust delay-dependent $\mathcal{H}_\infty$ control for uncertain T-S fuzzy systems with interval time-varying delay. IEEE Transactions on Fuzzy Systems 17(4), 890–900 (2009)
168. Petersen, I.R.: A stabilization algorithm for a class of uncertain linear systems. Systems and Control Letters 4(8), 351–357 (1987)

169. Qiu, J., Feng, G., Yang, J.: A new design of delay-dependent robust $\mathcal{H}_\infty$ filtering for discrete-time T-S fuzzy systems with time-varying delay. IEEE Transactions on Fuzzy Systems 17(5), 1044–1058 (2009)
170. Qiu, J., Feng, G., Gao, H.: Fuzzy-model-based piecewise $\mathcal{H}_\infty$ static-output-feedback controller design for networked nonlinear systems. IEEE Transactions on Fuzzy Systems 18(5), 919–934 (2010)
171. Qiu, J., Feng, G., Gao, H.: Static-output-feedback $\mathcal{H}_\infty$ control of continuous-time T-S fuzzy affine systems via piecewise Lyapunov functions. IEEE Transactions on Fuzzy Systems 21(2), 245–261 (2013)
172. Qiu, J., Tian, H., Lu, Q., Gao, H.: Nonsynchronized robust filtering design for continuous-time T-S fuzzy affine dynamic systems based on piecewise Lyapunov functions. IEEE Transactions on Systems, Man, and Cybernetics, Part B: Cybernetics 43(6), 1755–1766 (2013)
173. Richard, J.-P.: Time-delay systems: an overview of some recent advances and open problems. Automatica 39(10), 1667–1694 (2003)
174. Rojas, O.J., Bao, J., Lee, P.L.: On dissipativity, passivity and dynamic operability of nonlinear processes. Journal of Process Control 18(5), 515–526 (2008)
175. Sala, A., Arino, C.: Relaxed stability and performance conditions for Takagi-Sugeno fuzzy systems with knowledge on membership function overlap. IEEE Transactions on Systems, Man, and Cybernetics, Part B: Cybernetics 37(3), 727–732 (2007)
176. Sala, A., Arino, C.: Relaxed stability and performance LMI conditions for Takagi-Sugeno fuzzy systems with polynomial constraints on membership function shapes. IEEE Transactions on Fuzzy Systems 16(5), 1328–1336 (2008)
177. Schmidt, D.K.: Dynamics and control of hypersonic aeropropulsive/aeroelastic vehicles. AIAA Paper, pp. 1992–4326 (1992)
178. Shen, Q., Jiang, B., Cocquempot, V.: Fault-tolerant control for T-S fuzzy systems with application to near-space hypersonic vehicle with actuator faults. IEEE Transactions on Fuzzy Systems 20(4), 652–665 (2012)
179. Shen, Q., Jiang, B., Cocquempot, V.: Fuzzy logic system-based adaptive fault-tolerant control for near-space vehicle attitude dynamics with actuator faults. IEEE Transactions on Fuzzy Systems 21(2), 289–300 (2013)
180. Sigthorsson, D.O., Serrani, A., Yurkovich, S., Bolender, M.A., Doman, D.B.: Tracking control for an overactuated hypersonic air-breathing vehicle with steady state constraints. AIAA Paper, pp. 2006–6558 (2006)
181. Sigthorsson, D.O., Jankovsky, P., Serrani, A., Yurkovich, S., Bolender, M.A., Doman, D.B.: Robust linear output feedback control of an air-breathing hypersonic vehicle. Journal of Guidance, Control, and Dynamics 31(4), 1052–1066 (2008)
182. Sinha, P.K., Pechev, A.N.: Model reference adaptive control of a maglev system with stable maximum descent criterion. Automatica 35(8), 1457–1465 (1999)
183. Sinha, P.K., Pechev, A.N.: Nonlinear $\mathcal{H}_\infty$ controllers for electromagnetic suspension systems. IEEE Transactions on Automatic Control 49(4), 563–568 (2004)
184. Souza, F.O., Mozelli, L.A., Palhares, R.M.: On stability and stabilization of T-S fuzzy time-delayed systems. IEEE Transactions on Fuzzy Systems 17(6), 1450–1455 (2009)

185. Su, X., Shi, P., Wu, L., Song, Y.-D.: A novel approach to filter design for T-S fuzzy discrete-time systems with time-varying delay. IEEE Transactions on Fuzzy Systems 20(6), 1114–1129 (2012)
186. Su, X., Wu, L., Shi, P., Song, Y.-D.: $\mathcal{H}_\infty$ model reduction of Takagi-Sugeno fuzzy stochastic systems. IEEE Transactions on Systems, Man, and Cybernetics, Part B: Cybernetics 42(6), 1574–1585 (2012)
187. Su, X., Shi, P., Wu, L., Song, Y.-D.: A novel control design on discrete-time Takagi-Sugeno fuzzy systems with time-varying delays. IEEE Transaction on Fuzzy Systems 21(4), 655–671 (2013)
188. Su, X., Shi, P., Wu, L., Sing, K.N.: Induced ℓ_2 filtering of fuzzy stochastic systems with time-varying delays. IEEE Transactions on Cybernetics 43(4), 1251–1264 (2013)
189. Su, X., Shi, P., Wu, L., Basin, M.V.: Reliable filtering with strict dissipativity for T-S fuzzy time-delay systems. IEEE Transactions on Cybernetics (2014), doi:10.1109/TCYB.2014.2308983
190. Sun, C.H., Wang, W.J.: An improved stability criterion for T-S fuzzy discrete systems via vertex expression. IEEE Transactions on Systems, Man, and Cybernetics, Part B: Cybernetics 36(3), 672–678 (2006)
191. Sung, H.K., Lee, S.H., Bien, Z.: Design and implementation of a fault tolerant controller for EMS systems. Mechatronics 15(10), 1253–1272 (2005)
192. Takagi, T., Sugeno, M.: Fuzzy identification of systems and its applications to modeling and control. IEEE Transactions on Systems, Man, and Cybernetics SMC-15(1), 116–132 (1985)
193. Tanaka, K., Sugeno, M.: Stability analysis and design of fuzzy control systems. Fuzzy Sets and Systems 45(2), 135–156 (1992)
194. Tanaka, K., Wang, H.O.: Fuzzy Control Systems Design and Analysis: A Linear Matrix Inequality Approach. John Wiley and Sons, New York (2001)
195. Tanaka, K., Hori, T., Wang, H.O.: A multiple Lyapunov function approach to stabilization of fuzzy control systems. IEEE Transactions on Fuzzy Systems 11(4), 582–589 (2003)
196. Tanaka, K., Ohtake, H., Wang, H.O.: A descriptor system approach to fuzzy control system design via fuzzy Lyapunov functions. IEEE Transactions on Fuzzy Systems 15(3), 333–341 (2007)
197. Taniguchi, T., Sugeno, M.: Stabilization of nonlinear systems based on piecewise Lyapunov functions. In: Proceedings of the 13th IEEE International Conference on Fuzzy Systems, Budapest, Hungary, pp. 1607–1612 (July 2004)
198. Teixeira, M.C.M., Zak, S.H.: Stabilizaing controller design for uncertain nonlinear systems using fuzzy models. IEEE Transactions on Fuzzy Systems 7(2), 133–142 (1999)
199. Tognetti, E.S., Oliveira, R.C.L.F., Peres, P.L.D.: Selective $\mathcal{H}_2$ and $\mathcal{H}_\infty$ stabilization of Takagi-Sugeno fuzzy systems. IEEE Transactions on Fuzzy Systems 19(5), 890–900 (2011)
200. Tseng, C.-S.: Robust fuzzy filter design for nonlinear systems with persistent bounded disturbances. IEEE Transactions on Systems, Man, and Cybernetics, Part B: Cybernetics 36(4), 940–945 (2006)
201. Tseng, C.-S.: Robust fuzzy filter design for a class of nonlinear stochastic systems. IEEE Transactions on Fuzzy Systems 15(2), 261–274 (2007)
202. Tseng, C.-S.: A novel approach to $\mathcal{H}_\infty$ decentralized fuzzy-observer-based fuzzy control design for nonlinear interconnected systems. IEEE Transactions on Fuzzy Systems 16(5), 1337–1350 (2008)

203. Tuan, H.D., Apkarian, P., Narikiyo, T., Yamamoto, Y.: Parameterized linear matrix inequality techniques in fuzzy control system design. IEEE Transaction on Fuzzy Systems 9(2), 324–332 (2001)
204. van der Sande, T.P.J., Gysen, B.L.J., Besselink, I.J.M., Paulides, J.J.H., Lomonova, E.A., Nijmeijer, H.: Robust control of an electromagnetic active suspension system: simulations and measurements. Mechatronics 23(2), 204–212 (2013)
205. Veillette, R., Medanic, J.B., Perkins, W.: Design of reliable control systems. IEEE Transactions on Automatic Control 37(3), 290–304 (1992)
206. Wang, H.O., Tanaka, K., Griffin, M.F.: An approach to fuzzy control of nonlinear systems: stability and design issues. IEEE Transactions on Fuzzy Systems 4(1), 14–23 (1996)
207. Wang, L., Feng, G.: Piecewise $\mathcal{H}_\infty$ controller design of discrete time fuzzy systems. IEEE Transactions on Systems, Man, and Cybernetics, Part B: Cybernetics 34(1), 682–686 (2004)
208. Wang, W.J., Luoh, L.: Stability and stabilization of fuzzy large scale systems. IEEE Transactions on Fuzzy Systems 12(3), 309–315 (2004)
209. Wang, W.J., Wen, Y.J., Sun, C.H.: Relaxed stabilization criteria for discrete-time T-S fuzzy control systems based on a switching fuzzy model and piecewise Lyapunov function. IEEE Transactions on Systems, Man, and Cybernetics, Part B: Cybernetics 37(3), 551–559 (2007)
210. Wang, Y., Xie, L., de Souza, C.E.: Robust control of a class of uncertain nonlinear systems. Systems and Control Letters 19(2), 139–149 (1992)
211. Wang, Y., Sun, Z.Q., Sun, F.C.: Stability analysis and control of discrete-time fuzzy systems: a fuzzy Lyapunov function approach. In: Proceedings of the 5th Asian Control Conference, Melbourne, Australia, July 20-23, pp. 1855–(1860)
212. Wang, Z., Ho, D.W.C., Liu, X.: A note on the robust stability of uncertain stochastic fuzzy systems with time-delays. IEEE Transactions on Systems, Man and Cybernetics, Part A: Systems and Humans 34(4), 570–576 (2004)
213. Wei, G., Feng, G., Wang, Z.: Robust $\mathcal{H}_\infty$ control for discrete-time fuzzy systems with infinite-distributed delays. IEEE Transactions on Fuzzy Systems 17(1), 224–232 (2009)
214. Wilcox, Z.D., MacKunis, W., Bhat, S., Lind, R., Dixon, W.E.: Lyapunov-based exponential tracking control of a hypersonic aircraft with aerothermoelastic effects. Journal of Guidance, Control, and Dynamics 33(4), 1213–1224 (2010)
215. Willems, J.: Dissipative dynamical systems part I: general theory. Archive for Rational Mechanics and Analysis 45(5), 321–351 (1972)
216. Williams, T., Bolender, M.A., Doman, D.B., Morataya, O.: An aerothermal flexible mode analysis of a hypersonic vehicle. AIAA Paper, pp. 2006–6647 (2006)
217. Wu, H.: Delay-dependent stability analysis and stabilization for discrete-time fuzzy systems with state delay: a fuzzy Lyapunov-Krasovskii functional approach. IEEE Transactions on Systems, Man, and Cybernetics, Part B: Cybernetics 36(4), 954–962 (2006)
218. Wu, H., Zhang, H.: Reliable $\mathcal{H}_\infty$ fuzzy control for a class of discrete-time nonlinear systems using multiple fuzzy Lyapunov functions. IEEE Transactions on Circuits and Systems II: Express Briefs 54(42), 357–361 (2007)
219. Wu, L., Ho, D.W.C.: Fuzzy filter design for Itô stochastic systems with application to sensor fault detection. IEEE Transactions on Fuzzy Systems 17(1), 233–242 (2009)

220. Wu, L., Zheng, W.: $\mathcal{L}_2$-$\mathcal{L}_\infty$ control of nonlinear fuzzy Itô stochastic delay systems via dynamic output feedback. IEEE Transactions on Systems, Man, and Cybernetics, Part B: Cybernetics 39(5), 1308–1315 (2009)
221. Wu, L., Zheng, W.X.: Weighted $\mathcal{H}_\infty$ model reduction for linear switched systems with time-varying delay. Automatica 45(1), 186–193 (2009)
222. Wu, L., Su, X., Shi, P.: Model approximation for discrete-time state-delay systems in the T-S fuzzy framework. IEEE Transactions on Fuzzy Systems 19(2), 366–378 (2011)
223. Wu, L., Su, X., Shi, P., Qiu, J.: A new approach to stability analysis and stabilization of discrete-time T-S fuzzy time-varying delay systems. IEEE Transactions on Systems, Man, and Cybernetics, Part B: Cybernetics 41(1), 273–286 (2011)
224. Wu, M., He, Y., She, J.-H., Liu, G.-P.: Delay-dependent criteria for robust stability of time-varying delay systems. Automatica 40(8), 1435–1439 (2004)
225. Wu, Z.-G., Shi, P., Su, H., Chu, J.: Reliable $\mathcal{H}_\infty$ control for discrete-time fuzzy systems with infinite-distributed delay. IEEE Transactions on Fuzzy Systems 20(1), 22–31 (2012)
226. Xie, S., Xie, L., de Souza, C.E.: Robust dissipative control for linear systems with dissipative uncertainty. International Journal of Control 70(2), 169–191 (1998)
227. Xie, W.: Improved $\mathcal{L}_2$ gain performance controller synthesis for Takagi-Sugeno fuzzy system. IEEE Transactions on Fuzzy Systems 16(5), 1142–1150 (2008)
228. Xie, X., Ma, H., Zhao, Y., Ding, D.W., Wang, Y.: Control synthesis of discrete-time T-S fuzzy systems based on a novel non-PDC control scheme. IEEE Transactions on Fuzzy Systems 21(1), 147–157 (2013)
229. Xu, H., Mirmirani, M., Ioannou, P.: Robust neural adaptive control of a hypersonic aircraft. In: AIAA Guidance, Navigation, and Control Conference and Exhibit, pp. 1–8. AIAA, Austin (2003)
230. Xu, H., Mirmirani, M., Ioannou, P.: Adaptive sliding mode control design for a hypersonic flight vehicle. Journal of Guidance, Control, and Dynamics 27(5), 829–838 (2004)
231. Xu, J., Zhou, Y.: A nonlinear control method for the electromagnetic suspension system of the maglev train. Journal of Modern Transportation 19(3), 176–180 (2011)
232. Xu, S., Lam, J.: Improved delay-dependent stability criteria for time-delay systems. IEEE Transactions on Automatic Control 50(3), 384–387 (2005)
233. Xu, S., Lam, J.: Robust $\mathcal{H}_\infty$ control for uncertain discrete-time-delay fuzzy systems via output feedback controllers. IEEE Transactions on Fuzzy Systems 13(1), 82–93 (2005)
234. Xu, S., Lam, J.: A survey of linear matrix inequality techniques in stability analysis of delay systems. International Journal of Systems Science 39(12), 1095–1113 (2008)
235. Yan, W., Lam, J.: An approximate approach to $\mathcal{H}_2$ optimal model reduction. IEEE Transactions on Automatic Control 44(7), 1341–1358 (1999)
236. Yang, G., Dong, J.: $\mathcal{H}_\infty$ filtering for fuzzy singularly perturbed systems. IEEE Transactions on Systems, Man, and Cybernetics, Part B: Cybernetics 38(5), 1371–1389 (2008)
237. Yang, G., Dong, J.: Switching fuzzy dynamic output feedback $\mathcal{H}_\infty$ control for nonlinear systems. IEEE Transactions on Systems, Man, and Cybernetics, Part B: Cybernetics 40(2), 505–516 (2010)

238. Yang, D., Cai, K.-Y.: Reliable $\mathcal{H}_\infty$ non-uniform sampling fuzzy control for nonlinear systems with time delay. IEEE Transactions on Systems, Man, and Cybernetics, Part B: Cybernetics 38(6), 1606–1613 (2008)
239. Yang, H., Xia, Y., Liu, B.: Fault detection for T-S fuzzy discrete systems in finite-frequency domain. IEEE Transactions on Systems, Man, and Cybernetics, Part B: Cybernetics 41(4), 911–920 (2011)
240. Yang, R., Gao, H., Shi, P.: Novel robust stability criteria for stochastic Hopfield neural networks with time delays. IEEE Transactions on Systems, Man and Cybernetics, Part B: Cybernetics 39(2), 467–474 (2009)
241. Yoneyama, J., Nishikawa, M., Katayama, H., Ichikawa, A.: Output stabilization of Takagi-Sugeno fuzzy systems. Fuzzy Sets and Systems 111(2), 253–266 (2000)
242. Yoshida, K., Higeta, A.: Toward stochastic explanation of a neutrally stable delayed feedback model of human balance control. International Journal of Innovative Computing Information and Control 8(3(B)), 2249–2259 (2012)
243. Zadeh, L.: Outline of a new approach to the analysis of complex systems and decision processes. IEEE Transactions on Systems, Man, and Cybernetics 3(1), 28–44 (1973)
244. Zames, G.: On the input-output stability of time-varying nonlinear feedback systems – part one: conditions derived using concepts of loop gain, conicity, and positivity. IEEE Transactions on Automatic Control 11(2), 228–238 (1966)
245. Zhang, B., Xu, S., Zou, Y.: Improved stability criterion and its application in delayed controller design for discrete-time systems. Automatica 44(11), 2963–2967 (2008)
246. Zhang, B., Xu, S.: Delay-dependent robust $\mathcal{H}_\infty$ control for uncertain discrete-time fuzzy systems with time-varying delays. IEEE Transactions on Fuzzy Systems 17(4), 809–823 (2009)
247. Zhang, H., Li, C., Liao, X.: Stability analysis and $\mathcal{H}_\infty$ controller design of fuzzy large-scale systems based on piecewise Lyapunov functions. IEEE Transactions on Systems, Man, and Cybernetics, Part B: Cybernetics 36(3), 685–698 (2006)
248. Zhang, H., Lun, S., Liu, D.: Fuzzy $\mathcal{H}_\infty$ filter design for a class of nonlinear discrete-time systems with multiple time delays. IEEE Transactions on Fuzzy Systems 15(3), 453–469 (2007)
249. Zhang, H., Wang, Y., Liu, D.: Delay-dependent guaranteed cost control for uncertain stochastic fuzzy systems with multiple tire delays. IEEE Transactions on Systems, Man, and Cybernetics, Part B: Cybernetics 38(1), 126–140 (2008)
250. Zhang, H., Feng, G.: Stability analysis and $\mathcal{H}_\infty$ controller design of discrete-time fuzzy large-scale systems based on piecewise Lyapunov functions. IEEE Transactions on Systems, Man, and Cybernetics, Part B: Cybernetics 38(5), 1390–1401 (2008)
251. Zhang, H., Dang, C., Zhang, J.: Decentralized fuzzy $\mathcal{H}_\infty$ filtering for nonlinear interconnected systems with multiple time delays. IEEE Transactions on Systems, Man, and Cybernetics, Part B: Cybernetics 40(4), 1197–1203 (2010)
252. Zhang, H., Xie, X.: Relaxed stability conditions for continuous-time T-S fuzzy-control systems via augmented multi-indexed matrix approach. IEEE Transactions on Fuzzy Systems 19(3), 478–492 (2011)

253. Zhang, H., Zhong, H., Dang, C.: Delay-dependent decentralized $\mathcal{H}_\infty$ filtering for discrete-time nonlinear interconnected systems with time-varying delay based on the T-S fuzzy model. IEEE Transactions on Fuzzy Systems 20(3), 431–443 (2012)
254. Zhang, H., Shi, Y., Mehr, A.S.: On $\mathcal{H}_\infty$ filtering for discrete-time Takagi-Sugeno fuzzy systems. IEEE Transactions on Fuzzy Systems 20(2), 396–401 (2012)
255. Zhang, K., Jiang, B., Staroswiecki, M.: Dynamic output feedback-fault tolerant controller design for Takagi-Sugeno fuzzy systems with actuator faults. IEEE Transactions on Fuzzy Systems 18(1), 194–201 (2010)
256. Zhang, K., Jiang, B., Shi, P.: Fault estimation observer design for discrete-time Takagi-Sugeno fuzzy systems based on piecewise Lyapunov functions. IEEE Transactions on Fuzzy Systems 20(1), 192–200 (2012)
257. Zhao, J., Hill, D.J.: Dissipativity theory for switched systems. IEEE Transactions on Automatic Control 53(4), 941–953 (2008)
258. Zhao, L., Cao, H., Karimi, H.R.: Robust stability and stabilization of uncertain T-S fuzzy systems with time-varying delay: an input-output approach. IEEE Transactions on Fuzzy Systems 21(5), 883–897 (2013)
259. Zhao, Y., Gao, H., Lam, J., Du Stability, B.: stabilization of delayed T-S fuzzy systems: a delay partitioning approach. IEEE Transactions on Fuzzy Systems 17(4), 750–762 (2009)
260. Zhao, Y., Lam, J., Gao, H.: Fault detection for fuzzy systems with intermittent measurements. IEEE Transactions on Fuzzy Systems 17(2), 398–410 (2009)
261. Zhao, Y., Zhang, C., Gao, H.: A new approach to guaranteed cost control of T-S fuzzy dynamic systems with interval parameter uncertainties. IEEE Transactions on Systems, Man, and Cybernetics, Part B: Cybernetics 40(1), 1516–1527 (2009)
262. Zhao, Y., Gao, H.: Fuzzy-model-based control of an overhead crane with input delay and actuator saturation. IEEE Transactions on Fuzzy Systems 20(1), 181–186 (2012)
263. Zheng, Y., Fang, H., Wang, H.-O.: Takagi-Sugeno fuzzy-model-based fault detection for networked control systems with Markov delays. IEEE Transactions on Systems, Man, and Cybernetics, Part B: Cybernetics 36(4), 924–929 (2006)
264. Zhong, M., Ye, H., Shi, P., Wang, G.: Fault detection for Markovian jump systems. IEE Proceedings – Control Theory and Applications 152(4), 397–402 (2005)
265. Zhou, K., Doyle, J., Glover, K.: Robust and Optimal Control. Prentice Hall, Upper Saddle River (1996)
266. Zhou, S., Feng, G., Lam, J., Xu, S.: Robust $\mathcal{H}_\infty$ control for discrete fuzzy systems via basis-dependent Lyapunov functions. Information Sciences 174(3-4), 197–217 (2005)

Printed in the United States
By Bookmasters